T0313869

Mathematical Foundation of Railroad Vehicle Systems

Geometry and Mechanics

Mathematical Foundation of Railroad Vehicle Systems

Geometry and Mechanics

Ahmed A. Shabana
Richard and Loan Hill Professor of Engineering
University of Illinois at Chicago
Chicago, Illinois, USA

This edition first published 2021

Registered Offices
John Wiley & Sons, Inc., 111 River Street, Hoboken, NJ 07030, USA
John Wiley & Sons Ltd, The Atrium, Southern Gate, Chichester, West Sussex, PO19 8SQ, UK

Editorial Office
The Atrium, Southern Gate, Chichester, West Sussex, PO19 8SQ, UK

For details of our global editorial offices, customer services, and more information about Wiley products visit us at www .wiley.com.

Wiley also publishes its books in a variety of electronic formats and by print-on-demand. Some content that appears in standard print versions of this book may not be available in other formats.

Library of Congress Cataloging-in-Publication Data

Names: Shabana, Ahmed A., 1951- author.
Title: Mathematical foundation of railroad vehicle systems: geometry and mechanics / Ahmed A. Shabana, University of Illinois at Chicago Circle, IL, US.
Description: Hoboken, NJ : John Wiley & Sons, Inc., [2021] | Includes bibliographical references and index.
Identifiers: LCCN 2020020731 (print) | LCCN 2020020732 (ebook) | ISBN 9781119689041 (cloth) | ISBN 9781119689065 (adobe pdf) | ISBN 9781119689089 (epub)
Subjects: LCSH: Railroad engineering–Mathematics. | Railroad rails–Mathematical models. | Railroad trains–Dynamics. | Geometric analysis.
Classification: LCC TF205 .S47 2020 (print) | LCC TF205 (ebook) | DDC 625.1001/5118–dc23
LC record available at https://lccn.loc.gov/2020020731
LC ebook record available at https://lccn.loc.gov/2020020732

Cover Design: Wiley
Cover Image: © Xuanyu Han/Getty Images

Set in 9.5/12.5pt STIXTwoText by SPi Global, Chennai, India
Printed and bound by CPI Group (UK) Ltd, Croydon, CR0 4YY

10 9 8 7 6 5 4 3 2 1

Contents

Preface

The design of passenger and freight railroad vehicle systems, which are complex transportation systems, involves many areas of science and engineering, including mechanical, civil, electrical, and electronic engineering. Because a comprehensive treatment of such complex systems cannot be covered in a single book, the focus of this book is on developing the mathematical foundation of railroad vehicle systems with an emphasis on the integration of *geometry* and *mechanics*. Such a geometry/mechanics integration is necessary for developing a sound mathematical foundation, accurate formulation of nonlinear dynamic equations, and general computational algorithms that can be used effectively in the virtual prototyping, analysis, design, and performance evaluation of railroad vehicle systems. Geometry is particularly important in the formulation of the railroad system dynamic equations of the motion generated by contact between the wheel and rail surfaces. The surface geometry, therefore, plays a fundamental role in formulating the kinematics, forces, and equations of motion of such systems. The theory of curves is equally important, and it is central in the description of the track geometry. For this reason, the subject of differential geometry plays a fundamental role in formulating the equations that govern the motion of railroad vehicle systems.

In addition to basic geometry concepts, principles of mechanics are required in order to formulate railroad kinematic relationships and dynamic equations of motion. These mechanics principles allow for systematically modeling motion constraints resulting from mechanical joints and specified motion trajectories. Railroad vehicles have components that experience large displacements, including finite rotations; therefore, it is necessary to avoid linearization of the kinematic and dynamic equations when developing general computational algorithms. In the mechanics approaches used in this book, a fully nonlinear motion description is used in the formulation of dynamic equations of motion. As demonstrated in this book, applying mechanics principles can lead to different forms of dynamic equations of motion. Regardless of the form of the equations of motion obtained, the integration of geometry and mechanics is necessary.

This book is written to complement a previous book, *Railroad Vehicle Dynamics: A Computational Approach*. Both books are designed to be self contained, and therefore, overlap of some topics cannot be avoided. This book is designed for an introductory course on railroad vehicle system dynamics suitable for senior undergraduate and first-year graduate students. It can also be used as a reference book by researchers and practicing engineers. The book consists of seven chapters that are organized to introduce the reader to the basic concepts, formulations, and computational algorithms used in railroad vehicle system dynamics. Unlike other texts in the area of railroad vehicle dynamics, this book emphasizes geometry/mechanics integration throughout. It is shown how new mechanics-based approaches, such as the absolute nodal coordinate formulation (ANCF), can be used to achieve the geometry/mechanics integration necessary for developing accurate virtual prototyping algorithms.

Chapter 1 introduces the reader to several topics that are covered in subsequent chapters. Basic geometry concepts are discussed, and the need to integrate geometry and mechanics in the formulation of the railroad vehicle system equations is explained. The hunting oscillations that characterize the dynamic behavior of railroad vehicles and can lead to instabilities and derailments are discussed, and the methods used to describe wheel and track geometries are introduced. During curve negotiations, the effect of centrifugal forces must be taken into account to define the allowable balance speed that ensures the safe operation of the rail vehicle. This important topic, as well as the wheel/rail contact formulations and their integration with multibody system (MBS) computational algorithms, are introduced in Chapter 1. The chapter also discusses other important topics, including derailment criteria, high-speed rail, pantograph/catenary systems, and linear algebra, which can be conveniently used to formulate the kinematic and dynamic equations of railroad vehicle systems.

Chapter 2 covers topics in differential geometry that are fundamental in developing the nonlinear equations that govern the motion of the vehicle. Curve and surface geometries and their application to railroad vehicle systems are discussed. It is shown how the principal curvatures and principal directions of surfaces can be determined using the coefficients of the first and second fundamental forms of surfaces. The numerical representation of wheel and rail profile geometries using spline functions is discussed. The use of ANCF finite elements to describe the rail surface geometry is explained. It is shown that arbitrary surface geometry for the rail can be systematically described using the displacement field of fully parameterized ANCF finite elements.

Railroad vehicle motion and geometry descriptions are the subject of Chapter 3, which demonstrates how body motion equations and geometry are integrated. The coordinate systems used to describe rigid-body kinematics are first introduced, and different parameters for describing body orientation are presented. Among the orientation parameters discussed in this chapter are direction cosines, Euler angles, and Euler parameters. These orientation parameters are used to define the position, velocity, and acceleration equations in terms of the body generalized coordinates. The chapter explains the use of Euler angles in railroad vehicle system mechanics for two fundamentally different purposes: motion description and track geometry description. Using Euler angles as field variables to describe track geometry is explained. In Chapter 3, geometric motion constraints that result from mechanical joints used to connect vehicle components are introduced, and the trajectory coordinates used to develop specialized railroad vehicle system computer programs are defined.

While Chapters 2 and 3 deal, for the most part, with basic geometry and motion kinematic approaches and concepts, the focus of Chapter 4 is on railroad geometry. Wheel geometry equations are first defined, and it is shown how these geometry equations are used to define curvatures at an arbitrary point on the wheel surface. Two methods are introduced for describing rail geometry: semi-analytical and ANCF. Limitations of the semi-analytical approach are discussed to provide a justification for using the more general ANCF approach to describe rail geometry. Using the ANCF interpolation to define the tangents and normal at an arbitrary point on the rail surface is explained. As discussed in this chapter, such an ANCF approach, which employs position gradients, allows using higher-order interpolation for the position coordinates and avoids the interpolation of rotations. Fully parametrized ANCF finite elements can be used to systematically define surfaces by introducing a functional relationship between the element spatial coordinates. The structure of the track data used in computer simulations of railroad vehicle systems is discussed, and methods for describing track deviations that represent track irregularities are presented. The chapter concludes by providing a comparison between the semi-analytical and ANCF approaches used to describe rail geometry.

The contact problem is covered in Chapter 5, starting with a discussion of the wheel/rail contact mechanism in order to provide a better understanding of the need to account for the creep phenomenon in railroad vehicle system applications. The constraint contact formulation (CCF) and the elastic contact formulation (ECF), which are used to determine the locations of wheel/rail contact points online, are introduced. As explained in this chapter, the CCF approach does not allow wheel/rail separation, but the ECF approach does allow this separation. Therefore, when the ECF approach is used, no degrees of freedom are eliminated. Determining the normal contact force in both the CCF and ECF approaches is discussed in this chapter, and it is explained how the surface geometry and Hertz contact theory can be used to determine the dimensions of the contact area. To determine the tangential creep forces, velocity creepages used in wheel/rail creep force formulations are defined. Several creep force formulations are discussed in Chapter 5, and the assumptions used in these formulations are highlighted. The chapter concludes with a discussion of magnetically levitated (maglev) trains.

Methods for developing the nonlinear dynamic equations of motion of railroad vehicle systems are introduced in Chapter 6. The Newtonian and Lagrangian approaches are compared to highlight the basic concepts used in the two approaches. As discussed in this chapter, the Lagrangian approach does not require the use of free-body diagrams or making cuts at the joints because this approach is based on using connectivity conditions. The constraint forces can be defined systematically using algebraic constraint equations, and therefore, the Lagrangian approach lends itself to developing general computational algorithms for railroad vehicle systems. Using generalized coordinates, the inertia and applied and contact forces can be formulated and used to develop equations of motion. The form of the equations of motion can be defined using the augmented formulation or the embedding technique. The augmented formulation leads to a large, sparse system expressed in terms of redundant coordinates and constraint forces, while the embedding technique leads to a minimum number of equations of motion expressed in terms of the system degrees of freedom; therefore, these equations do not include any constraint forces. The chapter also discusses the formulation of other force elements used in railroad vehicle dynamics and explains using trajectory coordinates to develop longitudinal train dynamics (LTD) algorithms. Simple models that can be used to study hunting oscillations are also developed in Chapter 6. The chapter concludes with a discussion of MBS modeling of electromechanical systems.

Pantograph/catenary systems that provide the power supply for high-speed trains are the subject of Chapter 7. The chapter discusses in detail their design and the formulation of catenary equations of motion using ANCF finite elements. The formulation of pantograph/catenary contact forces is presented using the constraint and elastic contact formulations. It is shown how to account for lateral relative sliding between the pan-head and catenary when using the constraint and elastic contact formulations. Chapter 6 also discusses pantograph/catenary force control, the effect of aerodynamic forces, and the wear problem when such current-collection systems are used for high-speed trains.

The development of the materials presented in this book is based on research collaboration between the author and many colleagues and students. The author would like to acknowledge the contributions of his colleagues and students including UIC current and former students and visiting scholars Zhengfeng Bai, Issac Banes, Dario Bettamin, Mahmoud Elbakly, Ahmed Eldeek, Sibi Kandasamy, Hao Ling, Ramon Martinez, Huailong Shi, and Dayu Zhang for their help in the preparation and proofreading of the book manuscript. The author would like to thank Drs. Khaled Zaazaa and Hiroyuki Sugiyama for past collaboration on writing the book *Railroad Vehicle Dynamics: A Computational Approach*. The editorial and production staff of Wiley & Sons deserve special thanks for their cooperation and professional work in producing this book. Finally, the author would like to thank his family for their patience and understanding during the time spent preparing this book.

Ahmed A. Shabana
Chicago, Illinois, USA

Chapter 1

INTRODUCTION

Passenger and freight railroad vehicles are complex transportation systems whose design involves many areas of science and engineering, including mechanical, civil, electrical, and electronic engineering. Because, a comprehensive treatment of such railroad vehicle systems cannot be covered in a single volume book, the focus of this book is on developing the mathematical foundation with the emphasis placed on the integration of *geometry* and *mechanics*.

Integration of Geometry and Analysis As will be explained in this book, a sound mathematical foundation, the formulation of nonlinear dynamic equations, and the design of general computational algorithms of such complex systems require the integration of basic concepts of geometry and mechanics. Geometry is particularly important in formulating the railroad system dynamic equations that govern the vehicle motion produced by contact between the wheel and rail surfaces. The surface geometry, therefore, plays a fundamental role in formulating the kinematics, forces, and equations of motion of such transportation systems. Without an accurate description of the surface geometry, it is not possible to develop accurate formulations and computational algorithms that account for the nonlinear dynamic behavior of railroad vehicle systems. The theory of curves is equally important, and it is central to the description of track geometry. High-speed trains are electrically powered using pantograph/catenary systems. As discussed in this book, the catenary system consists of wires whose geometry and deformations are critical in ensuring high-quality electric current collection. These catenary wires also define space curves whose geometry must be accurately described in order to better understand their dynamic behavior and their response to contact and aerodynamic forces.

While geometry enters into formulating railroad kinematic relationships and dynamic equations of motion, these equations are formulated using the principles of mechanics, which allow for systematically modeling motion constraints resulting from mechanical joints and specified motion trajectories. Because railroad vehicles have components that experience large displacements, including finite rotations, it is important to avoid linearization of the kinematic and dynamic equations when developing general computational algorithms. In the mechanics approaches used in this book, fully nonlinear motion descriptions are adopted in formulating the dynamic equations of motion. Applying the principles of mechanics can lead to different forms of the dynamic equations of motion, as discussed in this book. Regardless of the form of the equations of motion obtained, the

Mathematical Foundation of Railroad Vehicle Systems: Geometry and Mechanics,
First Edition. Ahmed A. Shabana.

integration of geometry and mechanics is necessary in order to establish the foundation of those equations.

Passenger and Freight Trains The methods developed in this book are applicable to both passenger and freight railroad vehicle systems. These two types of systems can operate at significantly different speeds, have significantly different axle loads, and require the use of different track quality. Figure 1 shows a high-speed passenger train and a freight train that operate under different rules and guidelines and require different track-quality

(a)

(b)

Figure 1.1 Passenger and freight trains. Sources: (a) WANG SHIH-WEi/123 RF (2016); (b) Mike Danneman/Moment/Getty Images.

standards developed based on safety considerations. For both systems, however, regardless of the power supply needed for their operation, motion is generated by the force of interaction of wheels and rails like the ones shown in Figure 2. For the most part, freight trains use diesel engines and often run on lower-quality tracks, so their speeds cannot exceed a certain limit. High-speed trains, on the other hand, are electrically powered and can operate at a much higher speed as compared to freight trains. While the mathematical foundations developed in this book can be applied to both cases, it is important to recognize that practical issues must be taken into consideration when modeling these two different systems. While the basic geometric and mechanical formulations are the same, the analysis models, including dimensions, system configurations, number of rail cars, suspension characteristics, ride-comfort criteria, noise level, power supply, operating speeds, etc., can be significantly different. Developing accurate virtual prototyping models for both systems, based on proper integration of geometry and mechanics, is necessary in order to avoid deadly, costly, and environmentally damaging accidents. Passenger trains transport a large number of people around the world, while freight trains transport goods and hazardous materials. As the demand increases for higher-speed, heavier-axle loads; stricter operational and safety guidelines; less noise; a greater degree of comfort; and more robust transportation systems; more accurate virtual models with significant details are needed.

Figure 1.2 Wheel/rail contact.

Organization and Scope of This Chapter This chapter introduces the basic topics discussed in this book, starting with the *differential geometry* that covers the theory of curves and surfaces. The integration of *geometry* and *motion description* plays an important role in formulating the railroad vehicle equations of motion, as discussed in Section 1.1. In Section 1.2, the important topic of the *integration of geometry and mechanics* is discussed in more detail. Railroad wheelsets exhibit a type of vibration known as *hunting oscillations* during which the wheelset lateral displacement and yaw angle are related because of wheelset conicity. Hunting oscillations are discussed in Section 1.3 based on pure geometric considerations. Using a basic description of differential geometry and motion, *track and wheel geometries* can be defined, as explained in Section 1.4. Section 1.5 explains the effect of *centrifugal forces* during curve negotiations and presents a simple analysis for defining the *balance speed* that must not be exceeded by the vehicle when it travels on a curved section of track, to avoid derailment. Section 1.6 introduces the wheel/rail contact problem and the forces that influence vehicle stability. While the main focus in this book is on trains driven by the wheel/rail contact forces, maglev trains are also briefly discussed in Section 1.6. In Section 1.7, the multibody system (MBS) approach for formulating governing dynamic equations using the principles of mechanics is discussed. Section 1.8 discusses existing *derailment criteria* and the role of three-dimensional computational dynamics in developing more accurate and general derailment criteria. Section 1.9 reviews topics related to the operation of high-speed trains, including the *pantograph/catenary systems*. Section 1.10 discusses some mathematical preliminaries that are used throughout the chapters and the notations adopted in this book.

1.1 DIFFERENTIAL GEOMETRY

Figure 3 shows a train negotiating a curved track, the wheel and rail that come into contact, and the pantograph/catenary system used to supply the electric power necessary for the operation of high-speed trains. The description of the motion of the vehicle on the track, the layout of the track, the mathematical definition of the surfaces of the wheel and rail that come into contact to produce the train motion, and the geometry and deformations of the catenary cables that carry the electric current necessary for the operation of the high-speed rail system are examples that demonstrate the importance of geometry in railroad vehicle system dynamics. For this reason, understanding the differential-geometry theories of curves and surfaces is the first step in correctly formulating the dynamic equations that govern the motion of these complex systems (Do Carmo 1976; Goetz 1970; Kreyszig 1991).

For example, the track and wheel geometries must be accurately described in order to correctly predict the wheel/rail contact forces. Using both the differential-geometry *theory of curves* and the *theory of surfaces* is necessary in order to be able to formulate the wheel and rail kinematic and force equations. These theories are used to define nonlinear geometric equations that can be solved for the locations of the wheel/rail contact points, as discussed in this book. Furthermore, the spatial wheel and rail geometric representation is crucial in the study of railroad vehicle nonlinear dynamics in different motion scenarios, including curve negotiations and travel on tracks with irregularities and worn profiles that influence ride comfort, vehicle stability, and safe operation. Therefore, geometric concepts are an integral part of formulating nonlinear dynamic equations and computational algorithms that can be used effectively to develop credible virtual prototyping models that contribute to developing operational and safety guidelines and to reducing the possibility of train derailments and accidents.

Theory of Curves While a distinction is made between curve and surface geometry and dynamic motion descriptions, simple motion examples can be used to explain some concepts that are used in this book. In this section, the motion of a particle is used to explain basic concepts related to the geometry of curves and surfaces. In general, three coordinates (parameters) are required in order to define the position of a particle in space. These three coordinates can be selected to describe the location of the particle with respect to the origin of a coordinate system formed by three orthogonal axes, as shown in Figure 4. In this case, the position of the particle is defined by the vector $\mathbf{r} = [x \quad y \quad z]^T$, where x, y, and z are

(a)

(b)

Figure 1.3 Geometry of railroad vehicle systems. Sources: (a) serjiob74/Adobe Stock; (b) Susan Isakson/Alamy Stock Photo.

three independent parameters that define the position coordinates of the particle along the three orthogonal axes of the coordinate system XYZ. If the particle is constrained to trace a space curve like the one shown in Figure 4, the particle has the freedom to move only along the curve; the particle freedom to move along two directions perpendicular to the curve is eliminated. That is, the curve equation is completely defined in terms of one parameter, and the three coordinates of the particle are no longer independent.

To demonstrate this simple concept, consider a particle that traces the circle shown in Figure 5. The circle represents a curve with constant curvature defined by the radius of curvature a. If the motion of the particle is restricted to trace this circular curve, two constraints

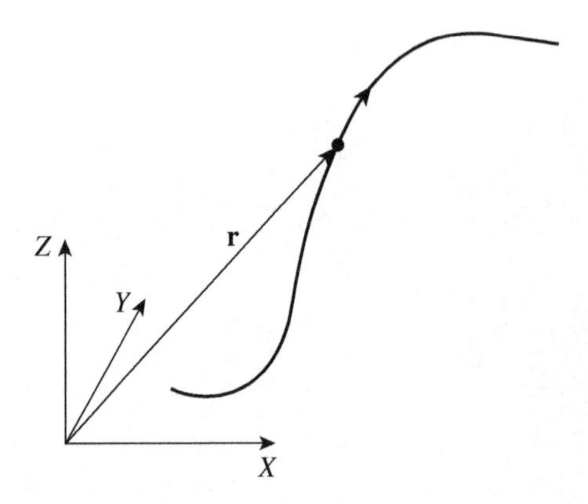

Figure 1.4 Curve geometry.

are imposed on that motion. These constraints, which relate the particle coordinates, can be described using the algebraic equations

$$(x)^2 + (y)^2 = (a)^2, \qquad z = 0 \tag{1.1}$$

The first equation relates the coordinates x and y, while the second equation implies that the motion on the circle is planar. Using the first equation, the coordinate y can be written in terms of the coordinate x as $y = \pm\sqrt{(a)^2 - (x)^2}$. Therefore, the particle position, or the position of an arbitrary point on the curve $\mathbf{r} = \begin{bmatrix} x & y & z \end{bmatrix}^T$, can be written as

$$\mathbf{r} = \begin{bmatrix} x & y & z \end{bmatrix}^T = \begin{bmatrix} x & \pm\sqrt{(a)^2 - (x)^2} & 0 \end{bmatrix}^T \tag{1.2}$$

This equation shows that points on the curve can be traced if the parameter (coordinate) x is given, and the curve equation can be written in terms of one parameter, which is x in this example. Given the parameter x, the location of an arbitrary point on the curve can be completely determined.

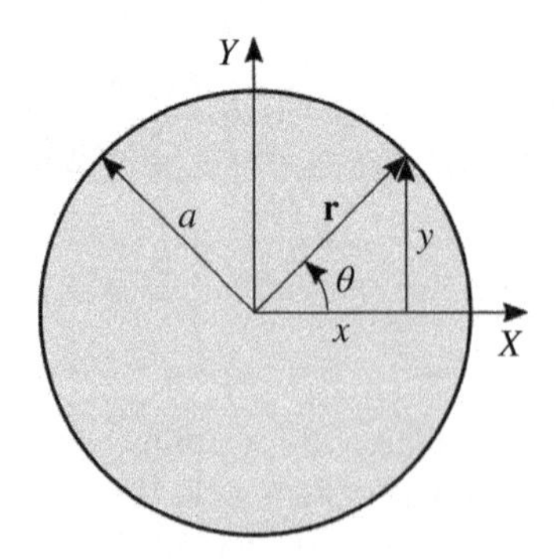

Figure 1.5 Circle geometry.

Curve Parameterization The curve parameter is not unique: that is, other parameters can be used to develop the curve equation. For example, the angle θ shown in Figure 5 can be used as the curve parameter. In this case, one must be able to write the curve Cartesian coordinates x, y, and z in terms of the parameter θ. It is clear from Figure 5 that the coordinates x and y can be written, respectively, in terms of the parameter θ as $x = a\cos\theta$

and $y = a \sin \theta$. Therefore, the curve equation can be written in terms of the parameter θ as

$$\mathbf{r} = \begin{bmatrix} x & y & z \end{bmatrix}^T = \begin{bmatrix} a\cos\theta & a\sin\theta & 0 \end{bmatrix}^T \tag{1.3}$$

Equations 2 and 3 clearly show that the same curve can have different parameterizations. The different parameters, however, are related by algebraic equations, as previously explained using the simple example considered in this section. Equation 2 or 3 is called the curve equation in its *parametric form*. The equation $(x)^2 + (y)^2 = (a)^2$, which defines the circle equation, is called the curve equation in its *implicit form*.

Arc Length It is clear from the simple analysis presented in this section that different parameters can be used to define the parametric form of the curve equation. As discussed earlier, these parameters are related, as evidenced by the fact that one can write the parameter x in terms of the parameter θ as $x = a \cos \theta$, and the parameter θ can be written in terms of the parameter x as $\theta = \cos^{-1}(x/a)$. Similarly, a curve can be parameterized by its *arc length* s. This can be easily demonstrated by writing $\theta = s/a$. This equation can be substituted into Eq. 3 to obtain

$$\mathbf{r} = \begin{bmatrix} x & y & z \end{bmatrix}^T = \begin{bmatrix} a\cos(s/a) & a\sin(s/a) & 0 \end{bmatrix}^T \tag{1.4}$$

Therefore, given any space curve, the curve parametric equation can be written in terms of one parameter as (Do Carmo, 1976; Goetz, 1970; Kreyszig, 1991)

$$\mathbf{r}(t) = \begin{bmatrix} x(t) & y(t) & z(t) \end{bmatrix}^T \tag{1.5}$$

where t is the curve parameter. While a curve can be parameterized using any scalar variable, the curve *arc length* that measures the distance from the curve starting point to an arbitrary point on the curve is often used as the curve parameter. Using the parametric form of the curve expressed in terms of its arc length, a spatial curve can be uniquely defined in terms of geometric invariants called the *curvature* and *torsion*, which appear in the curve *Serret–Frenet equations* presented in Chapter 2. A curve can also be defined in its *implicit form* by eliminating the parameter to obtain a single equation expressed in terms of the curve coordinates, as previously demonstrated by the circular curve example.

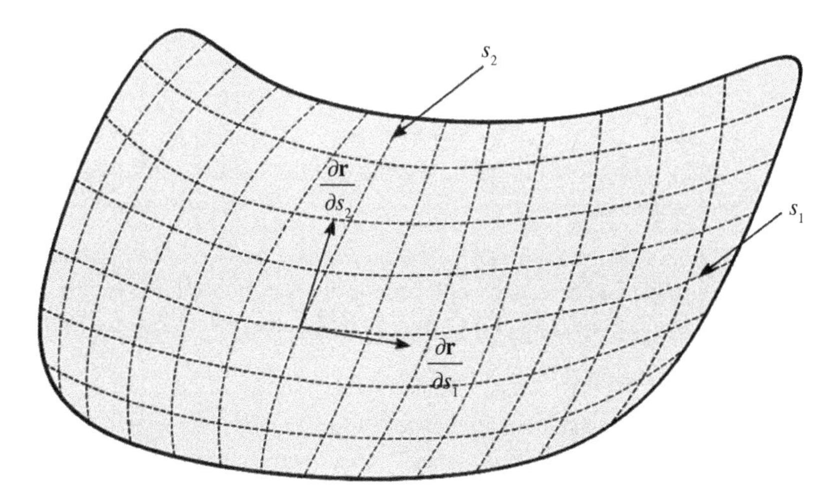

Figure 1.6 Surface geometry.

Theory of Surfaces Unlike curves, surface equations are defined in terms of two parameters, because on a surface, one has the freedom to move in two independent directions, as shown in Figure 6. For example, in the case of a rail surface, a wheel can roll and/or slide on the surface longitudinally or laterally. If the particle considered previously in this section is not constrained to move in the plane on the circular curve, the coordinate z is no longer constant, and such a coordinate can vary. In this case, there is only one constraint equation on the motion of the particle. This constraint equation is defined as $(x)^2 + (y)^2 = (a)^2$, and therefore the vector $\mathbf{r}$ can be written in terms of two independent coordinates x and z as

$$\mathbf{r}(x, z) = \begin{bmatrix} x & y & z \end{bmatrix}^T = \begin{bmatrix} x & \pm\sqrt{(a)^2 - (x)^2} & z \end{bmatrix}^T \tag{1.6}$$

It is clear that this equation, which defines the cylindrical surface shown in Figure 7, depends on the two independent *surface parameters* (coordinates) x and z. Once these two parameters are known, the coordinates of an arbitrary point on the surface can be determined. As in the case of curves, the surface parameterization is not unique. That is, other parameters such as θ and z can be used. In this case, the surface parametric equation can be written as

$$\mathbf{r}(\theta, z) = \begin{bmatrix} x & y & z \end{bmatrix}^T = \begin{bmatrix} a\cos\theta & a\sin\theta & z \end{bmatrix}^T \tag{1.7}$$

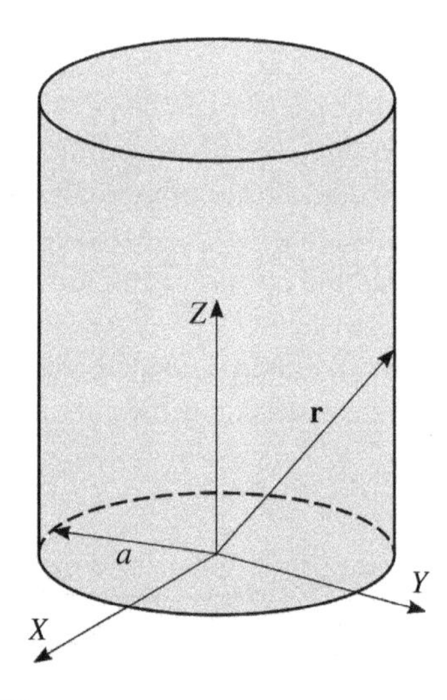

Figure 1.7 Cylindrical geometry.

In general, the surface equation can be written in its parametric form in terms of two independent parameters s_1 and s_2 as (Do Carmo, 1976; Goetz, 1970; Kreyszig, 1991)

$$\mathbf{r}(s_1, s_2) = \begin{bmatrix} x(s_1, s_2) & y(s_1, s_2) & z(s_1, s_2) \end{bmatrix}^T \tag{1.8}$$

Therefore, for a surface, one can define two independent tangent vectors $\partial\mathbf{r}/\partial s_1$ and $\partial\mathbf{r}/\partial s_2$ which define what is called a *tangent plane*. The geometric properties of a surface are defined using the *first* and *second fundamental forms* of surfaces, which are presented in

Chapter 2. The coefficients of these fundamental forms are used to define the *principal curvatures* and *principal directions* that enter into formulating the wheel/rail contact force equations.

Computational Approach for Geometric Representations To use the theory of differential geometry in practical applications, polynomial approximations are used to describe curves and surfaces. In railroad vehicle dynamics, using polynomials allows for defining arbitrary rail and wheel profile geometries. In Chapter 2, a finite element (FE) approach called the *absolute nodal coordinate formulation* (ANCF) is used to define the rail surface geometry. Starting with the polynomial interpolation, the ANCF finite elements are developed by replacing the polynomial coefficients with position and position gradient coordinates. This allows for describing the position of arbitrary points on a continuum using the equation $\mathbf{r}(x, y, z, t) = \mathbf{S}(x, y, z)\mathbf{e}(t)$, where $\mathbf{S}$ is a shape function matrix, t is time, and $\mathbf{e}$ is the vector of position and position gradient coordinates. If ANCF finite elements are used to describe the geometry of fixed rigid rail, one has $\mathbf{r}(x, y, z) = \mathbf{S}(x, y, z)\mathbf{e}$. Using this approach, which enables integrating geometry and analysis, allows for defining a surface systematically by writing an algebraic equation in which one coordinate (parameter) can be expressed in terms of the other two coordinates. For example, one can write $z = f(x, y)$ and use this functional relationship to define the rail surface equation as

$$\mathbf{r}(x,y) = \begin{bmatrix} x & y & f(x,y) \end{bmatrix}^T \tag{1.9}$$

In this surface equation, which can be conveniently defined using ANCF finite elements, only two parameters can be varied. Therefore, ANCF elements can be used to describe the geometries of curves and surfaces in their most general forms based on polynomial interpolations. The use of the ANCF position-gradient coordinates allows for conveniently describing complex shapes as well as the deformations in the case of flexible rails. The approach described in Chapter 2 also allows for using numerical or tabulated data to describe the surface geometry. The fact that one method can be used to define the geometry correctly and to accurately predict the deformation of the rail in the case of flexible rails allows for the systematic integration of geometry and the analysis of complex railroad vehicle systems, as discussed in more detail in the following section.

1.2 INTEGRATION OF GEOMETRY AND MECHANICS

The integration of geometry and mechanics represents the foundation for formulating the railroad vehicle system nonlinear dynamic equations of motion. The dynamic behavior and stability of the rail vehicle depend on the wheel/rail contact forces. These forces are functions of the geometry of the wheel and rail surfaces, which can be described using the techniques of differential geometry as well as computational geometric methods based on polynomial interpolations, as discussed in the preceding section. The track geometry also has a significant impact on rail-vehicle motion and stability. *Track irregularities* can influence vehicle dynamics and be a source of derailments and serious accidents when the vehicle negotiates curved and straight tracks. Therefore, the geometries of these irregularities need to be accurately represented in the simulation models in order to be able to predict their effect on overall vehicle behavior and nonlinear dynamics.

When the vehicle negotiates a curved track, the effect of *centrifugal forces* must be taken into account. To avoid derailments as the result of high centrifugal forces, the geometry of the track is altered by providing a track elevation that results in a lateral gravity-force component that opposes and balances the centrifugal forces, as discussed in this chapter. Accurate prediction of the effect of the centrifugal forces requires an accurate representation of the track geometry. Curved track sections can consist of curves with constant curvatures, and *spirals* that have curvatures that vary along the track. In the case of spirals, the radius of curvature is not constant, and consequently, the centrifugal force does not remain constant. Therefore, in railroad vehicle dynamics, geometry, motion descriptions, and force formulations are interrelated and cannot be separated.

General Displacement In the general case of unconstrained motion, the displacement of a rigid body in space can be described using six independent coordinates. Three coordinates define the global position of a point on the body, called the *body reference point*, and three coordinates define the orientation of the body with respect to the global coordinate system. The global position of the body reference point can be defined using three Cartesian coordinates. The orientation coordinates, on the other hand, can be introduced using three independent parameters that can represent angles or can be parameters that do not have an obvious physical meaning. Therefore, in spatial analysis, the orientation parameters are not unique, and different sets of parameters have been used in the literature and in developing computational *multibody system* (MBS) algorithms.

To define the configuration of a component (body) i in a vehicle system, two coordinate systems are first introduced, as shown in Figure 8. The first coordinate system is the global XYZ coordinate system, which is assumed fixed in time, while the second coordinate system $X^iY^iZ^i$ is the body coordinate system, which is assumed to be rigidly attached to the body *reference point* O^i. Using these two coordinate systems, the global position vector $\mathbf{r}^i$ of an arbitrary point on the rigid body i in the vehicle system can be written as

$$\mathbf{r}^i = \mathbf{R}^i + \mathbf{u}^i \tag{1.10}$$

where $\mathbf{R}^i = \begin{bmatrix} R_x^i & R_y^i & R_z^i \end{bmatrix}^T$ is the global position vector of the body reference point O^i, and $\mathbf{u}^i = \begin{bmatrix} u_x^i & u_y^i & u_z^i \end{bmatrix}^T$ defines the location of the arbitrary point with respect to the origin of the body coordinate system $X^iY^iZ^i$ in the global system: that is,

$$\mathbf{u}^i = \begin{bmatrix} u_x^i & u_y^i & u_z^i \end{bmatrix}^T = u_x^i\mathbf{i} + u_y^i\mathbf{j} + u_z^i\mathbf{k} \tag{1.11}$$

In this equation, $\mathbf{i}, \mathbf{j}$, and $\mathbf{k}$ are, respectively, unit vectors along the global axes X, Y, and Z. As discussed in Chapter 3, the vector $\mathbf{u}^i$ can be written in terms of constant components defined in the body coordinate system $X^iY^iZ^i$. This can be achieved by developing a transformation matrix that defines the body orientation. The columns of the transformation matrix define orthogonal unit vectors along the axes of the body coordinate system. While the body transformation matrix can be expressed in terms of three independent orientation parameters such as Euler angles or any other sets of parameters, the elements of the transformation matrix must assume the same numerical values regardless of the orientation parameters used. These elements of the transformation matrix, as discussed in Chapter 3, are the direction cosines of unit vectors along the axes of the body coordinate system $X^iY^iZ^i$.

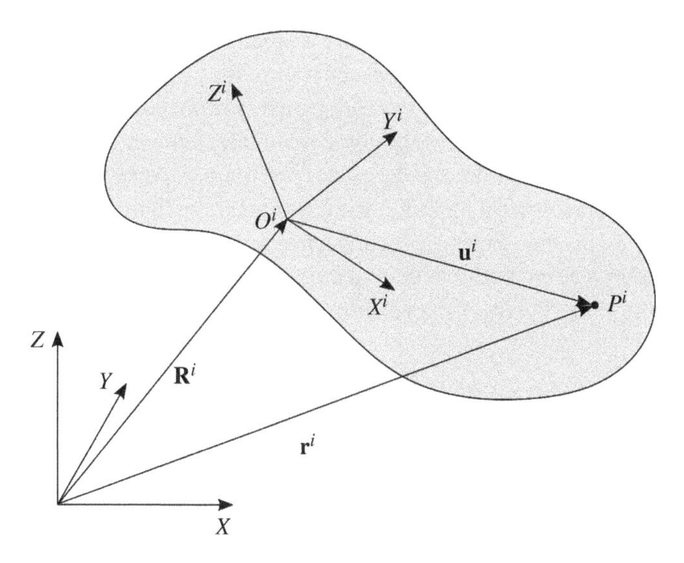

Figure 1.8 Coordinate systems.

Angular Velocity and Orientation Parameters By differentiating Eq. 10 once and twice with respect to time, the absolute velocity and acceleration vectors of the arbitrary point on the body can be defined. The derivative of the transformation matrix with respect to time can be used to define the angular velocity vector, as discussed in Chapter 3. In spatial analysis, the angular velocities are not *exact differentials*, and therefore they are not the time derivatives of orientation parameters. That is, the angular velocities cannot be directly integrated to determine the orientation parameters. Nonetheless, the angular velocities can always be written as linear functions of the time derivatives of the orientation parameters using a velocity transformation matrix. This velocity transformation matrix plays a fundamental role in determining the *generalized forces* associated with the orientation parameters since these orientation parameters serve as *generalized coordinates* and are not directly associated with the Cartesian moments applied to the bodies, as will be discussed in Chapter 6.

The kinematic description that will be used in this book to develop the equations of motion of the components of railroad vehicle systems is introduced in Chapter 3. It is shown in Chapter 3 that the use of three parameters, such as *Euler angles* (Greenwood, 1988; Huston, 1990; Roberson and Schwertassek, 1988; Rosenberg, 1977), to define the body orientation in space leads to kinematic singularities. Such singularities, however, can be avoided by using the four *Euler parameters* at the expense of adding an algebraic constraint equation that relates the four Euler parameters. Euler parameters, which are becoming more popular in developing general MBS algorithms, have many identities that can be used to simplify the kinematic and dynamic equations of the railroad vehicle system.

Euler Angles and Track Geometry In addition to using Euler angles to describe time-dependent motion by defining the orientation of bodies in space, these angles have also been used in railroad vehicle dynamics to define the geometry of the track based on given simple industry inputs. For the most part, track is constructed using three main segments: *tangent* (straight), *curve*, and *spiral*, as shown in Figure 9. The tangent segment has zero curvature, the curve segment has constant curvature, and the spiral segment, used to connect two segments with different curvature values, has a curvature that varies linearly along the track to ensure a smooth transition between the two segments connected by the spiral. The track geometry is often described using three inputs at points along the track at which the geometry changes. These three inputs are the *horizontal curvature*, *superelevation*, and *grade*, and they can be expressed in terms of three Euler angles that are used to construct the track and rail space curves. To this end, Euler angles are converted to *field variables* and used systematically to construct a curve with well-defined geometry based on the given simple track inputs. Unlike the three Euler angles used to describe the time-dependent motion of an

(a)

(b)

Figure 1.9 Track segments. Sources: (a) Dinodia Photos/Alamy Stock Photo. (b) Jens Teichmann/Adobe Stock.

unconstrained body in space, the three Euler angles used to describe the geometry of a curve are written in terms of one parameter that can be the arc length. Therefore, when Euler angles are used to describe curve geometry, these angles are converted to field variables expressed in terms of the curve arc length to ensure a unique definition of the geometry.

Therefore, it is important to recognize that Euler angles are used in this book for two fundamentally different purposes: (i) as motion-generalized coordinates to describe rigid body kinematics in space; and (ii) as geometric field variables to uniquely define the geometry of the track and rail space curves. The analysis presented in Chapter 3 is used as the basis for a computer procedure for developing the track geometry data required for nonlinear dynamic simulations of railroad vehicle systems. The data can be generated before the dynamic simulation at a preprocessing stage in a *track preprocessor* computer program, as will be explained in Chapter 4. The track preprocessor output file normally has data for three different space curves: the *track centerline space curve*, the *right rail space curve*, and the *left rail space curve*, as shown in Figure 10, in which R_H is the radius of curvature of the track centerline curve. These three curves can have different geometries. The right and left rail space curves are used in formulating the wheel/rail contact conditions, while the track space curve is used in the definition of the distance traveled and in the motion description of the coordinate systems of the vehicle components.

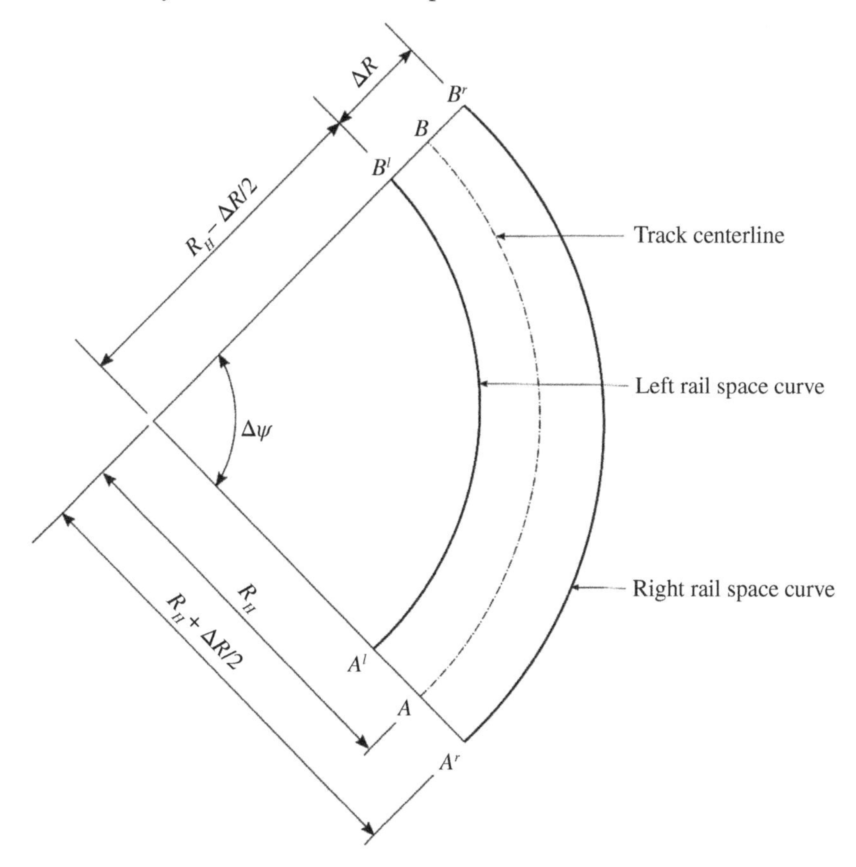

Figure 1.10 Track space curves.

1.3 HUNTING OSCILLATIONS

A simple analysis based on pure kinematics and geometric considerations can be used to shed light on the dynamics of railroad vehicle systems without consideration of the forces. In most railroad vehicle systems, a wheelset consists of two wheels connected by a stiff axle, as shown in Figure 11. The wheels are assumed to have conical profiles with the larger diameter close to the flange in order to achieve self-centering and minimize flange contact (Karnopp, 2004). Lateral wheelset oscillations with respect to the track centerline are referred to as *hunting*. During hunting oscillations, there is a relationship between the wheelset *lateral displacement* and the *yaw angle*, which represents the rotation of the wheelset about an axis normal to the track structure. In this section, a simple analysis based on pure geometry is used to demonstrate the relationship between the lateral displacement and yaw angle of the wheelset when it exhibits hunting oscillations. Such oscillations play a fundamental role in the stability of railroad vehicle systems. As will be demonstrated in this section, the hunting frequency is a function of the forward velocity of the wheelset as well as some other geometric parameters, including the wheel conicity, nominal rolling radius, and distance between the two rails.

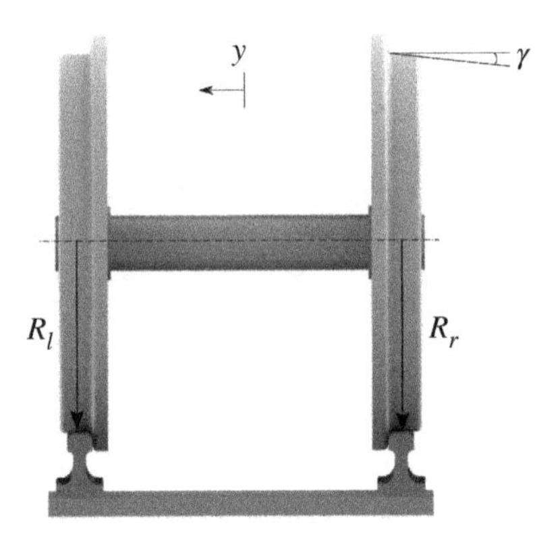

Figure 1.11 Railroad wheelset.

To provide an example of this simple geometric analysis, the wheelset shown in Figure 11 is considered. As shown in the figure, the wheelset conicity is denoted as γ, which defines the slope of the wheel profile curve. The lateral displacement of the wheelset center of mass is denoted as y. Before displacement, the wheelset is assumed to be centered and the displacement y is assumed to be zero: that is, $y = 0$. At this initial configuration, the radii of the two wheels at the points of contact with the rails are equal and denoted as R_o. As a result of disturbances that can be attributed to initial conditions or rail irregularities, the rolling radii of the two wheels will deviate from R_o as the wheelset starts to move forward. These rolling radii are denoted as R_r and R_l for the right and left wheels, respectively. As a result of a lateral displacement y, the rolling radii of the two wheels change, and such a change in the rolling radii is defined by $\Delta R = y\gamma$. It follows from simple geometry that $R_r = R_o - y\gamma$ and $R_l = R_o + y\gamma$. If the wheelset is assumed to rotate with a constant angular velocity ω,

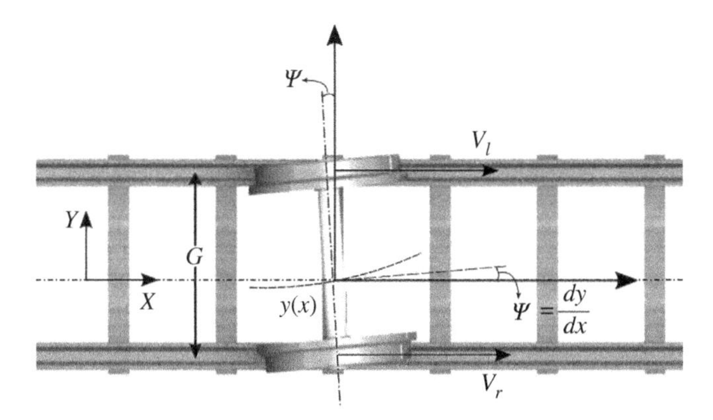

Figure 1.12 Hunting oscillations.

the forward velocities of the right and left wheels can be written, respectively, as

$$\left.\begin{aligned} V_r = \omega R_r = \omega\left(R_o - y\gamma\right) \\ V_l = \omega R_l = \omega\left(R_o + y\gamma\right) \end{aligned}\right\} \tag{1.12}$$

This equation shows that, during hunting oscillations, the two wheels have different forward velocities, and this gives rise to a yaw angle ψ, as shown in Figure 12. Nonetheless, the forward velocity of the wheelset center of mass remains constant and is always defined by the following equation:

$$V = \left(V_r + V_l\right)/2 = \omega R_o \tag{1.13}$$

Using the small oscillation assumption, one can write $\tan\psi = dy/dx \approx \psi$. Because one can write, using Eq. 12, $V_r - V_l = -2y\omega\gamma$, it follows that

$$\left.\begin{aligned} \dot{\psi} = \left(V_r - V_l\right)/G = -2y\omega\gamma/G, \\ \ddot{\psi} = -2\dot{y}\omega\gamma/G \end{aligned}\right\} \tag{1.14}$$

where G is the distance between the two rails. Furthermore, one can write, using the assumption of constant wheelset forward velocity V,

$$\left.\begin{aligned} \dot{y} = \frac{dy}{dt} = \frac{dy}{dx}\frac{dx}{dt} = \psi\, V = \psi\, R_o\omega, \\ \ddot{y} = \dot{\psi}\, V = \dot{\psi}\, R_0\omega \end{aligned}\right\} \tag{1.15}$$

Substituting the first equation of Eq. 14 in the second equation of Eq. 15; and substituting the first equation of Eq. 15 in the second equation of Eq. 14, one obtains, respectively, the following second-order homogeneous ordinary differential equations for the lateral displacement and yaw angle, respectively:

$$\ddot{y} + \left(\omega_h\right)^2 y = 0, \qquad \ddot{\psi} + \left(\omega_h\right)^2 \psi = 0 \tag{1.16}$$

where

$$\omega_h = \omega\sqrt{2R_o\gamma/G} \tag{1.17}$$

is the hunting frequency that can be defined only in the case of positive conicity. In the case of positive conicity, solutions of the preceding equations can be assumed in the forms $y = A_y \sin(\omega_h t + \phi_y)$ and $\psi = A_\psi \sin(\omega_h t + \phi_\psi)$, where A_y and A_ψ are the amplitudes, and ϕ_y and ϕ_ψ are phase angles that can be determined using the initial conditions. These solutions for the lateral displacement and yaw angle show that the frequencies of oscillation of the lateral and angular yaw displacements are the same and are defined by $\omega_h = \omega\sqrt{2R_o\gamma/G}$. Furthermore, by using these solutions, the first equation of Eq. 15, $\dot{y} = \psi R_o\omega$, can be used to prove that the amplitudes of the lateral displacement and yaw angles are related by the equation $A_y = R_o\omega A_\psi/\omega_h = VA_\psi/\omega_h$, and there is a phase angle $\pi/2$ between the lateral displacement y and the yaw angle ψ: that is, $\phi_y - \phi_\psi = \pi/2$. This difference in the phase angle and the relationship $\dot{y} = \psi R_o\omega$ show that the maximum and minimum values of the yaw angle ψ occur when the lateral displacement y is zero, and the maximum and minimum y occur when $\psi = 0$. The hunting oscillations in the case of positive conicity are shown in Figure 13a.

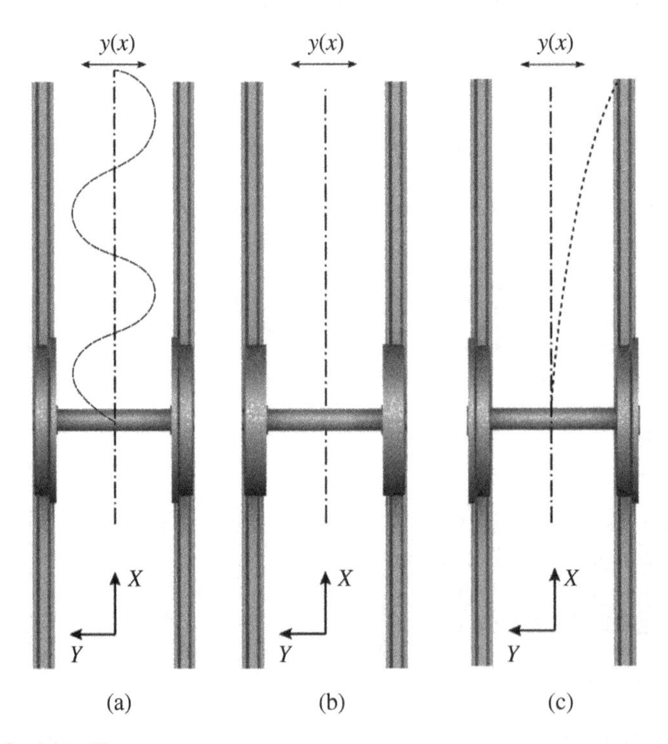

Figure 1.13 Conicity effect.

If, on the other hand, the conicity is equal to zero, $\gamma = 0$, which is the case of a cylindrical wheel, one has, from Eq. 15, $\ddot{y} = 0$ and $\ddot{\psi} = 0$. Integrating these two equations with respect to time shows that the solution is represented by straight lines and the motion is not oscillatory. If the initial conditions are different from zero, the solution will increase with time, leading to an unstable solution. In this case of a cylindrical wheel, the wheelset does not tend to self-center, as shown in Figure 13b.

In the case of negative conicity, $\gamma < 0$, the coefficient of y and ψ in Eq. 16 is negative, and the characteristic equations of the two equations have real roots that define an exponentially unstable solution, as shown in Figure 13c. The simple analysis presented in this section explains the effect of conicity on wheelset stability.

The hunting frequency can be written in terms of the forward velocity of the wheelset using the equation $V = \omega R_o$, which upon substitution into Eq. 17 leads to

$$\omega_h = V\sqrt{\frac{2\gamma}{R_o G}} \tag{1.18}$$

This equation defines the period of hunting oscillations T_h as

$$T_h = \frac{2\pi}{\omega_h} = \frac{2\pi}{V}\sqrt{\frac{R_o G}{2\gamma}} \tag{1.19}$$

The preceding two equations are called *Klingel's formulas* (Klingel 1883). Klingel's formula in Eq. 18 shows that the frequency of hunting oscillations increases as the wheelset forward velocity or the conicity increases, and the hunting frequency decreases as the nominal rolling radius or the distance between the two rails increases. While Klingel's formula is obtained in this section based on pure kinematic and geometric considerations without regard to the forces acting on the wheelset, computer simulations based on nonlinear dynamic formulations that account for all the forces acting on the system have demonstrated the accuracy of the hunting frequency predicted by Klingel's formula in Eq. 18.

1.4 WHEEL AND TRACK GEOMETRIES

The wheel and rail surface geometries enter into formulating the normal and tangential contact forces that influence the nonlinear dynamics and stability of railroad vehicle systems as well as the integrity of the track structure. Therefore, an accurate description of the geometry is necessary for developing credible virtual prototyping computer models for railroad vehicle systems. The wheel and rail surface geometries can be described using the theories of curves and surfaces that are discussed in more detail in Chapter 2. For example, to determine the dimensions of the wheel/rail contact area, the *principal curvatures* and *principal directions* of the wheel and rail surfaces in the contact region must be evaluated. While the wheel geometry can be described using a surface of revolution, the rail geometry can be defined by extrusion of the rail profile in the rail longitudinal direction. More complex surface geometries can be described in the mathematical models using numerical approximation methods in order to capture details that cannot be captured using analytical techniques that are more suited for simple or idealized geometries.

Wheel Surface Geometry A railroad wheel normally consists of two sections: the *tread* and the *flange*. The flange is used to limit the lateral motion of the wheel in order to avoid derailments. The tread is designed to have conicity that provides the tendency of self-centering and decreases the possibility of motion instability. Very high conicity can also lead to higher-frequency hunting oscillations, as discussed in the preceding section. The conicity γ, which depends on the particular combination of the wheel and rail used, normally varies from 1/20 to 1/40, with some values more popular in particular regions and countries. The conicity can also have an effect on the number of contact points between the wheel and rail; the frequency of the occurrence of two-point contacts between the wheel and rail depends on the conicity used.

The geometry of the unworn wheel can be described as a surface of revolution by rotating the profile curve about the wheel axis, as shown in Figure 14. While the profile can assume any shape, conical wheels are often used in order to improve vehicle stability and avoid derailments, as explained in the preceding section. Different profiles with different conicity values are used, depending on the type of vehicle, speed of operation, and loading conditions. The functions that define the profiles are not simple straight-line functions, and in most practical applications, a numerical description of the profile function is required for accurate computer modeling and virtual prototyping.

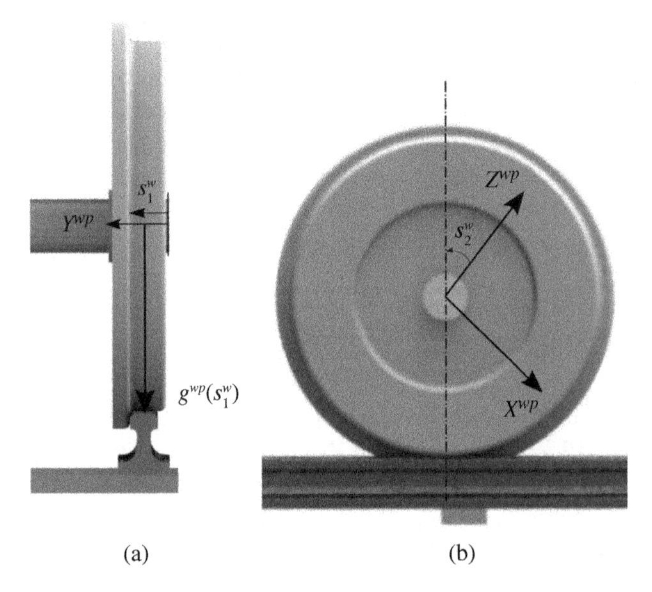

Figure 1.14 Wheel geometry.

The wheel surface can be described using two parameters, as discussed in the preceding section. Because the unworn wheel surface is a surface of revolution, it is convenient to use the parameterization $s_1^w = y_s^w$ and $s_2^w = \theta_s^w$, as shown in Figure 14, where subscript w refers to the wheel. To develop a mathematical definition of the wheel surface, a *profile frame* $X^{wp}Y^{wp}Z^{wp}$ is introduced, as shown in Figure 14, for the convenience of defining the profile curve. The angular surface parameter s_2^w is measured from the Z^{wp} axis. The position of the origin of the profile frame with respect to the wheel coordinate system $X^wY^wZ^w$ is defined by the Cartesian coordinates x_o^{wp}, y_o^{wp}, and z_o^{wp}, which form the elements of the vector: $\overline{\mathbf{R}}^{wp} = \begin{bmatrix} x_o^{wp} & y_o^{wp} & z_o^{wp} \end{bmatrix}^T$. The Y^w axis is assumed to coincide with the wheel axis of rotation. A profile function g^{wp} can be used to define the wheel profile curve in the wheel profile frame. In the case of *unworn wheels*, the profile function g^{wp} does not depend on the wheel angular surface parameter $s_2^w = \theta_s^w$, and in this special case, one has $g^{wp} = g^{wp}\left(s_1^w\right)$. In this case of unworn wheels, the coordinates of an arbitrary point on the wheel surface can be defined mathematically in the selected wheel coordinate system $X^w Y^w Z^w$ in terms of the surface parameters s_1^w and s_2^w as

$$\bar{\mathbf{u}}^w\left(s_1^w, s_2^w\right) = \overline{\mathbf{R}}^{wp} + \bar{\mathbf{u}}^{wp} = \begin{bmatrix} x_o^{wp} + g^{wp}\left(s_1^w\right)\sin s_2^w \\ y_o^{wp} + s_1^w \\ z_o^{wp} - g^{wp}\left(s_1^w\right)\cos s_2^w \end{bmatrix} \tag{1.20}$$

where, as previously defined, $\overline{\mathbf{R}}^{wp} = \left[x_o^{wp} \quad y_o^{wp} \quad z_o^{wp}\right]^T$ is the vector that defines the origin of the wheel profile frame $X^{wp}Y^{wp}Z^{wp}$ with respect to the coordinate system $X^w Y^w Z^w$ of the wheel or wheelset, and $\bar{\mathbf{u}}^{wp} = \left[g^{wp}\left(s_1^w\right)\sin s_2^w \quad s_1^w \quad -g^{wp}\left(s_1^w\right)\cos s_2^w\right]^T$ is the vector that defines the location of the arbitrary point in the profile frame. In the case of a single wheel, the coordinates x_o^{wp}, y_o^{wp}, and z_o^{wp} define the position of the origin of the profile frame $X^{wp}Y^{wp}Z^{wp}$ in the wheel coordinate system $X^w Y^w Z^w$. In the case of a rigid wheelset that has two wheels rigidly connected by an axle, the coordinate systems of the rigidly connected right and left wheels can be assumed the same and have origins located at the axle center point. Using the coordinates x_o^{wp}, y_o^{wp}, and z_o^{wp} of the origin of the wheel profile frame allows the systematic generalization of the geometric description used in this book to the case of deformable wheel axles or independent non-rigidly connected wheels. Furthermore, this description allows for using a numerical spline function representation of the profile function g^{wp}, and therefore, measured wheel profile data can be used. In the case of worn wheels, the profile function depends on both parameters s_1^w and s_2^w: that is, $g^{wp} = g^{wp}\left(s_1^w, s_2^w\right)$. In Chapter 4, a more detailed description of the wheel surface geometry is presented.

Track Geometry Developing an accurate description of the track geometry is one of the basic steps in formulating the rail vehicle nonlinear dynamic equations of motion and in the numerical solution of these equations. As previously discussed in this chapter and shown in Figure 9, for the most part, the track is constructed using three segment types with different geometries: tangent, curve, and spiral segments. A *tangent segment* is a straight section of track with zero curvature, which corresponds to a radius of curvature equal to infinity. A *curve segment* is a circular section of track that has a constant radius of curvature and constant curvature. To connect tangent and curve segments that have different curvatures, a *spiral segment* is used to ensure a smooth transition. When a spiral segment is used to connect tangent and curve sections of the track, the geometry of the spiral segment is designed such that the spiral has zero curvature at the end connected to the tangent segment and the value of the curve curvature at the end connected to the curve segment. This spiral design allows for smoothly varying the curvature and ensuring smooth operation of the rail vehicle during the transition between the tangent and curve sections and vice versa. These simple track segments (straight, curve, and spiral) represent the basic geometric elements for the construction of a track, but at intersections, the track can have a complex structure, as shown in Figure 9. This complex structure, which can include *switches* (*turnouts*) and *guardrails*, can still be constructed using basic rail segments.

The surface of each rail of the track can also be described using two surface parameters, which are selected in this book to be the longitudinal surface parameter s_1^r and the lateral surface parameter s_2^r, as shown in Figure 15, where superscript r refers to the rail. Using

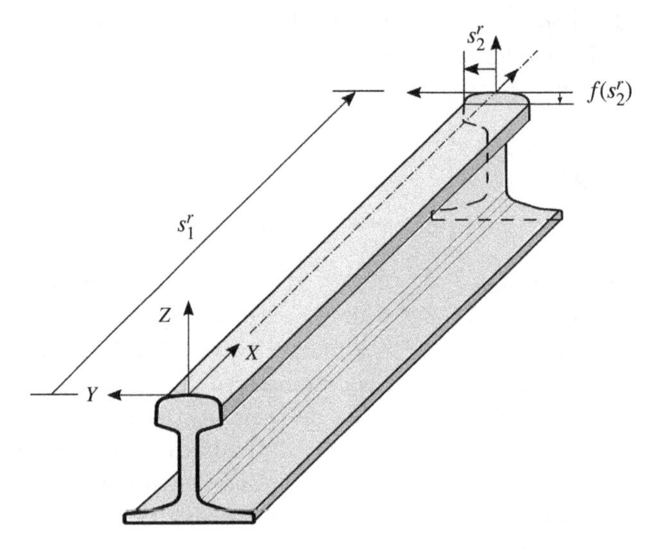

Figure 1.15 Rail geometry.

these surface parameters, the location of an arbitrary point on the rail surface can be defined in a selected rail coordinate system as explained in Chapter 4 as

$$\overline{\mathbf{u}}^r\left(s_1^r, s_2^r\right) = \overline{\mathbf{R}}^{rp} + \mathbf{A}^{rp}\overline{\overline{\mathbf{u}}}^{rp} \tag{1.21}$$

where $\overline{\mathbf{R}}^{rp} = \overline{\mathbf{R}}^{rp}\left(s_1^r\right) = \left[x_o^{rp}\left(s_1^r\right) \;\; y_o^{rp}\left(s_1^r\right) \;\; z_o^{rp}\left(s_1^r\right)\right]^T$ is the vector that defines the origin of the rail profile frame with respect to a rail coordinate system $X^rY^rZ^r$, $\mathbf{A}^{rp} = \mathbf{A}^{rp}\left(s_1^r\right)$ is a matrix that defines the orientation of the profile frame $X^{rp}Y^{rp}Z^{rp}$ with respect to the rail coordinate system $X^rY^rZ^r$, $\overline{\overline{\mathbf{u}}}^{rp} = \left[0 \;\; s_2^r \;\; g^{rp}\left(s_2^r\right)\right]^T$ is the vector that defines the location of the arbitrary point in the profile frame $X^{rp}Y^{rp}Z^{rp}$, and g^{rp} is a function that defines the geometry of the rail profile. In Chapter 4, it is shown how the transformation matrix $\mathbf{A}^{rp} = \mathbf{A}^{rp}\left(s_1^r\right)$ is defined using the techniques of differential geometry presented in Chapter 2.

Track Design Important definitions and terminologies are used in railroad vehicle dynamics, particularly in the layout of the track. These definitions are also important in developing computational algorithms in which numerical representation of the track is necessary. The gage G shown in Figure 16 is the lateral distance between two points on the heads of the right and left rails. In North America, these two points are located at

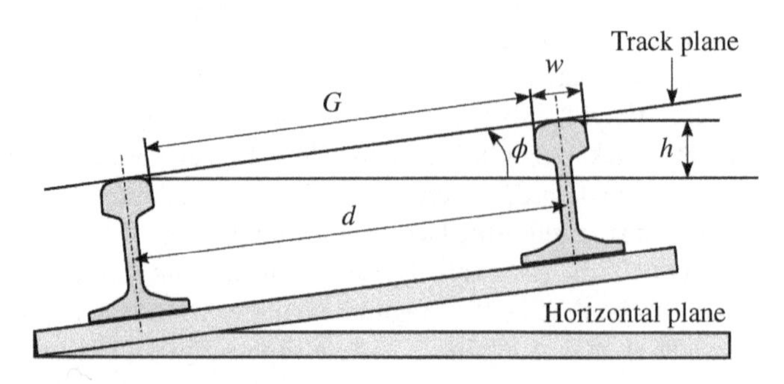

Figure 1.16 Gage and superelevation.

a distance 14 mm (5/8 in.) from the top of the railhead. The standard gage value used in North America varies from 56 to 57.25 in. The *superelevation h*, shown in Figure 16, is defined as the vertical distance between the right and left rails. This vertical distance defines the bank angle ϕ shown in the figure. In the case of a curved section of the track, the *curvature* is different from zero. As shown in Figure 17, the curvature is defined using the value of the angle ψ encompassed by a $100'$-length chord defined by the end points P_1 and P_2. The chord is assumed to have a constant radius R_H in the horizontal plan, as shown in Figure 17. The *grade* is defined as the ratio (percentage) between the vertical elevation and the longitudinal distance. The *cant angle* is defined as the rotation of each rail about its longitudinal axis, as shown in Figure 18.

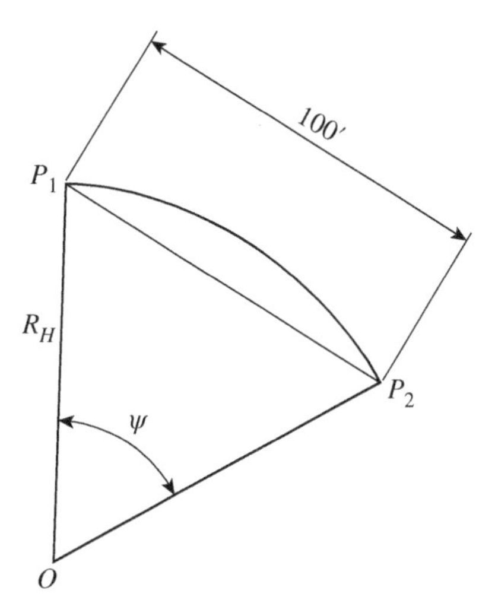

Figure 1.17 Curvature.

Track irregularities can be the source of derailments and serious accidents, and therefore, their effect must be evaluated. As discussed in Chapter 4, some standard track deviation functions can be used in computational models to test the vehicle stability. Measured track data are also often used in the computer simulations of railroad vehicle systems. Track deviations can be classified as *profile* or *alignment*. The *profile* is defined as the vertical deviation of the rail space curve, while the *alignment* is defined as the lateral deviation of the rail space curve, as shown in Figure 19. In Chapter 4, examples of different standard track deviations are provided.

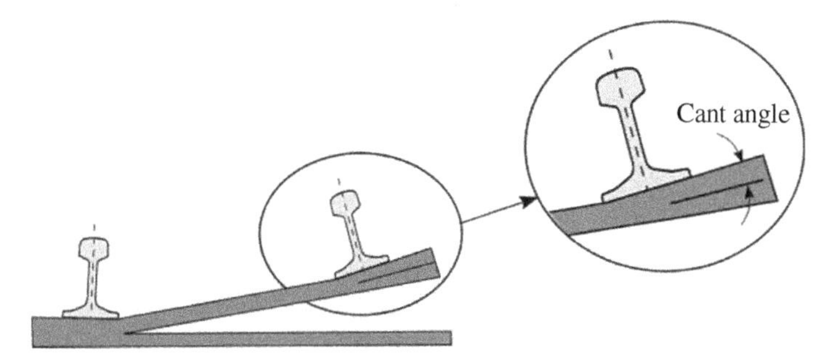

Figure 1.18 Cant angle.

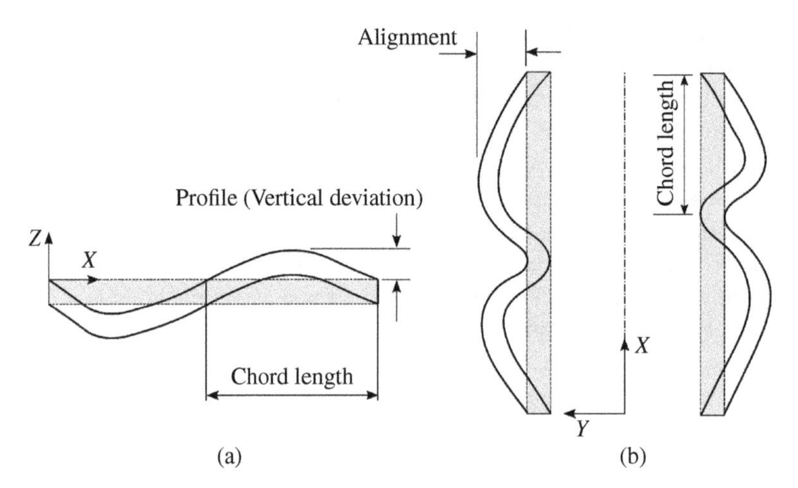

Figure 1.19 Track deviations.

Analytical and Computational Methods Chapter 4 discusses the geometries of the wheel and rail surfaces, which are fundamental in defining the kinematic and dynamic equations of the rail vehicle. The wheel and rail surface equations are defined in terms of two independent surface parameters, as previously mentioned. These surface parameters, which define the surface equations in their parametric form, can be used to determine the locations of arbitrary points on the wheel and rail surfaces. The parametric surface equations are necessary for developing the wheel/rail contact conditions, which are used to determine online the wheel/rail contact points. To describe the rail geometry, two different approaches are discussed in Chapter 4: the *semi-analytical approach* and the *ANCF interpolation approach*. The semi-analytical approach has two main disadvantages. The first is the need to evaluate the derivatives of angles with respect to the rail longitudinal surface parameter in order to determine the tangent, normal, and curvature vectors. The second is the low order of interpolation used to determine the position coordinates of arbitrary points on the rail with limited point data. These data, obtained as the output of a *track preprocessor* computer program, include three position coordinates of and three angles at discrete nodal points on the track. With these limited data at each point, the semi-analytical approach cannot be used with a higher order of interpolation. The ANCF interpolation approach is preferred because it does not have these disadvantages. Furthermore, the ANCF approach does not require differentiation of angles and allows for higher-order interpolation for the position coordinates of arbitrary points on the rail space curve.

1.5 CENTRIFUGAL FORCES AND BALANCE SPEED

When a rail vehicle negotiates a curve or a spiral section, the forces exerted on the vehicle can be significantly different from the forces that arise when the vehicle negotiates a straight segment. During curve and spiral negotiations, the centrifugal forces must be taken into account in order to ensure safe operation of the rail vehicle. In order to avoid derailments, the magnitude of the centrifugal force is used to enforce a limit on the vehicle speed; this limit is referred to as the *balance speed*. In the case of a curve which has a constant radius of curvature, the variations in the centrifugal forces acting on the vehicle are not as significant as in the case of a spiral which has variable curvature. When constructing a track, the geometry of the spiral sections is often designed to have a linearly varying curvature.

To balance the effect of the centrifugal force F_{ce} when the vehicle negotiates a curve, the curve must be designed to have a super-elevation h in order to produce a lateral gravity force component that balances the centrifugal force. The centrifugal force has the effect of pushing the vehicle away from the center of the curve in the direction of the *high rail*, while the lateral gravity force component resulting from the super-elevation has the effect of pushing the vehicle in the opposite direction towards the center of the curve. The direction of the centrifugal force depends on the direction of the normal to the curve traced by the center of mass of the vehicle. In order to demonstrate the interrelationship between the geometry and mechanics concepts, two different scenarios are considered in this section. In the first scenario, it is assumed that the vehicle center of mass traces a circular curve that lies in a plane parallel to the horizontal plane. In this case, the centrifugal force acting on the vehicle lies in the horizontal plane regardless of the amount of the track superelevation. In the second scenario, it is assumed that the vehicle experiences lateral motion, and therefore, the *motion-trajectory curve* is not in general circular and does not lie in a plane parallel to the horizontal plane. In this latter case, the vehicle center of mass can move vertically. Therefore, distinction must be made between the track geometry and the geometry of the curve the vehicle traces during its motion.

Circular Curve Equations If the curve is assumed to have a radius of curvature R, the arc length s that encompasses an angle θ can be written as $s = R\theta$, which implies that for a constant radius of curvature R, the forward velocity V of a vehicle negotiating the curve can be written as $V = \dot{s} = R\dot{\theta}$. If the vehicle, for simplicity, is assumed to be represented by a point mass, the position of the vehicle in the curve plane with respect to a coordinate system located at the center of the curve can be written as $\mathbf{r} = R\begin{bmatrix}\cos\theta & \sin\theta & 0\end{bmatrix}^T$, where the first and second elements in this vector represent, respectively, the lateral and forward components with the forward component in a direction tangent to the curve. Differentiating this position vector $\mathbf{r}$ with respect to time, the velocity of the vehicle is defined as $\dot{\mathbf{r}} = \dot{\theta}R\begin{bmatrix}-\sin\theta & \cos\theta & 0\end{bmatrix}^T$, which shows that the magnitude of the velocity vector is $V = \dot{\theta}R$. Differentiating the velocity with respect to time, one obtains the absolute acceleration vector defined as

$$\ddot{\mathbf{r}} = \ddot{\theta}R\begin{bmatrix}-\sin\theta & \cos\theta & 0\end{bmatrix}^T - (\dot{\theta})^2R\begin{bmatrix}\cos\theta & \sin\theta & 0\end{bmatrix}^T \tag{1.22}$$

If the vehicle is assumed to travel on the curve with a constant forward velocity $V = \dot{\theta}R$, $\dot{\theta}$ is constant, and $\ddot{\theta} = 0$. Therefore, the preceding acceleration equation reduces to $\ddot{\mathbf{r}} = -(\dot{\theta})^2R\begin{bmatrix}\cos\theta & \sin\theta & 0\end{bmatrix}^T$. Since $\theta = s/R$ and $V = \dot{s}$, where s is the arc length of the curve, one has in the case of constant forward velocity V

$$\ddot{\mathbf{r}} = -\frac{(V)^2}{R}\begin{bmatrix}\cos\theta & \sin\theta & 0\end{bmatrix}^T \tag{1.23}$$

This equation is used to determine the centrifugal inertia force.

Horizontal-Plane Curve During curve negotiation, if the vehicle is not allowed to move vertically, the vehicle center of mass traces a circular arc that lies in a plane parallel to the horizontal plane. The normal to this curve is in the horizontal direction, and therefore, the centrifugal force is in the direction shown in Figure 20 regardless of the amount of the superelevation defined by the bank angle ϕ_t. A simple force balance shows that $(mV^2/R)\cos\phi_t = mg\sin\phi_t$. This equation can be used to define the balance speed as $V = \sqrt{Rg\tan\phi_t}$. If the right and left rails are canted to have a gage value similar to the gage value of the track before the super-elevation, one can show that the balance speed obtained in this case does not differ significantly from the actual balance speed that is determined by including the effect of the cant. Because of the hunting oscillations, rolls, and suspension system, the assumption of zero vertical motion can be violated. Nonetheless, highway and railroad super-elevations are designed to limit the vertical motion of the vehicle during curve negotiations by proper vehicle steering or using flanged wheels as in the case of railroad vehicle systems.

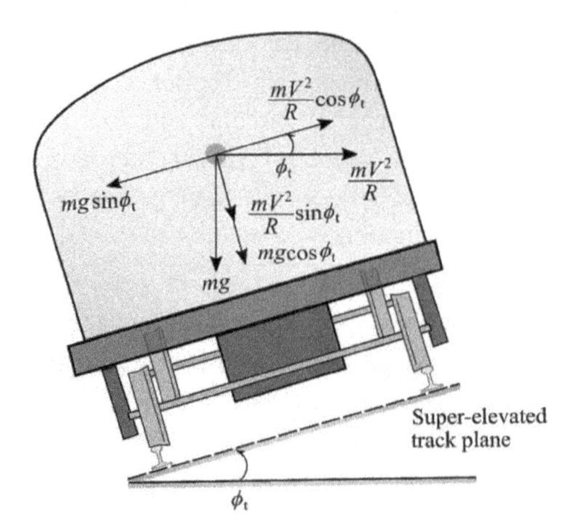

Figure 1.20 Centrifugal force in the case of a horizontal-plane curve.

General Motion-Trajectory Curve In the motion scenario considered above, the center of mass of the vehicle is assumed to negotiate a curve that lies in a plane parallel to the horizontal plane. In this case, the vector normal to the curve that defines the direction of the centrifugal force mV^2/R has no vertical component. In the case of railroad vehicle systems, the vehicle components can move laterally and vertically, and therefore, the assumption of a horizontal curve cannot be, in general, satisfied. In the case of more general motion scenario, the vector normal to the motion-trajectory curve has non-zero vertical component. Using the more general definition of the absolute acceleration $\ddot{\mathbf{r}}$, as will be explained in this section, one can show that the centrifugal force has the direction shown in Figure 21 and has magnitude $m|\ddot{\mathbf{r}}| = mV^2/R$, where m is the mass of the vehicle. In this figure, ϕ_t is the angle that defines the superelevation of the track and ϕ is the angle that defines the elevation of the *osculating plane* of the motion-trajectory curve. The osculating plane is the plane that contains the vectors tangent and normal to the motion-trajectory curve. In the case of the horizontal circular curve previously discussed in this section, $\phi = 0$. The centrifugal

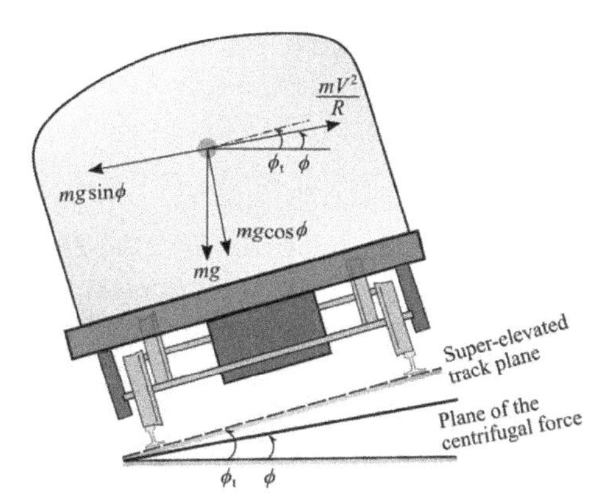

Figure 1.21 Centrifugal force in the case of a general motion-trajectory curve.

inertia force in the more general case in which $\phi \neq 0$ still tends to push the mass away from the center of the curve which can have curvature and radius of curvature that vary with the arc length. This centrifugal force can still be balanced by a component of the gravity force. Using Figure 21 and a simple force balance, one has

$$mV^2/R = mg\sin\phi \tag{1.24}$$

where g is the gravity constant, and ϕ is the angle that defines the direction of the normal to the curve. Because the angle ϕ is assumed to be small, one can write $\sin\phi \approx \tan\phi = h/G$, where G is the track gage. Using this approximation, Eq. 23 can be used to define an approximation of the *balance speed* as

$$V = \sqrt{gR\sin\phi} \approx \sqrt{gRh/G} = \sqrt{gR\tan\phi_t} \tag{1.25}$$

It is clear from this equation that the exact definition of the balance speed is derived using the angle ϕ and not the track superelevation angle ϕ_t. In practice, the difference between the two angles is small. Because the motion-trajectory curve is not a priori known, the use of the approximate value of the balance speed is justified.

Practical Considerations and Geometry If the vehicle is negotiating a curve with a velocity higher or lower than the balance speed, the vehicle is said to have *cant deficiency* or *cant excess*, respectively. Cant excess or cant deficiency is defined as the amount of super-elevation need to be reduced or increased, respectively, such that the given vehicle speed is equal to the balance speed. The super-elevation is normally kept below 7 inches, and therefore, the bank angle ϕ_t is small in most practical applications. While the two cases of general and horizontal-plane curves discussed in this section give results for the balance speed that do not differ significantly, it is important to recognize that because of the lateral oscillations of the railroad vehicle, the conditions of a horizontal-curve cannot be precisely met in practice, and the motion trajectory of the center of mass cannot be described by an exact circular curve. If the actual motion trajectory is defined by a curve $\mathbf{r}(s)$, where s is the arc length, a mass m tracing this curve has a velocity $\dot{\mathbf{r}}(s) = \mathbf{r}_s\dot{s}$, where $\mathbf{r}_s = \partial\mathbf{r}/\partial s$ is the unit tangent to the curve. The absolute acceleration of the mass is defined by the vector $\ddot{\mathbf{r}}(s) = \mathbf{r}_s\ddot{s} + \mathbf{r}_{ss}(\dot{s})^2$. As will be discussed in Chapter 2, the curvature vector $\mathbf{r}_{ss} = \partial^2\mathbf{r}/\partial s^2$ is along the unit vector $\mathbf{n}$ normal to the curve, which is defined from the equation $\mathbf{r}_{ss} = \kappa\mathbf{n}$, where $\kappa = 1/R$ is the curvature of the curve and R is the radius of curvature. Therefore, the acceleration vector can be written as $\ddot{\mathbf{r}}(s) = \ddot{s}\mathbf{r}_s + \left((\dot{s})^2/R\right)\mathbf{n}$, which shows that the centrifugal acceleration $(\dot{s})^2/R = (V)^2/R$ is along the normal to the curve. Therefore, if the trajectory of motion does not lie in the horizontal plane, the direction of the inertia force $m(V)^2/R$ will be the same as the direction of the unit vector normal to the curve. Understanding these basic differential-geometry equations and concepts is important in properly defining the direction of the centrifugal force. For example, regardless of how the high rail is elevated, if a wheelset with flanged wheels is placed on a super-elevated track plane which has two rails to support the two wheels, the center of the wheelset will follow the centerline curve of the track, but not precisely because of any lateral motion. Therefore, the motion trajectory of the wheelset deviates from the centerline curve of the track due to the lateral motion. The motion trajectory of the center of the wheelset is represented by a curve which defines the direction of the centrifugal force and lies in the osculating plane of the curve.

In order to provide an example for the difference between the balance speeds evaluated using the exact and approximate definitions, it is important to recognize that $\phi \le \phi_t$. That is, the maximum value of ϕ cannot exceed ϕ_t. For example, in the case of a bank angle $\phi_t = 6.370°$, the maximum balance speed obtained using the exact definition cannot exceed $V = \sqrt{Rg\sin\phi} = 0.3333\sqrt{Rg}$, while using the assumption of a horizontal-plane curve leads to a balance speed $V = \sqrt{Rg\tan\phi_t} = 0.3344\sqrt{Rg}$. That is, the error is less than 0.11%, which is not significant, as previously mentioned.

1.6 CONTACT FORMULATIONS

Another basic step in the study of railroad vehicle system dynamics and stability is formulating the dynamic interaction forces between the wheel and rail. Accurate prediction of these forces is necessary in order to obtain credible results that shed light on the system dynamic behavior. When the wheel is pressed against the rail, a contact region, referred to as the *contact area* or *contact patch*, is formed. The shape of the region of contact between two solids depends on many factors that include the material properties, the contact pressure, the solid geometries in the contact area, etc. If the contact region covers a finite area that cannot be approximated by a point or a line, one has the case of *conformal contact*. In the case of wheel/rail contact, the dimensions of the contact area are small compared to the dimensions of the wheel and rail, and therefore, a localized concentrated contact is often assumed when the wheel/rail contact forces are evaluated. This assumption of *non-conformal contact* can be justified because of the shapes of the wheel and rail surfaces in the contact area.

Creep Forces When two solids come into contact and are subjected to external pressure, some points on the contact surfaces may slip while other points may stick, and therefore, the contact region consists of slip and adhesion areas. The small relative slip and spin between the two solids can be the result of the difference between the deformations in the contact region, and they lead to creep forces and spin moments, respectively.

The wheel/rail contact forces are the result of a combination of relative *sliding* and *rolling*. As the vehicle attains a certain speed, and because of the effect of friction forces, the motion of the wheel with respect to the rail becomes predominantly rolling with small amount of slipping; this gives rise to *tangential creep forces* as well as *spin moments* that have a significant effect on the vehicle dynamics and stability. In the case of traction and braking, for example, the relative velocity between the wheel and rail increases, causing significant sliding, a case known as *full saturation* in which the tangential forces can be approximated using the *Coulomb friction law*. Below a certain relative velocity value, slipping as the result of the creep phenomenon produces creep forces that can be expressed in terms of normalized velocities called *creepages*. The creep phenomenon, which is due to the elasticity of the two solids in the contact area, is the source of tangential creep forces that are functions of the creepages. Different linear and nonlinear wheel/rail tangential contact force and spin moment models expressed in terms of velocity creepages are used in the railroad vehicle system literature, as discussed in Chapter 5.

Contact Formulations The accuracy of normal and tangential contact force calculations depends on the accuracy of predicting the locations of the wheel/rail contact points. Predicting the locations of the contact points online is necessary for the generality of the wheel/rail dynamic model and for accurately formulating the wheel/rail interaction forces and moments. Fundamentally different approaches are used in formulating the wheel/rail dynamic interaction. In some of these approaches, referred to as *constraint contact formulations* (CCFs), the wheel is assumed to remain in contact with the rail: that is, wheel/rail separation is not allowed. In other formulations, referred to in this book as *elastic contact formulations* (ECFs), wheel/rail separation is allowed. Both formulations can be used to determine the normal contact force, which is the wheel/rail interaction force in the direction normal to the surfaces of contact. To develop efficient computational algorithms for predicting wheel/rail contact forces, it is often assumed that wheel/rail contact is *non-conformal*: that is, the contact is assumed to cover a very small region such that the use of a point contact can be justified. Once the normal contact force is determined, tangential creep forces can be computed using linear or nonlinear models, as discussed in Chapter 5.

The CCF and ECF approaches discussed in Chapter 5 lead to different models with different numbers of degrees of freedom. Because the CCF approach uses contact-constraint equations that must be satisfied at position, velocity, and acceleration levels, the CCF solution procedure is more complex than the ECF solution procedure in which wheel/rail contact force is represented using a compliant force model. The fact that assumptions of non-conformal contact are used in both approaches (CCF and ECF) does not imply that conformal contact is not encountered in railroad vehicle system dynamics. Developing an efficient and accurate method for modeling conformal contact in railroad vehicle system applications remains a challenging problem, and it has been the subject of several research investigations.

Hertz Theory As previously mentioned, in the case of non-conformal contact, the wheel/rail contact area is assumed small in comparison with the wheel and rail dimensions. When the constraint contact formulation is used, the normal contact force is determined as a reaction force. In the elastic contact formulation, on the other hand, normal contact forces are determined using a compliant force model with assumed stiffness and damping coefficients. Experimental observations have shown that the wheel/rail contact area can be approximated using an elliptical shape. The dimensions of the *contact ellipse*, which enter into formulating wheel/rail tangential creep forces, can be determined using *Hertz contact theory* (Hertz 1882). The contact ellipse dimensions, wheel and rail material properties, creepages, and normal contact force predicted using the constraint or elastic contact formulation can be used to formulate the tangential creep force and spin moment. In Hertz contact theory, the geometry of the two surfaces in contact is used to determine the principal curvatures, which are used to determine the dimensions of the contact ellipse. Chapter 5 discusses Hertz contact theory, wheel/rail contact formulations, and creep force models.

Maglev Trains While most of Chapter 5 is devoted to formulating the wheel/rail contact problem, discussions of the forces used in *magnetically levitated* (maglev) trains are also presented. In the case of maglev trains, in which there is no contact between the vehicle and the guideway, the vehicle is lifted using magnetic forces. This allows for a significant increase in train speeds because it eliminates the limitations that are the result of wheel/rail and pantograph/catenary contacts.

Maglev trains are currently being used, for the most part, for short-distance transportation. Many issues need to be addressed before using this technology for long-distance transportation. It is also not clear whether this technology will be useful for freight trains that consist of a very large number of cars.

1.7 COMPUTATIONAL MBS APPROACHES

Railroad vehicles are complex systems that consist of many components, joints, and force elements. Their motion is three-dimensional and is composed of both large translations and finite rotations (Greenwood 1988; Huston 1990; Roberson and Schwertassek 1988; Rosenberg 1977). Some railroad vehicle components, such as wheelsets, spin at very high speeds. Wheel/rail contact must be formulated using a three-dimensional, fully nonlinear approach in order to accurately represent vehicle motion. Simplified planar and linearized approaches are not suited for the analysis of modern railroad vehicle systems, particularly when such systems operate at speeds that do not justify simplifications or linearization. For this reason, using a fully nonlinear MBS approach is necessary for the analysis, design, and performance evaluation of modern railroad vehicle systems.

Complexity of Railroad Vehicles Railroad vehicles consist of interconnected components that can experience large relative displacements with respect to each other. One rail car has a large number of components that have distributed inertia that cannot be modeled using discrete elements. Figure 22 shows an example of a rail car that consists of several subsystems. The *car body* of this rail car, as shown in the figure, is mounted on two *bogies* (also referred to as *trucks*). The car body has distributed inertia, and in accurate simulation models, the car inertia cannot be represented by a concentrated mass. In many investigations, the distributed elasticity of the car body must also be accounted for using a continuum-based approach, such as the finite element (FE) method. Figure 23 shows an

Figure 1.22 Rail car. Source: rruntsch/Adobe Stock.

example of a bogie on which the wheelsets are mounted. The bogie consists of several components with distributed inertia that include two wheelsets, two *equalizers*, a *frame*, and a *bolster*. The car body is mounted on the bogie at the *center plate*, and this connection is often modeled using a pin (revolute) joint. In addition to this pin-joint connection, *secondary suspensions* are used between the car body and the bogie bolster. This secondary suspension, as shown in the figure, has springs and dampers in order to create a vibration-isolation system and reduce the effect of the force transmitted from the bogie to the car body. The *primary suspensions*, shown in the figure, are used to connect the wheelsets to the bogie frame. The primary suspensions serve the purpose of reducing frame vibration that results from the wheel/rail contact forces.

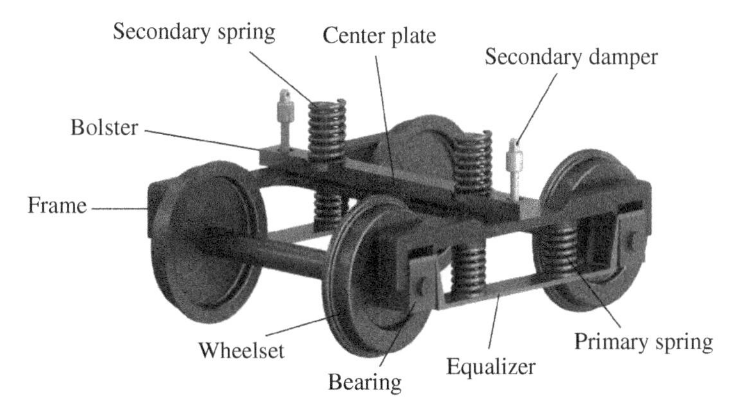

Figure 1.23 Bogie components.

The vehicle components described in this section represent the basic components that must be included in a MBS railroad vehicle model. In addition to these basic components, a large number of bushings and bearing elements must be included in order to develop realistic simulation models. As discussed in this book, a simple rigid body model of a railroad vehicle system must also take into consideration the track geometry and three-dimensional wheel/rail contact. In the case of high-speed trains, a pantograph/catenary system model need also be included, as discussed in a later section. The flexibility of the vehicle components can have a significant effect on vehicle dynamics and stability. Accounting for the car body, track, and catenary deformations using a three-dimensional continuum-based approach may be necessary in order to obtain accurate results in analysis, performance evaluation, and accident investigations.

Computational Approaches The dynamics of complex mechanical systems such as railroad vehicle systems is governed by a system of *differential/algebraic equations* (DAE). The differential equations represent the second-order ordinary differential equations of motion, while the algebraic equations describe the constraints imposed on the motion of the system. The constraint equations represent mechanical joints and specified motion trajectories. In this book, the motion of a multibody vehicle system is described using n coordinates defined by the vector $\mathbf{q} = \begin{bmatrix} q_1 & q_2 & \dots & q_n \end{bmatrix}^T$. These coordinates are related by kinematic constraints because of mechanical joints and specified motion trajectories; therefore, the coordinates are not independent. As explained in Chapter 6, using the principles of mechanics, the system differential equations of motion can be written as

$$\mathbf{M}\ddot{\mathbf{q}} = \mathbf{Q}_e + \mathbf{Q}_v + \mathbf{F}_c \tag{1.26}$$

where $\mathbf{M}$ is the system mass matrix, $\mathbf{Q}_e$ is the vector of applied forces, $\mathbf{Q}_v$ is the vector of Coriolis and centrifugal inertia forces, and $\mathbf{F}_c$ is the vector of constraint forces. As explained in Chapter 6, the constraint equations that represent algebraic relationships between the system coordinates $\mathbf{q} = \begin{bmatrix} q_1 & q_2 & \dots & q_n \end{bmatrix}^T$ and describe the system mechanical joints and specified motion trajectories can be written in a vector form as

$$\mathbf{C}(\mathbf{q}, t) = \begin{bmatrix} C_1 & C_2 & \dots & C_{n_c} \end{bmatrix}^T = \mathbf{0} \tag{1.27}$$

where n_c is the number of algebraic constraint equations. Because each algebraic equation can be used to write a dependent coordinate in terms of the independent coordinates, the vector of coordinates can be written in a partitioned form as $\mathbf{q} = \begin{bmatrix} \mathbf{q}_i^T & \mathbf{q}_d^T \end{bmatrix}^T$, where $\mathbf{q}_i$ is the vector of independent coordinates or *system degrees of freedom*, and $\mathbf{q}_d$ is the vector of dependent coordinates. The number of dependent coordinates is equal to the number of algebraic constraint equations n_c, while the number of degrees of freedom is $n_d = n - n_c$. Each constraint equation in Eq. 27 introduces an independent constraint force, and therefore, for given applied forces, the number of unknowns in Eq. 26 is $n + n_c$. These unknowns are n accelerations $\ddot{\mathbf{q}}$ and n_c independent constraint forces that appear in vector $\mathbf{F}_c$ in Eq. 26. Therefore, the following $n + n_c$ DAE system is sufficient for solving for all unknowns:

$$\left.\begin{aligned} \mathbf{M}\ddot{\mathbf{q}} &= \mathbf{Q}_e + \mathbf{Q}_v + \mathbf{F}_c, \\ \mathbf{C}(\mathbf{q}, t) &= \begin{bmatrix} C_1 & C_2 & \dots & C_{n_c} \end{bmatrix}^T = \mathbf{0} \end{aligned}\right\} \tag{1.28}$$

Chapter 6 presents procedures for solving the DAE system defined by the preceding equations. These procedures lead to different forms of the dynamic equations of motion. One of the procedures used for solving the DAE system is the *augmented formulation,* which leads to a large system of equations that has a sparse matrix structure. Another approach is to use the *embedding technique,* which leads to a number of equations equal to the system number of degrees of freedom; these equations do not include any constraint forces.

Coordinate Selection While different approaches can be used to formulate the dynamic equations of motion of constrained dynamical systems, the *Lagrangian approach* lends itself easily for developing general algorithms for the computer-aided analysis of railroad vehicle systems. In the Lagrangian approach, the concepts of *virtual displacement, virtual work, generalized coordinates*, and *generalized forces* are fundamental. As previously discussed and explained in more detail in Chapter 3, the general unconstrained spatial motion of a rigid body is described using six independent coordinates. Three of these coordinates describe the translation of the body reference point, and three rotation coordinates define the body orientation. The orientation coordinates can be three Euler angles or four Euler parameters, as discussed in Chapter 3. In Chapter 6, the Cartesian translation and orientation coordinates are referred to as the *absolute generalized coordinates*, which define, respectively, the body translation and orientation with respect to a selected global coordinate system. Another set of coordinates that is also used in formulating the dynamic equations of railroad vehicle systems is the set of *trajectory coordinates* as which includes coordinates that define the body translation and orientation with respect to a *body-track coordinate system* that follows the body motion. This body-track coordinate system is referred to in this book as the *trajectory body coordinate system.* The trajectory coordinates will be briefly discussed later in this section and are discussed in more detail in Chapter 6.

Newtonian and Lagrangian Approaches Generalized coordinates are used in this book to define the global position of the origin and the orientation of the body coordinate system. Because the formulations presented in Chapter 6 are based on the *Newton–Euler equations* of motion, the body reference point, which defines the origin of the body coordinate system, is assumed to be attached to the body center of mass in order to eliminate the inertia coupling between the body translation and rotation. The *virtual work principle* used in the Lagrangian formulation is introduced and used with expressions for the absolute velocity and acceleration vectors of an arbitrary point on the body to formulate the body equations of motion. Chapter 6 also explains using *contact conditions* to formulate the wheel/rail interaction forces and discusses hunting oscillations using a constrained wheelset model that accounts for the coupling between the lateral and yaw displacements.

In formulating the MBS constrained dynamic equations of motion, two approaches are commonly used: the *Newtonian approach* and the *Lagrangian approach.* The Newtonian approach, which is based on vector mechanics, requires the use of free-body diagrams by making cuts at the joints. On these free-body diagrams, the joint reaction forces, as well as the inertia and applied forces, are shown. This approach is not well-suited for developing general MBS algorithms and can be used for developing the dynamic equations of motion of relatively simple systems. For the analysis of complex systems, such as railroad vehicles, the Lagrangian approach, which has its roots in D'Alembert's principle and employs scalar quantities such as virtual work and kinetic and potential energies, is often used for developing general-purpose MBS algorithms. When the Lagrangian approach is used, there is no need for free-body diagrams or for the study of the equilibrium of the bodies separately. This is mainly because the Lagrangian approach is based on connectivity conditions, which describe mechanical joints in the system and can be formulated using nonlinear algebraic *constraint equations*. In the Lagrangian approach, joint forces take a standard form expressed in terms of the *constraint Jacobian matrix* and multipliers called *Lagrange multipliers*. The fact that algebraic constraint equations can be used to systematically define joint reaction forces allows for developing general computational procedures for the computer-aided analysis of a wide class of physics and engineering systems without the need for using free-body diagrams.

Trajectory Coordinates Some specialized railroad vehicle formulations do not use the absolute Cartesian coordinates adopted in the development of general MBS algorithms. Instead, another set of coordinates is adopted: *trajectory coordinates* as described in Chapter 3. The detailed formulation of the equations of motion in terms of trajectory coordinates is also presented in Chapter 6. Trajectory coordinates are also used in developing *longitudinal train dynamics* (LTD) algorithms in which each vehicle is normally represented by a single body with one degree of freedom that defines the distance traveled along the track centerline.

As explained in Chapter 3, the trajectory and Cartesian coordinates are related by a velocity transformation, thereby allowing for systematically converting the equations of motion expressed in terms of Cartesian coordinates to equations expressed in terms of trajectory coordinates. As shown in Figure 24, the general displacement of a rigid body i in a vehicle system can be described using six trajectory coordinates

$$\mathbf{p}^i = \begin{bmatrix} s^i & y^{ir} & z^{ir} & \psi^{ir} & \phi^{ir} & \theta^{ir} \end{bmatrix}^T \tag{1.29}$$

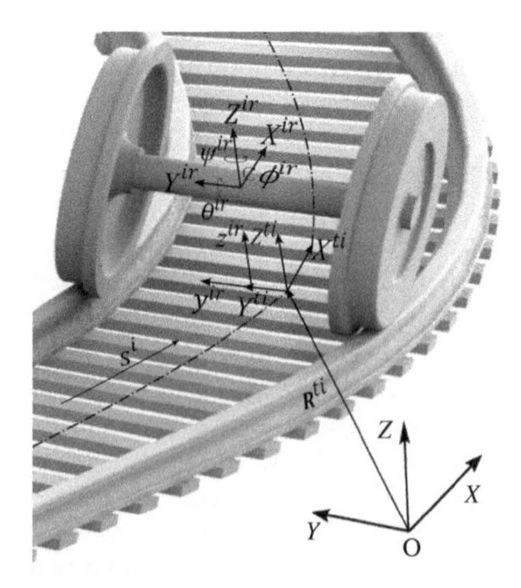

Figure 1.24 Trajectory coordinates.

where s^i is the arc length coordinate of the track space curve; y^{ir} and z^{ir} are, respectively, the lateral and vertical displacements of the origin O^i of the body coordinate system $X^iY^iZ^i$ relative to the trajectory body coordinate system $X^{ti}Y^{ti}Z^{ti}$ that follows the body as shown in Figure 24; and ψ^{ir}, ϕ^{ir}, and θ^{ir} are, respectively, the *yaw*, *roll*, and *pitch* angles that define the orientation of the body coordinate system $X^iY^iZ^i$ with respect to the trajectory body coordinate system $X^{ti}Y^{ti}Z^{ti}$. As discussed in Chapter 3, if the arc length parameter s^i is known, the location of the origin and the matrix that defines the orientation of the trajectory body coordinate system $X^{ti}Y^{ti}Z^{ti}$ can be defined and written in terms of the arc length parameter as $\mathbf{R}^{ti} = \mathbf{R}^{ti}(s^i)$ and $\mathbf{A}^{ti} = \mathbf{A}^{ti}(s^i)$, respectively. Using trajectory coordinates has the advantage of simplifying the specified-motion constraints. However, it has the disadvantage of making the formulation and the computational algorithm more complex, less user-friendly, and more difficult to generalize for the analysis of flexible bodies.

1.8 DERAILMENT CRITERIA

Railroad *derailments* involve complex three-dimensional dynamic behavior and forces that cannot be captured using simplified planar analysis or linearized models. Three-dimensional wheel/rail contact formulations must be adopted in order to accurately predict the spatial motion of the wheel with respect to the rail. Using three-dimensional analysis requires a full parameterization for the wheel and rail surfaces because the theory of curves alone is not sufficient to capture the complexity of some derailment and *wheel climb* scenarios. For example, wheel climb can occur at a low velocity with a relatively large *angle of attack* ($<3°$). The angle of attack α_a is defined as the angle between the forward velocity of the wheel and the longitudinal tangent $\mathbf{t}_1^r$ of the rail, as shown in Figure 25.

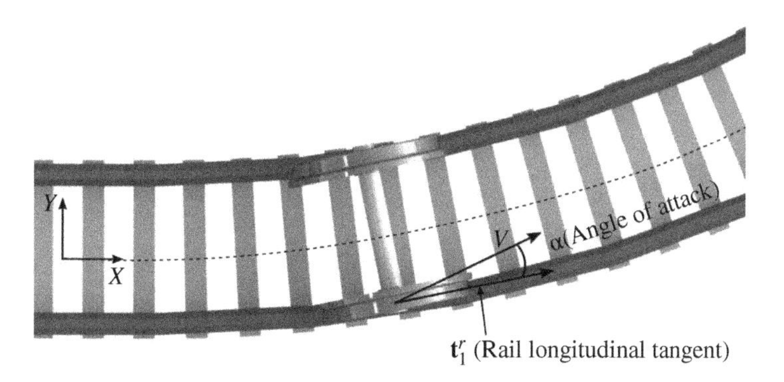

Figure 1.25 Angle of attack.

A large number of investigations have been devoted to the development and validation of derailment criteria (Blader 1989; Elkins and Wu 2000; Marquis and Grief 2011; Shust et al. 1997; Wu and Elkins 1999; Wilson et al. 2004). These criteria have been widely used for developing safety and operation guidelines as well as in accident investigations. Some of these criteria are based on the ratio L/V, where L and V are, respectively, the lateral and vertical forces acting on the wheel, as shown in Figure 26. Examples of derailment criteria are

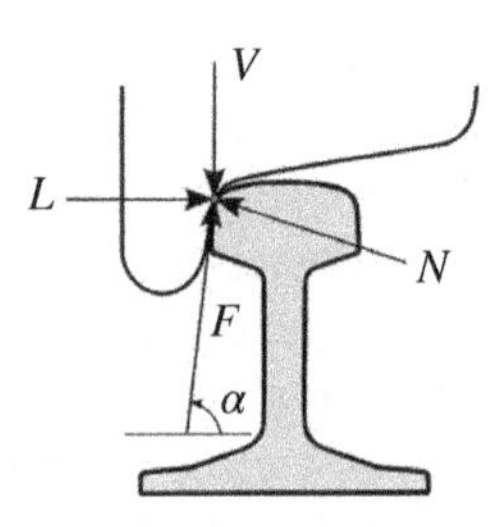

Figure 1.26 Lateral and vertical forces in Nadal's criterion.

the Nadal single-wheel L/V limit criterion, Weinstock axle-sum L/V limit criterion, Federal Railroad Administration (FRA) high-speed passenger distance limit (5 ft), Association of American Railroads (AAR) Chapter 11 50-ms time limit, Japanese National Railway (JNR) L/V time duration criterion, Electro-Motive Diesel (EMD) L/V time duration criterion, and Transportation Technology Center, Inc. (TTCI) wheel climb distance criterion.

General computational MBS approaches can be used to investigate derailment scenarios using more realistic railroad vehicle models that include significant details. These computational algorithms can be used to develop derailment criteria based on three-dimensional analysis and check the accuracy of these criteria using fully nonlinear spatial models. The ratio between the lateral force L and the vertical force V acting on the wheel, as shown in Figure 26, is used as the basis for developing several existing derailment criteria, as previously mentioned. In these criteria, it is assumed that if the lateral force exceeds a certain limit, derailment can occur. The L/V ratio can be more accurately predicted based on fully nonlinear MBS models, and the results can be compared with those obtained using simplified approaches such as *Nadal's formula* (Nadal 1908), which is often adopted for determining the limit of the L/V ratio.

In the simple analysis used to develop the L/V ratio of Nadal's formula, it is assumed that the lateral and vertical forces L and V acting on the rail are balanced by the forces N and F that apply on the wheel, where N is the normal reaction force, $F = \mu N$ is the tangential friction force, and μ is the coefficient of friction. If the wheel *flange angle* is α, a simple force analysis can be used to show that the L/V ratio can be written as

$$\frac{L}{V} = \frac{\tan\alpha - \mu}{1 + \mu\tan\alpha} \tag{1.30}$$

In some of the existing derailment criteria, if the L/V ratio exceeds the right side of Eq. 30, wheel climb is assumed to occur. Because of the shape of the flange, the flange angle is not constant as the wheel climbs the rail. In Nadal's formula, defined by the preceding equation, the maximum wheel-flange contact angle is used. Furthermore, in some criteria, a positive angle of attack is assumed when applying the preceding equation, despite the fact this equation is not a function of the angle of attack.

Because of the complexity of wheel/rail interaction forces, it is difficult to have a universal limit for the L/V ratio in all motion scenarios. For example, Blader (1989) suggested that, in order to avoid derailments when the wheel/rail contact point is located at the rail gage point, the L/V ratio must be within the range 0.66–0.73, depending on the shape of the rail profile. A high value for the lateral force acting on the track can be the source of other problems that can lead to derailments. For example, a high-magnitude lateral force not only

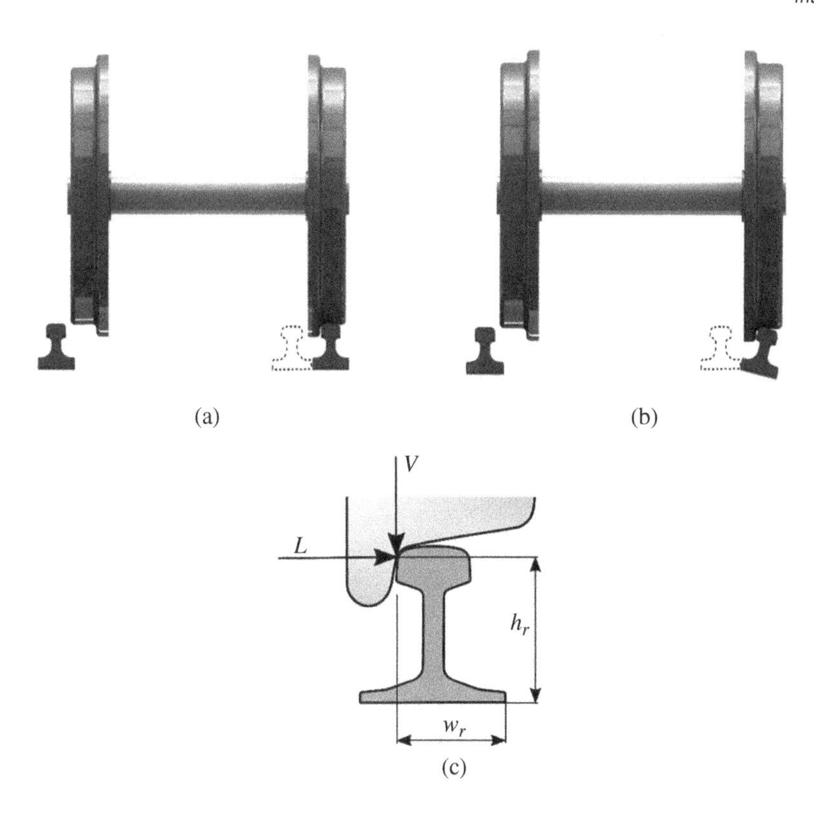

Figure 1.27 Gage widening and rail rollover.

causes wheel climbs but also can lead to *gage widening* and *rail rollover*. Gage widening can lead to wheel/rail separation, as shown in Figure 27a. Rail rollover, shown in Figure 27b, is also a common source of accidents that occur when the vehicle negotiates a spiral section of track. It is clear that if the L/V ratio is higher than the ratio w_r/h_r, where w_r and h_r are shown in the figure, the moment $L \times h_r$ generated by the lateral force L becomes higher than the moment $V \times w_r$ generated by the vertical force V, and this can lead to rail rollover as the result of the rail rotation about its corner.

While a simple analysis can give insight on the issues encountered in the analysis of railroad vehicle derailments and accident investigations, wheel/rail interaction forces, as discussed in this book, are better understood by developing more detailed three-dimensional models that account for displacement modes that cannot be represented using a simple planar analysis model. For instance, wheel climb at an angle of attack can be better understood using a spatial analysis, despite the fact that the yaw angle remains small compared with the pitch angle of rotation of the wheelset about its axis. Because finite and infinitesimal rotations do not commute, using linearized equations to investigate derailments can lead to wrong force prediction, inaccurate velocity results, and wrong conclusions regarding the root causes of derailments of complex railroad vehicles.

Example 1.1 The derivation of Nadal's formula of Eq. 30 is based on simple force balance. Using Figure 26, one can write

$$L = N \sin \alpha - F \cos \alpha$$
$$V = N \cos \alpha + F \sin \alpha$$

The flange tangential force F can be expressed in terms of the normal force as $F = \mu N$. Substituting this equation into the expressions of L and V, one obtains

$$L = N(\sin \alpha - \mu \cos \alpha)$$
$$V = N(\cos \alpha + \mu \sin \alpha)$$

It follows that

$$\frac{L}{V} = \frac{N(\sin \alpha - \mu \cos \alpha)}{N(\cos \alpha + \mu \sin \alpha)} = \frac{\tan \alpha - \mu}{1 + \mu \tan \alpha}$$

It is recommended to use a flange angle α above 70° in order to increase the limiting value of the L/V ratio. Old rail systems used a wheel flange angle $\alpha \approx$ 63–65°, while for modern rail system, the wheel flange angle $\alpha \approx$ 72–75°. In the case of dry wheel/rail contact conditions, the friction coefficient μ can be assumed equal to 0.5, that is $\mu = 0.5$. Using this value of the friction coefficient, one can show that the limiting value for the L/V ratio is 1.1276 for $\alpha =$ 72–75°; and this L/V ratio is 0.7382 for $\alpha = 63°$. This shows that larger flange angle allows for supporting higher lateral force, and this can contribute to the vehicle stability by avoiding derailments.

1.9 HIGH-SPEED RAIL SYSTEMS

As the demand for higher rail-transportation speed increases, more scientific challenges emerge. Modern high-speed rail transportation systems are being designed to operate at speeds of around 370 km/h (220 mph). Nonetheless, research and experimentation are currently focused on trains that can have speeds close to or above 500 km/h (298 mph). At these high speeds, contact and friction introduce many challenging problems that cannot be scientifically examined using some of the contact formulations currently being used. In addition to the wheel/rail contact problem previously discussed in this chapter, which is a subject of more detailed discussion in later chapters, there are challenging problems that are particular to high-speed trains. High-speed trains are electrically powered using a pantograph/catenary system like the one shown in Figure 28. Electrification is not normally used for freight trains, which operate at much lower speeds, have higher axle loads, can consist of a much larger number of cars, and use lower-quality tracks. Freight trains are normally driven by diesel locomotives, which have a speed limit of approximately 238 km/h (148 mph). As a result of this speed limitation, diesel engines cannot be used for high-speed passenger trains. Extensive experimentation is currently under way to significantly increase the train speed to a level that allows minimizing travel time, and at such high speeds, the pantograph/catenary technology presents itself as the only viable power-supply option. Electrification systems, which were used in the 1800s to provide the electric power supply for other transportation systems such as trams, trolleys, and buses, are quieter, more reliable, and less costly, and can provide the power needed for higher speeds.

Figure 1.28 Pantograph/catenary system. Source: hpgruesen/Pixabay.

Electric Power Collection As shown in Figure 28, the *pantograph* is often mounted on the top of one of the train cars. Multiple pantographs can be used for a single train in order to draw sufficient power as the train speed increases. However, there has to be a minimum distance between two pantographs. The voltage used for electric train operations, which can reach 25 kV AC at 50–60 Hz, is generated using electric feeder stations at several locations along the track. The pantograph mechanism collects power through contact with an overhead power line, called the *catenary,* which receives electric power from feeder stations. The pantograph consists of several components connected by joints, while the catenary consists of several cables and can be treated as a structural system. Electric current is collected by the train using the pantograph, which normally has a *carbon strip* that maintains contact with a catenary wire called the *contact wire*. Contact is maintained with the catenary by applying uplift force using pneumatic actuators attached to the pantograph. The combined pantograph/catenary system is also called the *current collection system*. Current collection systems allow electric current to flow through to the train and back to the feeder station through wheel/rail contact. This arrangement for the flow of current represents a closed electric circuit.

Pantograph/Catenary Contact To achieve an uninterrupted power supply to the train, loss of contact between the pantograph carbon strip and catenary contact wire should be minimized. This loss of contact can lead to electric arcing that can cause damage to the carbon strip and contact wire. In addition to the pantograph uplift force, maintaining continuous contact between the pantograph carbon strip and catenary contact wire can be achieved by reducing contact wire vibration. This uninterrupted contact can be achieved by increasing the bending stiffness of the catenary wires using *pretension* as well as using other cables such as a *messenger wire* and *droppers* to support the contact wire and reduce its oscillations resulting from contact and aerodynamic forces.

Pretension of the catenary wires, which leads to higher bending stiffness as a result of the coupling between axial and bending deformations, is necessary to maintain catenary stability and ensure stable contact with the pantograph carbon strip. As the stiffness of the contact wire increases, the speed of the propagation of elastic waves resulting from the pantograph/catenary interaction also increases to a level that is higher than the vehicle operating speed. To avoid resonance, the frequency of the wire oscillations and the speed of the elastic waves must be much higher than the train operating speed, and this can be achieved using pretension of the catenary wires. The pretension can be made independent of weather conditions by using balance weights or hydraulic tensioners, as described in Chapter 7.

Pantograph/Catenary Mathematical Formulations A large number of investigations have been devoted to studying pantograph/catenary system dynamics and vibration. Some of these investigations are based on linear models, while others propose nonlinear models with varying degrees of complexity and assumptions. In some of these models, the pantograph mechanism is modeled as a system of masses and springs, and the catenary is modeled using simple discrete spring-damper elements. Such simplified models do not account for the effect of pantograph joint articulation or distributed inertia and elasticity of the catenary cables. For this reason, simplified models may not accurately predict the system dynamics and stresses that are necessary for a credible evaluation of mechanical pantograph design and catenary structural integrity and durability. Such credible assessment can be made using MBS computational approaches for modeling pantograph dynamics, and continuum-based approaches that account for the distributed inertia and elasticity of catenary wires, as discussed in Chapter 7. MBS approaches for modeling a pantograph system allow for the accurate representation of the rotations of pantograph components, mathematical description of joints between these components, systematic computation of joint and actuator forces, and accurate prediction of contact forces. Continuum-based models of the catenary wire allow for a more accurate evaluation of the speed of elastic wave propagation and the implementation of different material models as well as accounting for geometric nonlinearities that result from large-amplitude oscillations.

ANCF Catenary Model As previously mentioned, the train operating speed should be kept lower than the speed of the propagation of catenary elastic waves due to safety considerations (Kumaniecka and Snamina 2008; Pappalardo et al. 2016). Therefore, it is recommended to use a continuum-based approach, such as the *absolute nodal coordinate formulation* (ANCF) for catenary models, as discussed in Chapter 7. Discrete spring models do not allow for modeling the propagation of elastic waves or assessing the wear and durability of catenary wires. ANCF finite elements can be systematically integrated with general MBS computational algorithms to develop detailed models of the pantograph/catenary system. Formulating the catenary equations of motion using two different ANCF finite elements is discussed in Chapter 7. These elements are the *gradient-deficient cable element* and the *fully parameterized beam element*. While these two elements use different numbers of parameters, a unified approach for modeling contact with the pantograph is presented in Chapter 7. Using ANCF finite elements avoids the use of *incremental-rotation* and *co-simulation* approaches, since the resulting dynamic equations can be solved using a non-incremental-rotation solution procedure.

Pantograph/Catenary Contact Formulations As in the case of wheel/rail contact previously discussed in this section, two fundamentally different formulations can be used to develop pantograph/catenary interaction forces: the *constraint contact formulation* (CCF) and the *elastic contact formulation* (ECF). In the constraint contact formulation, the pantograph carbon strip is assumed to remain in contact with the catenary wire, and therefore, separations or penetrations are not allowed. In this case, pantograph/catenary loss of contact, which is a source of arcing, cannot be modeled. Nonetheless, a constraint contact formulation can be developed to model the longitudinal and lateral relative motion between the pantograph and catenary. In the CCF approach, the normal contact force is determined as a constraint (reaction) force. In the elastic contact formulation, on the other hand, pantograph/catenary separations and penetrations are allowed, and therefore, the loss of contact can be modeled. In the ECF approach, no constraints are imposed on the motion of the pantograph carbon strip with respect to the catenary wire, and the pantograph/catenary normal contact force is described using a compliant force model with assumed stiffness and damping coefficients. The constraint contact formulation is recommended in simulation scenarios in which it is desirable to have a smoother solution by avoiding the oscillations and high frequencies that result from using an elastic contact force formulation. Avoiding these oscillations and high frequencies makes it easier to observe and assess some dynamic behaviors and phenomena that can be difficult to observe and assess when high-frequency oscillations are superimposed on solutions. Therefore, it is recommended to implement both the constraint and elastic contact formulations in computer software developed for nonlinear dynamic simulations of railroad vehicle systems, to be able to conduct research investigations and perform computer simulations of practical motion scenarios.

Pantograph/Catenary Contact Force Control Maintaining pantograph/catenary contact and ensuring system stability at high speeds are necessary for smooth train operations. This is a challenging problem because of disturbances that can result from aerodynamics forces, rail car vibrations, and track irregularities. These disturbances can have an adverse effect on current collection quality as a result of variations in the contact force between the pantograph strip and catenary wire. For this reason, investigations have been conducted to study the effectiveness of developing control algorithms that ensure the stability of pantograph/catenary As previously mentioned, contact and ensure smooth train operation, particularly at high speeds.

The design of effective control systems requires accurate modeling of train dynamics (Poetsch et al. 1997). The stability of the current collection system depends on the dynamic behavior of the pantograph/catenary system and its response to disturbances. As previously mentioned, contact between the pantograph carbon strip and catenary contact wire is normally maintained using an *uplift force* exerted by actuators on the pantograph lower arm. The magnitude of the uplift force should not be very high, to avoid a high contact force between the pantograph strip and catenary wire that can lead to increased wear as a result of high friction forces. Low uplift forces, on the other hand, can lead to loss of contact, which increases the probability of electrical arcing. Arcing, which is the result of electric current flow in an air gap between the catenary contact wire and pantograph contact strip, leads to electric sparks with intense light that can cause damage to the contact wire and contact strip (Hsiao 2010). One solution to this problem is to use an *active control system* by placing an actuator between the pantograph components, with the goal of reducing contact force variations, as discussed in Chapter 7 and the literature (Pappalardo et al. 2016).

Effect of Aerodynamic Forces Environmental conditions such as temperature and wind can have a significant impact on pantograph/catenary interaction forces. For example, high temperature can alter the static equilibrium position and pretension in the catenary wires, while cold weather conditions can result in the formation of ice on the wires, leading to undesirable deformations and poor contact and current collection quality. Aerodynamic forces, on the other hand, can result in severe vibrations of the catenary cables as well as undesirable forces acting on the components of the pantograph system that negatively influence its functional operation. Aerodynamic drag and lift forces can cause variations in contact forces, wear, and loss of contact. Wear can generate asymmetric drag and lift forces, leading to *galloping catenary motion* (Stickland Scanlon 2001; Stickland et al. 2003), while high cross-wind loads can also cause severe vehicle vibrations that directly influence pantograph/catenary interaction (Bacciolone et al. 2008b; Cheli et al. 2010). Aerodynamics drag and lift force components alter the uplift force exerted on the pantograph mechanism. For double-arm pantographs, aerodynamic forces can cause an imbalance, resulting in faster wear of one of the collector strips (Bacciolone et al. 2006a; Carnevale et al. 2016; Pombo et al. 2009). Furthermore, boundary layer turbulence near the car body roof due to vortex shedding can excite pantograph components, generate high frequencies, and increase the sparking level; this, in turn, can negatively impact the current collection quality (Bacciolone et al. 2006a,2006b; Ikeda et al.). For these reasons, evaluating the effect of aerodynamic forces on pantograph/catenary dynamics is an important design consideration. As pointed out by Kulkarni et al. (2017), in general, two main approaches can be used to evaluate the effect of aerodynamic forces on pantograph/catenary systems. In the first approach, *computational fluid dynamics* (CFD) is used to calculate the effect of aerodynamic forces on pantograph system components and to examine the contribution of these forces to total uplift pantograph force (Bacciolone et al. 2007; Carnevale et al. 2016). In the second approach, *drag and lift coefficients* are obtained from experimental studies, and aerodynamic forces are computed using simpler equations. Developing an aerodynamic force model that can be applied to pantograph/catenary systems in which the catenary wires are modeled using continuum-based ANCF finite elements is described in Chapter 7.

Wear Effect Developing a general computational MBS algorithm for predicting pantograph/catenary wear allows incorporating the effects of vehicle vibration, wheel/rail contact forces, and track irregularities. Such an algorithm also allows for predicting the wear rate in the case of different motion scenarios that require the use of nonlinear models, including curve negotiations, accelerations, and braking (Daocharoenporn et al. 2019). In these motion scenarios, the effects of contact forces resulting from vehicle dynamics on the wear rates of the catenary wire can be significant. Severe environmental and operating conditions may cause arcing, high wear rates, or even failure of the pantograph/catenary system to provide desired uninterrupted electric power necessary for train operation. Arcing, for example, may increase significantly the wear rate of the pantograph contact strip, leading to variations of contact forces and a deterioration in pantograph/catenary system performance.

Continuous localized contact between the pantograph carbon strip and catenary contact wire can cause significant wear to the carbon strip inserted on the pantograph top and used for electric current collection. Between two overhead line supports, the contact wire segment is straight; therefore, during curve negotiations, the contact wire sweeps laterally over the whole carbon strip, resulting in uniform wear. When the vehicle negotiates a tangent track, it is also desirable to obtain a pattern of contact in which the contact wires sweep laterally over the surface of the carbon strip to achieve uniform (instead of localized and more severe) wear. Therefore, in the case of a tangent track, the contact catenary wire is slightly zigzagged around the centerline of the track to produce a lateral sweep that leads to uniform wear.

As discussed in Chapter 7, several important factors can have a direct effect on pantograph/catenary wear, including the design of the contact wire, pretension in the catenary cable, and uplift force of the pantograph mechanism. For example, using a high uplift force can lead to a significant increase in the magnitude of the contact force, and this in turn can lead to an increased wear rate. Therefore, the wear rate can be reduced by controlling this uplift force of the pantograph mechanism, as previously mentioned. The wear rate can also be reduced by properly designing the pantograph/catenary system to better handle aerodynamic forces, staggering the contact wire, controlling the intensity of the collection current, using the proper materials for the contact wire and contact strip to reduce friction, and properly adjusting the pretension of the catenary cables (Bucca and Collina 2009; Bucca and Collina 2015; Daocharoenporn et al. 2019). A wear model that accounts for electric and mechanical effects such as the electric arcing effect, *Joule effect* of the electrical current, and effect of contact friction forces is presented in Chapter 7 based on the model developed by Bucca and Collina (2015).

1.10 LINEAR ALGEBRA AND BOOK NOTATIONS

Vector, matrix, and tensor algebra are used to develop static, kinematic, and dynamic equations of physics and engineering systems in compact forms. They are also used to perform addition and multiplication operations required for the proofs of many important identities. Therefore, vector, matrix, and tensor algebra are integral parts of engineering and physics courses, including courses at the undergraduate level. Therefore, it is assumed that the reader has some familiarity with the subject of linear algebra. This section reviews some specific topics of linear algebra that are frequently used in this book. The notations used in the book are also described.

Cross Product and Skew-Symmetric Matrices The relationship between the cross product and skew-symmetric matrix representation is used in this book to define the angular velocity vector. A square matrix $\tilde{\mathbf{A}}$ is said to be *skew-symmetric* if its elements a_{ij}, $i, j = 1, 2, \ldots, n$, where n is the order of the matrix, satisfy the relationship $a_{ij} = -a_{ji}$. This implies that all the diagonal elements of a skew-symmetric matrix are equal to zero, since zero is the only number that is equal to its negative. An example of a 3×3 skew-symmetric matrix $\tilde{\mathbf{A}}$ can be written as

$$\tilde{\mathbf{A}} = \begin{bmatrix} 0 & -a_3 & a_2 \\ a_3 & 0 & -a_1 \\ -a_2 & a_1 & 0 \end{bmatrix} \tag{1.31}$$

It is clear that this form of skew-symmetric matrix can be constructed using the elements of the vector $\mathbf{a} = \begin{bmatrix} a_1 & a_2 & a_3 \end{bmatrix}^T$. Therefore, given any three-dimensional vector $\mathbf{a}$, a unique skew-symmetric matrix $\tilde{\mathbf{A}}$ is associated with the vector. In this book, the skew-symmetric matrix $\tilde{\mathbf{A}}$ associated with the vector $\mathbf{a} = \begin{bmatrix} a_1 & a_2 & a_3 \end{bmatrix}^T$ is assumed to be in the form given by the preceding equation. It is also important to note that a matrix that is equal to the negative of its transpose must be skew-symmetric: that is, if $\mathbf{A} = -\mathbf{A}^T$, then $\mathbf{A}$ must be a skew-symmetric matrix.

The cross product between two three-dimensional vectors $\mathbf{a} = \begin{bmatrix} a_1 & a_2 & a_3 \end{bmatrix}^T$ and $\mathbf{b} = \begin{bmatrix} b_1 & b_2 & b_3 \end{bmatrix}^T$ is defined as

$$\mathbf{a} \times \mathbf{b} = \begin{vmatrix} \mathbf{i} & \mathbf{j} & \mathbf{k} \\ a_1 & a_2 & a_3 \\ b_1 & b_2 & b_3 \end{vmatrix} = (a_2 b_3 - a_3 b_2)\ \mathbf{i} + (a_3 b_1 - a_1 b_3)\ \mathbf{j} + (a_1 b_2 - a_2 b_1)\ \mathbf{k} \tag{1.32}$$

where $\mathbf{i}$, $\mathbf{j}$, and $\mathbf{k}$ are unit vectors along the axes of the coordinate system in which the components of vectors $\mathbf{a}$ and $\mathbf{b}$ are defined. It is clear that the cross product of the preceding equation can be written as

$$\mathbf{a} \times \mathbf{b} = \begin{bmatrix} a_2 b_3 - a_3 b_2 \\ a_3 b_1 - a_1 b_3 \\ a_1 b_2 - a_2 b_1 \end{bmatrix} \tag{1.33}$$

This equation can also be written as

$$\mathbf{a} \times \mathbf{b} = \begin{bmatrix} a_2 b_3 - a_3 b_2 \\ a_3 b_1 - a_1 b_3 \\ a_1 b_2 - a_2 b_1 \end{bmatrix} = \begin{bmatrix} 0 & -a_3 & a_2 \\ a_3 & 0 & -a_1 \\ -a_2 & a_1 & 0 \end{bmatrix} \begin{bmatrix} b_1 \\ b_2 \\ b_3 \end{bmatrix} = \tilde{\mathbf{A}}\mathbf{b} \tag{1.34}$$

This equation shows that the cross product of two vectors $\mathbf{a}$ and $\mathbf{b}$ can be written in terms of the skew-symmetric matrix $\tilde{\mathbf{A}}$ associated with the vector $\mathbf{a} = \begin{bmatrix} a_1 & a_2 & a_3 \end{bmatrix}^T$. Similarly, one can show that $\mathbf{a} \times \mathbf{b} = -\mathbf{b} \times \mathbf{a} = -\tilde{\mathbf{B}}\mathbf{a}$, where $\tilde{\mathbf{B}}$ is the skew-symmetric matrix associated with vector $\mathbf{b}$. The skew-symmetric form of the cross product is repeatedly used in this book and is used to obtain a general definition of the angular velocity vector in the spatial analysis.

Orthogonal Matrices Two n-dimensional vectors $\mathbf{a} = \begin{bmatrix} a_1 & a_2 & \ldots & a_n \end{bmatrix}^T$ and $\mathbf{b} = \begin{bmatrix} b_1 & b_2 & \ldots & b_n \end{bmatrix}^T$ are said to be *orthogonal* if their dot product is equal to zero: that is,

$$\mathbf{a} \cdot \mathbf{b} = \mathbf{a}^T \mathbf{b} = a_1 b_1 + a_2 b_2 + \cdots + a_n b_n = \sum_{i=1}^{n} a_i b_i = 0 \tag{1.35}$$

A square matrix $\mathbf{A}$ is said to be orthogonal if

$$\mathbf{A}\mathbf{A}^T = \mathbf{A}^T \mathbf{A} = \mathbf{I} \tag{1.36}$$

where $\mathbf{I}$ is the identity matrix. It follows that the transpose of an orthogonal matrix is equal to its inverse: that is, $\mathbf{A}^T = \mathbf{A}^{-1}$. Orthogonal matrices are encountered in formulating the dynamic equations of motion of mechanical systems. The transformation matrices that define the orientations of coordinate systems in space are orthogonal matrices. The columns of these transformation matrices define orthogonal unit vectors along the axes of coordinate systems. Therefore, for such orthogonal matrices, no element of the transformation matrix should be greater than one.

The orthogonality condition of Eq. 36 is used in this book to obtain a general definition of the angular velocity vector. In this case, the orthogonal matrix that defines the orientation of the body coordinate system is written in terms of a set of orientation parameters that depend on time. If $\mathbf{A}$ is an orthogonal matrix, and the identity $\mathbf{A}\mathbf{A}^T = \mathbf{I}$ is differentiated with respect to time, one obtains

$$\frac{d}{dt}\left(\mathbf{A}\mathbf{A}^T\right) = \dot{\mathbf{A}}\mathbf{A}^T + \mathbf{A}\dot{\mathbf{A}}^T = \mathbf{0} \tag{1.37}$$

Keeping in mind that $(\mathbf{AB})^T = \mathbf{B}^T\mathbf{A}^T$, then for any arbitrary matrices $\mathbf{A}$ and $\mathbf{B}$ that have the right number of rows and columns for a valid matrix product, the preceding equation can be written as

$$\dot{\mathbf{A}}\mathbf{A}^T = -\mathbf{A}\dot{\mathbf{A}}^T = -\left(\dot{\mathbf{A}}\mathbf{A}^T\right)^T \tag{1.38}$$

This equation shows that $\dot{\mathbf{A}}\mathbf{A}^T$ is equal to the negative of its transpose, and therefore, it is a skew-symmetric matrix that can be written as $\dot{\mathbf{A}}\mathbf{A}^T = \tilde{\boldsymbol{\omega}}$, where $\tilde{\boldsymbol{\omega}}$ can be used to define a vector $\boldsymbol{\omega} = \begin{bmatrix}\omega_1 & \omega_2 & \omega_3\end{bmatrix}^T$. Following a similar procedure, one can use the identity $\mathbf{A}^T\mathbf{A} = \mathbf{I}$ to show that $\dot{\mathbf{A}}^T\mathbf{A} + \mathbf{A}^T\dot{\mathbf{A}} = \mathbf{0}$, which leads to $\dot{\mathbf{A}}^T\mathbf{A} = -\mathbf{A}^T\dot{\mathbf{A}} = -\left(\dot{\mathbf{A}}^T\mathbf{A}\right)^T = \tilde{\overline{\boldsymbol{\omega}}}$, where $\tilde{\overline{\boldsymbol{\omega}}}$ can be used to define a vector $\overline{\boldsymbol{\omega}} = \begin{bmatrix}\overline{\omega}_1 & \overline{\omega}_2 & \overline{\omega}_3\end{bmatrix}^T$. The procedure used to define the two vectors $\boldsymbol{\omega} = \begin{bmatrix}\omega_1 & \omega_2 & \omega_3\end{bmatrix}^T$ and $\overline{\boldsymbol{\omega}} = \begin{bmatrix}\overline{\omega}_1 & \overline{\omega}_2 & \overline{\omega}_3\end{bmatrix}^T$ is followed in Chapter 3 to develop a general expression for the angular velocity vector.

Differentiation of Vector Functions As previously discussed in this chapter, the configuration of a railroad vehicle system is defined using the generalized coordinates $\mathbf{q} = \begin{bmatrix}q_1 & q_2 & \dots & q_n\end{bmatrix}^T$. These coordinates are not, in general, independent, because of mechanical joints and specified motion trajectories. The kinematic constraint equations imposed on system motion can be written in a vector form as $\mathbf{C}(\mathbf{q}, t) = \begin{bmatrix}C_1 & C_2 & \dots & C_{n_c}\end{bmatrix}^T = \mathbf{0}$, where $\mathbf{C}$ is the vector of constraint functions and n_c is the number of algebraic constraint equations. In formulating the equations of motion presented in this book, it is necessary to evaluate the derivatives of the constraint equations with respect to time. For a given constraint function $C_k(\mathbf{q}, t) = C_k(q_1, q_2, \dots, q_n, t)$, the total derivative of the function can be written using the chain rule of differentiation as

$$\begin{aligned}\frac{dC_k}{dt} = \dot{C}_k &= \frac{\partial C_k}{\partial q_1}\dot{q}_1 + \frac{\partial C_k}{\partial q_2}\dot{q}_2 + \dots + \frac{\partial C_k}{\partial q_n}\dot{q}_n + \frac{\partial C_k}{\partial t} \\ &= \sum_{j=1}^{n}\frac{\partial C_k}{\partial q_j}\dot{q}_j + \frac{\partial C_k}{\partial t}\end{aligned} \tag{1.39}$$

This equation can be written as

$$\begin{aligned}\frac{dC_k}{dt} &= \begin{bmatrix}\dfrac{\partial C_k}{\partial q_1} & \dfrac{\partial C_k}{\partial q_2} & \cdots & \dfrac{\partial C_k}{\partial q_n}\end{bmatrix}\begin{bmatrix}\dot{q}_1 \\ \dot{q}_2 \\ \vdots \\ \dot{q}_n\end{bmatrix} + \frac{\partial C_k}{\partial t} \\ &= \frac{\partial C_k}{\partial \mathbf{q}}\dot{\mathbf{q}} + \frac{\partial C_k}{\partial t}\end{aligned} \tag{1.40}$$

where

$$\frac{\partial C_k}{\partial \mathbf{q}} = \begin{bmatrix} \frac{\partial C_k}{\partial q_1} & \frac{\partial C_k}{\partial q_2} & \cdots & \frac{\partial C_k}{\partial q_n} \end{bmatrix} \tag{1.41}$$

and $\partial C_k/\partial t$ is the partial derivative of $C_k(\mathbf{q}, t)$ with respect to time. The preceding equation shows that the partial derivative of a scalar function with respect to the coordinates is a row vector. Therefore, in the case of the vector functions $\mathbf{C}(\mathbf{q}, t) = \begin{bmatrix} C_1 & C_2 & \dots & C_{n_c} \end{bmatrix}^T = \mathbf{0}$, one has

$$\frac{d\mathbf{C}}{dt} = \begin{bmatrix} dC_1/dt \\ dC_2/dt \\ \vdots \\ dC_{n_c}/dt \end{bmatrix} = \begin{bmatrix} \partial C_1/\partial q_1 & \partial C_1/\partial q_2 & \cdots & \partial C_1/\partial q_n \\ \partial C_2/\partial q_1 & \partial C_2/\partial q_2 & \cdots & \partial C_2/\partial q_n \\ \vdots & \vdots & \vdots & \ddots \quad \vdots \\ \partial C_{n_c}/\partial q_1 & \partial C_{n_c}/\partial q_2 & \cdots & \partial C_{n_c}/\partial q_n \end{bmatrix} \begin{bmatrix} \dot{q}_1 \\ \dot{q}_2 \\ \vdots \\ \dot{q}_n \end{bmatrix} + \begin{bmatrix} \partial C_1/\partial t \\ \partial C_2/\partial t \\ \vdots \\ \partial C_{n_c}/\partial t \end{bmatrix}$$

$$= \mathbf{C_q}\dot{\mathbf{q}} + \mathbf{C}_t \tag{1.42}$$

where

$$\mathbf{C_q} = \begin{bmatrix} \partial C_1/\partial q_1 & \partial C_1/\partial q_2 & \cdots & \partial C_1/\partial q_n \\ \partial C_2/\partial q_1 & \partial C_2/\partial q_2 & \cdots & \partial C_2/\partial q_n \\ \vdots & \vdots & \vdots & \vdots \\ \partial C_{n_c}/\partial q_1 & \partial C_{n_c}/\partial q_2 & \cdots & \partial C_{n_c}/\partial q_n \end{bmatrix}, \quad \mathbf{C}_t = \frac{\partial \mathbf{C}}{\partial t} = \begin{bmatrix} \partial C_1/\partial t \\ \partial C_2/\partial t \\ \vdots \\ \partial C_{n_c}/\partial t \end{bmatrix} \tag{1.43}$$

The $n_c \times n$ matrix $\mathbf{C_q}$ is the Jacobian matrix, and the vector $\mathbf{C}_t$ is the partial derivative of the vector functions $\mathbf{C}$ with respect to time. It is clear from the preceding equation that each row in the Jacobian matrix corresponds to a function, and each column corresponds to a coordinate. The Jacobian matrix of the constraint functions plays a fundamental role in the dynamic formulations considered in this book. This matrix can be used to identify the degrees of freedom in the case of complex systems, and it can also be used to define the constraint forces in the Lagrangian formulation discussed in this book.

Book Notations In this book, scalar, vectors, and matrices are used. Italic letters are used for scalar symbols, while bold-face letters are used for vectors and matrices. Some of these scalars, vectors, and matrices are associated with certain bodies in multibody vehicle systems. A scalar a, a vector $\mathbf{a}$, and a matrix $\mathbf{A}$ associated with a body i are referred to as a^i, $\mathbf{a}^i$, and $\mathbf{A}^i$, respectively. That is, superscript i is used to refer to the body number. If a scalar is raised to a certain power, parentheses are used. For example, if a^i is raised to the power 3, it is written as $(a^i)^3$. Using this notation makes clear the difference between the body number and the power. Scalars, vectors, and matrices can also be defined in different coordinate systems. If scalar a, vector $\mathbf{a}$, and matrix $\mathbf{A}$ are defined in the coordinate system of body i, a bar is used, and the scalar, vector, and matrix are written as $\bar{a}^i$, $\bar{\mathbf{a}}^i$, and $\bar{\mathbf{A}}^i$, respectively.

Chapter 2

DIFFERENTIAL GEOMETRY

The geometric description is a fundamental step in the formulation of the equations that govern the mechanics of railroad vehicle systems. The track and wheel geometries must be accurately described in order to correctly predict the wheel/rail contact forces. Both the differential geometry *theory of curves* and *theory of surfaces* are required in order to be able to formulate the wheel and rail kinematic and force equations. These theories are used to define nonlinear geometric equations that can be solved for the location of the wheel/rail contact points, as will be explained in a later chapter of this book. Furthermore, the spatial wheel and rail geometric representations are crucial in the study of railroad vehicle nonlinear dynamics in different motion scenarios, including curve negotiations and travel on tracks with irregularities and worn profiles that influence ride comfort, vehicle stability, safe operations, and the possibility of train derailments.

Theory of Curves Curve equations are defined in terms of only one parameter (Do Carmo 1976; Goetz 1970; Kreyszig 1991). To trace a curve, one has to move in only one direction along the curve because motion is not allowed in any other directions perpendicular to the tangent to the curve at an arbitrary point. Because of this restriction, the location of an arbitrary point on the curve in a three-dimensional space can be defined completely in terms of one parameter. If the value of this parameter is given at an arbitrary point, the three Cartesian coordinates that define the location of the point can be determined. While a curve can be parameterized using any scalar variable, the curve *arc length* that measures the distance from the curve starting point is often used as the curve parameter. Using the parametric form of the curve, a spatial curve can be uniquely defined in terms of geometric invariants called the *curvature* and *torsion*, which appear in the curve *Serret–Frenet equations*. A curve can also be given in an *implicit form* by eliminating the parameter to obtain a single equation expressed in terms of the curve coordinates.

Theory of Surfaces Unlike curve equations, surface equations are defined in terms of two parameters because, on a surface, one has the freedom to move in two independent directions. For example, in the case of a rail surface, a wheel can slide and/or rotate on the surface longitudinally or laterally. Therefore, for a surface, one can define two independent tangent vectors that define what is called a *tangent plane*. The geometric properties of a surface are defined using the *first* and *second fundamental forms* of surfaces. The coefficients

Mathematical Foundation of Railroad Vehicle Systems: Geometry and Mechanics,
First Edition. Ahmed A. Shabana.

of these fundamental forms are used to define the *principal curvatures* that enter into the formulation of the contact force equations.

Numerical Description of Geometry To use the theory of differential geometry in practical applications, polynomial approximations are used for curves and surfaces. In railroad vehicle dynamics, using polynomials allows for defining arbitrary rail and wheel profile geometries. In this chapter, a finite element (FE) approach, called the *absolute nodal coordinate formulation* (ANCF), is used to define the rail geometry. Using this approach enables integrating geometry and analysis, particularly if the rail deformation is of concern and must be predicted. ANCF elements can be used to conveniently describe the geometries of both curves and surfaces. Using the ANCF position-gradient coordinates allows for describing complex shapes as well as the deformations in the case of flexible rails. The first two sections of this chapter follow, for the most part, the first two sections of Chapter 3 of a previous book on railroad vehicle dynamics (Shabana et al. 2008).

2.1 CURVE GEOMETRY

A curve can be described using a single implicit-form equation or in a parametric form in which the coordinates of an arbitrary point on the curve are written in terms of one parameter. In the *implicit form*, the single equation relates the coordinates of an arbitrary point on the curve. For example, in the case of a planar parabolic curve, one can write $y=(x)^2$, where x and y are the Cartesian coordinates of a point on the curve. When using the *parametric form* of a spatial curve, on the other hand, the three coordinates of a point on the curve in a Cartesian coordinate system XYZ, as shown in Figure 1, can be written as $\mathbf{r}=[x \; y \; z]^T$, where $x=x(t)$, $y=y(t)$, and $z=z(t)$ are the Cartesian coordinates of the point on the curve, and t is the parameter. Therefore, a curve defined over the interval $a \leq t \leq b$ can, in general, be written in the parametric form

$$\mathbf{r}(t) = \begin{bmatrix} x(t) & y(t) & z(t) \end{bmatrix}^T \tag{2.1}$$

Given the value of the parameter t, the location of a point on the curve can be determined. The coordinates $x=x(t)$, $y=y(t)$, and $z=z(t)$ must be differentiable over the interval on

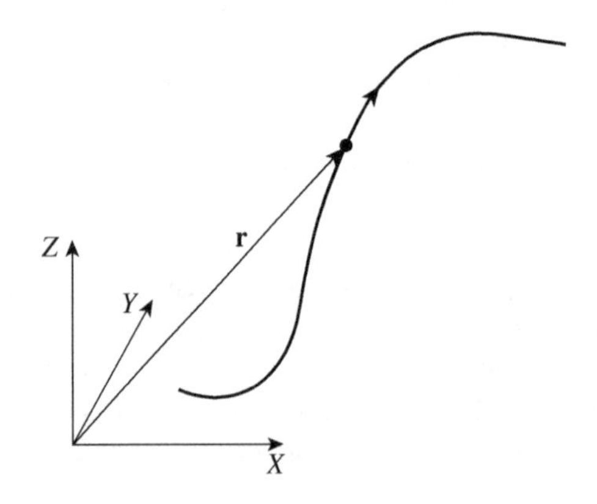

Figure 2.1 Spatial curve.

which t is defined. In the case of a curve, one can trace points in only one direction, and motion in a plane perpendicular to this direction is not permitted; so, Eq. 1 is completely defined once the parameter t is given.

Example 2.1 The simple equation of a parabolic curve, shown in Figure 2, defined in its implicit form, is $y=(x)^2$. The parametric form of the parabolic curve can be written as

$$x(t) = t, \qquad y(t) = (t)^2$$

where t is the parameter. Note that by eliminating the parameter t from the parametric representation, the single equation $y=(x)^2$, which defines the implicit curve form, can be obtained. Note also that in this case, the parameter $t=x$: that is, the coordinate x is used as the curve parameter. This choice of the curve parameter is not unique.

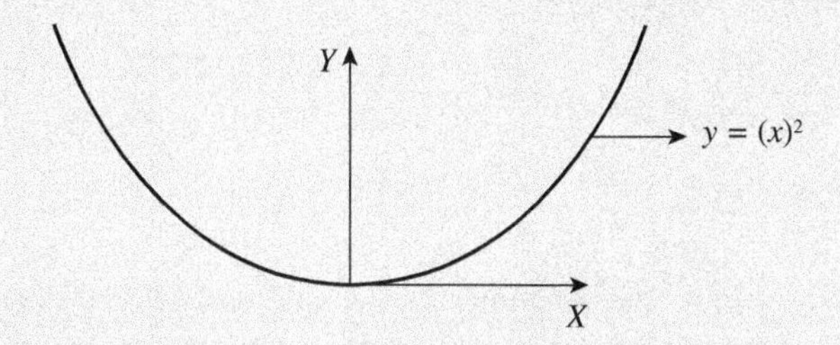

Figure 2.2 Parabolic curve.

Example 2.2 The equation of a circle, shown in Figure 3, with radius a and center at the origin, is defined by the parametric equations

$$x(t) = a\cos t, \qquad y(t) = a\sin t$$

These equations, in which t is the parameter, can be used to obtain the implicit form $(x)^2+(y)^2=(a)^2$ by simply eliminating t by squaring and adding the two parametric equations and using the trigonometric identity $\cos^2 t+\sin^2 t=1$. The parameter t represents the angle that defines the point on the circle, as shown in Figure 3.

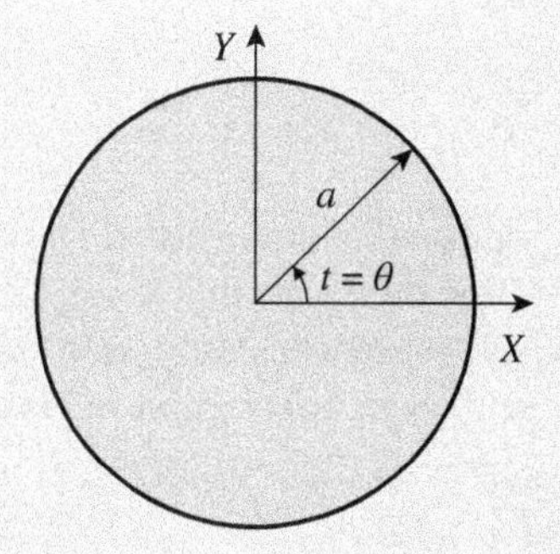

Figure 2.3 Circle geometry.

(Continued)

If the center of the circle is not at the origin of the Cartesian coordinate system and is defined by the coordinates x_c and y_c, the parametric representation of the curve can be written as

$$x(t) = x_c + a\cos t, \qquad y(t) = y_c + a\sin t$$

These two equations can be used to define the implicit form of the circle, by eliminating the parameter t, as $(x - x_c)^2 + (y - y_c)^2 = a^2$.

In this book, polynomials will be used to define the curve and surface geometries. Non-rational polynomials can lead to an accurate representation of the geometry despite the fact that some geometries cannot be described exactly using non-rational polynomials. A circle is an example of a geometry that can only be described exactly using *rational polynomials*. One can show that the parametric curve equations $x(t) = a\cos t$ and $y(t) = a\sin t$ of a circle with center at the origin can be written using the following rational polynomials:

$$x(t) = a\left(\frac{1-(t)^2}{1+(t)^2}\right), \qquad y(t) = a\left(\frac{2t}{1+(t)^2}\right)$$

Example 2.3 The parametric equation of a helix of radius a like the one shown in Figure 4 is $\mathbf{r}(t) = \begin{bmatrix} a\cos t & a\sin t & \alpha t \end{bmatrix}^T$, where a and α are constants. One can show that the trace of this curve is a helix of pitch $2\pi\alpha$ on the cylinder $(x)^2 + (y)^2 = (a)^2$ (Do Carmo 1976).

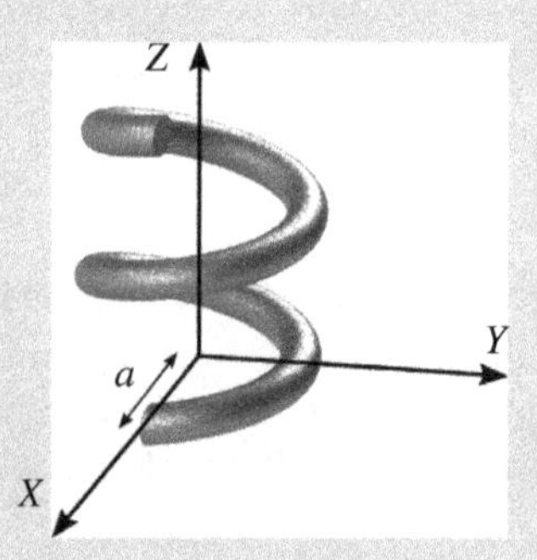

Figure 2.4 Helix geometry.

Tangent Vector Given two points on the curve defined by the parameter values t_1 and t_2 as shown in Figure 5, one can define the vector difference $\Delta\mathbf{r} = \mathbf{r}(t_2) - \mathbf{r}(t_1)$. In the limit as t_1 approaches t_2: that is, $\Delta t = t_2 - t_1$ approaches zero, $\Delta\mathbf{r}/\Delta t$ becomes tangent to the curve. Therefore, for a given t, the tangent vector to the curve at t is defined as

$$\frac{d\mathbf{r}}{dt} = \begin{bmatrix} \frac{dx(t)}{dt} & \frac{dy(t)}{dt} & \frac{dz(t)}{dt} \end{bmatrix}^T \tag{2.2}$$

If $|d\mathbf{r}(t)/dt| = 0$ at a point t, the point is called a *singular point*. The *arc length s* of a curve between two arbitrary points t_o and t can be defined using the norm of the tangent vector as

$$s = \int_{t_o}^{t} \left| \frac{d\mathbf{r}}{dt} \right| dt \tag{2.3}$$

If a curve is parameterized by its arc length s – that is, $\mathbf{r} = \mathbf{r}(s)$ – one can show using Eq. 3 that the tangent vector, $\mathbf{t}(s) = d\mathbf{r}/ds$, is a unit vector: that is,

$$\left| \frac{d\mathbf{r}}{ds} \right| = |\mathbf{t}(s)| = 1 \tag{2.4}$$

The proof of Eq. 3 is straightforward and can be established by considering two points on the curve defined by the parameters t_1 and t_2, as shown in Figure 5. As t_1 approaches t_2, the infinitesimal arc length ds can be written as $(ds)^2 = d\mathbf{r}^T d\mathbf{r}$, where $d\mathbf{r} = \mathbf{r}(t_2) - \mathbf{r}(t_1)$. Since $d\mathbf{r} = (d\mathbf{r}/dt)dt$, one has $(ds)^2 = (d\mathbf{r}/dt)^T(d\mathbf{r}/dt)(dt)^2$, which yields $ds = \sqrt{(d\mathbf{r}/dt)^T (d\mathbf{r}/dt)}\ dt = |d\mathbf{r}/dt|\, dt$. Equation 3 follows by integrating this equation from t_o to t. It is also clear from Eq. 3 that if $t = s$, then $|d\mathbf{r}/ds| = 1$, which provides a proof that $\mathbf{t}(s) = d\mathbf{r}/ds$ is a unit vector. This result is also intuitively obvious since as t_1 approaches t_2, it is clear that $\Delta s = |\Delta \mathbf{r}|$, which shows that $(\Delta s)^2 = \Delta \mathbf{r}^T \Delta \mathbf{r}$. This equation yields $(\Delta \mathbf{r}/\Delta s)^T(\Delta \mathbf{r}/\Delta s) = 1$, which shows that the tangent to the curve evaluated by differentiation with respect to the arc length is indeed a unit vector.

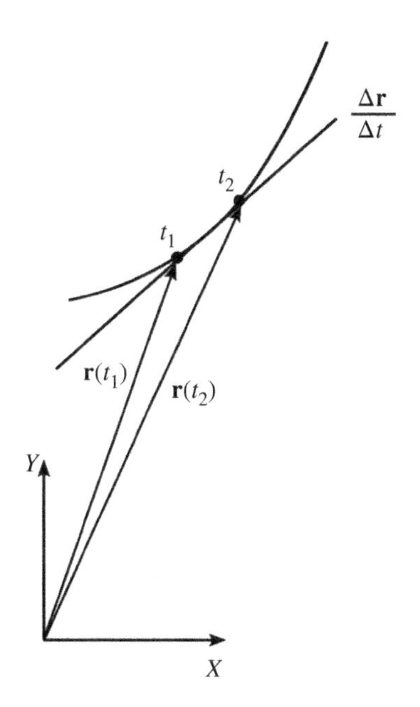

Figure 2.5 Parametric derivative.

While any parameter can be used to define the curve, the tangent vector is a unit vector only when the curve is parameterized by its arc length and the differentiation is carried out with respect to the arc length. For an arbitrary parameter t, one can write the relationship

$\mathbf{dr}/dt = (\mathbf{dr}/ds)(ds/dt)$. Therefore, without any loss of generality, it is assumed in the remainder of this section that the curve is parameterized by its arc length s.

Example 2.4 The equation of a circle with radius a and center at the origin is defined in Example 2 by the parametric equation $\mathbf{r}(t) = \begin{bmatrix} a\cos t & a\sin t \end{bmatrix}^T$. The tangent vector is defined as

$$\frac{\mathbf{dr}(t)}{dt} = \begin{bmatrix} -a\sin t & a\cos t \end{bmatrix}^T$$

It is clear that the length of this tangent vector is a, and therefore, this vector is not a unit vector. It is also clear, in this example, that the parameter t represents an angle θ : that is, $t = \theta$, and θ varies from 0 to 2π. Therefore, the length of the circle can be evaluated using Eq. 3 as

$$s = \int_{t_0}^{t} \left| \frac{\mathbf{dr}}{dt} \right| dt = \int_{0}^{2\pi} a d\theta = 2\pi a$$

which defines the expected length of the circle perimeter. Furthermore, because $\mathbf{dr}/d\theta = (\mathbf{dr}/ds)(ds/d\theta)$, $\mathbf{dr}/d\theta = \begin{bmatrix} -a\sin\theta & a\cos\theta \end{bmatrix}^T$, and $\mathbf{dr}/ds$ is the unit tangent vector $\mathbf{dr}/ds = \begin{bmatrix} -\sin\theta & \cos\theta \end{bmatrix}^T$, one has $(ds/d\theta) = a$ or, equivalently, $ds = ad\theta = adt$, which implies that $t = s/a$. Substituting this equation in the curve equations and differentiating with respect to s, one can show that the resulting tangent vector is indeed a unit vector. The equation $ds = ad\theta$, which defines the relationship between the infinitesimal arc length ds and the angle $d\theta$ encompassed by a curve segment with a radius of curvature a, is general and is applicable to circular and noncircular curves.

Curvature Vector and Serret–Frenet Equations Assuming that a curve is parameterized by its arc length, the *curvature vector* is obtained by differentiating the unit tangent vector with respect to the arc length as

$$\mathbf{r}''(s) = \frac{d^2\mathbf{r}}{ds^2} = \frac{\mathbf{dt}}{ds} \tag{2.5}$$

In this equation, $\mathbf{r}'' = d^2\mathbf{r}/ds^2$ denotes twice differentiation with respect to s. Since the tangent vector resulting from differentiation with respect to the arc length is a unit vector, the tangent and curvature vectors are orthogonal: that is, $\mathbf{r}''^T\mathbf{t} = 0$. This can be proven by differentiating the equation $\mathbf{r}'^T\mathbf{r}' = 1$ with respect to s, leading to $2\mathbf{r}'^T\mathbf{r}'' = 0$, which demonstrates the orthogonality of the unit tangent vector and the curvature vector.

The *curvature* of the curve $\kappa(s)$ at a point s is a scalar defined as the magnitude of the curvature vector: that is,

$$\kappa(s) = \left|\mathbf{r}''(s)\right| = \left|\mathbf{t}'(s)\right| \tag{2.6}$$

Because the tangent vector $\mathbf{t}(s)$ has unit length, the norm of its derivative defined by the scalar curvature $\kappa(s)$ measures the rate of change of the tangent vector orientation. The *radius of curvature* at a point s is defined as $R(s) = 1/\kappa(s)$. Because the tangent and curvature vectors are orthogonal vectors, the normal to the curve $\mathbf{n}$ is defined as a unit vector along the curvature vector: that is,

$$\mathbf{n}(s) = \frac{\mathbf{r}''(s)}{\kappa(s)} = \frac{\mathbf{t}'(s)}{\kappa(s)} \tag{2.7}$$

The *osculating plane* at a given point s is defined as the plane formed by the unit tangent and normal vectors. The *binormal vector* at a point s on the curve is a vector normal to the osculating plane and is defined by the cross product between the orthogonal tangent and normal unit vectors as

$$\mathbf{b}(s) = \mathbf{t}(s) \times \mathbf{n}(s) \tag{2.8}$$

Because $\mathbf{t}$, $\mathbf{n}$, and $\mathbf{b}$ are orthogonal unit vectors, they form a frame called the *Frenet frame*, whose orientation is defined by the orthogonal matrix $\mathbf{A}_f = \begin{bmatrix}\mathbf{t} & \mathbf{n} & \mathbf{b}\end{bmatrix}$. The $\mathbf{t}$-$\mathbf{b}$ plane is called the *rectifying plane*, and the $\mathbf{n}$-$\mathbf{b}$ plane is called the *normal plane*.

Keeping in mind that the two vectors $\mathbf{t}'(s)$ and $\mathbf{n}(s)$ are parallel vectors, the derivative of the binormal vector $\mathbf{b}$ with respect to s can be written as

$$\mathbf{b}'(s) = \mathbf{t}'(s) \times \mathbf{n}(s) + \mathbf{t}(s) \times \mathbf{n}'(s) = \mathbf{t}(s) \times \mathbf{n}'(s) \tag{2.9}$$

This equation demonstrates that vector $\mathbf{b}'(s)$ is normal to the tangent vector $\mathbf{t}(s)$, and because $\mathbf{b}(s)$ is a unit vector, $\mathbf{b}'(s)$ is also normal to $\mathbf{b}(s)$; consequently, $\mathbf{b}'(s)$ is parallel to $\mathbf{n}$. The parallelism of $\mathbf{b}'(s)$ and $\mathbf{n}$ can be used to write $\mathbf{b}'(s)$ in the following form, which defines the *torsion* τ of the curve as

$$\mathbf{b}'(s) = -\tau(s)\,\mathbf{n}(s) \tag{2.10}$$

The curve geometry is completely defined by the curvature and torsion in the neighborhood of s. In summary, one can write the following equations (Kreyszig 1991; Shabana et al. 2008):

$$\left.\begin{aligned}\mathbf{t}' &= \kappa\mathbf{n}\\ \mathbf{n}' &= -\kappa\mathbf{t} + \tau\mathbf{b}\\ \mathbf{b}' &= -\tau\mathbf{n}\end{aligned}\right\} \tag{2.11}$$

These equations are the *Serret–Frenet formulas*.

Example 2.5 In Example 4, a circle with radius a and center at the origin defined by the parametric equations $\mathbf{r}(t) = \begin{bmatrix}a\cos t & a\sin t\end{bmatrix}^T$ was considered. It was shown in Example 4 that $t = s/a$, where s is the arc length. The circle equation can then be written, using s as a parameter, as $\mathbf{r}(s) = \begin{bmatrix}a\cos(s/a) & a\sin(s/a)\end{bmatrix}^T$. The tangent vector is defined as

$$\mathbf{t}(s) = \frac{d\mathbf{r}(s)}{ds} = \begin{bmatrix}-\sin(s/a) & \cos(s/a)\end{bmatrix}^T$$

It is clear from this equation that the tangent vector obtained by differentiation with respect to the arc length s is a unit vector. The curvature vector is defined as

$$\mathbf{r}''(s) = \mathbf{t}'(s) = \frac{d^2\mathbf{r}(s)}{ds^2} = \frac{1}{a}\begin{bmatrix}-\cos(s/a) & -\sin(s/a)\end{bmatrix}^T$$

(Continued)

This equation defines the curvature as $\kappa(s) = |\mathbf{r}''(s)| = 1/a$ and the normal vector as

$$\mathbf{n}(s) = \begin{bmatrix} -\cos(s/a) & -\sin(s/a) \end{bmatrix}^T$$

The binormal vector is defined as

$$\mathbf{b}(s) = \mathbf{t}(s) \times \mathbf{n}(s) = \begin{bmatrix} 0 & 0 & 1 \end{bmatrix}^T$$

This equation shows that $\mathbf{b}'(s) = \mathbf{0}$; therefore, the torsion $\tau = 0$ since the circular curve is planar and is not twisted. The matrix that defines the orientation of the Serret–Frenet frame at a given point s is defined as

$$\mathbf{A}_f = \begin{bmatrix} \mathbf{t} & \mathbf{n} & \mathbf{b} \end{bmatrix} = \begin{bmatrix} -\sin(s/a) & -\cos(s/a) & 0 \\ \cos(s/a) & -\sin(s/a) & 0 \\ 0 & 0 & 1 \end{bmatrix}$$

One can show that this matrix is an orthogonal matrix: that is, $\mathbf{A}_f^T \mathbf{A}_f = \mathbf{I}$, where $\mathbf{I}$ is the 3×3 identity matrix.

Example 2.6 The parametric equation of the helix in Example 3 is $\mathbf{r}(t) = \begin{bmatrix} a\cos t & a\sin t & \alpha t \end{bmatrix}^T$, where a and α are constants. The tangent vector for this curve is $\mathbf{r}'(t) = d\mathbf{r}/dt = \begin{bmatrix} -a\sin t & a\cos t & \alpha \end{bmatrix}^T$, which shows that $|\mathbf{r}'(t)| = \sqrt{(a)^2 + (\alpha)^2}$. It follows that $dt = \left(1/\sqrt{(a)^2 + (\alpha)^2}\right) ds$. Since a and α are constants, one has $t = \left(1/\sqrt{(a)^2 + (\alpha)^2}\right) s$. The unit tangent vector is, therefore, defined as $\mathbf{r}'(s) = d\mathbf{r}/ds = \left(1/\sqrt{(a)^2 + (\alpha)^2}\right) \begin{bmatrix} -a\sin t & a\cos t & \alpha \end{bmatrix}^T$. Using the equation $t = \left(1/\sqrt{(a)^2 + (\alpha)^2}\right) s$, one can then write

$$\mathbf{r}(s) = \begin{bmatrix} a\cos\left(s/\sqrt{(a)^2 + (\alpha)^2}\right) & a\sin\left(s/\sqrt{(a)^2 + (\alpha)^2}\right) & \alpha\left(s/\sqrt{(a)^2 + (\alpha)^2}\right) \end{bmatrix}^T$$

which defines the unit tangent vector obtained by differentiation with respect to the arc length s as

$$\mathbf{t}(s) = d\mathbf{r}/ds = \frac{1}{\sqrt{(a)^2 + (\alpha)^2}} \begin{bmatrix} -a\sin\left(\frac{s}{\sqrt{(a)^2 + (\alpha)^2}}\right) & a\cos\left(\frac{s}{\sqrt{(a)^2 + (\alpha)^2}}\right) & \alpha \end{bmatrix}^T$$

The curvature vector is defined as

$$\begin{aligned} \mathbf{r}''(s) = d^2\mathbf{r}/ds^2 &= \frac{a}{\left((a)^2 + (\alpha)^2\right)} \begin{bmatrix} -\cos\left(\frac{s}{\sqrt{(a)^2 + (\alpha)^2}}\right) & -\sin\left(\frac{s}{\sqrt{(a)^2 + (\alpha)^2}}\right) & 0 \end{bmatrix}^T \\ &= \frac{a}{\left((a)^2 + (\alpha)^2\right)} \begin{bmatrix} -\cos t & -\sin t & 0 \end{bmatrix}^T \end{aligned}$$

This equation defines the helix curvature at s as $\kappa(s) = |\mathbf{r}''(s)| = |a|/((a)^2 + (\alpha)^2)$, which shows that the helix has a constant curvature. It follows that the normal vector is defined as

$$\mathbf{n}(s) = \left[-\cos\left(\frac{s}{\sqrt{(a)^2+(\alpha)^2}}\right) \quad -\sin\left(\frac{s}{\sqrt{(a)^2+(\alpha)^2}}\right) \quad 0\right]^T$$
$$= \left[-\cos t \quad -\sin t \quad 0\right]^T$$

The binormal vector is

$$\mathbf{b}(s) = \mathbf{t}(s) \times \mathbf{n}(s) = \left(\frac{1}{\sqrt{(a)^2+(\alpha)^2}}\right)$$
$$\times \left[\alpha \sin\left(\frac{s}{\sqrt{(a)^2+(\alpha)^2}}\right) \quad -\alpha\cos\left(\frac{s}{\sqrt{(a)^2+(\alpha)^2}}\right) \quad a\right]^T$$
$$= \left(\frac{1}{\sqrt{(a)^2+(\alpha)^2}}\right)\left[\alpha \sin t \quad -\alpha\cos t \quad a\right]^T$$

Differentiating the binormal vector with respect to the arc length, one obtains

$$\mathbf{b}'(s) = \left(\frac{\alpha}{(a)^2+(\alpha)^2}\right)\left[\cos\left(\frac{s}{\sqrt{(a)^2+(\alpha)^2}}\right) \quad \sin\left(\frac{s}{\sqrt{(a)^2+(\alpha)^2}}\right) \quad 0\right]^T$$
$$= \left(\frac{\alpha}{(a)^2+(\alpha)^2}\right)\left[\cos t \quad \sin t \quad 0\right]^T$$

which shows that the helix has a constant torsion defined as $\tau(s) = \alpha/((a)^2+(\alpha)^2)$: that is, the curvature of the helix can be written in terms of the torsion as $\kappa(s) = (|a|/\alpha)\tau$. The matrix that defines the orientation of the Serret–Frenet frame at a given point defined by s can then be written in terms of the curvature and torsion as

$$\mathbf{A}_f = \begin{bmatrix} -\sqrt{\kappa a}\sin\left(s\sqrt{\tau/\alpha}\right) & -\cos\left(s\sqrt{\tau/\alpha}\right) & \sqrt{\tau a}\sin\left(s\sqrt{\tau/\alpha}\right) \\ \sqrt{\kappa a}\cos\left(s\sqrt{\tau/\alpha}\right) & -\sin\left(s\sqrt{\tau/\alpha}\right) & -\sqrt{\tau a}\cos\left(s\sqrt{\tau/\alpha}\right) \\ \sqrt{\tau \alpha} & 0 & a\sqrt{\tau/\alpha} \end{bmatrix}$$

Railroad Application of the Theory of Curves The theory of curves is extensively used in developing the equations that govern the dynamics of railroad vehicle systems. The *profiles* of the wheel and rail are first defined using planar curves, as shown in Figure 6. Using the planar wheel profile curve, the wheel surface of revolution can be constructed, as will be discussed in a later chapter of this book. Similarly, the rail surface can be extruded using the planar rail profile. The wheel and rail profiles must be measured frequently in order to determine the influence of the wear and irregularities on the profile curve characteristics as the result of the wheel/rail dynamic interaction.

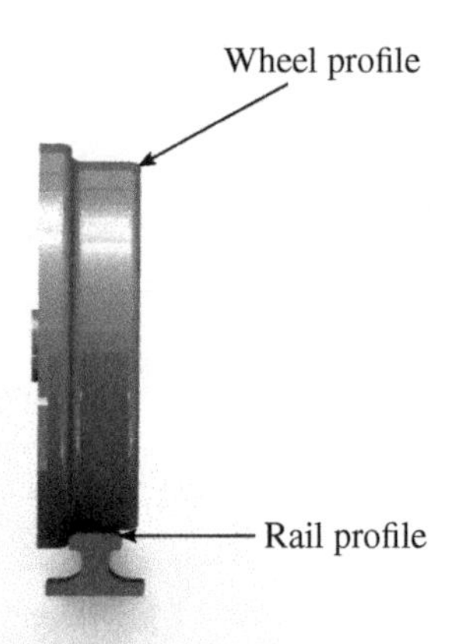

Figure 2.6 Wheel and rail profiles.

Additionally, as shown in Figure 7, the track is constructed using three different segments for the longitudinal *rail space curve*. The first is a straight segment called a *tangent*, which has zero curvature; the second is a circular segment called a *curve*, which has constant curvature; and the third segment, called a *spiral*, connects the tangent segment to the curve or connects two curves with different curvature values. The spiral segment, when connecting tangent and curve segments, is designed such that it starts with zero (or constant) curvature and ends with constant (or zero curvature). The spiral curvature is assumed to vary linearly, as will be discussed in Chapter 4, in order to ensure continuity of the curvature. Rail derailments occur when a vehicle negotiates spiral sections that have varying geometric characteristics and, consequently, the balance speed is not constant during the spiral negotiation.

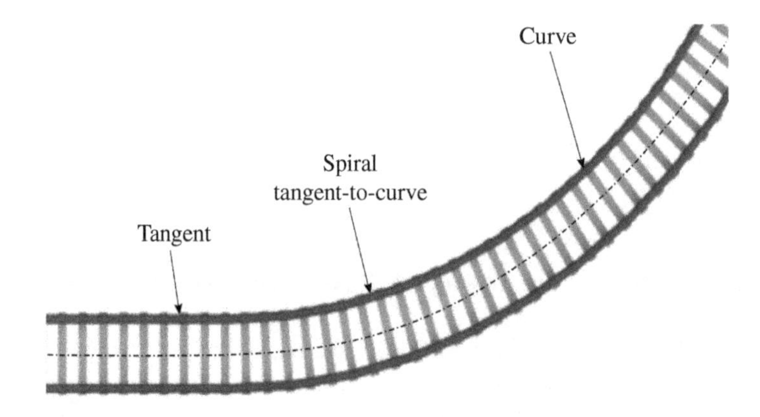

Figure 2.7 Rail space curve.

2.2 SURFACE GEOMETRY

While in the case of a curve, one can only trace points along one direction, in the case of a surface, the coordinates of the points can be defined in terms of two independent surface

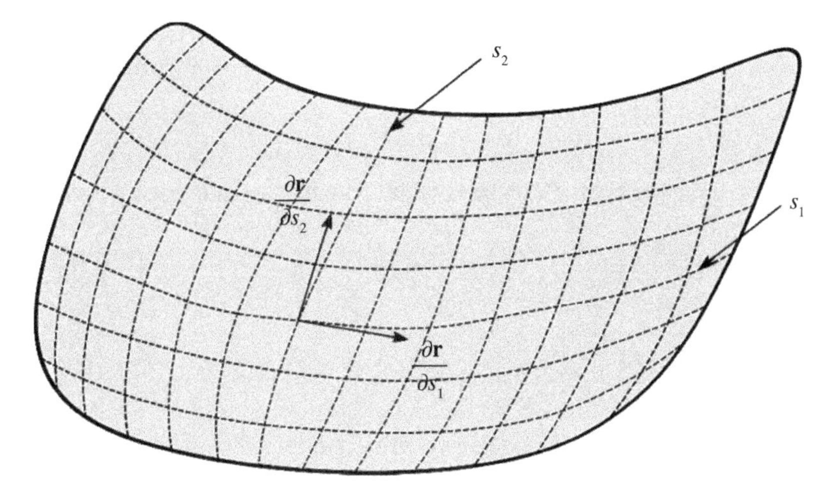

Figure 2.8 Surface geometry.

parameters s_1 and s_2, as shown in Figure 8. Using these parameters and the Cartesian coordinates x, y, and z, an arbitrary point on the surface has a unique position defined by vector $\mathbf{r}$, which can be written in terms of the two independent parameters s_1 and s_2 as follows:

$$\mathbf{r}(s_1, s_2) = \begin{bmatrix} x(s_1, s_2) & y(s_1, s_2) & z(s_1, s_2) \end{bmatrix}^T \tag{2.12}$$

That is, the parameters s_1 and s_2 completely describe the surface geometry. The parameters s_1 and s_2 are called the *surface parameters.* The surface representation of Eq. 12 is assumed to satisfy continuity and differentiability requirements. Specifically, the following two conditions must be met (Kreyszig 1991; Goetz 1970):

1. The mapping in Eq. 12 must be one-to-one: that is, each point on the surface corresponds to a unique set of the surface parameters s_1 and s_2.
2. For a given set of s_1 and s_2, the Jacobian matrix

$$\mathbf{J} = \begin{bmatrix} \frac{\partial \mathbf{r}}{\partial s_1} & \frac{\partial \mathbf{r}}{\partial s_2} \end{bmatrix} = \begin{bmatrix} \frac{\partial x}{\partial s_1} & \frac{\partial x}{\partial s_2} \\ \frac{\partial y}{\partial s_1} & \frac{\partial y}{\partial s_2} \\ \frac{\partial z}{\partial s_1} & \frac{\partial z}{\partial s_2} \end{bmatrix} \tag{2.13}$$

must have a rank of two, which means the two tangent vectors $\partial \mathbf{r}/\partial s_1$ and $\partial \mathbf{r}/\partial s_2$ are linearly independent. This implies mathematically that $(\partial \mathbf{r}/\partial s_1) \times (\partial \mathbf{r}/\partial s_2) \neq 0$: that is, the two tangent vectors are not parallel.

The condition that $(\partial \mathbf{r}/\partial s_1) \times (\partial \mathbf{r}/\partial s_2) \neq 0$ is necessary in order to uniquely define the normal to the surface at an arbitrary point defined by the parameters s_1 and s_2. While the two tangent vectors $\partial \mathbf{r}/\partial s_1$ and $\partial \mathbf{r}/\partial s_2$ must be linearly independent, these tangent vectors are not necessarily orthogonal or unit vectors. The definitions of the two independent tangents to the coordinate lines s_1 and s_2 are required for the formulation of the surface fundamental forms.

Example 2.7 The parametric equations of the cylinder shown in Figure 9, which has an axis along the X axis and a radius a, are defined in terms of the two parameters x and θ as $\mathbf{r}(x,\theta) = \begin{bmatrix} x & y & z \end{bmatrix}^T = \begin{bmatrix} x & a\cos\theta & a\sin\theta \end{bmatrix}^T$. The two tangent vectors are defined as

$$\frac{\partial \mathbf{r}}{\partial x} = \begin{bmatrix} 1 \\ 0 \\ 0 \end{bmatrix}, \qquad \frac{\partial \mathbf{r}}{\partial \theta} = a \begin{bmatrix} 0 \\ -\sin\theta \\ \cos\theta \end{bmatrix}$$

These two tangent vectors are orthogonal vectors, and therefore, they are independent for any values of the two parameters x and θ. Note that for the cylinder and for a given x, one has $(y)^2 + (z)^2 - (a)^2$.

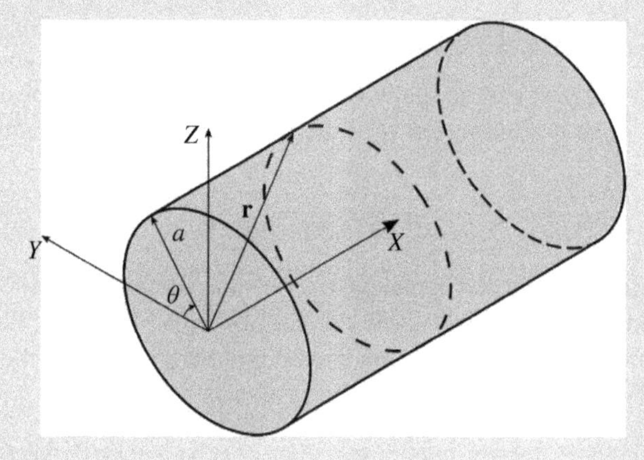

Figure 2.9 Cylindrical surface.

Example 2.8 The parametric equations of the conic surface shown in Figure 10, which has an apex at the origin, has an aperture 2γ, and opens along the X axis, are defined in terms of the two parameters x and θ as $\mathbf{r}(x,\theta) = \begin{bmatrix} x & y & z \end{bmatrix}^T = x\begin{bmatrix} 1 & \tan\gamma\cos\theta & \tan\gamma\sin\theta \end{bmatrix}^T$. The two tangent vectors are defined as

$$\frac{\partial \mathbf{r}}{\partial x} = \begin{bmatrix} 1 \\ \tan\gamma\cos\theta \\ \tan\gamma\sin\theta \end{bmatrix}, \qquad \frac{\partial \mathbf{r}}{\partial \theta} = x\tan\gamma \begin{bmatrix} 0 \\ -\sin\theta \\ \cos\theta \end{bmatrix}$$

These two tangent vectors are not orthogonal vectors, and singularity occurs at the cone apex when $x = 0$. For the conic surface, one has $(y)^2 + (z)^2 = (x\tan\gamma)^2$.

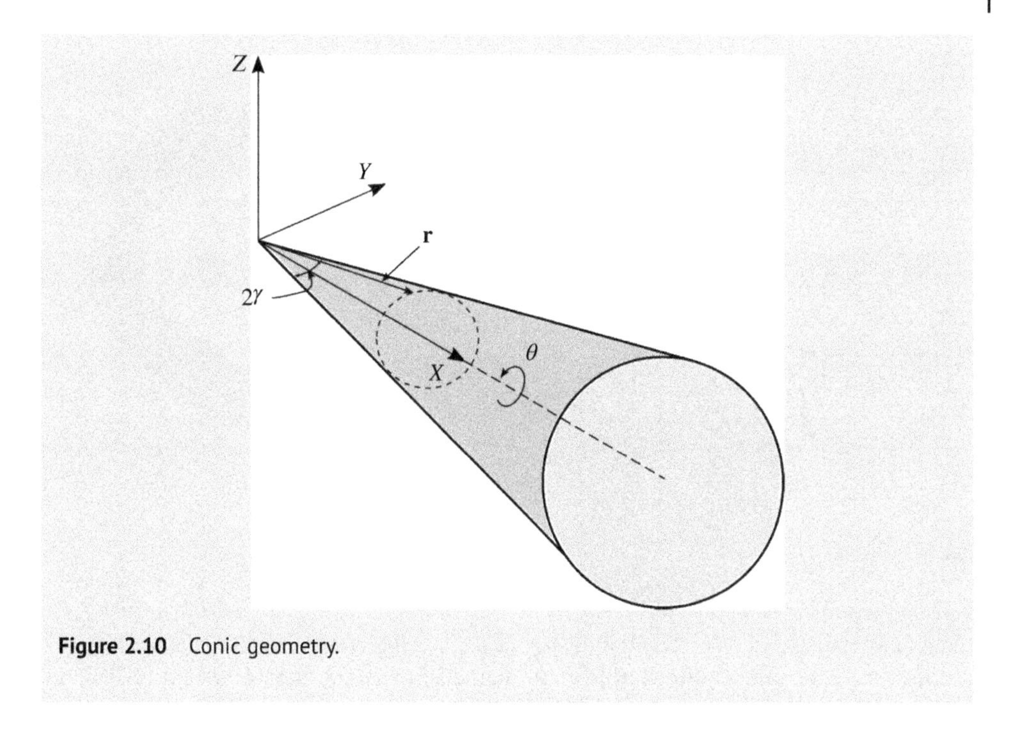

Figure 2.10 Conic geometry.

2.3 APPLICATION TO RAILROAD GEOMETRY

In railroad vehicle system dynamics, one of the main steps in solving the wheel/rail contact problem is to determine online the locations of the contact points on the wheel and rail surfaces. The surface parameters s_1 and s_2 are used to define the locations of these contact points. Nonlinear algebraic contact equations can be formulated and solved for the surface parameters of the wheel and rail, as will be discussed in this book. These surface parameters are used in the surface parametric equations to define the location of the contact points. Knowing the locations of the contact points, the velocities at these points can be evaluated and used to define velocity variables called *creepages* that enter into the formulation of the wheel/rail tangential contact forces. Knowing the surface parameters also allows for defining the tangent and normal vectors as well as the curvatures at the contact points. The *principal curvatures* and *principal directions* are determined using the surface parameters in order to determine the dimensions of the wheel/rail contact area.

To explain the use of the surface parameterization in railroad vehicle dynamics, the rail shown in Figure 11a is considered. In railroad vehicle dynamics, the surface geometry of the rail is defined in a profile coordinate system $X^{rp}Y^{rp}Z^{rp}$, shown in Figure 11a, where superscript r refers to the rail and superscript p refers to the profile. For the purpose of demonstration, the rail is assumed to have a cylindrical surface with radius a. Since the surface parameterization is not unique, one can use the parameters $s_1 = x$ and $s_2 = y_s$, as shown in Figure 11 where y_s measures the lateral distance from the origin of the rail profile frame $X^{rp}Y^{rp}Z^{rp}$. This $x - y_s$ parameterization, often used in railroad

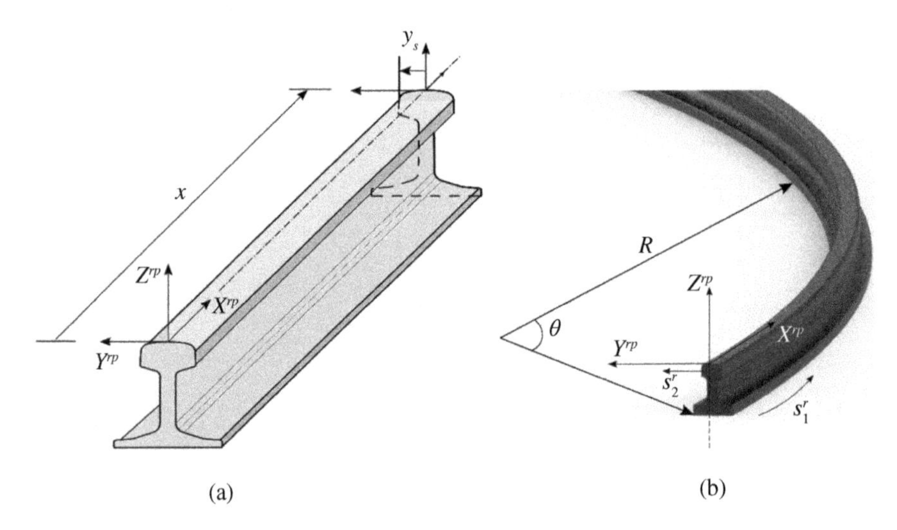

Figure 2.11 Rail surface.

vehicle dynamics, is slightly different from the parameterization used in Example 7 for the cylindrical surface. The position of points on a cylindrical surface, translated by the distance d in the lateral direction, can be defined in the coordinate system $X^rY^rZ^r$ as $\mathbf{r} = \begin{bmatrix} x & y & z \end{bmatrix}^T = \begin{bmatrix} x & -d + y_s & \sqrt{(a)^2 - (y_s)^2} \end{bmatrix}^T$, which can be written as $\mathbf{r} = \begin{bmatrix} x & y & z \end{bmatrix}^T = \begin{bmatrix} x & -d + y_s & f^r(y) \end{bmatrix}^T$, where $f^r(y) = \sqrt{(a)^2 - (y_s)^2}$ defines the profile geometry in the profile coordinate system $X^{rp}Y^{rp}Z^{rp}$. The rail profile can also change as a function of the parameter x that represents the distance traveled. In this case, one can use the function $f^r = f^r(x, y_s)$ to account for the change in the profile shape as a function of the parameter x. Therefore, for a rail with arbitrary geometry, the following general representation can be used:

$$\mathbf{r} = \begin{bmatrix} x & y & z \end{bmatrix}^T = \begin{bmatrix} x & y & f^r(x, y_s) \end{bmatrix}^T \tag{2.14}$$

Keeping in mind that $y = -d + y_s$, in the general case of Eq. 14, the tangent vectors are defined as

$$\frac{\partial \mathbf{r}}{\partial s_1} = \frac{\partial \mathbf{r}}{\partial x} = \begin{bmatrix} 1 \\ 0 \\ \partial f^r / \partial x \end{bmatrix}, \qquad \frac{\partial \mathbf{r}}{\partial s_2} = \frac{\partial \mathbf{r}}{\partial y_s} = \begin{bmatrix} 0 \\ 1 \\ \partial f^r / \partial y_s \end{bmatrix} \tag{2.15}$$

It is clear from this equation that if the profile function f^r does not depend on the parameter x, using x and y_s as surface parameters leads to orthogonal tangents. It is also clear that in the special case of a cylindrical rail in which $f^r(y_s) = \sqrt{(a)^2 - (y_s)^2}$, the tangent and normal vectors are defined as $\partial \mathbf{r}/\partial x = \begin{bmatrix} 1 & 0 & 0 \end{bmatrix}^T$, $\partial \mathbf{r}/\partial y_s = \begin{bmatrix} 0 & 1 & \left(-y_s/\sqrt{(a)^2 - (y_s)^2}\right) \end{bmatrix}^T$, and $\mathbf{n} = \begin{bmatrix} 0 & \left(y_s/\sqrt{(a)^2 - (y_s)^2}\right) & 1 \end{bmatrix}^T$.

In the case of curved rail, the longitudinal parameter, shown in Figure 11b, can be considered as the arc length s of the rail space curve, and therefore, the two parameters s and y_s can be used instead of x and y_s. In this case, one can write Eq. 14 as

$$\mathbf{r} = \begin{bmatrix} x & y & z \end{bmatrix}^T = \begin{bmatrix} g_1^r(s, y_s) & g_2^r(s, y_s) & f^r(s, y_s) \end{bmatrix}^T \tag{2.16}$$

In this equation, $g_1^r(s, y_s)$ and $g_2^r(s, y_s)$ are functions that, respectively, define the Cartesian coordinates x and y of points on the rail surface in terms of the surface parameters s and y_s. For example, in the case of a circular rail space curve, the functions $g_1^r(s, y_s)$ and $g_2^r(s, y_s)$ are defined, respectively, as $g_1^r(s) = d_s \sin(s/R)$ and $g_2^r(s) = -d_s \cos(s/R)$, where $d_s = -R + y_s$ and R is the radius of curvature of the rail space curve. Using simple geometry, the angle θ shown in Figure 11b is defined as $\theta = s/R$. Therefore, for a circular rail space curve with a radius of curvature R and a cylindrical profile with a radius a, one has $\mathbf{r} = \begin{bmatrix} x & y & z \end{bmatrix}^T = \begin{bmatrix} d_s \sin(s/R) & -d_s \cos(s/R) & \sqrt{(a)^2 - (y_s)^2} \end{bmatrix}^T$. It follows that the tangent vectors are $\partial \mathbf{r}/\partial s = \begin{bmatrix} (d_s/R)\cos(s/R) & (d_s/R)\sin(s/R) & 0 \end{bmatrix}^T$ and $\partial \mathbf{r}/\partial y_s = \begin{bmatrix} \sin(s/R) & -\cos(s/R) & -\left(y_s/\sqrt{(a)^2 - (y_s)^2} \right) \end{bmatrix}^T$. Using these two tangent vectors, the normal vector can be defined using the cross product $(\partial \mathbf{r}/\partial s) \times (\partial \mathbf{r}/\partial y_s)$.

A similar procedure can be used for the definition of the parametric form of the wheel surface. The surface of an unworn wheel is a surface of revolution generated by rotating the wheel profile curve about the wheel axle. To define the wheel surface geometry in a parametric form, a coordinate system $X^w Y^w Z^w$ is assigned to the wheel, where superscript w refers to the wheel. This coordinate system is rigidly attached to the wheel axle, and therefore, it rotates with the wheel. The wheel surface parametric form can be defined by the lateral surface parameter $s_1^w = y_s$ and the angular surface parameter $s_2^w = \theta$. The angle θ, measured here from the Z^w axis, as shown in Figure 12, is a surface parameter and has nothing to do with the angle that defines the wheel rotation as it rolls. As shown in Figure 12a, the wheel profile curve can, in general, be defined by the function $f^w(y_s, \theta)$ in a wheel profile frame $X^{wp} Y^{wp} Z^{wp}$ using the surface parameters y_s and θ. Using this function, and assuming that the origin of the wheel profile coordinate system $X^{wp} Y^{wp} Z^{wp}$ is defined in a wheel coordinate system by the coordinates x_o, y_o, and z_o, the parametric form that defines the coordinates of the points on the wheel surface can be written as

$$\mathbf{r} = \begin{bmatrix} x & y & z \end{bmatrix}^T = \begin{bmatrix} g_1^w(y_s, \theta) & y_o + y_s & g_2^w(y_s, \theta) \end{bmatrix}^T \tag{2.17}$$

In this equation, $g_1^w(y_s, \theta) = x_o + f^w(y_s, \theta)\sin\theta$ and $g_2^w(y_s, \theta) = z_o + f^w(y_s, \theta)\cos\theta$ are functions that, respectively, define the Cartesian coordinates x and z of points on the wheel surface in terms of the surface parameters y_s and θ. The parametric form

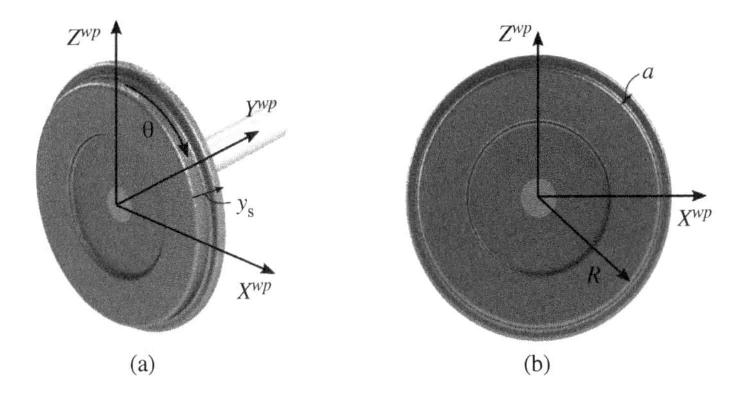

Figure 2.12 Wheel surface geometry.

of Eq. 17 is general and allows for changing the profile geometry as a function of the angular surface parameter θ. In the special case of a wheel with radius R and cylindrical profile of radius a, as shown in Figure 12b, one can select the origin of the profile coordinate system such that $x_o = R\sin\theta$, $y_o = 0$, $z_o = R\cos\theta$, and $f^w(y_s, \theta) = f^w(y_s) = \sqrt{(a)^2 - (y_s)^2}$. It follows that $\mathbf{r} = \begin{bmatrix} x & y & z \end{bmatrix}^T = \begin{bmatrix} (R + f^w)\sin\theta & y_s & (R + f^w)\cos\theta \end{bmatrix}^T$. The two tangent vectors are $\partial\mathbf{r}/\partial y_s = \begin{bmatrix} (\partial f^w/\partial y_s)\sin\theta & 1 & (\partial f^w/\partial y_s)\cos\theta \end{bmatrix}^T$ and $\partial\mathbf{r}/\partial\theta = (R + f^w)\begin{bmatrix} \cos\theta & 0 & -\sin\theta \end{bmatrix}^T$, where $\partial f^w/\partial y_s = -y_s/\sqrt{(a)^2 - (y_s)^2}$.

The profiles of the wheel and rail in railroad vehicle system applications are designed using more complex geometry as compared to the simple cylindrical profiles considered in the examples used in this section. The actual profiles can be measured using a device called a *MiniProf*, which generates the $y_s - z$ table of spline data that can be used in dynamic simulations. Using spline functions to define the wheel and rail profiles, which allows for describing arbitrary profile geometries, and using the *absolute nodal coordinate formulation* for the definition of the rail geometry, will be discussed in later sections of this chapter.

2.4 SURFACE TANGENT PLANE AND NORMAL VECTOR

The *tangent plane* to the surface at a point P is the plane defined by the two tangent vectors $\mathbf{t}_1 = \partial\mathbf{r}/\partial s_1$ and $\mathbf{t}_2 = \partial\mathbf{r}/\partial s_2$, where s_1 and s_2 are the surface parameters that define point P. A curve can be defined on the surface, as shown in Figure 13, by assuming that the two surface parameters s_1 and s_2 can be written in terms of one parameter t. In this case, one can write $s_1 = s_1(t)$ and $s_2 = s_2(t)$, where t is the curve parameter: that is, s_1 and s_2 are related and are no longer independent. Using this curve parameterization, Eq. 12 can be written as

$$\mathbf{r}\left(s_1(t),\ s_2(t)\right) = \mathbf{y}(t) \tag{2.18}$$

where $\mathbf{y}(t)$ is a regular curve on the surface that has a tangent vector defined by the equation

$$\frac{d\mathbf{y}}{dt} = \frac{\partial\mathbf{r}}{\partial s_1}\frac{ds_1}{dt} + \frac{\partial\mathbf{r}}{\partial s_2}\frac{ds_2}{dt} \tag{2.19}$$

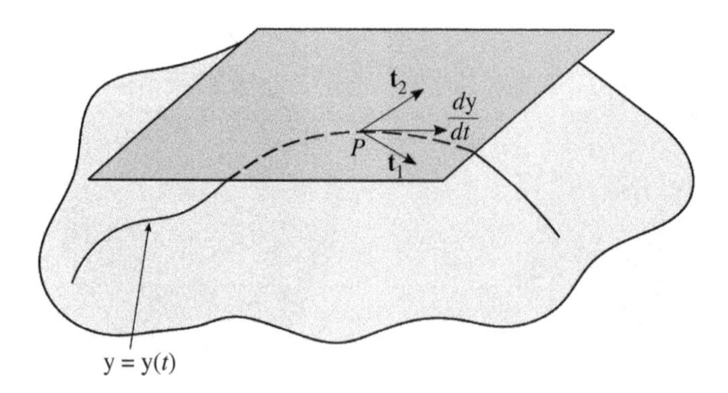

Figure 2.13 Tangent plane.

If $(\partial\mathbf{r}/\partial s_1)\times(\partial\mathbf{r}/\partial s_2)\neq 0$ and ds_1/dt and ds_2/dt are assumed to be different from zero – that is, $(ds_1/dt)\neq 0$ and/or $(ds_2/dt)\neq 0$ – the curve tangent vector of Eq. 19 is well defined. It is clear from Eq. 19 that the curve tangent vector $\mathbf{dy}(t)/dt$ is a linear combination of the two surface tangent vectors $\partial\mathbf{r}/\partial s_1$ and $\partial\mathbf{r}/\partial s_2$, and therefore, $\mathbf{dy}(t)/dt$ lies in the tangent plane and is also tangent to the surface, as shown in Figure 13. The unit vector normal to the surface at point P is defined as the normal to the tangent plane and can be written as follows:

$$\mathbf{n}=\frac{\mathbf{t}_1\times\mathbf{t}_2}{|\mathbf{t}_1\times\mathbf{t}_2|}=\frac{((\partial\mathbf{r}/\partial s_1)\times(\partial\mathbf{r}/\partial s_2))}{|(\partial\mathbf{r}/\partial s_1)\times(\partial\mathbf{r}/\partial s_2)|} \tag{2.20}$$

This vector is, therefore, normal to the tangent vectors $\partial\mathbf{r}/\partial s_1$ and $\partial\mathbf{r}/\partial s_2$ as well as $\mathbf{dy}(t)/dt$ at point P. For example, for general rail geometry, the parametric representation of the surface was found to be $\mathbf{r}=\begin{bmatrix}x & y & z\end{bmatrix}^T=\begin{bmatrix}x & y & f^r(x,y_s)\end{bmatrix}^T$ (Eq. 14). The two tangent vectors are defined in Eq. 15 as $\partial\mathbf{r}/\partial s_1=\partial\mathbf{r}/\partial x=\begin{bmatrix}1 & 0 & \partial f^r/\partial x\end{bmatrix}^T$ and $\partial\mathbf{r}/\partial s_2=\partial\mathbf{r}/\partial y_s=\begin{bmatrix}0 & 1 & \partial f^r/\partial y_s\end{bmatrix}^T$. In this case, the normal vector is $(\partial\mathbf{r}/\partial s_1)\times(\partial\mathbf{r}/\partial s_2)=\begin{bmatrix}-\partial f^r/\partial x & -\partial f^r/\partial y_s & 1\end{bmatrix}^T$. Therefore, the unit normal to the surface at a point defined by the surface parameters $s_1=x$ and $s_2=y_s$ is given by $\mathbf{n}=(\partial\mathbf{r}/\partial s_1)\times(\partial\mathbf{r}/\partial s_2)/|(\partial\mathbf{r}/\partial s_1)\times(\partial\mathbf{r}/\partial s_2)|=(1/\alpha)\begin{bmatrix}-\partial f^r/\partial x & -\partial f^r/\partial y_s & 1\end{bmatrix}^T$, where $\alpha=|\mathbf{n}|=\sqrt{(\partial f^r/\partial x)^2+(\partial f^r/\partial y_s)^2+1}$.

Example 2.9 The parametric equation of the cylindrical surface in Example 7, which has an axis along the X axis and radius a, is defined in terms of the two parameters x and θ as $\mathbf{r}(x,\theta)=\begin{bmatrix}x & y & z\end{bmatrix}^T=\begin{bmatrix}x & a\cos\theta & a\sin\theta\end{bmatrix}^T$. The two tangent vectors were defined as $\partial\mathbf{r}/\partial x=\begin{bmatrix}1 & 0 & 0\end{bmatrix}^T$ and $\partial\mathbf{r}/\partial\theta=a\begin{bmatrix}0 & -\sin\theta & \cos\theta\end{bmatrix}^T$. The unit normal vector is defined as $\mathbf{n}=\begin{bmatrix}0 & -\cos\theta & -\sin\theta\end{bmatrix}^T$.

Example 2.10 It was shown in Example 8 that the parametric equation of the conic surface shown in Figure 10, which opens along the X axis, is defined in terms of the two parameters x and θ by $\mathbf{r}(x,\theta)=\begin{bmatrix}x & y & z\end{bmatrix}^T=x\begin{bmatrix}1 & \tan\gamma\cos\theta & \tan\gamma\sin\theta\end{bmatrix}^T$. The two tangent vectors were defined as $\partial\mathbf{r}/\partial x=\begin{bmatrix}1 & \tan\gamma\cos\theta & \tan\gamma\sin\theta\end{bmatrix}^T$ and $\partial\mathbf{r}/\partial\theta=x\begin{bmatrix}0 & -\tan\gamma\sin\theta & \tan\gamma\cos\theta\end{bmatrix}^T$. The normal vector is defined as $(\partial\mathbf{r}/\partial x)\times(\partial\mathbf{r}/\partial\theta)=x\tan\gamma\begin{bmatrix}\tan\gamma & -\cos\theta & -\sin\theta\end{bmatrix}^T$. The norm of this vector is $x\tan\gamma/\cos\gamma$, which shows that the unit normal is given by $\mathbf{n}=\cos\gamma\begin{bmatrix}\tan\gamma & -\cos\theta & -\sin\theta\end{bmatrix}^T$. It is clear that there is a singularity at $x=0$, where the normal vector is not defined.

The definition of the unit normal vector to the wheel and rail surfaces at the point of contact is necessary in order to be able to determine the direction of the normal contact force. The normal contact force, relative velocity, and wheel and rail material properties and surface geometry at the contact point are used to determine the tangential creep and friction forces, which play a detrimental role in railroad vehicle dynamics and stability.

A surface, as in the case of a curve, can be defined using local properties like those determined by differentiating the curve equations with respect to the surface parameters. Gauss introduced two forms that can be used to determine the local surface properties, which are discussed next: the *first* and *second fundamental forms* of the surfaces.

2.5 SURFACE FUNDAMENTAL FORMS

This section introduces the first and second fundamental forms of surfaces. These fundamental forms allow measuring distances and defining surface curvatures. The coefficients of the surface fundamental forms are used to define the principal curvatures and principal directions that enter into the formulation of wheel/rail interaction forces.

First Fundamental Form The first fundamental form of a surface is defined as

$$I = d\mathbf{r} \cdot d\mathbf{r} = d\mathbf{r}^T d\mathbf{r} \tag{2.21}$$

Because $d\mathbf{r} = (\partial\mathbf{r}/\partial s_1)ds_1 + (\partial\mathbf{r}/\partial s_2)ds_2$, the first fundamental form I can be written as

$$I = \left(\left(\partial\mathbf{r}/\partial s_1\right) ds_1 + \left(\partial\mathbf{r}/\partial s_2\right) ds_2\right)^T \left(\left(\partial\mathbf{r}/\partial s_1\right) ds_1 + \left(\partial\mathbf{r}/\partial s_2\right) ds_2\right) \tag{2.22}$$

which can be written as

$$\begin{aligned} I &= E\left(ds_1\right)^2 + 2F ds_1 ds_2 + G\left(ds_2\right)^2 \\ &= \begin{bmatrix} ds_1 & ds_2 \end{bmatrix} \begin{bmatrix} E & F \\ F & G \end{bmatrix} \begin{bmatrix} ds_1 \\ ds_2 \end{bmatrix} \end{aligned} \tag{2.23}$$

where E, F, and G are called the *coefficients of the first fundamental form* and are defined as

$$\left.\begin{aligned} E &= \mathbf{t}_1^T \mathbf{t}_1 = \left(\partial\mathbf{r}/\partial s_1\right)^T \left(\partial\mathbf{r}/\partial s_1\right), \\ F &= \mathbf{t}_1^T \mathbf{t}_2 = \left(\partial\mathbf{r}/\partial s_1\right)^T \left(\partial\mathbf{r}/\partial s_2\right), \\ G &= \mathbf{t}_2^T \mathbf{t}_2 = \left(\partial\mathbf{r}/\partial s_2\right)^T \left(\partial\mathbf{r}/\partial s_2\right) \end{aligned}\right\} \tag{2.24}$$

The first fundamental form of Eq. 23, which is a homogenous function of second degree in ds_1 and ds_2, can be used to measure distances, angles, and areas on the surface. To demonstrate this, consider a curve on the surface defined by the parametric equation $\mathbf{y}(t) = \mathbf{r}(s_1(t), s_2(t))$, where t is the curve parameter. The length of the curve $\mathbf{y}(t) = \mathbf{r}(s_1(t), s_2(t))$ over the domain $t_1 \le t \le t_2$ is given by $l = \int_{t_1}^{t_2} |d\mathbf{y}/dt|\, dt = \int_{t_1}^{t_2} \sqrt{|(d\mathbf{r}/dt) \cdot (d\mathbf{r}/dt)|}dt$, which can be written in terms of the coefficients of the first fundamental form I as $l = \int_{t_1}^{t_2} \sqrt{E\left(ds_1/dt\right)^2 + 2F\left(ds_1/dt\right)\left(ds_2/dt\right) + G\left(\left(ds_2/dt\right)\right)^2}dt$, which shows that the length of a curve can be evaluated using the square root of the first fundamental form. Furthermore, the angle between two tangent vectors $\mathbf{t}_1 = \partial\mathbf{r}/\partial s_1$ and $\mathbf{t}_2 = \partial\mathbf{r}/\partial s_2$ is defined by the equation $\cos\beta = \mathbf{t}_1^T\mathbf{t}_2 / \left(|\mathbf{t}_1|\,|\mathbf{t}_2|\right) = \left(\partial\mathbf{r}/\partial s_1\right)^T\left(\partial\mathbf{r}/\partial s_2\right) / \left(|\partial\mathbf{r}/\partial s_1|\,|\partial\mathbf{r}/\partial s_2|\right)$, which can be written in terms of the coefficients of the first fundamental form as $\cos\beta = F/\sqrt{EG}$. Clearly, the two tangent vectors $\mathbf{t}_1 = \partial\mathbf{r}/\partial s_1$ and $\mathbf{t}_2 = \partial\mathbf{r}/\partial s_2$ are orthogonal at a point if $F = 0$ at that point.

It was previously shown in this chapter that in the general case in which the rail profile changes as a function of the longitudinal surface parameter x that represents the distance traveled, the rail profile curve can be written as $f^r = f^r(x, y_s)$, where y_s is the lateral rail surface parameter. In this case, the surface parametric equation of a rail with arbitrary geometry can, in general, be written as $\mathbf{r} = \begin{bmatrix} x & y & z \end{bmatrix}^T = \begin{bmatrix} x & y & f^r\left(x, y_s\right) \end{bmatrix}^T$. While it is assumed here that $y = y_s$, one can also assume that $y = y_o + y_s$, where y_o is a constant shift. The longitudinal surface parameter x is often chosen to be the arc length of the rail space curve. Using the parametric form of the rail surface, it was shown that the two tangent vectors are given by $\mathbf{t}_1 = \partial\mathbf{r}/\partial s_1 = \partial\mathbf{r}/\partial x = \begin{bmatrix} 1 & 0 & \partial f^r/\partial x \end{bmatrix}^T$ and $\mathbf{t}_2 = \partial\mathbf{r}/\partial s_2 = \partial\mathbf{r}/\partial y_s = \begin{bmatrix} 0 & 1 & \partial f^r/\partial y_s \end{bmatrix}^T$. One can then show that for this rail surface description, the coefficients of the first fundamental form are

$$\left.\begin{aligned} E &= \mathbf{t}_1^T\mathbf{t}_1 = (\partial\mathbf{r}/\partial x)^T(\partial\mathbf{r}/\partial x) = 1 + (\partial f^r/\partial x)^2, \\ F &= \mathbf{t}_1^T\mathbf{t}_2 = (\partial\mathbf{r}/\partial x)^T\left(\partial\mathbf{r}/\partial y_s\right) = (\partial f^r/\partial x)\left(\partial f^r/\partial y_s\right), \\ G &= \mathbf{t}_2^T\mathbf{t}_2 = \left(\partial\mathbf{r}/\partial y_s\right)^T\left(\partial\mathbf{r}/\partial y_s\right) = 1 + \left(\partial f^r/\partial y_s\right)^2 \end{aligned}\right\} \tag{2.25}$$

If the profile function f^r does not vary as a function of the longitudinal surface parameter x – that is, $f^r = f^r(y_s)$ – the two tangent vectors are orthogonal vectors since $F = 0$, which shows that $\cos\beta = \mathbf{t}_1^T\mathbf{t}_2/\left(|\mathbf{t}_1|\,|\mathbf{t}_2|\right) = \left(\partial\mathbf{r}/\partial s_1\right)^T\left(\partial\mathbf{r}/\partial s_2\right)/\left(|\partial\mathbf{r}/\partial s_1|\,|\partial\mathbf{r}/\partial s_2|\right) = 0$.

Example 2.11 In the special case of a straight cylindrical rail in which $f^r\left(y_s\right) = \sqrt{(a)^2 - \left(y_s\right)^2}$, where a is the cylinder radius, the two tangent vectors are defined as $\partial\mathbf{r}/\partial x = \begin{bmatrix} 1 & 0 & 0 \end{bmatrix}^T$, and $\partial\mathbf{r}/\partial y_s = \begin{bmatrix} 0 & 1 & \left(-y_s/\sqrt{(a)^2 - \left(y_s\right)^2}\right) \end{bmatrix}^T$. In this case, the coefficients of the first fundamental form are given by

$$E = \mathbf{t}_1^T\mathbf{t}_1 = (\partial\mathbf{r}/\partial x)^T(\partial\mathbf{r}/\partial x) = 1, \quad F = \mathbf{t}_1^T\mathbf{t}_2 = (\partial\mathbf{r}/\partial x)^T\left(\partial\mathbf{r}/\partial y_s\right) = 0,$$
$$G = \mathbf{t}_2^T\mathbf{t}_2 = \left(\partial\mathbf{r}/\partial y_s\right)^T\left(\partial\mathbf{r}/\partial y_s\right) = 1 + \left(\left(y_s\right)^2/\left((a)^2 - \left(y_s\right)^2\right)\right)$$

As previously discussed, the surface of the wheel is described by the profile curve $f^w(y_s, \theta)$, defined in the profile frame $X^{wp}Y^{wp}Z^{wp}$, where y_s and θ are, respectively, the wheel lateral and angular surface parameters. Assuming that the origin of the wheel profile coordinate system $X^{wp}Y^{wp}Z^{wp}$ is defined in the $X^wY^wZ^w$ coordinate system by the coordinates x_o, y_o, and z_o, the parametric form of the wheel surface is $\mathbf{r} = \begin{bmatrix} x & y & z \end{bmatrix}^T = \begin{bmatrix} g_1^w\left(y_s, \theta\right) & y_o + y_s & g_2^w\left(y_s, \theta\right) \end{bmatrix}^T$, where $g_1^w\left(y_s, \theta\right) = x_o + f^w\left(y_s, \theta\right)\sin\theta$ and $g_2^w\left(y_s, \theta\right) = z_o + f^w\left(y_s, \theta\right)\cos\theta$. The two tangent vectors are defined as $\partial\mathbf{r}/\partial y_s = \begin{bmatrix} \partial g_1^w/\partial y_s & 1 & \partial g_2^w/\partial y_s \end{bmatrix}^T$ and $\partial\mathbf{r}/\partial\theta = \begin{bmatrix} \partial g_1^w/\partial\theta & 0 & \partial g_2^w/\partial\theta \end{bmatrix}^T$. The coefficients of the first fundamental form are defined in this case as

$$\left.\begin{aligned} E &= \mathbf{t}_1^T\mathbf{t}_1 = \left(\partial\mathbf{r}/\partial y_s\right)^T\left(\partial\mathbf{r}/\partial y_s\right) = 1 + \left(\partial g_1^w/\partial y_s\right)^2 + \left(\partial g_2^w/\partial y_s\right)^2, \\ F &= \mathbf{t}_1^T\mathbf{t}_2 = \left(\partial\mathbf{r}/\partial y_s\right)^T(\partial\mathbf{r}/\partial\theta) = \left(\partial g_1^w/\partial y_s\right)\left(\partial g_1^w/\partial\theta\right) + \left(\partial g_2^w/\partial y_s\right)\left(\partial g_2^w/\partial\theta\right), \\ G &= \mathbf{t}_2^T\mathbf{t}_2 = (\partial\mathbf{r}/\partial\theta)^T(\partial\mathbf{r}/\partial\theta) = \left(\partial g_1^w/\partial\theta\right)^2 + \left(\partial g_2^w/\partial\theta\right)^2 \end{aligned}\right\} \tag{2.26}$$

The condition of the orthogonality of the two vectors tangent to the wheel surface is $F = 0$, which implies that $\left(\partial g_1^w/\partial y_s\right)\left(\partial g_1^w/\partial\theta\right) + \left(\partial g_2^w/\partial y_s\right)\left(\partial g_2^w/\partial\theta\right) = 0$: that is, $\left(\partial g_1^w/\partial y_s\right)\left(\partial g_1^w/\partial\theta\right) = -\left(\partial g_2^w/\partial y_s\right)\left(\partial g_2^w/\partial\theta\right)$. If the wheel profile is not a function of the angular surface parameter θ, one has $f^w = f^w(y_s)$. In this special case, one has $\left(\partial g_1^w/\partial y_s\right)\left(\partial g_1^w/\partial\theta\right) = f^w\left(\partial f^w/\partial y_s\right)\cos\theta\sin\theta$ and $\left(\partial g_2^w/\partial y_s\right)\left(\partial g_2^w/\partial\theta\right) = -f^w\left(\partial f^w/\partial y_s\right)\cos\theta\sin\theta$, demonstrating that the two tangent vectors at an arbitrary point defined by the parameters y_s and θ remain orthogonal vectors. This is, in general, the case for unworn wheels. In this case, the coefficients of the first fundamental form reduce to

$$E = 1 + \left(\partial f^w/\partial y_s\right)^2, \quad F = 0, \quad G = \left(f^w\right)^2 \tag{2.27}$$

These coefficients have much simpler forms as compared to the more general expressions given by Eq. 26.

Example 2.12 In the special case of the wheel shown in Figure 12b, which has radius R and cylindrical profile of radius a, the origin of the profile coordinate system $X^{wp}Y^{wp}Z^{wp}$ is defined in the wheel coordinate system $X^wY^wZ^w$ by $x_o = R\sin\theta, y_o = 0, z_o = R\cos\theta$. The profile function is defined as $f^w(y_s,\theta) = f^w(y_s) = \sqrt{(a)^2 - (y_s)^2}$, and the parametric form of the surface is defined as $\mathbf{r} = \begin{bmatrix} x & y & z\end{bmatrix}^T = \begin{bmatrix}(R+f^w)\sin\theta & y_s & (R+f^w)\cos\theta\end{bmatrix}^T$. The two tangent vectors are $\mathbf{t}_1 = \partial\mathbf{r}/\partial y_s = \begin{bmatrix}(\partial f^w/\partial y_s)\sin\theta & 1 & (\partial f^w/\partial y_s)\cos\theta\end{bmatrix}^T$ and $\mathbf{t}_2 = \partial\mathbf{r}/\partial\theta = (R+f^w)\begin{bmatrix}\cos\theta & 0 & -\sin\theta\end{bmatrix}^T$, where $\partial f^w/\partial y_s = -y_s/\sqrt{(a)^2 - (y_s)^2}$. The coefficients of the first fundamental form are defined as

$$
\begin{aligned}
E &= \mathbf{t}_1^T\mathbf{t}_1 = (\partial\mathbf{r}/\partial y_s)^T(\partial\mathbf{r}/\partial y_s) = 1 + (\partial f^w/\partial y_s)^2 = 1 + (y_s)^2/\left((a)^2 - (y_s)^2\right),\\
F &= \mathbf{t}_1^T\mathbf{t}_2 = (\partial\mathbf{r}/\partial y_s)^T(\partial\mathbf{r}/\partial\theta) = 0,\\
G &= \mathbf{t}_2^T\mathbf{t}_2 = (\partial\mathbf{r}/\partial\theta)^T(\partial\mathbf{r}/\partial\theta) = (R+f^w)^2
\end{aligned}
$$

Example 2.13 The parametric equation of the conical wheel surface shown in Figure 14, which is assumed to have aperture 2γ and opens along the Y, axis are defined in terms of the two parameters y_s and θ as $\mathbf{r}(y_s,\theta) = \begin{bmatrix} x & y & z\end{bmatrix}^T = \begin{bmatrix} y_s\tan\gamma\sin\theta & y_s & y_s\tan\gamma\cos\theta\end{bmatrix}^T$. The two tangent vectors are defined as $\mathbf{t}_1 = \partial\mathbf{r}/\partial y_s = \begin{bmatrix}\tan\gamma\sin\theta & 1 & \tan\gamma\cos\theta\end{bmatrix}^T$ and $\mathbf{t}_2 = \partial\mathbf{r}/\partial\theta = y_s\begin{bmatrix}\tan\gamma\cos\theta & 0 & -\tan\gamma\sin\theta\end{bmatrix}^T$. It is clear that these two tangent vectors are orthogonal because $\mathbf{t}_1^T\mathbf{t}_2 = (\partial\mathbf{r}/\partial y_s)^T(\partial\mathbf{r}/\partial\theta) = 0$. The coefficients of the first fundamental form of this conical surface are given by

$$
\begin{aligned}
E &= (\partial\mathbf{r}/\partial y_s)^T(\partial\mathbf{r}/\partial y_s) = 1 + \tan^2\gamma,\\
F &= (\partial\mathbf{r}/\partial y_s)^T(\partial\mathbf{r}/\partial\theta) = 0,\\
G &= (\partial\mathbf{r}/\partial\theta)^T(\partial\mathbf{r}/\partial\theta) = (y_s\tan\gamma)^2
\end{aligned}
$$

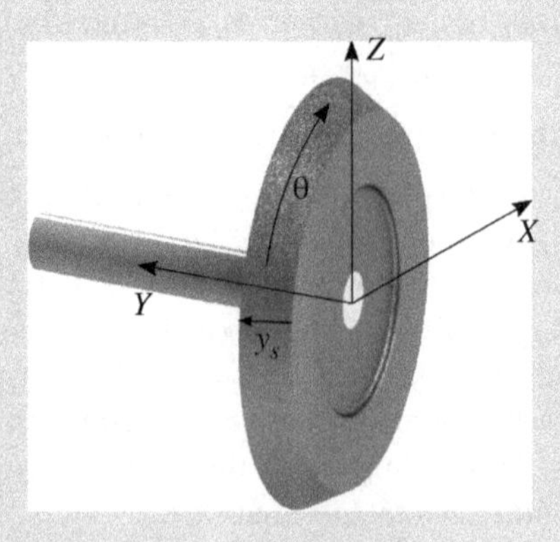

Figure 2.14 Conical wheel surface.

Second Fundamental Form The vector $\mathbf{n}$ normal to the surface was defined as $\mathbf{n}=\mathbf{t}_1\times\mathbf{t}_2/|\mathbf{t}_1\times\mathbf{t}_2|=((\partial\mathbf{r}/\partial s_1)\times(\partial\mathbf{r}/\partial s_2))/|(\partial\mathbf{r}/\partial s_1)\times(\partial\mathbf{r}/\partial s_2)|$. Therefore, the unit normal vector $\mathbf{n}$ is a function in the surface parameters s_1 and s_2, and its differential can be written as

$$d\mathbf{n} = \left(\partial\mathbf{n}/\partial s_1\right) ds_1 + \left(\partial\mathbf{n}/\partial s_2\right) ds_2 \tag{2.28}$$

Because $d\mathbf{r}=(\partial\mathbf{r}/\partial s_1)ds_1+(\partial\mathbf{r}/\partial s_2)ds_2$ is tangent to the surface, one has $d\mathbf{r}\cdot\mathbf{n}=\mathbf{0}$, which upon differentiation yields $d^2\mathbf{r}\cdot\mathbf{n}=-d\mathbf{r}\cdot d\mathbf{n}$. In this equation, $d^2\mathbf{r}=\left(\partial^2\mathbf{r}/\partial s_1^2\right)\left(ds_1\right)^2+2\left(\partial^2\mathbf{r}/\partial s_1\partial s_2\right)ds_1ds_2+\left(\partial^2\mathbf{r}/\partial s_2^2\right)\left(ds_2\right)^2$. The second fundamental form of a surface is defined as

$$\begin{aligned} II &= d^2\mathbf{r}\cdot\mathbf{n} = -d\mathbf{r}\cdot d\mathbf{n} \\ &= -\left(\left(\partial\mathbf{r}/\partial s_1\right) ds_1 + \left(\partial\mathbf{r}/\partial s_2\right) ds_2\right)^T\left(\left(\partial\mathbf{n}/\partial s_1\right) ds_1 + \left(\partial\mathbf{n}/\partial s_2\right) ds_2\right) \end{aligned} \tag{2.29}$$

This equation shows that the second fundamental form defines the projection of the second derivative of the vector that represents the relative position of two neighboring points on the surface normal vector. The second fundamental form of Eq. 29 can also be written as

$$\begin{aligned} II &= L\left(ds_1\right)^2 + 2M ds_1 ds_2 + N\left(ds_2\right)^2 \\ &= \begin{bmatrix} ds_1 & ds_2 \end{bmatrix}\begin{bmatrix} L & M \\ M & N \end{bmatrix}\begin{bmatrix} ds_1 \\ ds_2 \end{bmatrix} \end{aligned} \tag{2.30}$$

where L, M, and N are called the *coefficients of the second fundamental form*, defined as

$$\left.\begin{aligned} L &= -\left(\partial\mathbf{r}/\partial s_1\right)^T\left(\partial\mathbf{n}/\partial s_1\right), \\ M &= -\frac{1}{2}\left(\left(\partial\mathbf{r}/\partial s_1\right)^T\left(\partial\mathbf{n}/\partial s_2\right) + \left(\partial\mathbf{r}/\partial s_2\right)^T\left(\partial\mathbf{n}/\partial s_1\right)\right), \\ N &= -\left(\partial\mathbf{r}/\partial s_2\right)^T\left(\partial\mathbf{n}/\partial s_2\right) \end{aligned}\right\} \tag{2.31}$$

Equation 29 or 30 shows that the second fundamental form II is a homogenous function of second degree in ds_1 and ds_2. Because the normal vector $\mathbf{n}$ is orthogonal to the tangent plane formed by vectors $\mathbf{t}_1=\partial\mathbf{r}/\partial s_1$ and $\mathbf{t}_2=\partial\mathbf{r}/\partial s_2$, the following identities can be written:

$$\left.\begin{aligned} \frac{\partial}{\partial s_1}\left(\mathbf{t}_1^T\mathbf{n}\right) &= \left(\partial^2\mathbf{r}/\partial s_1^2\right)^T\mathbf{n} + \left(\partial\mathbf{r}/\partial s_1\right)^T\left(\partial\mathbf{n}/\partial s_1\right) = 0 \\ \frac{\partial}{\partial s_2}\left(\mathbf{t}_1^T\mathbf{n}\right) &= \left(\partial^2\mathbf{r}/\partial s_1\partial s_2\right)^T\mathbf{n} + \left(\partial\mathbf{r}/\partial s_1\right)^T\left(\partial\mathbf{n}/\partial s_2\right) = 0 \\ \frac{\partial}{\partial s_1}\left(\mathbf{t}_2^T\mathbf{n}\right) &= \left(\partial^2\mathbf{r}/\partial s_2\partial s_1\right)^T\mathbf{n} + \left(\partial\mathbf{r}/\partial s_2\right)^T\left(\partial\mathbf{n}/\partial s_1\right) = 0 \\ \frac{\partial}{\partial s_2}\left(\mathbf{t}_2^T\mathbf{n}\right) &= \left(\partial^2\mathbf{r}/\partial s_2^2\right)^T\mathbf{n} + \left(\partial\mathbf{r}/\partial s_2\right)^T\left(\partial\mathbf{n}/\partial s_2\right) = 0 \end{aligned}\right\} \tag{2.32}$$

These identities can also be written in the following forms:

$$\left.\begin{aligned} \left(\partial^2\mathbf{r}/\partial s_1^2\right)^T\mathbf{n} &= -\left(\partial\mathbf{r}/\partial s_1\right)^T\left(\partial\mathbf{n}/\partial s_1\right), \\ \left(\partial^2\mathbf{r}/\partial s_1\partial s_2\right)^T\mathbf{n} &= -\left(\partial\mathbf{r}/\partial s_1\right)^T\left(\partial\mathbf{n}/\partial s_2\right), \\ \left(\partial^2\mathbf{r}/\partial s_1\partial s_2\right)^T\mathbf{n} &= -\left(\partial\mathbf{r}/\partial s_2\right)^T\left(\partial\mathbf{n}/\partial s_1\right), \\ \left(\partial^2\mathbf{r}/\partial s_2^2\right)^T\mathbf{n} &= -\left(\partial\mathbf{r}/\partial s_2\right)^T\left(\partial\mathbf{n}/\partial s_2\right) \end{aligned}\right\} \tag{2.33}$$

Using these identities, the coefficients of the second fundamental form can be written in alternate forms in terms of second derivatives as

$$L = \left(\partial^2\mathbf{r}/\partial s_1^2\right)^T\mathbf{n}, \quad M = \left(\partial^2\mathbf{r}/\partial s_1\partial s_2\right)^T\mathbf{n}, \quad N = \left(\partial^2\mathbf{r}/\partial s_2^2\right)^T\mathbf{n} \tag{2.34}$$

As will be explained in this chapter, $\partial^2\mathbf{r}/\partial s_i\partial s_j$, $i, j = 1, 2$ represent *curvature vectors*, and therefore, the coefficients of the second fundamental form represent the projections or components of these curvature vectors at a point along the normal to the surface at this point. These coefficients of the second fundamental form determine the nature of the surface in the neighborhood of a point. Based on the values of these coefficients, one has the following surface classifications:

(1) If $LN - (M)^2 > 0$ at a point, the surface is called *elliptic* at that point.
(2) If $LN - (M)^2 < 0$ at a point, the surface is called *hyperbolic* at that point.
(3) If $LN - (M)^2 = 0$ at a point, the surface is called *parabolic* at that point.
(4) If $L = N = M = 0$ at a point, the surface is called *planar* at that point.

Figure 15 shows examples of elliptic, hyperbolic, parabolic, and planar surfaces.

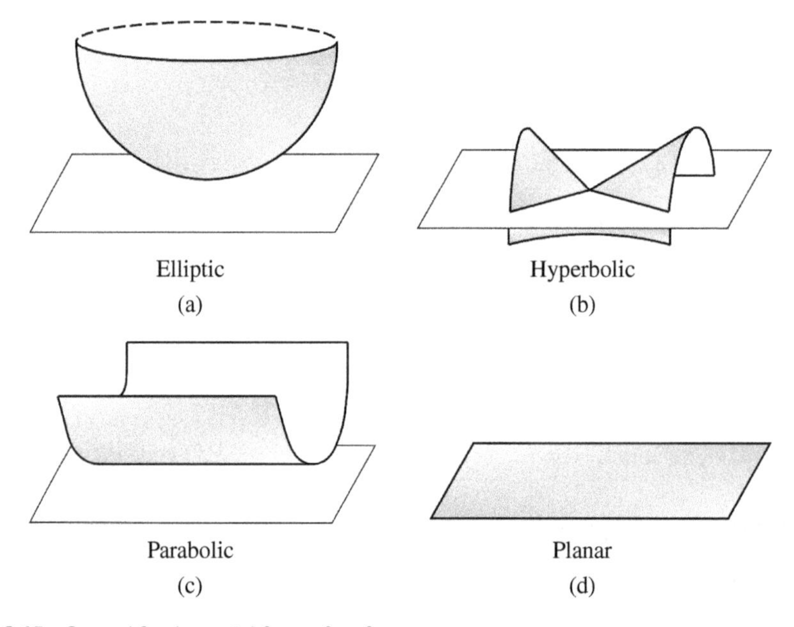

Figure 2.15 Second fundamental form of surfaces.

Example 2.14 In the special case of a straight cylindrical rail in which $f^r(y_s) = \sqrt{(a)^2 - (y_s)^2}$, where a is the cylinder radius, the two tangent vectors are defined in Example 11 as $\mathbf{t}_1 = \partial\mathbf{r}/\partial x = \begin{bmatrix}1 & 0 & 0\end{bmatrix}^T$ and $\mathbf{t}_2 = \partial\mathbf{r}/\partial y_s = \left[0 \; 1 \; \left(-y_s/\sqrt{(a)^2 - (y_s)^2}\right)\right]^T$. In this case, the normal vector is defined as $\mathbf{n} = \left(1/\sqrt{1 + \left(f_{y_s}^r\right)^2}\right)\begin{bmatrix}0 & -f_{y_s}^r & 1\end{bmatrix}^T$, where

$f^r_{y_s} = -y_s/\sqrt{(a)^2 - (y_s)^2}$. The following vectors can be defined:

$$\partial^2\mathbf{r}/\partial s_1^2 = \partial^2\mathbf{r}/\partial x^2 = \begin{bmatrix} 0 & 0 & 0 \end{bmatrix}^T$$
$$\partial^2\mathbf{r}/\partial s_1\partial s_2 = \partial^2\mathbf{r}/\partial x\partial y_s = \begin{bmatrix} 0 & 0 & 0 \end{bmatrix}^T$$
$$\partial^2\mathbf{r}/\partial s_2^2 = \partial^2\mathbf{r}/\partial y_s^2 = \begin{bmatrix} 0 & 0 & \partial f^r_{y_s}/\partial y_s \end{bmatrix}^T$$

where $\partial f^r_{y_s}/\partial y_s = -(a)^2/\left((a)^2 - (y_s)^2\right)^{3/2}$. One can, therefore, define the coefficients of the second fundamental form as

$$L = \left(\partial^2\mathbf{r}/\partial s_1^2\right)^T\mathbf{n} = 0,$$
$$M = \left(\partial^2\mathbf{r}/\partial s_1\partial s_2\right)^T\mathbf{n} = 0,$$
$$N = \left(\partial^2\mathbf{r}/\partial s_2^2\right)^T\mathbf{n} = \left(\partial f^r_{y_s}/\partial y_s\right)/\sqrt{1 + \left(f^r_{y_s}\right)^2}$$

To give a geometric interpretation of the coefficients of the second fundamental form, the straight cylindrical rail of the preceding example is considered. The surface parametrization can be changed from x and y_s to x and s, where s is the arc length of the profile curve, as shown in Figure 16. Using this parametrization, the equation of the cylindrical rail in its parametric form can be written as $\mathbf{r} = \begin{bmatrix} x & a\cos(s/a) & a\sin(s/a) \end{bmatrix}^T$. The two tangent vectors are defined as $\mathbf{t}_1 = \partial\mathbf{r}/\partial x = \begin{bmatrix} 1 & 0 & 0 \end{bmatrix}^T$ and $\mathbf{t}_2 = \partial\mathbf{r}/\partial s = \begin{bmatrix} 0 & -\sin(s/a) & \cos(s/a) \end{bmatrix}^T$. The normal vector is defined as $\mathbf{n} = (\partial\mathbf{r}/\partial x) \times (\partial\mathbf{r}/\partial s) = \begin{bmatrix} 0 & -\cos(s/a) & -\sin(s/a) \end{bmatrix}^T$. The second derivatives of vector $\mathbf{r}$ with respect to the new parameters are $\partial^2\mathbf{r}/\partial x^2 = \begin{bmatrix} 0 & 0 & 0 \end{bmatrix}^T$, $\partial^2\mathbf{r}/\partial x\partial s = \begin{bmatrix} 0 & 0 & 0 \end{bmatrix}^T$, and $\partial^2\mathbf{r}/\partial s^2 = (1/a)\begin{bmatrix} 0 & -\cos(s/a) & -\sin(s/a) \end{bmatrix}^T$. Using these definitions, the coefficients of the second fundamental form, when using the arc length parameter for the profile curve, are $L = (\partial^2\mathbf{r}/\partial x^2)^T\mathbf{n} = 0$, $M = (\partial^2\mathbf{r}/\partial x\partial s)^T\mathbf{n} = 0$, and $N = (\partial^2\mathbf{r}/\partial s^2)^T\mathbf{n} = 1/a$, which shows that in this case, the coefficient N represents the constant curvature of the rail profile circular curve. Note also that in this example, the curvature vector $\partial^2\mathbf{r}/\partial s^2 = (1/a)\begin{bmatrix} 0 & -\cos(s/a) & -\sin(s/a) \end{bmatrix}^T$ is parallel to the unit normal vector to the surface $\mathbf{n} = \begin{bmatrix} 0 & -\cos(s/a) & -\sin(s/a) \end{bmatrix}^T$.

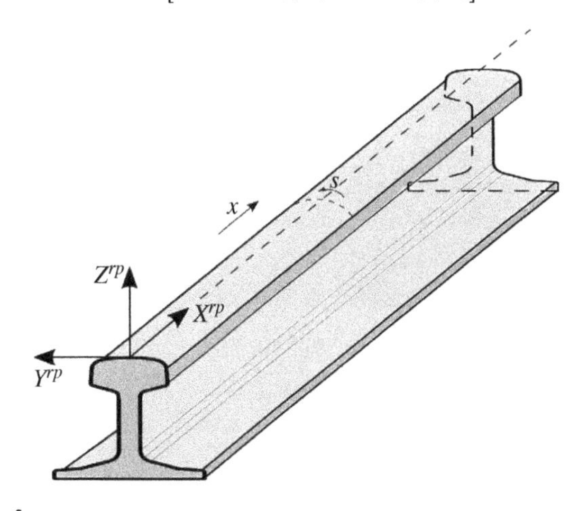

Figure 2.16 Rail surface.

Example 2.15 In the special case of the wheel shown in Figure 12b, which has radius R and cylindrical profile of radius a, the origin of the profile coordinate system $X^{wp}Y^{wp}Z^{wp}$ is defined in the wheel coordinate system $X^wY^wZ^w$ by $x_o = R\sin\theta$, $y_o = 0$, $z_o = R\cos\theta$. The profile function is defined as $f^w(y_s, \theta) = f^w(y_s) = \sqrt{(a)^2 - (y_s)^2}$, and the parametric form of the surface is defined as $\mathbf{r} = [x \;\; y \;\; z]^T = [(R + f^w)\sin\theta \;\; y_s \;\; (R + f^w)\cos\theta]^T$. The two tangent vectors were defined in Example 12 as $\mathbf{t}_1 = \partial\mathbf{r}/\partial y_s = [(\partial f^w/\partial y_s)\sin\theta \;\; 1 \;\; (\partial f^w/\partial y_s)\cos\theta]^T$ and $\mathbf{t}_2 = \partial\mathbf{r}/\partial\theta = (R + f^w)[\cos\theta \;\; 0 \;\; -\sin\theta]^T$, where $f^w_{y_s} = \partial f^w/\partial y_s = -y_s/\sqrt{(a)^2 - (y_s)^2}$. The normal vector is defined as $\mathbf{n} = (\partial\mathbf{r}/\partial y_s) \times (\partial\mathbf{r}/\partial\theta)/\left|(\partial\mathbf{r}/\partial y_s) \times (\partial\mathbf{r}/\partial\theta)\right| = \beta[-\sin\theta \;\; -f^w_{y_s} \;\; -\cos\theta]^T$, where $\beta = (R + f^w)/\sqrt{1 + \left(f^w_{y_s}\right)^2}$. The second derivatives of $\mathbf{r}$ are $\partial^2\mathbf{r}/\partial y_s^2 = \left(\partial f^w_{y_s}/\partial y_s\right)[\sin\theta \;\; 0 \;\; \cos\theta]^T$, $\partial^2\mathbf{r}/\partial y_s\partial\theta = -(\partial f^w/\partial y_s)[\cos\theta \;\; 0 \;\; -\sin\theta]^T$, and $\partial^2\mathbf{r}/\partial\theta^2 = -(R + f^w)[\sin\theta \;\; 0 \;\; \cos\theta]^T$. Therefore, the coefficients of the second fundamental form are

$$L = \left(\partial^2\mathbf{r}/\partial y_s^2\right)\cdot\mathbf{n} = -\beta\left(\partial f^w_{y_s}/\partial y_s\right),$$
$$M = \left(\partial^2\mathbf{r}/\partial y_s\partial\theta\right)\cdot\mathbf{n} = 0,$$
$$N = \left(\partial^2\mathbf{r}/\partial\theta^2\right)\cdot\mathbf{n} = \beta\left(R + f^w\right)$$

As in the case of the cylindrical rail, one can parameterize the cylindrical wheel profile using the arc length s. In this case, the parametric equation of the cylindrical wheel surface can be written in terms of the surface parameters s and θ as $\mathbf{r} = [x \;\; y \;\; z]^T = [(R + a\sin(s/a))\sin\theta \;\; a\cos(s/a) \;\; (R + a\sin(s/a))\cos\theta]^T$. The two tangent vectors can be defined in this case as $\mathbf{t}_1 = \partial\mathbf{r}/\partial s = [\cos(s/a)\sin\theta \;\; -\sin(s/a) \;\; \cos(s/a)\cos\theta]^T$ and $\mathbf{t}_2 = \partial\mathbf{r}/\partial\theta = (R + a\sin(s/a))[\cos\theta \;\; 0 \;\; -\sin\theta]^T$. The normal vector is defined as $\mathbf{n} = ((\partial\mathbf{r}/\partial s) \times (\partial\mathbf{r}/\partial\theta))/|(\partial\mathbf{r}/\partial s) \times (\partial\mathbf{r}/\partial\theta)| = \beta[\sin(s/a)\sin\theta \;\; \cos(s/a) \;\; \sin(s/a)\cos\theta]^T$, where $\beta = R + a\sin(s/a)$. The second derivatives of $\mathbf{r}$ are $\partial^2\mathbf{r}/\partial s^2 = -(1/a)[\sin(s/a)\sin\theta \;\; \cos(s/a) \;\; \sin(s/a)\cos\theta]^T$, $\partial^2\mathbf{r}/\partial s\partial\theta = [\cos(s/a)\cos\theta \;\; 0 \;\; -\cos(s/a)\sin\theta]^T$, and $\partial^2\mathbf{r}/\partial\theta^2 = -\beta[\sin\theta \;\; 0 \;\; \cos\theta]^T$. Therefore, the coefficients of the second fundamental form are $L = (\partial^2\mathbf{r}/\partial s^2)\cdot\mathbf{n} = -\beta/a$, $M = (\partial^2\mathbf{r}/\partial s\partial\theta)\cdot\mathbf{n} = 0$, and $N = (\partial^2\mathbf{r}/\partial\theta^2)\cdot\mathbf{n} = -\beta\sin(s/a)$.

Example 2.16 The parametric equation of the conical wheel surface shown in Figure 14, which opens along the Y axis, is defined in Example 13 in terms of the two parameters y_s and θ as $\mathbf{r}(y_s, \theta) = [x \;\; y \;\; z]^T = [y_s\tan\gamma\sin\theta \;\; y_s \;\; y_s\tan\gamma\cos\theta]^T$. The two tangent vectors are defined as $\mathbf{t}_1 = \partial\mathbf{r}/\partial y_s = [\tan\gamma\sin\theta \;\; 1 \;\; \tan\gamma\cos\theta]^T$ and $\mathbf{t}_2 = \partial\mathbf{r}/\partial\theta = y_s[\tan\gamma\cos\theta \;\; 0 \;\; -\tan\gamma\sin\theta]^T$. The unit normal vector is defined as $\mathbf{n} = \cos\gamma[-\sin\theta \;\; \tan\gamma \;\; -\cos\theta]^T$. The second derivatives of vector $\mathbf{r}$ are $\partial^2\mathbf{r}/\partial y_s^2 = [0 \;\; 0 \;\; 0]^T$, $\partial^2\mathbf{r}/\partial y_s\partial\theta = \tan\gamma[\cos\theta \;\; 0 \;\; -\sin\theta]^T$, and $\partial^2\mathbf{r}/\partial\theta^2 = -y_s\tan\gamma[\sin\theta \;\; 0 \;\; \cos\theta]^T$. The coefficients of the second fundamental form can then be defined as $L = \left(\partial^2\mathbf{r}/\partial y_s^2\right)\cdot\mathbf{n} = 0$, $M = (\partial^2\mathbf{r}/\partial y_s\partial\theta)\cdot\mathbf{n} = 0$, and $N = (\partial^2\mathbf{r}/\partial\theta^2)\cdot\mathbf{n} = y_s\sin\gamma$.

2.6 NORMAL CURVATURE

As explained in Section 1, the normal to a curve is a unit vector along the curvature vector. In general, the normal of a regular curve drawn on a surface is not the same as the surface normal $\mathbf{n}$. This fact can be demonstrated by considering the simple example of a planar surface parameterized by the Cartesian coordinates x and y, as shown in Figure 17. The parametric equation of the planar surface in terms of these Cartesian parameters can be written as $\mathbf{r} = \begin{bmatrix} x & y & 0 \end{bmatrix}^T$. The two tangents to the planar surface are $\mathbf{t}_1 = \partial \mathbf{r}/\partial x = \begin{bmatrix} 1 & 0 & 0 \end{bmatrix}^T$ and $\mathbf{t}_2 = \partial \mathbf{r}/\partial y = \begin{bmatrix} 0 & 1 & 0 \end{bmatrix}^T$. The normal to the surface is defined as $\mathbf{n} = (\partial \mathbf{r}/\partial x) \times (\partial \mathbf{r}/\partial y) = \begin{bmatrix} 0 & 0 & 1 \end{bmatrix}^T$. Consider a circular curve $\mathbf{C}$ drawn on the planar surface. The circular curve is assumed to have radius a, and it is parameterized by its arc length s. By tracing the curve on the surface, the surface parameters x and y are related because $x = a\cos(s/a)$ and $y = a\sin(s/a)$. Therefore, the parametric equation of the circular curve on the planar surface can be written as $\mathbf{C} = \mathbf{C}(s) = \begin{bmatrix} a\cos(s/a) & a\sin(s/a) & 0 \end{bmatrix}^T$. The tangent to the curve and the curvature vector are defined, respectively, as $\partial \mathbf{C}(s)/\partial s = \begin{bmatrix} -\sin(s/a) & \cos(s/a) & 0 \end{bmatrix}^T$ and $\partial^2 \mathbf{C}(s)/\partial s^2 = -(1/a)\begin{bmatrix} \cos(s/a) & \sin(s/a) & 0 \end{bmatrix}^T$, which shows the curve normal vector $\mathbf{n}^c$ is $\left(\partial^2 \mathbf{C}(s)/\partial s^2\right) / \left|\partial^2 \mathbf{C}(s)/\partial s^2\right| = -\begin{bmatrix} \cos(s/a) & \sin(s/a) & 0 \end{bmatrix}^T$, which is a vector that lies on the surface; therefore, in this example, the vectors normal to the curve and the surface are orthogonal vectors.

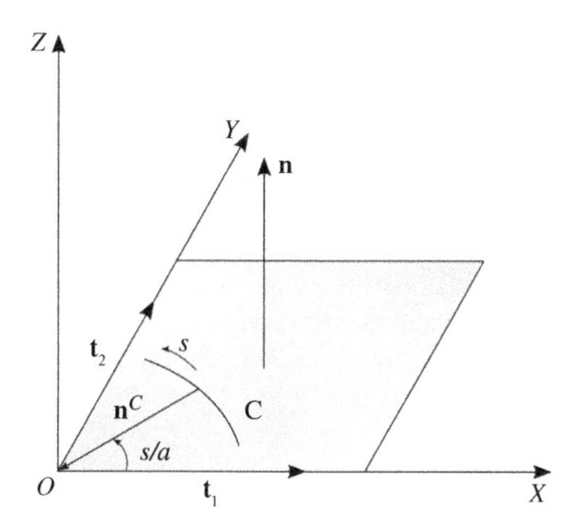

Figure 2.17 Surface normal versus curve normal.

While for the planar surface, the vectors normal to the curves on the surface are orthogonal to the surface normal, for other surfaces, the vectors normal to the curves do not lie in the same plane. For example, consider the cylindrical surface previously discussed in this chapter in Example 9 and shown in Figure 18. The parametric equation of the cylindrical surface, which has an axis along the X axis and radius a, are defined in terms of the two parameters x and θ as $\mathbf{r}(x,\theta) = \begin{bmatrix} x & y & z \end{bmatrix}^T = \begin{bmatrix} x & a\cos\theta & a\sin\theta \end{bmatrix}^T$. The two tangent vectors were defined as $\mathbf{t}_1 = \partial \mathbf{r}/\partial x = \begin{bmatrix} 1 & 0 & 0 \end{bmatrix}^T$ and $\mathbf{t}_2 = \partial \mathbf{r}/\partial \theta = a\begin{bmatrix} 0 & -\sin\theta & \cos\theta \end{bmatrix}^T$. The unit vector normal to the surface is defined as $(\partial \mathbf{r}/\partial x) \times$

$(\partial \mathbf{r}/\partial \theta) / |(\partial \mathbf{r}/\partial x) \times (\partial \mathbf{r}/\partial \theta)| = -\begin{bmatrix} 0 & \cos\theta & \sin\theta \end{bmatrix}^T$. For a given $x = x_s$, and keeping in mind that $\theta = s/a$ where s is the curve arc length, one has a profile curve defined by the parametric equation $\mathbf{C}(s) = \begin{bmatrix} x_s & a\cos(s/a) & a\sin(s/a) \end{bmatrix}^T$. The tangent and curvature vectors of this curve are, respectively, given by $\partial \mathbf{C}(s)/\partial s = \begin{bmatrix} 0 & -\sin(s/a) & \cos(s/a) \end{bmatrix}^T$ and $\partial^2 \mathbf{C}(s)/\partial s^2 = -(1/a)\begin{bmatrix} 0 & \cos(s/a) & \sin(s/a) \end{bmatrix}^T$, which defines the unit normal to the curve as $\left(\partial^2 \mathbf{C}(s)/\partial s^2\right) / \left|\partial^2 \mathbf{C}(s)/\partial s^2\right| = -\begin{bmatrix} 0 & \cos(s/a) & \sin(s/a) \end{bmatrix}^T$; this can also be written in terms of θ as $\left(\partial^2 \mathbf{C}(s)/\partial s^2\right) / \left|\partial^2 \mathbf{C}(s)/\partial s^2\right| = -\begin{bmatrix} 0 & \cos\theta & \sin\theta \end{bmatrix}^T$, which is the same as the normal to the surface defined by vector $\mathbf{n}$. Therefore, on a surface with general geometry, different curves can be constructed. These curves may intersect at a given point and can have curvature vectors that have different orientations with respect to the normal to the surface at this point.

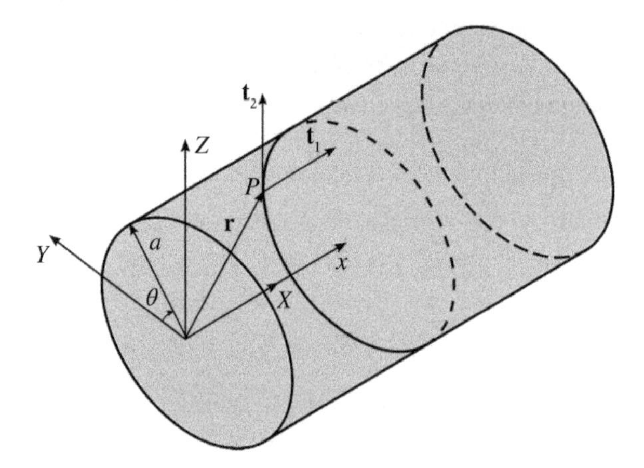

Figure 2.18 Normals of cylindrical curves.

Let $\mathbf{C} = \mathbf{C}(s_1(t), s_2(t))$ be a regular curve defined on the surface $\mathbf{r} = \mathbf{r}(s_1, s_2)$, and let $\mathbf{K}$ be the curvature vector of the curve that defines the curve normal direction. The *normal curvature vector* $\mathbf{K}_n$ at a point on the curve is defined as the projection of the curvature vector $\mathbf{K}$ on the normal to the surface $\mathbf{n}$ at this point. That is,

$$\mathbf{K}_n = (\mathbf{K} \cdot \mathbf{n})\mathbf{n} \tag{2.35}$$

The norm of this vector, called the *normal curvature*, is defined as

$$\kappa_n = \mathbf{K} \cdot \mathbf{n} \tag{2.36}$$

The sign of κ_n can be positive or negative, depending on the direction of the normal $\mathbf{n}$, with a positive sign implying that the normal vector $\mathbf{n}$ is directed toward the center of the radius of curvature. Furthermore, because the surface normal $\mathbf{n}$ is determined using the cross product, $\mathbf{n} = ((\partial \mathbf{r}/\partial s_1) \times (\partial \mathbf{r}/\partial s_2))/|(\partial \mathbf{r}/\partial s_1) \times (\partial \mathbf{r}/\partial s_2)|$, the effect of the choice of the order of multiplication must be observed.

If $\mathbf{t}$ is the tangent vector to curve $\mathbf{C}$ and s and t are two different parameters, with s representing the curve arc length, then the curvature vector of curve $\mathbf{C}$ at a point on surface $\mathbf{r}$ can be written as

$$\mathbf{K} = \frac{d^2\mathbf{C}}{ds^2} = \frac{d\mathbf{t}(s)}{ds} = \frac{d\mathbf{t}}{dt}\frac{dt}{ds} = \frac{(d\mathbf{t}/dt)}{|d\mathbf{r}/dt|} \tag{2.37}$$

In this equation, $d\mathbf{r}$ is an infinitesimal line element between two points on the curve that also correspond to two points on the surface. Furthermore, the fact that $ds = |d\mathbf{r}/dt|dt$ was utilized. Because $\mathbf{t}$ and $\mathbf{n}$ are orthogonal vectors ($\mathbf{t}^T\mathbf{n} = 0$), one has $d(\mathbf{t}^T\mathbf{n})/dt = 0$, which leads to $(d\mathbf{t}/dt)\cdot\mathbf{n} = -\mathbf{t}\cdot(d\mathbf{n}/dt)$. Using this equation in Eq. 37 after multiplying by $\mathbf{n}$, and given the fact that $\mathbf{t} = (d\mathbf{r}/dt)(dt/ds) = (d\mathbf{r}/dt)/|d\mathbf{r}/dt|$ because the curve lies on the surface, one obtains the following expression for the normal curvature κ_n:

$$\kappa_n = \mathbf{K}\cdot\mathbf{n} = \frac{(d\mathbf{t}/dt)\cdot\mathbf{n}}{|d\mathbf{r}/dt|} = -\frac{\mathbf{t}\cdot(d\mathbf{n}/dt)}{|d\mathbf{r}/dt|} = -\frac{(d\mathbf{r}/dt)\cdot(d\mathbf{n}/dt)}{(|d\mathbf{r}/dt|)^2} \tag{2.38}$$

It is important to note that by dividing by $(dt)^2$, this equation can also be written as $\kappa_n = \mathbf{K}\cdot\mathbf{n} = -(d\mathbf{r}\cdot d\mathbf{n})/(|d\mathbf{r}|)^2$. One can show that the normal curvature given in Eq. 38 depends on the ratio ds_1/ds_2, which defines direction. Because the tangent and normal vectors to the surface at a given point are independent of the curve drawn on the surface, all curves that intersect at a given point on the surface have the same normal curvature $\kappa_n = \mathbf{K}\cdot\mathbf{n} = -(d\mathbf{r}\cdot d\mathbf{n})/(|d\mathbf{r}|)^2$ at the intersection point for the same ratio ds_1/ds_2. Equation 38 for the normal curvature κ_n can also be written in terms of the coefficients of the first and second fundamental forms of surfaces as

$$\kappa_n = \frac{L(ds_1/dt)^2 + 2M(ds_1/dt)(ds_2/dt) + N(ds_2/dt)^2}{E(ds_1/dt)^2 + 2F(ds_1/dt)(ds_2/dt) + G(ds_2/dt)^2} \tag{2.39}$$

This equation or the equation $\kappa_n = \mathbf{K}\cdot\mathbf{n} = -(d\mathbf{r}\cdot d\mathbf{n})/(|d\mathbf{r}|)^2$ can also be used to write the normal curvature as the ratio of the second and first fundamental forms as

$$\kappa_n = \frac{L(ds_1)^2 + 2M ds_1 ds_2 + N(ds_2)^2}{E(ds_1)^2 + 2F ds_1 ds_2 + G(ds_2)^2} = \frac{II}{I} \tag{2.40}$$

In deriving Eqs. 39 and 40, it is assumed that $(ds_1)^2 + (ds_2)^2 \neq 0$. Furthermore, because the first fundamental form I measures the square of the length of a line element $d\mathbf{r}$ and consequently is positive definite, the sign of the normal curvature κ_n depends on the sign of the second fundamental form II. For a planar surface, κ_n is zero everywhere; for an elliptic point on the surface, $\kappa_n \neq 0$ and assumes the sign as ds_1/ds_2; for a hyperbolic point, κ_n can be positive, negative, or zero, depending on the value and sign of ds_1/ds_2; and for a parabolic point, κ_n has the same sign and is zero for a zero value of II.

Example 2.17 A cylindrical surface that has an axis along the X axis and radius a was previously discussed in Example 9 and shown in Figure 18. The parametric equation of this surface is defined in terms of the two parameters x and θ by $\mathbf{r}(x,\theta) = [x \;\; y \;\; z]^T = [x \;\; a\cos\theta \;\; a\sin\theta]^T$. The two tangent vectors were defined as $\mathbf{t}_1 = \partial\mathbf{r}/\partial x = [1 \;\; 0 \;\; 0]^T$ and $\mathbf{t}_2 = \partial\mathbf{r}/\partial\theta = a[0 \;\; -\sin\theta \;\; \cos\theta]^T$. The unit vector normal to the surface is defined using the cross product $\mathbf{n} = (\partial\mathbf{r}/\partial x)\times(\partial\mathbf{r}/\partial\theta)\,/\,|(\partial\mathbf{r}/\partial x)\times(\partial\mathbf{r}/\partial\theta)| = -[0 \;\; \cos\theta \;\; \sin\theta]^T$. As previously discussed in this section, for a given $x = x_s$, keeping in mind that $\theta = s/a$ where s is the curve arc length, one has a profile curve defined by the parametric equation $\mathbf{C}(s) = [x_s \;\; a\cos(s/a) \;\; a\sin(s/a)]^T$. The tangent and curvature vector of this curve

(Continued)

are, respectively, given by

$$\partial \mathbf{C}(s)/\partial s = \begin{bmatrix} 0 & -\sin(s/a) & \cos(s/a) \end{bmatrix}^T$$
$$\partial^2 \mathbf{C}(s)/\partial s^2 = -(1/a)\begin{bmatrix} 0 & \cos(s/a) & \sin(s/a) \end{bmatrix}^T$$

The normal curvature is given by

$$\kappa_n = \left(\partial^2 \mathbf{C}(s)/\partial s^2\right)\cdot \mathbf{n} = -(1/a)\begin{bmatrix} 0 & \cos(s/a) & \sin(s/a) \end{bmatrix}\begin{bmatrix} 0 \\ -\cos\theta \\ -\sin\theta \end{bmatrix} = (1/a)$$

2.7 PRINCIPAL CURVATURES AND DIRECTIONS

The fact that the expression of the normal curvature at a point depends on a direction defined by the ratio ds_1/ds_2 suggests that there are directions in which the normal curvature assumes maximum and minimum values. The normal curvature is called the *principal curvature* if its value is this maximum or minimum value. From differential calculus, the maximum and minimum values of the normal curvature can be obtained by equating its derivatives with respect to s_1 and s_2 to zero: that is, $\partial\kappa_n/\partial(ds_1)=0$ and $\partial\kappa_n/\partial(ds_2)=0$. Using these two equations and the expression of the normal curvature given by Eq. 40, one can write $\partial\kappa_n/\partial(ds_k)=(I(\partial II/\partial s_k)-II(\partial I/\partial s_k))/(I)^2=0,\ k=1,\ 2$. This equation, upon using the definition $\kappa_n=II/I$, leads to $(\partial II/\partial s_k)-\kappa_n(\partial I/\partial s_k)=0$, where $\partial I/\partial s_1=2Eds_1+2Fds_2$, $\partial I/\partial s_2=2Fds_1+2Gds_2$, $\partial II/\partial s_1=2Lds_1+2Mds_2$, and $\partial II/\partial s_2=2Mds_1+2Nds_2$. Therefore, the derivatives of κ_n with respect to s_1 and s_2 lead to the two equations $Lds_1+Mds_2=\kappa_n(Eds_1+Fds_2)$ and $Mds_1+Nds_2=\kappa_n(Fds_1+Gds_2)$, respectively. These two scalar equations can be written in a matrix form as

$$\begin{bmatrix} L & M \\ M & N \end{bmatrix}\begin{bmatrix} ds_1 \\ ds_2 \end{bmatrix} = \kappa_n \begin{bmatrix} E & F \\ F & G \end{bmatrix}\begin{bmatrix} ds_1 \\ ds_2 \end{bmatrix} \tag{2.41}$$

This is an eigenvalue problem with symmetric coefficient matrices. In this equation, κ_n is the eigenvalue and $\mathbf{y} = \begin{bmatrix} ds_1 & ds_2 \end{bmatrix}^T$ is the eigenvector. This eigenvalue problem can be written as

$$\begin{bmatrix} L-\kappa_n E & M-\kappa_n F \\ M-\kappa_n F & N-\kappa_n G \end{bmatrix}\begin{bmatrix} ds_1 \\ ds_2 \end{bmatrix} = \begin{bmatrix} 0 \\ 0 \end{bmatrix} \tag{2.42}$$

This system of homogenous equations in ds_1 and ds_2 has a nontrivial solution if and only if the determinant of the coefficient matrix is equal to zero. This yields the condition

$$\begin{vmatrix} L-\kappa_n E & M-\kappa_n F \\ M-\kappa_n F & N-\kappa_n G \end{vmatrix} = \left(EG-(F)^2\right)\left(\kappa_n\right)^2 - (EN+GL-2FM)\,\kappa_n + LN-(M)^2 = 0 \tag{2.43}$$

This is the *characteristic equation,* which is quadratic in the normal curvature κ_n. This equation has the following two roots:

$$\kappa_{1,2} = \frac{-\bar{b} \pm \sqrt{\left(\bar{b}\right)^2 - 4\bar{a}\,\bar{c}}}{2\bar{a}} \tag{2.44}$$

In this equation, $\bar{a} = EG - (F)^2$, $\bar{b} = -(EN + GL - 2FM)$, and $\bar{c} = LN - (M)^2$. The two roots κ_1 and κ_2 determine the principal curvatures that define the maximum and minimum curvatures. The principal directions $\left[(ds_1)_i \quad (ds_2)_i\right]^T$, $i = 1, 2$ are determined using Eq. 42 as

$$\begin{bmatrix} L - \kappa_i E & M - \kappa_i F \\ M - \kappa_i F & N - \kappa_i G \end{bmatrix} \begin{bmatrix} (ds_1)_i \\ (ds_2)_i \end{bmatrix} = \begin{bmatrix} 0 \\ 0 \end{bmatrix}, \quad i = 1, 2 \tag{2.45}$$

Because Eq. 45 is a homogeneous system of algebraic equations with a singular coefficient matrix, the principal directions $\left[(ds_1)_i \quad (ds_2)_i\right]^T$, $i = 1, 2$ can be determined to within an arbitrary constant. Furthermore, because of the symmetry of the coefficient matrices of Eq. 41, the eigenvalues that represent the principal curvatures and the eigenvectors that represent the principal directions are guaranteed to be real non-complex numbers.

Two important definitions are often made in the differential geometry of surfaces. These are the *mean curvature* κ_m and the *Gaussian curvature* κ_G. These two curvatures are defined using the principal curvatures as

$$\kappa_m = \frac{1}{2}\left(\kappa_1 + \kappa_2\right), \qquad \kappa_G = \kappa_1 \kappa_2 \tag{2.46}$$

It can be proven that the mean curvature is half the trace of a matrix **H**, while the Gaussian curvature is the determinant of the same matrix **H**. Matrix **H**, formed from the coefficient matrices of Eq. 41, is a function of the coefficients of the first and second fundamental forms and can be written as follows:

$$\mathbf{H} = \begin{bmatrix} E & F \\ F & G \end{bmatrix}^{-1} \begin{bmatrix} L & M \\ M & N \end{bmatrix} = \frac{1}{EG - (F)^2} \begin{bmatrix} (GL - FM) & (GM - FN) \\ (EM - FL) & (EN - FM) \end{bmatrix} \tag{2.47}$$

That is, the mean curvature κ_m, which represents the average of the principal curvatures, and the Gaussian curvature κ_G, which is the product of the principal curvatures, are invariants of matrix **H**. Note that the eigenvalue problem in Eq. 42 can be simply written as $\mathbf{H} - \kappa_n \mathbf{I} = \mathbf{0}$, where **I** is the 2×2 identity matrix.

Example 2.18 Examples 11 and 14 considered the special case of a straight cylindrical rail that has the parametric surface equation $\mathbf{r} = \begin{bmatrix} x & y & z \end{bmatrix}^T = \begin{bmatrix} x & y_s & f^r(x, y_s) \end{bmatrix}^T$, where x and y_s are, respectively, the longitudinal and lateral surface parameters, and $f^r(y_s) = \sqrt{(a)^2 - (y_s)^2}$, where a is the cylinder radius. The coefficients of the first fundamental form were determined in Example 11 as

$$E = (\partial \mathbf{r}/\partial x)^T (\partial \mathbf{r}/\partial x) = 1, \quad F = (\partial \mathbf{r}/\partial x)^T (\partial \mathbf{r}/\partial y_s) = 0,$$

$$G = (\partial \mathbf{r}/\partial y_s)^T (\partial \mathbf{r}/\partial y_s) = 1 + \left((y_s)^2 / \left((a)^2 - (y_s)^2\right)\right)$$

In Example 14, the normal vector was defined as $\mathbf{n} = \left(1/\sqrt{1 + \left(f_{y_s}^r\right)^2}\right)\begin{bmatrix} 0 & -f_{y_s}^r & 1 \end{bmatrix}^T$, and the coefficients of the second fundamental form were determined as

$$L = \left(\partial^2 \mathbf{r}/\partial s_1^2\right)^T \mathbf{n} = 0, \quad M = \left(\partial^2 \mathbf{r}/\partial s_1 \partial s_2\right)^T \mathbf{n} = 0,$$

$$N = \left(\partial^2 \mathbf{r}/\partial s_2^2\right)^T \mathbf{n} = \left(1/\sqrt{1 + \left(f_{y_s}^r\right)^2}\right)\left(\partial f_{y_s}^r / \partial y_s\right)$$

where $\partial f_{y_s}^r / \partial y_s = -(a)^2 / \left((a)^2 - (y_s)^2\right)^{3/2}$.

(Continued)

Consider first the case in which $a = 0.05$ m and $y_s = 0$ m. Using these values, one has $f^r = 0.05$, $f^r_{y_s} = 0$, and $\partial f^r_{y_s}/\partial y_s = -20$; the coefficients of the first fundamental form are $E = 1$, $F = 0$, and $G = 1$; and the coefficients of the second fundamental form are $L = 0$, $M = 0$, and $N = -20$. In this case, the characteristic equation $(EG - (F)^2)(\kappa_n)^2 - (EN + GL - 2FM)\kappa_n + LN - (M)^2 = 0$ reduces to $(\kappa_n)^2 + 20\kappa_n = 0$, which has the roots $\kappa_1 = 0$ and $\kappa_2 = -20$. The principal directions can be determined using Eq. 45, which shows that $(ds_2)_i/(ds_1)_i = -(L - \kappa_i E)/(M - \kappa_i F)$ or $(ds_2)_i/(ds_1)_i = -(M - \kappa_i F)/(N - \kappa_i G)$. For $\kappa_1 = 0$, and considering $s_1 = x$ and $s_2 = y_s$, one has $(dy_s)_1/(dx)_1 = 0$, which shows that the first principal direction associated with κ_1, which can be determined to within an arbitrary constant, can be written as $\begin{bmatrix}(dx)_1 & (dy_s)_1\end{bmatrix}^T = \begin{bmatrix}1 & 0\end{bmatrix}^T$. For $\kappa_2 = -20$, one has $(dx)_2/(dy_s)_2 = 0$, which shows that the second principal direction can be written as $\begin{bmatrix}(dx)_1 & (dy_s)_1\end{bmatrix}^T = \begin{bmatrix}0 & 1\end{bmatrix}^T$. One can also show that matrix $\mathbf{H}$ is given in this case as

$$\mathbf{H} = \frac{1}{EG - (F)^2}\begin{bmatrix}(GL - FM) & (GM - FN) \\ (EM - FL) & (EN - FM)\end{bmatrix} = \begin{bmatrix}0 & 0 \\ 0 & -20\end{bmatrix}$$

This matrix has a trace equal to -20, which is equal to $\kappa_1 + \kappa_2$, and a zero determinant that is equal to $\kappa_1\kappa_2$. Recall that the mean curvature is defined as $\kappa_m = (\kappa_1 + \kappa_2)/2$, and the Gaussian curvature is defined as $\kappa_G = \kappa_1\kappa_2$. Note also that because $LN - (M)^2 < 0$, the surface is hyperbolic at the given point.

Consider second the case in which $a = 0.05$ m and $y_s = 0.025$ m. Using these values, one has $f^r = 0.0433$, $f^r_{y_s} = -0.5774$, and $\partial f^r_{y_s}/\partial y_s = -30.7920$; the coefficients of the first fundamental form are $E = 1$, $F = 0$, and $G = 1.3333$; and the coefficients of the second fundamental form are $L = 0$, $M = 0$, and $N = -26.6667$. In this case, the characteristic equation $(EG - (F)^2)(\kappa_n)^2 - (EN + GL - 2FM)\kappa_n + LN - (M)^2 = 0$ reduces to $1.3333(\kappa_n)^2 + 26.6667\kappa_n = 0$, which has the roots $\kappa_1 = 0$ and $\kappa_2 = -20$. These are the same results obtained in the case in which $y_s = 0$. Furthermore, because $LN - (M)^2 < 0$, the surface is hyperbolic at the point defined by $y_s = 0.025$ m.

Example 2.19 A surface is defined in its parametric form by the equation

$$\mathbf{r} = \begin{bmatrix}x & 4xy & y\end{bmatrix}^T$$

Determine the type of the surface and the principal curvatures and directions at the surface point whose coordinates are $x = 1$ and $y = 1$.

Solution Using the surface parametric equation, one can write

$$\mathbf{t}_1 = \partial\mathbf{r}/\partial x = \begin{bmatrix}1 \\ 4y \\ 0\end{bmatrix}, \quad \mathbf{t}_2 = \partial\mathbf{r}/\partial y = \begin{bmatrix}0 \\ 4x \\ 1\end{bmatrix}, \quad \mathbf{n} = \frac{1}{\sqrt{1 + 16\left((x)^2 + (y)^2\right)}}\begin{bmatrix}4y \\ -1 \\ 4x\end{bmatrix},$$

$$\partial^2\mathbf{r}/\partial x^2 = \partial^2\mathbf{r}/\partial y^2 = \mathbf{0}, \quad \partial^2\mathbf{r}/\partial xy = \begin{bmatrix}0 \\ 4 \\ 0\end{bmatrix}$$

Using these definitions for the surface, the coefficients of the first fundamental form are given by

$$E = (\partial \mathbf{r}/\partial x) \cdot (\partial \mathbf{r}/\partial x) = 1 + 16(y)^2$$

$$F = (\partial \mathbf{r}/\partial x) \cdot (\partial \mathbf{r}/\partial y) = 16xy$$

$$G = (\partial \mathbf{r}/\partial y) \cdot (\partial \mathbf{r}/\partial y) = 1 + 16(x)^2$$

The coefficients of the second fundamental form are

$$L = \left(\partial^2 \mathbf{r}/\partial x^2\right) \cdot \mathbf{n} = 0, \qquad N = \left(\partial^2 \mathbf{r}/\partial y^2\right) \cdot \mathbf{n} = 0,$$

$$M = \left(\partial^2 \mathbf{r}/\partial x \partial y\right) \cdot \mathbf{n} = \frac{-4}{\sqrt{1 + 16\left((x)^2 + (y)^2\right)}}$$

At the point defined by the parameters $x = 1$ and $y = 1$, the coefficients of the first and second fundamental forms are given by

$$E = 17, \quad F = 16, \quad G = 17, \quad L = 0, \quad M = -4/\sqrt{33}, \quad N = 0$$

Because $LN - (M)^2 < 0$ at the given point, the surface is hyperbolic at this point. In fact, the surface is hyperbolic everywhere because the condition $LN - (M)^2 < 0$ is satisfied at every point on the surface. The principal curvatures at the given point can be defined using Eq. 44 as

$$\kappa_{1,2} = \frac{-\bar{b} \pm \sqrt{\left(\bar{b}\right)^2 - 4\bar{a}\,\bar{c}}}{2\bar{a}}$$

where $\bar{a} = EG - (F)^2 = 33$, $\bar{b} = -(EN + GL - 2FM) = -128/\sqrt{33}$ and $\bar{c} = LN - (M)^2 = -16/\sqrt{33}$. Using these values, the principal curvatures can be evaluated as

$$\kappa_1 = 0.0211, \qquad \kappa_2 = -0.6963$$

The principal directions can be determined using Eq. 45, which shows that $(ds_2)_i/(ds_1)_i = -(L - \kappa_i E)/(M - \kappa_i F)$ or $(ds_2)_i/(ds_1)_i = -(M - \kappa_i F)/(N - \kappa_i G)$. Using these ratios, and considering $s_1 = x$ and $s_2 = y$, the principal directions are defined, to within an arbitrary constant, as

$$\begin{bmatrix} (dx)_1 \\ (dy)_1 \end{bmatrix} = \begin{bmatrix} 1 \\ 0.3469 \end{bmatrix}, \qquad \begin{bmatrix} (dx)_2 \\ (dy)_2 \end{bmatrix} = \begin{bmatrix} 1 \\ 1.1712 \end{bmatrix}$$

To check the results of the principal curvatures, one can write matrix $\mathbf{H}$ as

$$\mathbf{H} = \frac{1}{EG - (F)^2} \begin{bmatrix} (GL - FM) & (GM - FN) \\ (EM - FL) & (EN - FM) \end{bmatrix} = \begin{bmatrix} 0.3376 & -0.3587 \\ -0.3587 & 0.3376 \end{bmatrix}$$

This equation shows that the trace of $\mathbf{H}$ is given by $\mathrm{tr}(\mathbf{H}) = -0.6752 = \kappa_1 + \kappa_2 = 2\kappa_m$, and the determinant of $\mathbf{H}$ is $|\mathbf{H}| = -0.01469 = \kappa_G$.

2.8 NUMERICAL REPRESENTATION OF THE PROFILE GEOMETRY

The wheel and rail profiles that define the geometry of the contact surfaces play a significant role in railroad vehicle dynamics and stability. The wheel and rail profile geometries, like the ones shown in Figure 19, are designed to ensure the stability of the railroad vehicle as it negotiates both tangent and curved tracks. These profiles are not described using a single analytical function. The rail profile, for example, is constructed using a series of arcs that are combined together to form the profile geometry. For both new and worn profiles, the profile geometry can be measured using a simple device called a *MiniProf*, like the one shown in Figure 20. By sweeping the MiniProf over the wheel and rail profiles, it generates $x-y$ tabulated data that can be used with interpolating functions to define the profile numerically. The tabulated data can be used in numerical computer simulations to define the wheel and rail contact surface geometries.

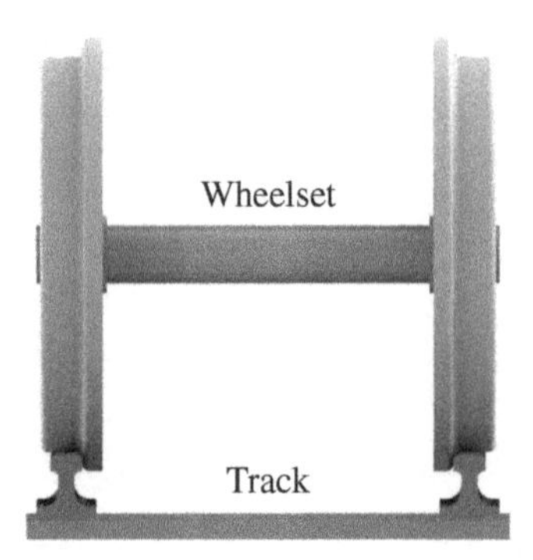

Figure 2.19 Wheel and rail profiles.

Cubic Spline Interpolation Given the wheel or rail profile tabulated data in the form $y_i = y(x_i)$, $i = 0, 1, \ldots, n$, where $x_0, x_1, \ldots, x_n$ are not necessarily equally spaced and are

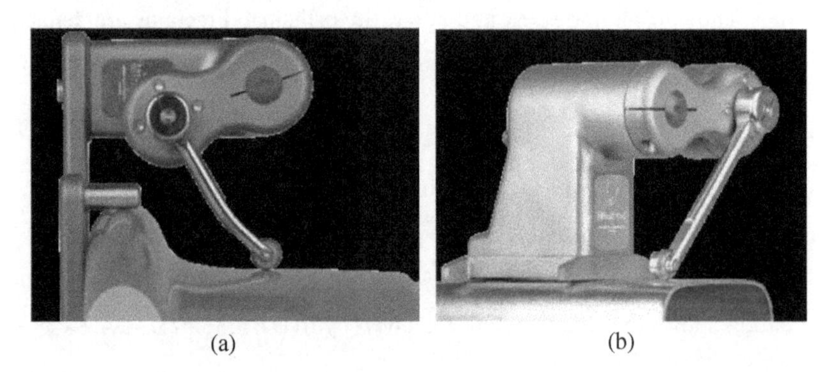

Figure 2.20 MiniProf profile measurements; (a) Wheel, Source: Greenwood Engineering (b) Rail.

not overlapping – that is, $x_0 \leq x_1 \leq \cdots \leq x_n$ – different interpolation functions can be used to describe the profile geometry. One of the most popular methods for the interpolation functions used to approximate the profile geometry is the *cubic spline* interpolation. Cubic spline functions are considered the most popular spline functions because they are smooth functions and do not exhibit the oscillatory behavior that characterizes higher-order interpolation. They are simple to develop and use and can lead to very efficient implementation. Consider a profile segment over the interval defined by the two coordinates x_{i-1} and x_i. The goal is to be able to describe the profile geometry over this interval using the cubic interpolation (Atkinson 1978)

$$y(x) = a_i + b_i x + c_i(x)^2 + d_i(x)^3, \quad x_{i-1} \leq x \leq x_i, \quad i = 1, \ldots, n \tag{2.48}$$

Because there are n profile segments, there are $4n$ unknown coefficients when the cubic interpolation is used for each segment. Therefore, one needs $4n$ conditions in order to be able to determine these unknown coefficients and uniquely define the profile geometry.

To develop the conditions that are required to determine the $4n$ unknown coefficients in Eq. 48, one may require that the cubic spline interpolation must pass through the data points $x_0, x_1, \cdots, x_n$ and must satisfy the continuity of the coordinates, first derivatives, and second derivatives: that is,

$$\left.\begin{aligned} & y_i = y\left(x_i\right), && i = 0, 1, \ldots, n \\ & y_{i+1}\left(x_i\right) = y_i\left(x_i\right), && i = 1, \ldots, n-1 \\ & \left(\frac{\partial y_{i+1}}{\partial x}\right)_{x_i} = \left(\frac{\partial y_i}{\partial x}\right)_{x_i}, && i = 1, \ldots, n-1 \\ & \left(\frac{\partial^2 y_{i+1}}{\partial x^2}\right)_{x_i} = \left(\frac{\partial^2 y_i}{\partial x^2}\right)_{x_i}, && i = 1, \ldots, n-1 \end{aligned}\right\} \tag{2.49}$$

where y_i refers to the interpolating function over the segment defined by the non-overlapping coordinates x_{i-1} and x_i. It is clear that Eq. 49 defines $n+1+3(n-1)=4n-2$ conditions; therefore, two coefficients can be varied arbitrarily, thereby defining two families of solutions. To have a unique cubic spline representation, two additional conditions must be enforced. Typically, these two additional conditions, which are not unique, are enforced as boundary conditions at x_0 and x_n. Among the most common choices for these boundary conditions are the following:

1. Assuming that the derivatives y'_0 and y'_n are known at the endpoints x_0 and x_n, respectively, one can require that $y(x)$ satisfies the two endpoint conditions $(\partial y/\partial x)_{x_0} = y'_0$ and $(\partial y/\partial x)_{x_n} = y'_n$. The resulting cubic spline, in this case, is called the *complete cubic spline interpolation* (Atkinson 1978).
2. If the derivatives at the two endpoints are not known, one may set the second derivatives equal to zero at the endpoints: that is, $\left(\partial^2 y/\partial x^2\right)_{x_0} = 0$ and $\left(\partial^2 y/\partial x^2\right)_{x_n} = 0$. The resulting interpolation, in this case, is called the *natural cubic spline*.

Construction of the Cubic Spline Interpolation To explain how the cubic spline is constructed, the notation $\kappa_i = \left(\partial^2 y/\partial x^2\right)_{x_i}$, $i = 0, 1, \ldots, n$ is used, where the constants κ_i, $i = 0, 1, \ldots, n$ will be determined. It is clear that in the case of the cubic interpolation of

Eq. 48, one has the following linear interpolation for the second derivative over the interval $x_i \leq x \leq x_{i+1}$:

$$\frac{\partial^2 y}{\partial x^2} = \frac{\left(x_{i+1} - x\right) \kappa_i + \left(x - x_i\right) \kappa_{i+1}}{\Delta x_i}, \quad i = 0, 1, \ldots, n-1 \tag{2.50}$$

where $\Delta x_i = x_{i+1} - x_i$. Using Eq. 50 ensures that $\partial^2 y/\partial x^2$ is continuous over the entire domain of the spline function. Integrating Eq. 50 twice, one can show that the cubic spline function can be written as (Atkinson 1978)

$$y(x) = \frac{\left(x_{i+1} - x\right)^3 \kappa_i + \left(x - x_i\right)^3 \kappa_{i+1}}{6\Delta x_i} + C_i\left(x_{i+1} - x\right) + D_i\left(x - x_i\right) \tag{2.51}$$

where the arbitrary constants of integration C_i and D_i can be determined using the first condition in Eq. 49, $y_i = y(x_i)$, $i = 0, 1, \ldots, n$, as

$$C_i = \frac{6y_i - \left(\Delta x_i\right) \kappa_i}{6\Delta x_i}, \qquad D_i = \frac{6y_{i+1} - \left(\Delta x_i\right) \kappa_{i+1}}{6\Delta x_i} \tag{2.52}$$

Substituting these constants into Eq. 51 yields

$$\begin{aligned} y(x) = {} & \frac{\left(x_{i+1} - x\right)^3 \kappa_i + \left(x - x_i\right)^3 \kappa_{i+1}}{6\Delta x_i} + \frac{\left(x_{i+1} - x\right) y_i + \left(x - x_i\right) y_{i+1}}{\Delta x_i} \\ & - \frac{\Delta x_i}{6}\left[\left(x_{i+1} - x\right) \kappa_i + \left(x - x_i\right) \kappa_{i+1}\right], \quad x_i \leq x \leq x_{i+1}, \quad 0 \leq i \leq n-1 \end{aligned} \tag{2.53}$$

This interpolation ensures the continuity of $y(x)$ over the entire domain of the spline function and also ensures that the spline function passes by all the data points: that is, $y_i = y(x_i)$, $i = 0, 1, \ldots, n$. To determine the constants $\kappa_0, \ldots, \kappa_n$, the third condition of Eq. 49, $\left(\partial y_{i+1}/\partial x\right)_{x_i} = \left(\partial y_i/\partial x\right)_{x_i}$, $i = 1, \ldots, n-1$, which implies the continuity of the first derivatives at $x_1, \ldots, x_{n-1}$, can be applied. Enforcing this continuity condition and using some manipulations, by equating the derivatives of the two cubic functions over the two segments $x_i \leq x \leq x_{i+1}$ and $x_{i-1} \leq x \leq x_i$, one obtains

$$\frac{\Delta x_{i-1}}{6} \kappa_{i-1} + \frac{\Delta x_i + \Delta x_{i-1}}{3} \kappa_i + \frac{\Delta x_i}{6} \kappa_{i+1} = \frac{y_{i+1} - y_i}{\Delta x_i} - \frac{y_i - y_{i-1}}{\Delta x_{i-1}}, \quad i = 1, \ldots, n-1 \tag{2.54}$$

which defines $n - 1$ equations, while there are $n + 1$ unknown constants $\kappa_0, \ldots, \kappa_n$. Therefore, two additional end conditions, as previously discussed, need to be introduced. One can specify the first derivatives or assume zero curvatures at the endpoints x_0 and x_n in order to be able to determine all the constants $\kappa_0, \ldots, \kappa_n$. Using either of the two choices described previously in this section for the end conditions at x_0 and x_n ensures obtaining a system of algebraic equations with a non-singular coefficient matrix that is diagonally dominant. This system of algebraic equations can be efficiently solved in order to determine the coefficients $\kappa_0, \ldots, \kappa_n$ and have a unique cubic spline representation (Atkinson 1978).

2.9 NUMERICAL REPRESENTATION OF SURFACE GEOMETRY

One effective and general method that can be used to describe surface geometry in railroad vehicle system applications is to use the *absolute nodal coordinate formulation*

(*ANCF*) finite elements (Shabana 2018). ANCF finite elements have been used effectively to integrate the geometry and analysis in railroad vehicle system applications (Berzeri et al. 2000). These elements, which have been used to describe both rail and catenary geometries, are well-suited for the description of arbitrary geometry, and their displacement field is related to computer-aided design (CAD) computational geometry methods by a linear mapping. Therefore, such ANCF elements allow for developing a solid model geometry and using it in the analysis without any adjustments. For example, ANCF fully parameterized beam elements can be used to develop the geometry of the rail space curve and superimpose on the curve geometry the profile geometry to create the desired surface geometry, as will be described in this section.

ANCF Element Kinematics In the case of a fully parameterized three-dimensional ANCF element, the displacement field can be written as

$$\mathbf{r}(x, y, z, t) = \mathbf{S}(x, y, z)\,\mathbf{e}(t) \tag{2.55}$$

In this equation, x, y, and z are the element volume parameters, t is time, $\mathbf{r}$ is the position vector of an arbitrary point on the element, $\mathbf{S}$ is the element shape function matrix, and $\mathbf{e}$ is the element vector of the nodal coordinates. If ANCF elements are used to describe only the geometry without considering deformations or displacements, as in the case of a fixed rigid rail, then the preceding equation becomes independent of time, and the vector of nodal coordinates becomes constant. In this special case, the preceding equation reduces to

$$\mathbf{r}(x, y, z) = \mathbf{S}(x, y, z)\,\mathbf{e} \tag{2.56}$$

Since the focus in this chapter is on the numerical description of the geometry, Eq. 56 will be used. In the case of the fully parameterized ANCF beam element shown in Figure 21, the shape function matrix $\mathbf{S}$ and vector of nodal coordinates $\mathbf{e}$ can be defined by starting with the following interpolating polynomials:

$$\mathbf{r} = \begin{bmatrix} r_1 \\ r_2 \\ r_3 \end{bmatrix} = \begin{bmatrix} a_0 + a_1x + a_2y + a_3z + a_4xy + a_5xz + a_6(x)^2 + a_7(x)^3 \\ b_0 + b_1x + b_2y + b_3z + b_4xy + b_5xz + b_6(x)^2 + b_7(x)^3 \\ c_0 + c_1x + c_2y + c_3z + c_4xy + c_5xz + c_6(x)^2 + c_7(x)^3 \end{bmatrix} \tag{2.57}$$

In this equation, $a_i, b_i, c_i, i = 0, 1, \ldots, 7$ are the coefficients of the interpolating polynomials. The interpolation in the preceding equation is cubic in x and linear in y and z. Using polynomials, which are cubic in x, allows for accurately describing the curvature of the rail space curve. The 24 coefficients $a_i, b_i, c_i, i = 0, 1, \ldots, 7$ can be replaced by coordinates that have physical meaning. For ANCF elements, these are position and position-gradient coordinates. As shown in Figure 22, the position and gradient coordinates are introduced at two endpoints called *nodes*. For a node k, $k = 1, 2$, the following 12 coordinates are

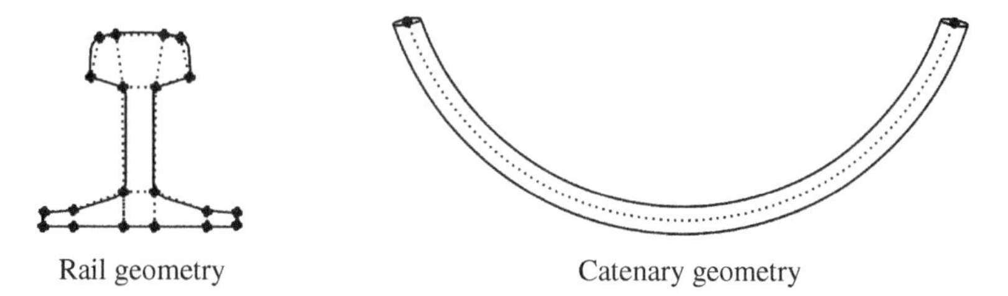

Figure 2.21 ANCF geometry.

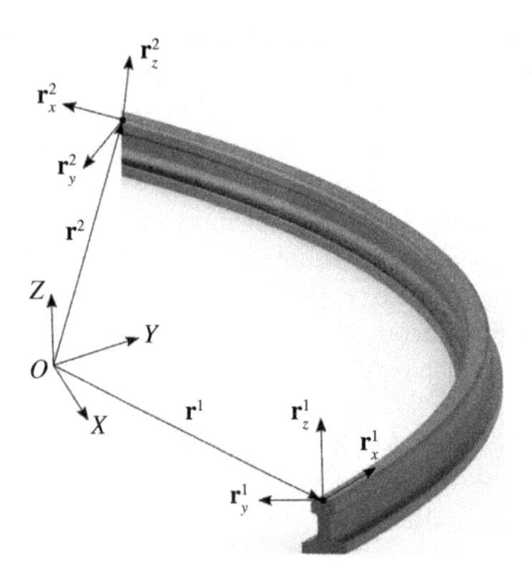

Figure 2.22 ANCF elements.

introduced: $\mathbf{r}^k, \mathbf{r}_x^k = \partial \mathbf{r}^k/\partial x$, $\mathbf{r}_y^k = \partial \mathbf{r}^k/\partial y$, and $\mathbf{r}_z^k = \partial \mathbf{r}^k/\partial z$. Therefore, there are 24 new coordinates that can be used to replace the polynomial coefficients. To this end, the following conditions are applied at the first node, $k = 1$:

$$\left.\begin{aligned} \mathbf{r}^1 = \mathbf{r}(0,0,0) = \begin{bmatrix} e_1 \\ e_2 \\ e_3 \end{bmatrix}, \quad \frac{\partial \mathbf{r}^1}{\partial x} = \mathbf{r}_x(0,0,0) = \begin{bmatrix} e_4 \\ e_5 \\ e_6 \end{bmatrix}, \\ \frac{\partial \mathbf{r}^1}{\partial y} = \mathbf{r}_y(0,0,0) = \begin{bmatrix} e_7 \\ e_8 \\ e_9 \end{bmatrix}, \quad \frac{\partial \mathbf{r}^1}{\partial z} = \mathbf{r}_z(0,0,0) = \begin{bmatrix} e_{10} \\ e_{11} \\ e_{12} \end{bmatrix} \end{aligned}\right\} \tag{2.58}$$

and for the second node, $k = 2$, one has

$$\left.\begin{aligned} \mathbf{r}^2 = \mathbf{r}(l,0,0) = \begin{bmatrix} e_{13} \\ e_{14} \\ e_{15} \end{bmatrix}, \quad \frac{\partial \mathbf{r}^2}{\partial x} = \mathbf{r}_x(l,0,0) = \begin{bmatrix} e_{16} \\ e_{17} \\ e_{18} \end{bmatrix}, \\ \frac{\partial \mathbf{r}^2}{\partial y} = \mathbf{r}_y(l,0,0) = \begin{bmatrix} e_{19} \\ e_{20} \\ e_{21} \end{bmatrix}, \quad \frac{\partial \mathbf{r}^2}{\partial z} = \mathbf{r}_z(l,0,0) = \begin{bmatrix} e_{22} \\ e_{23} \\ e_{24} \end{bmatrix} \end{aligned}\right\} \tag{2.59}$$

where l is the length of the element. The nodal coordinates defined in the preceding two equations can be used to define a system of algebraic equations. These algebraic equations can be used to write the polynomial coefficients in terms of the ANCF coordinate vector $\mathbf{e} = \begin{bmatrix} e_1 & e_2 & \cdots & e_{24} \end{bmatrix}^T$. This leads to the equation $\mathbf{r}(x, y, z) = \mathbf{S}(x, y, z)\mathbf{e}$, where

$$\mathbf{S} = \begin{bmatrix} \bar{s}_1\mathbf{I} & \bar{s}_2\mathbf{I} & \bar{s}_3\mathbf{I} & \bar{s}_4\mathbf{I} & \bar{s}_5\mathbf{I} & \bar{s}_6\mathbf{I} & \bar{s}_7\mathbf{I} & \bar{s}_8\mathbf{I} \end{bmatrix} \tag{2.60}$$

and the shape functions $\bar{s}_i$, $i = 1, 2, \ldots, 8$ are defined as (Yakoub and Shabana 2001)

$$\left.\begin{aligned} &\bar{s}_1 = 1 - 3\xi^2 + 2\xi^3, \quad \bar{s}_2 = l\left(\xi - 2\xi^2 + \xi^3\right), \quad \bar{s}_3 = l(\eta - \xi\eta), \quad \bar{s}_4 = l(\varsigma - \xi\varsigma), \\ &\bar{s}_5 = 3\xi^2 - 2\xi^3, \quad \bar{s}_6 = l\left(-\xi^2 + \xi^3\right), \quad \bar{s}_7 = l\xi\eta, \quad \bar{s}_8 = l\xi\varsigma \end{aligned}\right\} \tag{2.61}$$

where $\xi = x/l$, $\eta = y/l$ and $\varsigma = z/l$. Using the position vector gradients as nodal coordinates allows for representing complex railroad geometries using a small number of ANCF elements.

Surface Parameterization The surface of an element can assume an arbitrary shape by writing the coordinate z as a function of the other two coordinates x and y. To this end, one can write $z = f(x, y)$, where f is a function that defines the shape of the element surface. In the case of the rail surface, for example, one has $z = f(y)$ if the rail profile does not depend on the longitudinal parameter x. Using the equation $z = f(x, y)$ implies that the element surface is defined using the two parameters $s_1 = x$ and $s_2 = y$. That is, on the surface of the element, one has

$$\mathbf{r}(x, y, z) = \mathbf{r}(x, y, f(x, y)) = \mathbf{r}(x, y) = \mathbf{r}(s_1, s_2) = \mathbf{S}(s_1, s_2)\,\mathbf{e} \tag{2.62}$$

The function $z = f(x, y)$ can assume any form and can also be represented numerically in a tabulated form. In the special and important case $z = f(y)$, the cubic spline function can be used to describe this function based on tabulated data that can be obtained, in the case of a rail profile, using the MiniProf device.

Tangent and Normal Vectors Using Eq. 62, one can write $d\mathbf{r} = (\partial\mathbf{r}/\partial x)dx + (\partial\mathbf{r}/\partial y)dy + (\partial\mathbf{r}/\partial z)dz$. Using the functional relationship $z = f(x, y)$, one has $dz = (\partial f/\partial x)dx + (\partial f/\partial y)dy$. It follows that

$$\begin{aligned} d\mathbf{r} &= (\partial\mathbf{r}/\partial x)\,dx + (\partial\mathbf{r}/\partial y)\,dy + (\partial\mathbf{r}/\partial z)\,dz \\ &= ((\partial\mathbf{r}/\partial x) + (\partial\mathbf{r}/\partial z)(\partial f/\partial x))\,dx + ((\partial\mathbf{r}/\partial y) + (\partial\mathbf{r}/\partial z)(\partial f/\partial y))\,dy \end{aligned} \tag{2.63}$$

This equation defines the two tangent vectors at an arbitrary point on the surface of the element as

$$\left.\begin{aligned} (\partial\mathbf{r}/\partial s_1) &= (\partial\mathbf{r}/\partial x) + (\partial\mathbf{r}/\partial z)(\partial f/\partial x) \\ (\partial\mathbf{r}/\partial s_2) &= (\partial\mathbf{r}/\partial y) + (\partial\mathbf{r}/\partial z)(\partial f/\partial y) \end{aligned}\right\} \tag{2.64}$$

It is important to recognize the difference between x and y when used without the relationship $z = f(x, y)$. In this case, the volume parameters x, y, and z are independent, and the three tangent vectors at any point x, y, and z of the element $(\partial\mathbf{r}/\partial x)$, $(\partial\mathbf{r}/\partial y)$, and $(\partial\mathbf{r}/\partial z)$ are independent vectors. However, if the surface of the element is specified by the function $z = f(x, y)$, one has only two independent parameters $s_1 = x$ and $s_2 = y$, and the interpretation of parameters x and y in this case is different from their interpretation when z is not specified on the surface. This is clear from the definition of the tangent vectors given by Eq. 64, in which $(\partial\mathbf{r}/\partial x) = (\partial\mathbf{r}/\partial s_1)$ with $x = s_1$ and $(\partial\mathbf{r}/\partial y) = (\partial\mathbf{r}/\partial s_2)$ with $y = s_2$ have different directions, magnitudes, and interpretations as compared to the case when the relationship $z = f(x, y)$ is not used.

Using Eq. 64, the normal vector to the surface of the element can be defined as

$$\begin{aligned} \mathbf{n} &= \frac{(\partial\mathbf{r}/\partial s_1) \times (\partial\mathbf{r}/\partial s_2)}{\left|(\partial\mathbf{r}/\partial s_1) \times (\partial\mathbf{r}/\partial s_2)\right|} \\ &= \frac{((\partial\mathbf{r}/\partial x) + (\partial\mathbf{r}/\partial z)(\partial f/\partial x)) \times ((\partial\mathbf{r}/\partial y) + (\partial\mathbf{r}/\partial z)(\partial f/\partial y))}{|((\partial\mathbf{r}/\partial x) + (\partial\mathbf{r}/\partial z)(\partial f/\partial x)) \times ((\partial\mathbf{r}/\partial y) + (\partial\mathbf{r}/\partial z)(\partial f/\partial y))|} \end{aligned} \tag{2.65}$$

The tangent vectors $(\partial \mathbf{r}/\partial x) = (\partial \mathbf{r}/\partial s_1)$ and $(\partial \mathbf{r}/\partial y) = (\partial \mathbf{r}/\partial s_2)$ with $x = s_1$ and $y = s_2$ can be used to define the coefficients of the first fundamental form. The definition of the normal vector given by Eq. 65 and the second derivatives of vector $\mathbf{r}$ with respect to $x = s_1$ and $y = s_2$ can be used to define the coefficients of the second fundamental form. These derivatives can be evaluated conveniently using the element shape function matrix and vector of nodal coordinates in Eq. 62. The coefficients of the first and second fundamental forms of surfaces obtained using the ANCF geometry can be used to systematically define the principal curvatures and principal directions of the surfaces at the wheel/rail contact points using the procedure previously discussed in this chapter.

Chapter 3

MOTION AND GEOMETRY DESCRIPTIONS

In railroad vehicle dynamics, the motion of vehicle components strongly depends on the wheel and rail geometries. The dynamics and vibration behavior, ride comfort, noise, and forces of a passenger or freight train are determined by the track quality and geometry and are also heavily influenced by the wheel and rail profile geometries. In general, in the case of unconstrained motion, the displacement of a rigid body in space can be described using six independent coordinates. Three coordinates define the global position of a point on the body, called the *body reference point*; and three coordinates define the orientation of the body with respect to the global coordinate system. The global position of the body reference point can be defined using three Cartesian coordinates. The orientation coordinates can be introduced using three independent parameters that can represent angles or can be parameters that do not have an obvious physical meaning. Therefore, in spatial analysis, orientation parameters are not unique, and different sets of parameters have been used in the literature and in developing computational multibody system (MBS) algorithms. Furthermore, in spatial analysis, angular velocities are not *exact differentials*; and therefore, they are not the time derivatives of orientation parameters. That is, angular velocities cannot be directly integrated to determine orientation parameters. Nonetheless, angular velocities can always be written as linear functions of the derivatives of the orientation parameters using a velocity transformation matrix. This velocity transformation plays a fundamental role in determining the *generalized forces* associated with the orientation parameters since these orientation parameters serve as *generalized coordinates* and are not directly associated with the Cartesian moments applied to the bodies, as discussed in Chapter 6.

The kinematic description that will be used in this book to develop the equations of motion of the components of railroad vehicles are introduced in this chapter. As discussed in this chapter, using three parameters, such as *Euler angles*, to define a body orientation in space leads to kinematic singularities. This singularity can be avoided by using the four *Euler parameters* at the expense of adding an algebraic constraint equation that relates them. Euler parameters, which are becoming more popular in developing general MBS algorithms, have many identities that can be used to simplify kinematic and dynamic equations.

Mathematical Foundation of Railroad Vehicle Systems: Geometry and Mechanics,
First Edition. Ahmed A. Shabana.

In addition to defining the orientation of bodies in space, Euler angles have been also used in railroad vehicle dynamics to define the geometry of the track based on given simple industry inputs. For the most part, track is constructed using three main segments: *tangent* (straight), *curve*, and *spiral*. The tangent segment has zero curvature; the curve segment has constant curvature; and the spiral segment, used to connect two segments with different curvature values, has a curvature that varies linearly in order to ensure a smooth transition between the two segments. Track geometry is often described using three inputs at points along the track at which the geometry changes. These three inputs – *horizontal curvature*, *superelevation*, and *grade* – can be used to define uniquely three Euler angles that are used to construct the track and rail space curves. To this end, Euler angles are converted to *field variables* and used systematically to construct a curve with well-defined geometry based on given simple track inputs. Because the three Euler angles are in general independent when used as time-dependent motion coordinates for unconstrained bodies, while the geometry of a curve is uniquely defined using one parameter that can be the arc length, converting Euler angles to field variables expressed in terms of the curve arc length ensures unique definition of the curve geometry.

Therefore, it is important to recognize that Euler angles are used in this chapter for two fundamentally different purposes: (i) as motion-generalized coordinates to describe rigid body kinematics in space; and (ii) as geometry field variables to uniquely define the geometry of track and rail space curves. The analysis presented in this chapter is used as the foundation for a computer procedure designed to develop the track geometry data required for nonlinear dynamic simulations of railroad vehicle systems. The data can be generated once before the dynamic simulation at a preprocessing stage in a *track preprocessor* computer program, as discussed in Chapter 4. The track preprocessor output file normally has data for three different space curves: the *track centerline space curve*, *right rail space curve*, and *left rail space curve*. These three curves can have different geometries. The right and left rail space curves are used in the formulation of wheel/rail contact conditions, while the track space curve is used in the motion description of the coordinate systems of vehicle components.

3.1 RIGID-BODY KINEMATICS

An arbitrary rigid body in a railroad vehicle/track system will be referred to as body i. The unconstrained motion of this body in spatial analysis is described using six independent coordinates, called body *generalized coordinates*. These coordinates are three independent translation coordinates of a selected reference point on the body and three rotation coordinates that define the body orientation in space. As shown in Figure 1, translational coordinates can be defined in the global coordinate system XYZ using the position vector $\mathbf{R}^i = \begin{bmatrix} R_x^i & R_y^i & R_z^i \end{bmatrix}^T$ of the body reference point O^i, which is assumed to be rigidly attached to the body in rigid body dynamics. The body orientation can be defined using orientation parameters that define the direction cosines of the axes of the body coordinate system $X^iY^iZ^i$ in the global XYZ coordinate system. It is clear from Figure 1 that one can write the global position vector $\mathbf{r}^i$ of an arbitrary point on the body as

$$\mathbf{r}^i = \mathbf{R}^i + \mathbf{u}^i \tag{3.1}$$

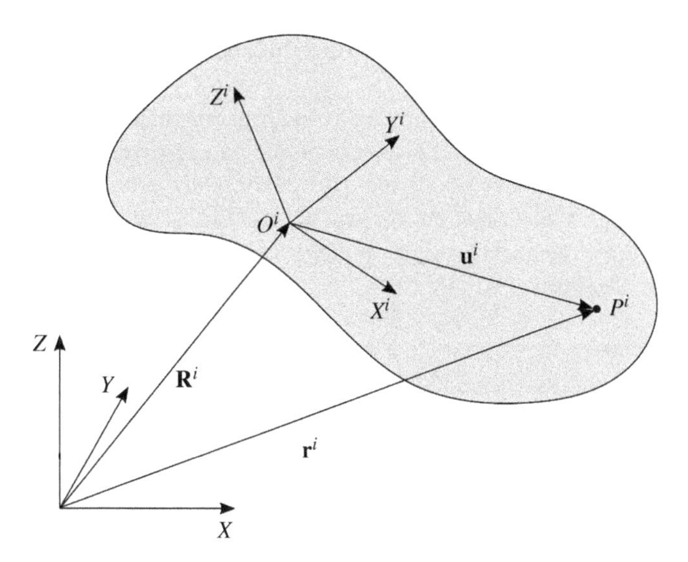

Figure 3.1 Rigid body coordinates.

where the vector $\mathbf{u}^i = \begin{bmatrix} u_x^i & u_y^i & u_z^i \end{bmatrix}^T$ defines the location of the arbitrary point with respect to the origin of the body coordinate system $X^iY^iZ^i$ in the global system: that is,

$$\mathbf{u}^i = \begin{bmatrix} u_x^i & u_y^i & u_z^i \end{bmatrix}^T = u_x^i\mathbf{i} + u_y^i\mathbf{j} + u_z^i\mathbf{k} \tag{3.2}$$

In this equation, $\mathbf{i}, \mathbf{j}$, and $\mathbf{k}$ are, respectively, unit vectors along the global axes X, Y, and Z. Alternatively, vector $\mathbf{u}^i$ can be written in terms of components defined in the body coordinate system $X^iY^iZ^i$ as

$$\begin{aligned} \mathbf{u}^i &= \overline{u}_x^i\mathbf{i}^i + \overline{u}_y^i\mathbf{j}^i + \overline{u}_z^i\mathbf{k}^i \\ &= \begin{bmatrix} \mathbf{i}^i & \mathbf{j}^i & \mathbf{k}^i \end{bmatrix} \begin{bmatrix} \overline{u}_x^i \\ \overline{u}_y^i \\ \overline{u}_z^i \end{bmatrix} = \mathbf{A}^i\overline{\mathbf{u}}^i \end{aligned} \tag{3.3}$$

where $\mathbf{i}^i, \mathbf{j}^i$, and $\mathbf{k}^i$ are, respectively, unit vectors along the axes of the body coordinate system X^i, Y^i, and Z^i defined in the global system; $\overline{u}_x^i, \overline{u}_y^i$, and $\overline{u}_z^i$ are the component of vector $\mathbf{u}^i$ defined in the body coordinate system $X^iY^iZ^i$; $\overline{\mathbf{u}}^i = \begin{bmatrix} \overline{u}_x^i & \overline{u}_y^i & \overline{u}_z^i \end{bmatrix}^T = \begin{bmatrix} x^i & y^i & z^i \end{bmatrix}^T$; and $\mathbf{A}^i = \begin{bmatrix} \mathbf{i}^i & \mathbf{j}^i & \mathbf{k}^i \end{bmatrix}$ is the 3×3 transformation matrix whose columns define the axes (orientation) of the body coordinate system $X^iY^iZ^i$. The components of vector $\overline{\mathbf{u}}^i$ are constant for rigid body dynamics. Substituting Eq. 3 into Eq. 1, the global position vector of an arbitrary point on the rigid body can be written as

$$\mathbf{r}^i = \mathbf{R}^i + \mathbf{A}^i\overline{\mathbf{u}}^i \tag{3.4}$$

While there is no restriction on the choice of body reference point O^i, selecting the center of mass of the body as the reference point leads to significant simplifications in the form of the equations of motion. Using the body center of mass as the reference point eliminates inertia coupling between the body translation and rotation coordinates and leads to the definition of *Newton–Euler equations* widely used in rigid body dynamics.

3.2 DIRECTION COSINES AND SIMPLE ROTATIONS

This section introduces a general form of transformation matrix $\mathbf{A}^i$ that defines the orientation of the body in space and is used to define transformation matrices as a result of performing simple rotations about the axes of the body coordinate system $X^iY^iZ^i$. These simple-rotation transformation matrices are used in a later section of this chapter to introduce the three independent Euler angles widely used in the motion description of three-dimensional bodies.

Direction Cosines It is clear that the columns of the transformation matrix $\mathbf{A}^i = [\mathbf{i}^i \;\; \mathbf{j}^i \;\; \mathbf{k}^i]$ are unit vectors along the axes of the body coordinate system $X^iY^iZ^i$. The elements of vectors $\mathbf{i}^i, \mathbf{j}^i$, and $\mathbf{k}^i$ represent the projection of these unit vectors along the axes of the global coordinate system defined by three unit vectors $\mathbf{i}$, $\mathbf{j}$, and $\mathbf{k}$. For example, as shown in Figure 2, the components of unit vector $\mathbf{i}^i$ along the X, Y, and Z axes are defined, respectively, by $\alpha_{11}^i = \mathbf{i}^i \cdot \mathbf{i} = \cos\beta_1$, $\alpha_{12}^i = \mathbf{i}^i \cdot \mathbf{j} = \cos\beta_2$, and $\alpha_{13}^i = \mathbf{i}^i \cdot \mathbf{k} = \cos\beta_3$, where β_1, β_2, and β_3 are, respectively, the angles between the body axis X^i and the global axes X, Y, and Z. Similar definitions for the components of unit vectors $\mathbf{j}^i$ and $\mathbf{k}^i$ can be made. Therefore, one can write

$$\left.\begin{aligned} \mathbf{i}^i &= \begin{bmatrix}\mathbf{i}^i\cdot\mathbf{i} & \mathbf{i}^i\cdot\mathbf{j} & \mathbf{i}^i\cdot\mathbf{k}\end{bmatrix}^T = \begin{bmatrix}\alpha_{11}^i & \alpha_{12}^i & \alpha_{13}^i\end{bmatrix}^T, \\ \mathbf{j}^i &= \begin{bmatrix}\mathbf{j}^i\cdot\mathbf{i} & \mathbf{j}^i\cdot\mathbf{j} & \mathbf{j}^i\cdot\mathbf{k}\end{bmatrix}^T = \begin{bmatrix}\alpha_{21}^i & \alpha_{22}^i & \alpha_{23}^i\end{bmatrix}^T, \\ \mathbf{k}^i &= \begin{bmatrix}\mathbf{k}^i\cdot\mathbf{i} & \mathbf{k}^i\cdot\mathbf{j} & \mathbf{k}^i\cdot\mathbf{k}\end{bmatrix}^T = \begin{bmatrix}\alpha_{31}^i & \alpha_{32}^i & \alpha_{33}^i\end{bmatrix}^T \end{aligned}\right\} \tag{3.5}$$

Elements α_{jl}^i, $j, l = 1, 2, 3$, which represent the components of orthogonal unit vectors $\mathbf{i}^i$, $\mathbf{j}^i$, and $\mathbf{k}^i$ along the global X, Y, and Z axes are called *direction cosines*. Therefore, the transformation matrix that defines the orientation of the body coordinate system with respect to the global system can be written in terms of direction cosines as

$$\mathbf{A}^i = \begin{bmatrix}\mathbf{i}^i & \mathbf{j}^i & \mathbf{k}^i\end{bmatrix} = \begin{bmatrix}\alpha_{11}^i & \alpha_{21}^i & \alpha_{31}^i \\ \alpha_{12}^i & \alpha_{22}^i & \alpha_{32}^i \\ \alpha_{13}^i & \alpha_{23}^i & \alpha_{33}^i\end{bmatrix} \tag{3.6}$$

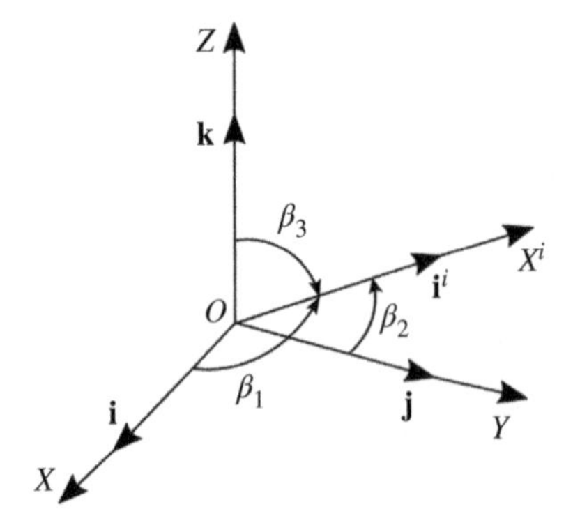

Figure 3.2 Direction cosines.

Direction cosines are not independent because the columns of the transformation matrix $\mathbf{A}^i$ are orthogonal unit vectors. Therefore, direction cosines are related by the following six constraint equations:

$$\alpha_{k1}^i\alpha_{l1}^i + \alpha_{k2}^i\alpha_{l2}^i + \alpha_{k3}^i\alpha_{l3}^i = \delta_{kl}, \quad k, l = 1, 2, 3 \tag{3.7}$$

where δ_{kl} is the *Kronecker delta*, which is equal to one if $k = l$ and equal to zero if $k \neq l$. Using the transformation matrix of Eq. 6 and the six constraint equations of Eq. 7, one can show that transformation matrix $\mathbf{A}^i$ is an orthogonal matrix: that is, $\mathbf{A}^{i^T}\mathbf{A}^i = \mathbf{A}^i\mathbf{A}^{i^T} = \mathbf{I}$, where $\mathbf{I}$ is the 3×3 identity matrix. It is also important to note that since there are nine direction cosines related by six algebraic constraint equations (Eq. 7), only three independent parameters are required in order to describe the orientation of the body in space. These three independent parameters will be introduced in a later section of this chapter.

Example 3.1

Axes X^i and Z^i of the coordinate system of a rigid body i are defined, respectively, in the global coordinate system by the vectors $\begin{bmatrix}0.0 & 2.0 & 2.0\end{bmatrix}^T$ and $\begin{bmatrix}-2.0 & -1.0 & 1.0\end{bmatrix}^T$. Obtain the transformation matrix that defines the orientation of body i with respect to the global system.

Solution The norm of the vector $\begin{bmatrix}0.0 & 2.0 & 2.0\end{bmatrix}^T$ is $\sqrt{(0)^2 + (2.0)^2 + (2.0)^2} = 2\sqrt{2}$, and the norm of the vector $\begin{bmatrix}-2.0 & -1.0 & 1.0\end{bmatrix}^T$ is $\sqrt{(-2.0)^2 + (-1.0)^2 + (1.0)^2} = \sqrt{6}$. Therefore, unit vectors $\mathbf{i}^i$ and $\mathbf{k}^i$ along the body axes X^i and Z^i can be determined, respectively, as

$$\mathbf{i}^i = \frac{1}{2\sqrt{2}}\begin{bmatrix}0.0\\2.0\\2.0\end{bmatrix} = \begin{bmatrix}0.0\\0.7071\\0.7071\end{bmatrix}, \qquad \mathbf{k}^i = \frac{1}{\sqrt{6}}\begin{bmatrix}-2.0\\-1.0\\1.0\end{bmatrix} = \begin{bmatrix}-0.8165\\-0.4082\\0.4082\end{bmatrix}$$

A unit vector $\mathbf{j}^i$ along axis Y^i can be determined using the cross product $\mathbf{j}^i = \mathbf{k}^i \times \mathbf{i}^i$, which leads to

$$\mathbf{j}^i = \mathbf{k}^i \times \mathbf{i}^i = \begin{bmatrix}-0.5774\\0.5774\\-0.5774\end{bmatrix}$$

The three vectors $\mathbf{i}^i, \mathbf{j}^i$, and $\mathbf{k}^i$ represent the columns of the transformation matrix that defines the orientation of the coordinate system $X^iY^iZ^i$ with respect to the global coordinate system XYZ as

$$\mathbf{A}^i = \begin{bmatrix}\mathbf{i}^i & \mathbf{j}^i & \mathbf{k}^i\end{bmatrix} = \begin{bmatrix}0.0 & -0.5774 & -0.8165\\0.7071 & 0.5774 & -0.4082\\0.7071 & -0.5774 & 0.4082\end{bmatrix}$$

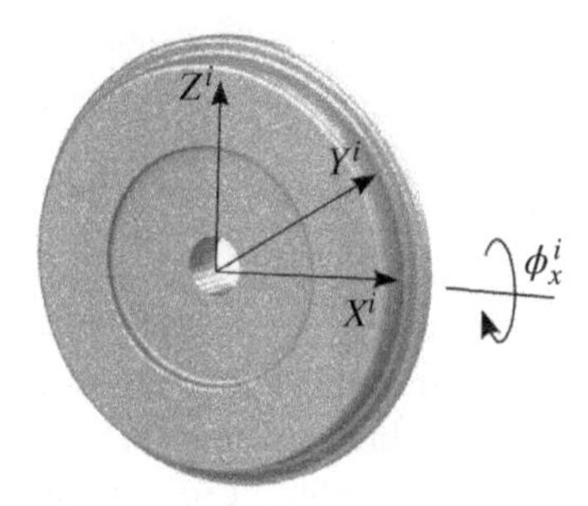

Figure 3.3 Simple rotations.

Simple Rotations Using direction cosines, one can develop the form of the transformation matrices that result from simple rotations about the axes of the body coordinate system. In the development presented in this section, it is assumed that the global *XYZ* and body $X^iY^iZ^i$ coordinate systems initially coincide before performing a simple rotation.

For a simple rotation ϕ_x about the X^i axis, shown in Figure 3, one can show that $\left[\alpha_{11}^i\ \alpha_{12}^i\ \alpha_{13}^i\right]^T = \left[\mathbf{i}^i\cdot\mathbf{i}\ \ \mathbf{i}^i\cdot\mathbf{j}\ \ \mathbf{i}^i\cdot\mathbf{k}\right]^T = \left[1\ \ 0\ \ 0\right]^T$, $\left[\alpha_{21}^i\ \alpha_{22}^i\ \alpha_{23}^i\right]^T = \left[\mathbf{j}^i\cdot\mathbf{i}\ \ \mathbf{j}^i\cdot\mathbf{j}\ \ \mathbf{j}^i\cdot\mathbf{k}\right]^T = \left[0\ \ \cos\phi_x\ \ \sin\phi_x\right]^T$, and $\left[\alpha_{31}^i\ \ \alpha_{32}^i\ \ \alpha_{33}^i\right]^T = \left[\mathbf{k}^i\cdot\mathbf{i}\ \ \mathbf{k}^i\cdot\mathbf{j}\ \ \mathbf{k}^i\cdot\mathbf{k}\right]^T = \left[0\ \ -\sin\phi_x\ \ \cos\phi_x\right]^T$. It follows that the transformation matrix in Eq. 6 can be written as a result of this simple rotation as

$$\mathbf{A}^i = \begin{bmatrix} 1 & 0 & 0 \\ 0 & \cos\phi_x & -\sin\phi_x \\ 0 & \sin\phi_x & \cos\phi_x \end{bmatrix} \tag{3.8}$$

Similarly, a simple rotation ϕ_y about the Y^i axis is defined by the following rotation matrix:

$$\mathbf{A}^i = \begin{bmatrix} \cos\phi_y & 0 & \sin\phi_y \\ 0 & 1 & 0 \\ -\sin\phi_y & 0 & \cos\phi_y \end{bmatrix} \tag{3.9}$$

And a simple rotation ϕ_z about the Z^i axis defines the following transformation matrix:

$$\mathbf{A}^i = \begin{bmatrix} \cos\phi_z & -\sin\phi_z & 0 \\ \sin\phi_z & \cos\phi_z & 0 \\ 0 & 0 & 1 \end{bmatrix} \tag{3.10}$$

The forms of the simple-rotation matrices in Eqs. 8–10 can be used to develop a more general transformation matrix expressed in terms of three independent rotations. This general transformation matrix can describe any orientation of the body in space. The three angles used to form this general transformation are called *Euler angles.*

3.3 EULER ANGLES

In railroad vehicle dynamics, Euler angles are used for two fundamentally different purposes. First, they are used as rotation parameters for the definition of a body orientation in space. In this case, transformation matrix $\mathbf{A}^i$ in Eq. 6 is expressed in terms of three independent angles that represent motion-generalized coordinates that vary with time as the body moves and changes its orientation. Second, Euler angles are used as field variables to define the geometry of rail and track space curves. In this case, Euler angles are no longer independent, and they represent geometric variables that vary with the rail arc length. Using Euler angles to define the curve geometry allows for developing a simple procedure based on simple inputs used by the rail industry to create a data file in a track preprocessor computer program that defines space curves in terms of nodes. This discretized form of the space curve is used during dynamic simulations to define the tangent, normal, and curvature vectors required to solve the wheel/rail contact problem, as discussed in later chapters.

Euler Angle Transformation Matrix As discussed in the preceding section, the transformation matrix that defines the body orientation can be written in terms of three independent parameters. These three independent parameters can be three angles, called *Euler angles*, performed about three independent axes. When Euler angles are used, three simple successive rotations are performed about three axes of the body coordinate system $X^iY^iZ^i$. Different rotation sequences can be used to define Euler angles. For example, Euler used the sequence Z^i, X^i, and Z^i to study gyroscopic motion. In this sequence used by Euler, the first axis Z^i is different from the third axis Z^i as a result of rotation about the second axis X^i. In railroad vehicle dynamics, a different sequence of Euler angles is used. First a rotation ψ^i (*yaw*) about the Z^i axis is performed, followed by a rotation ϕ^i (*roll*) about the new X^i axis, followed by a rotation θ^i (*pitch*) about the new Y^i body axis, as shown in Figure 4. These three simple successive rotations define the following simple rotation matrices:

$$\mathbf{A}_\psi^i = \begin{bmatrix} \cos\psi^i & -\sin\psi^i & 0 \\ \sin\psi^i & \cos\psi^i & 0 \\ 0 & 0 & 1 \end{bmatrix}, \quad \mathbf{A}_\phi^i = \begin{bmatrix} 1 & 0 & 0 \\ 0 & \cos\phi^i & -\sin\phi^i \\ 0 & \sin\phi^i & \cos\phi^i \end{bmatrix}, \quad \mathbf{A}_\theta^i = \begin{bmatrix} \cos\theta^i & 0 & \sin\theta^i \\ 0 & 1 & 0 \\ -\sin\theta^i & 0 & \cos\theta^i \end{bmatrix} \tag{3.11}$$

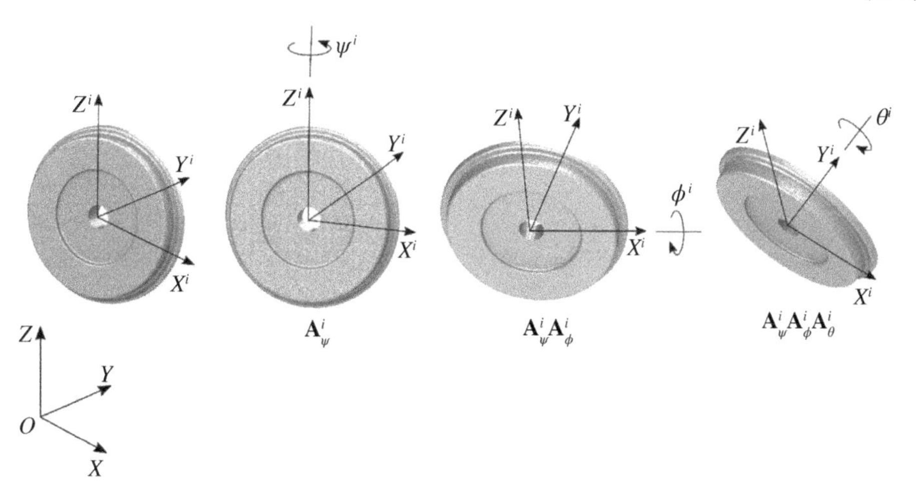

Figure 3.4 Euler angles.

This sequence, as explained later, is used to avoid kinematic singularities in most railroad vehicle simulations in which the yaw and roll angles are always small and the pitch angle can be very large due to wheel rotation. It is clear from Figure 4 that the orthogonal transformation matrix $\mathbf{A}^i$ can be defined as the product of the three simple rotation matrices in Eq. 11 as $\mathbf{A}^i = \mathbf{A}^i_\psi \mathbf{A}^i_\phi \mathbf{A}^i_\theta$, that is,

$$\mathbf{A}^i = \mathbf{A}^i_\psi \mathbf{A}^i_\phi \mathbf{A}^i_\theta$$
$$= \begin{bmatrix} \cos\psi^i \cos\theta^i - \sin\psi^i \sin\phi^i \sin\theta^i & -\sin\psi^i \cos\phi^i & \cos\psi^i \sin\theta^i + \sin\psi^i \sin\phi^i \cos\theta^i \\ \sin\psi^i \cos\theta^i + \cos\psi^i \sin\phi^i \sin\theta^i & \cos\psi^i \cos\phi^i & \sin\psi^i \sin\theta^i - \cos\psi^i \sin\phi^i \cos\theta^i \\ -\cos\phi^i \sin\theta^i & \sin\phi^i & \cos\phi^i \cos\theta^i \end{bmatrix} \tag{3.12}$$

The columns of this transformation matrix are orthogonal unit vectors along the axes of the body coordinate system; therefore, the magnitude of any of the elements in the matrix of Eq. 12 should not exceed one. These elements are simply the direction cosines of the axes of the body coordinate system X^i, Y^i, and Z^i. It is also important to note that because of the non-commutativity of finite rotations, $\mathbf{A}^i_z \mathbf{A}^i_x \mathbf{A}^i_y \neq \mathbf{A}^i_y \mathbf{A}^i_x \mathbf{A}^i_z$, and the order in which Euler angles are performed must be observed.

Kinematic Singularity All three-parameter representations of finite rotations suffer from singularities. This is a major drawback that can lead to serious numerical problems in computer simulations of railroad vehicle systems (Roberson and Schwertassek 1988; Shabana 2010). When Euler angles are used, the singularity arises when the three axes about which Euler angles are performed are not independent. The Euler-angle singular configuration depends on the sequence of rotation used. For the sequence Z^i, X^i, and Y^i used in this section, the singularity occurs when the roll angle ϕ^i is equal to $\pm\pi/2$ because the axes of rotations for angles ψ^i and θ^i become parallel, and the yaw angle ψ^i and pitch angle θ^i are not independent. In this case, as demonstrated in a later section of this chapter, one cannot write the derivatives of the angles in terms of the angular velocity vector when the singular configuration is encountered. As previously mentioned, this type of kinematic singularity is encountered when any known three-parameter method is used to define the body orientation in space. This kinematic singularity can be avoided by using four *Euler parameters* at the expense of adding an algebraic constraint equation.

Example 3.2 As previously mentioned, in his study of gyroscopic motion, Euler used the sequence $Z^i - X^i - Z^i$. This sequence of Euler angles can be defined by a rotation ϕ^i about the body Z^i axis, a rotation θ^i about the body X^i axis, and a rotation ψ^i about the body Z^i axis. These three simple rotations lead to the three transformation matrices

$$\mathbf{A}^i_\phi = \begin{bmatrix} \cos\phi^i & -\sin\phi^i & 0 \\ \sin\phi^i & \cos\phi^i & 0 \\ 0 & 0 & 1 \end{bmatrix}, \quad \mathbf{A}^i_\theta = \begin{bmatrix} 1 & 0 & 0 \\ 0 & \cos\theta^i & -\sin\theta^i \\ 0 & \sin\theta^i & \cos\theta^i \end{bmatrix}, \quad \mathbf{A}^i_\psi = \begin{bmatrix} \cos\psi^i & -\sin\psi^i & 0 \\ \sin\psi^i & \cos\psi^i & 0 \\ 0 & 0 & 1 \end{bmatrix}$$

Using these matrices, the orthogonal transformation matrix that defines the orientation of the body $X^iY^iZ^i$ coordinate system with respect to the XYZ global coordinate system is

$$\mathbf{A}^i = \mathbf{A}^i_\phi \mathbf{A}^i_\theta \mathbf{A}^i_\psi$$

$$= \begin{bmatrix} \cos\psi^i\cos\phi^i - \cos\theta^i\sin\phi^i\sin\psi^i & -\sin\psi^i\cos\phi^i - \cos\theta^i\sin\phi^i\cos\psi^i & \sin\theta^i\sin\phi^i \\ \cos\psi^i\sin\phi^i + \cos\theta^i\cos\phi^i\sin\psi^i & -\sin\psi^i\sin\phi^i + \cos\theta^i\cos\phi^i\cos\psi^i & -\sin\theta^i\cos\phi^i \\ \sin\theta^i\sin\psi^i & \sin\theta^i\cos\psi^i & \cos\theta^i \end{bmatrix}$$

3.4 EULER PARAMETERS

The four *Euler parameters* can be used to avoid singular configurations at the expense of using a nonlinear algebraic constraint equation. Euler parameters, however, are becoming more popular in MBS algorithms because of the singularity problem associated with using three rotation parameters. This singularity problem can be the source of numerical problems in the dynamic simulation of spinning bodies or bodies rotating at high speeds, as is the case with rail wheelsets. Euler parameters can be systematically defined using Rodrigues' formula, which defines the transformation matrix in terms of the angle of rotation and the components of a unit vector along the instantaneous *axis of rotation*. A general three-dimensional rotation is equivalent to a single rotation about this instantaneous axis of rotation.

Rodrigues' Formula In order to derive Rodrigues' formula, vector $\bar{\mathbf{r}}^i$ on body i is assumed to rotate with an angle γ^i about the axis of rotation defined by unit vector $\mathbf{v}^i = \begin{bmatrix} v_1^i & v_2^i & v_3^i \end{bmatrix}^T$, as shown in Figure 5. Vector $\bar{\mathbf{r}}^i$ is assumed to make an angle α with the axis of rotation. It follows that radius a of the circle shown in the figure is defined as $a = |\bar{\mathbf{r}}^i| \sin\alpha = |\mathbf{v}^i \times \bar{\mathbf{r}}^i|$. As a result of this rotation, vector $\bar{\mathbf{r}}^i$ occupies a new position defined by vector $\mathbf{r}^i$, as shown in Figure 5. It is clear from the figure that $\mathbf{r}^i = \bar{\mathbf{r}}^i + \Delta\mathbf{r}^i$. Vector $\Delta\mathbf{r}^i$ has two components $(\Delta\mathbf{r}^i)_1$ and $(\Delta\mathbf{r}^i)_2$ that are perpendicular to the axis of rotation defined by unit vector $\mathbf{v}^i$. It is clear from Figure 5 that $(\Delta\mathbf{r}^i)_1$ has a magnitude $a\sin\gamma$ and a direction defined by unit vector $(\mathbf{v}^i \times \bar{\mathbf{r}}^i)/|\mathbf{v}^i \times \bar{\mathbf{r}}^i|$, while $(\Delta\mathbf{r}^i)_2$ has a magnitude $a(1-\cos\gamma^i) = 2a\sin^2(\gamma^i/2)$ and a direction $\mathbf{v}^i \times (\mathbf{v}^i \times \bar{\mathbf{r}}^i)/\left|\mathbf{v}^i \times (\mathbf{v}^i \times \bar{\mathbf{r}}^i)\right| = \mathbf{v}^i \times (\mathbf{v}^i \times \bar{\mathbf{r}}^i)/a$. Therefore, one has

$$\left.\begin{aligned} (\Delta\mathbf{r}^i)_1 &= a\sin\gamma^i \frac{(\mathbf{v}^i \times \bar{\mathbf{r}}^i)}{|\mathbf{v}^i \times \bar{\mathbf{r}}^i|} = (\mathbf{v}^i \times \bar{\mathbf{r}}^i)\sin\gamma^i \\ (\Delta\mathbf{r}^i)_2 &= 2a\sin^2\left(\frac{\gamma^i}{2}\right)\frac{\mathbf{v}^i \times (\mathbf{v}^i \times \bar{\mathbf{r}}^i)}{a} = 2\left(\mathbf{v}^i \times (\mathbf{v}^i \times \bar{\mathbf{r}}^i)\right)\sin^2\left(\frac{\gamma^i}{2}\right) \end{aligned}\right\} \tag{3.13}$$

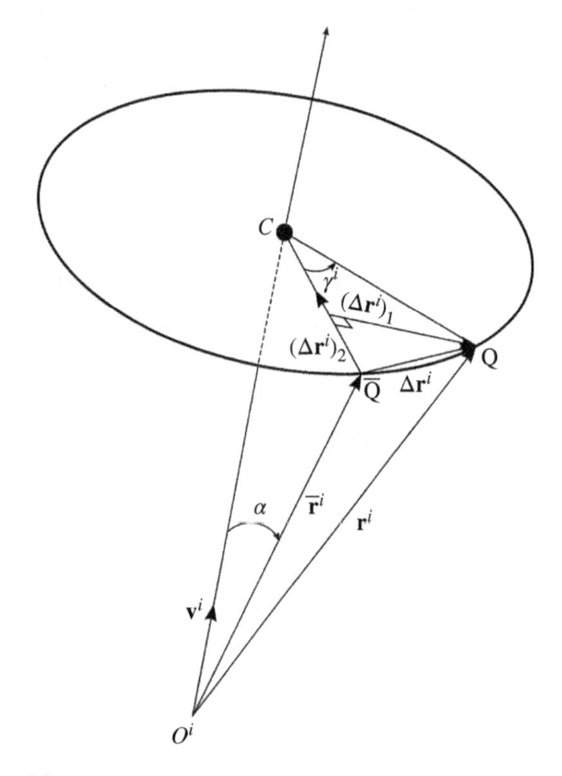

Figure 3.5 Rodrigues' formula.

Using this equation, one can write $\mathbf{r}^i = \overline{\mathbf{r}}^i + \Delta\mathbf{r}^i$ in a more explicit form as

$$\mathbf{r}^i = \overline{\mathbf{r}}^i + \left(\mathbf{v}^i \times \overline{\mathbf{r}}^i\right)\sin\gamma^i + 2\left(\mathbf{v}^i \times \left(\mathbf{v}^i \times \overline{\mathbf{r}}^i\right)\right)\sin^2\left(\frac{\gamma^i}{2}\right) \tag{3.14}$$

The cross products in Eq. 14 can be written using matrix notation as $\mathbf{v}^i \times \overline{\mathbf{r}}^i = \tilde{\mathbf{v}}^i\overline{\mathbf{r}}^i$ and $\mathbf{v}^i \times \left(\mathbf{v}^i \times \overline{\mathbf{r}}^i\right) = \left(\tilde{\mathbf{v}}^i\right)^2\overline{\mathbf{r}}^i$, where $\tilde{\mathbf{v}}^i$ is the skew-symmetric matrix associated with vector $\mathbf{v}^i$ and is defined as

$$\tilde{\mathbf{v}}^i = \begin{bmatrix} 0 & -v_3^i & v_2^i \\ v_3^i & 0 & -v_1^i \\ -v_2^i & v_1^i & 0 \end{bmatrix} \tag{3.15}$$

Using skew-symmetric matrix notation, Eq. 14 can be written as

$$\mathbf{r}^i = \mathbf{A}^i\overline{\mathbf{r}}^i = \left[\mathbf{I} + \tilde{\mathbf{v}}^i\sin\gamma^i + 2\left(\tilde{\mathbf{v}}^i\right)^2\sin^2\left(\frac{\gamma^i}{2}\right)\right]\overline{\mathbf{r}}^i \tag{3.16}$$

where $\mathbf{I}$ is the 3×3 identity matrix, and transformation matrix $\mathbf{A}^i$ is defined as

$$\mathbf{A}^i = \mathbf{I} + \tilde{\mathbf{v}}^i\sin\gamma^i + 2\left(\tilde{\mathbf{v}}^i\right)^2\sin^2\left(\frac{\gamma^i}{2}\right) \tag{3.17}$$

This equation, called *Rodrigues' formula*, defines the transformation matrix in terms of four parameters v_1^i, v_2^i, v_3^i, and γ^i. These parameters are not totally independent because

$\sum_{k=1}^{3}\left(v_k^i\right)^2 = 1$, demonstrating again that general three-dimensional rotation can be described using only three independent parameters. Rodrigues' formula, in which matrix $\left(\tilde{\mathbf{v}}^i\right)^2$ is a symmetric matrix, can be used to determine the transformation matrix as a result of rotation about a vector that has arbitrary orientation. It can also be used to obtain the simple rotation matrices previously presented in this chapter, as demonstrated by the following example.

Example 3.3 Consider the simple rotation ϕ^i about the Z^i axis, which is defined by unit vector $\mathbf{v}^i = \begin{bmatrix} 0 & 0 & 1 \end{bmatrix}^T$. In this case,

$$\tilde{\mathbf{v}}^i = \begin{bmatrix} 0 & -1 & 0 \\ 1 & 0 & 0 \\ 0 & 0 & 0 \end{bmatrix}, \qquad \left(\tilde{\mathbf{v}}^i\right)^2 = \begin{bmatrix} -1 & 0 & 0 \\ 0 & -1 & 0 \\ 0 & 0 & 0 \end{bmatrix}$$

The transformation matrix can be determined using Rodrigues' formula as

$$\begin{aligned} \mathbf{A}^i = \mathbf{A}^i &= \mathbf{I} + \tilde{\mathbf{v}}^i \sin\phi^i + 2\left(\tilde{\mathbf{v}}^i\right)^2 \sin^2\left(\frac{\phi^i}{2}\right) \\ &= \begin{bmatrix} 1 & 0 & 0 \\ 0 & 1 & 0 \\ 0 & 0 & 1 \end{bmatrix} + \begin{bmatrix} 0 & -1 & 0 \\ 1 & 0 & 0 \\ 0 & 0 & 0 \end{bmatrix} \sin\phi^i + \begin{bmatrix} -1 & 0 & 0 \\ 0 & -1 & 0 \\ 0 & 0 & 0 \end{bmatrix} \left(1 - \cos\phi^i\right) \\ &= \begin{bmatrix} \cos\phi^i & -\sin\phi^i & 0 \\ \sin\phi^i & \cos\phi^i & 0 \\ 0 & 0 & 1 \end{bmatrix} \end{aligned}$$

which is the same matrix previously obtained as a result of a simple rotation about the Z^i axis.

Euler Parameters *Euler parameters* are defined in terms of the angle of rotation γ^i and the three components v_1^i, v_2^i, and v_3^i of unit vector $\mathbf{v}^i$ along the axis of rotation as

$$\theta_0^i = \cos\frac{\gamma^i}{2}, \quad \theta_1^i = v_1^i \sin\frac{\gamma^i}{2}, \quad \theta_2^i = v_2^i \sin\frac{\gamma^i}{2}, \quad \theta_3^i = v_3^i \sin\frac{\gamma^i}{2} \tag{3.18}$$

The four Euler parameters can be written in a vector form as $\boldsymbol{\theta}^i = \begin{bmatrix} \theta_0^i & \theta_1^i & \theta_2^i & \theta_3^i \end{bmatrix}^T$. These parameters are not totally independent because they are related by the algebraic constraint equation $\sum_{k=0}^{3}\left(\theta_k^i\right)^2 = 1$. Using Rodrigues' formula of Eq. 17 and the definitions of Euler parameters of Eq. 18, one can show that matrix $\mathbf{A}^i$ can be written in terms of Euler parameters as

$$\mathbf{A}^i = \mathbf{I} + 2\tilde{\boldsymbol{\theta}}_s^i\left(\theta_0^i \mathbf{I} + \tilde{\boldsymbol{\theta}}_s^i\right) = \mathbf{E}^i \overline{\mathbf{E}}^{iT} \tag{3.19}$$

where $\boldsymbol{\theta}_s^i = \begin{bmatrix}\theta_1^i & \theta_2^i & \theta_3^i\end{bmatrix}^T$, $\tilde{\boldsymbol{\theta}}_s^i$ is the skew-symmetric matrix associated with vector $\boldsymbol{\theta}_s^i$, and the two matrices $\mathbf{E}^i$ and $\overline{\mathbf{E}}^i$ are the 3×4 matrices defined as (Nikravesh 1988; Shabana 2005)

$$\mathbf{E}^i = \begin{bmatrix} -\theta_1^i & \theta_0^i & -\theta_3^i & \theta_2^i \\ -\theta_2^i & \theta_3^i & \theta_0^i & -\theta_1^i \\ -\theta_3^i & -\theta_2^i & \theta_1^i & \theta_0^i \end{bmatrix}, \quad \overline{\mathbf{E}}^i = \begin{bmatrix} -\theta_1^i & \theta_0^i & \theta_3^i & -\theta_2^i \\ -\theta_2^i & -\theta_3^i & \theta_0^i & \theta_1^i \\ -\theta_3^i & \theta_2^i & -\theta_1^i & \theta_0^i \end{bmatrix} \tag{3.20}$$

Because of the Euler parameter constraint $\sum_{k=0}^{3} \left(\theta_k^i\right)^2 = 1$, which allows writing one parameter in terms of the other three, the transformation matrix written in terms of Euler parameters can take different forms. One of these forms is given as

$$\mathbf{A}^i = \begin{bmatrix} 2(\theta_0^i)^2 + 2(\theta_1^i)^2 - 1 & 2\left(\theta_1^i\theta_2^i - \theta_0^i\theta_3^i\right) & 2\left(\theta_1^i\theta_3^i + \theta_0^i\theta_2^i\right) \\ 2\left(\theta_1^i\theta_2^i + \theta_0^i\theta_3^i\right) & 2(\theta_0^i)^2 + 2(\theta_2^i)^2 - 1 & 2\left(\theta_2^i\theta_3^i - \theta_0^i\theta_1^i\right) \\ 2\left(\theta_1^i\theta_3^i - \theta_0^i\theta_2^i\right) & 2\left(\theta_2^i\theta_3^i + \theta_0^i\theta_1^i\right) & 2(\theta_0^i)^2 + 2(\theta_3^i)^2 - 1 \end{bmatrix} \tag{3.21}$$

This transformation matrix is quadratic in Euler parameters, while the transformation matrix expressed in terms of Euler angles has trigonometric functions that have infinite order. Furthermore, using Euler parameters allows eliminating the singularity associated with Euler angles.

Rodrigues' formula and Euler parameters can be used to define the transformation matrix resulting from body rotation about an axis that has any orientation in space. Therefore, they are not restricted to simple rotations about the axes of coordinate systems. Additionally, Euler parameters have many useful identities that can be used to simplify the formulation of the equations of motion. In Table 1, which shows some of the identities, $\boldsymbol{\theta}^i = \begin{bmatrix}\theta_0^i & \theta_1^i & \theta_2^i & \theta_3^i\end{bmatrix}^T$, $\mathbf{I}$ is the 3×3 identity matrix, and $\mathbf{I}_4$ is the 4×4 identity matrix.

Table 3.1 Euler parameter identities.

Euler parameters	$\mathbf{E}\mathbf{E}^{\mathrm{T}} = \overline{\mathbf{E}}\overline{\mathbf{E}}^{\mathrm{T}} = \mathbf{I}$, $\mathbf{E}^{\mathrm{T}}\mathbf{E} = \overline{\mathbf{E}}^{\mathrm{T}}\overline{\mathbf{E}} = \mathbf{I}_4 - \boldsymbol{\theta}\boldsymbol{\theta}^{\mathrm{T}}$, $\mathbf{E}\boldsymbol{\theta} = \overline{\mathbf{E}}\boldsymbol{\theta} = \mathbf{0}$, $\boldsymbol{\theta}^{\mathrm{T}}\boldsymbol{\theta} = 1$
Euler parameter derivatives	$\dot{\mathbf{E}}\boldsymbol{\theta} = \dot{\overline{\mathbf{E}}}\boldsymbol{\theta} = \mathbf{0}$, $\mathbf{E}\dot{\overline{\mathbf{E}}}^{\mathrm{T}} = \dot{\mathbf{E}}\overline{\mathbf{E}}^{\mathrm{T}}$, $\dot{\boldsymbol{\theta}}^{\mathrm{T}}\boldsymbol{\theta} = 0$, $\mathbf{E}\dot{\boldsymbol{\theta}} = -\dot{\mathbf{E}}\boldsymbol{\theta}$, $\overline{\mathbf{E}}\dot{\boldsymbol{\theta}} = -\dot{\overline{\mathbf{E}}}\boldsymbol{\theta}$

Example 3.4

A body rotates an angle $\gamma^i = \pi/2$ about an axis of rotation defined by vector $\begin{bmatrix}1.0 & 0 & -1.0\end{bmatrix}^T$. Use Rodrigues' formula to determine the transformation matrix as a result of this rotation, and determine the values of the four Euler parameters.

Solution A unit vector along the axis of rotation is defined as $\mathbf{v}^i = \left(1/\sqrt{2}\right)\begin{bmatrix}1.0 & 0 & -1.0\end{bmatrix}^T$. Using this unit vector, one has

$$\tilde{\mathbf{v}}^i = \frac{1}{\sqrt{2}}\begin{bmatrix} 0 & 1 & 0 \\ -1 & 0 & -1 \\ 0 & 1 & 0 \end{bmatrix}, \quad \left(\tilde{\mathbf{v}}^i\right)^2 = -\frac{1}{2}\begin{bmatrix} 1 & 0 & 1 \\ 0 & 2 & 0 \\ 1 & 0 & 1 \end{bmatrix}$$

For the given angle $\gamma^i = \pi/2$, one has $\sin\gamma^i = 1$ and $2\sin^2(\gamma^i/2) = 1 - \cos\gamma^i = 1$. Therefore, Rodrigues' formula can be written as

$$\mathbf{A}^i = \mathbf{I} + \tilde{\mathbf{v}}^i \sin\gamma^i + 2(\tilde{\mathbf{v}}^i)^2 \sin^2\frac{\gamma^i}{2}$$

$$= \begin{bmatrix} 1 & 0 & 0 \\ 0 & 1 & 0 \\ 0 & 0 & 1 \end{bmatrix} + \frac{1}{\sqrt{2}} \begin{bmatrix} 0 & 1 & 0 \\ -1 & 0 & -1 \\ 0 & 1 & 0 \end{bmatrix} (1) - \frac{1}{2} \begin{bmatrix} 1 & 0 & 1 \\ 0 & 2 & 0 \\ 1 & 0 & 1 \end{bmatrix} (1)$$

$$= \begin{bmatrix} 0.5 & 0.7072 & -0.5 \\ -0.7072 & 0 & -0.7072 \\ -0.5 & 0.7072 & 0.5 \end{bmatrix}$$

Using the angle $\gamma^i = \pi/2$ and the unit vector $\mathbf{v}^i = \left(1/\sqrt{2}\right)\begin{bmatrix}1.0 & 0 & -1.0\end{bmatrix}^T$ along the axis of rotation, the four Euler parameters can be defined as

$$\theta_0^i = \cos\left(\gamma^i/2\right) = \left(1/\sqrt{2}\right)$$

$$\begin{bmatrix}\theta_1^i & \theta_2^i & \theta_3^i\end{bmatrix}^T = \mathbf{v}^i \sin\left(\gamma^i/2\right) = \begin{bmatrix}0.5 & 0 & -0.5\end{bmatrix}^T$$

3.5 VELOCITY AND ACCELERATION EQUATIONS

Equation 4 shows that the global position of an arbitrary point on a body i is defined as $\mathbf{r}^i = \mathbf{R}^i + \mathbf{A}^i\bar{\mathbf{u}}^i$. When using Euler angles, transformation matrix $\mathbf{A}^i$ is a function of the Euler angles defined by vector $\boldsymbol{\theta}^i = \begin{bmatrix}\psi^i & \phi^i & \theta^i\end{bmatrix}^T$. Therefore, the time-dependent coordinates of the body can be written as

$$\mathbf{q}^i = \begin{bmatrix}\mathbf{R}^{i^T} & \boldsymbol{\theta}^{i^T}\end{bmatrix}^T = \begin{bmatrix}R_x^i & R_y^i & R_z^i & \psi^i & \phi^i & \theta^i\end{bmatrix}^T \tag{3.22}$$

For Euler parameters, this is the seven-dimensional vector defined as $\mathbf{q}^i = \begin{bmatrix}\mathbf{R}^{i^T} & \boldsymbol{\theta}^{i^T}\end{bmatrix}^T = \begin{bmatrix}R_x^i & R_y^i & R_z^i & \theta_0^i & \theta_1^i & \theta_2^i & \theta_3^i\end{bmatrix}^T$. In this case, the constraint on the Euler parameters $\sum_{k=0}^{3}\left(\theta_k^i\right)^2 = 1$ ensures that only three rotational parameters are independent.

Angular Velocity Vectors In three-dimensional motion analysis, angular velocities are not time derivatives of orientation parameters. Nonetheless, angular velocities can be written as linear functions of time derivatives of orientation coordinates. In this section, a general expression for the angular velocity vector is obtained using the orthogonality property of the transformation matrix.

Because transformation matrix $\mathbf{A}^i$ is an orthogonal matrix, one has $\mathbf{A}^i\mathbf{A}^{i^T} = \mathbf{I}$. It follows that $\dot{\mathbf{A}}^i\mathbf{A}^{i^T} + \mathbf{A}^i\dot{\mathbf{A}}^{i^T} = \mathbf{0}$. This equation shows that $\dot{\mathbf{A}}^i\mathbf{A}^{i^T} = -\mathbf{A}^i\dot{\mathbf{A}}^{i^T} = -\left(\dot{\mathbf{A}}^i\mathbf{A}^{i^T}\right)^T$. A matrix that is equal to the negative of its transpose must be a skew-symmetric matrix. Therefore, one can write $\dot{\mathbf{A}}^i\mathbf{A}^{i^T} = \tilde{\boldsymbol{\omega}}^i$, where $\tilde{\boldsymbol{\omega}}^i$ is a skew-symmetric matrix associated with vector $\boldsymbol{\omega}^i$, which is the *angular velocity vector* defined in the global coordinate system. It is clear from the equation $\tilde{\boldsymbol{\omega}}^i = \dot{\mathbf{A}}^i\mathbf{A}^{i^T}$ that if the transformation matrix is known, the angular velocity vector can be defined.

Alternatively, one can use the orthogonality of the transformation matrix to write $\mathbf{A}^{i^T}\mathbf{A}^i = \mathbf{I}$. By differentiating this equation, one can show that $\mathbf{A}^{i^T}\dot{\mathbf{A}}^i = -\left(\mathbf{A}^{i^T}\dot{\mathbf{A}}^i\right)^T$, which demonstrates that $\mathbf{A}^{i^T}\dot{\mathbf{A}}^i$ is a skew-symmetric matrix that can be written as $\mathbf{A}^{i^T}\dot{\mathbf{A}}^i = \tilde{\overline{\boldsymbol{\omega}}}^i$, where $\tilde{\overline{\boldsymbol{\omega}}}^i$ is a skew-symmetric matrix associated with vector $\overline{\boldsymbol{\omega}}^i$, which is the angular velocity vector defined in the body coordinate system. Therefore, one can write the derivative of the transformation matrix using one of the following forms:

$$\dot{\mathbf{A}}^i = \tilde{\boldsymbol{\omega}}^i\mathbf{A}^i, \qquad \dot{\mathbf{A}}^i = \mathbf{A}^i\tilde{\overline{\boldsymbol{\omega}}}^i \tag{3.23}$$

This equation shows that $\tilde{\boldsymbol{\omega}}^i = \mathbf{A}^i\tilde{\overline{\boldsymbol{\omega}}}^i\mathbf{A}^{i^T}$. The two representations of Eq. 23 can be used to obtain two different expressions for the absolute velocity vector of an arbitrary point on the body. Equation 23 can be used to show that $\dot{\mathbf{A}}^i\overline{\boldsymbol{\omega}}^i = \mathbf{0}$

Absolute Velocity Vector In order to obtain the absolute velocity vector of an arbitrary point on rigid body i, Eq. 4, $\mathbf{r}^i = \mathbf{R}^i + \mathbf{A}^i\overline{\mathbf{u}}^i$, can be differentiated with respect to time to obtain $\dot{\mathbf{r}}^i = \dot{\mathbf{R}}^i + \dot{\mathbf{A}}^i\overline{\mathbf{u}}^i$, where $\dot{\mathbf{R}}^i$ is the absolute velocity of the reference point. This equation, upon using the first equation in Eq. 23, can be written as $\dot{\mathbf{r}}^i = \dot{\mathbf{R}}^i + \tilde{\boldsymbol{\omega}}^i\mathbf{A}^i\overline{\mathbf{u}}^i$. Because $\mathbf{u}^i = \mathbf{A}^i\overline{\mathbf{u}}^i$ and $\tilde{\boldsymbol{\omega}}^i\mathbf{u}^i = \boldsymbol{\omega}^i \times \mathbf{u}^i$, the absolute velocity vector of the arbitrary point can be written in terms of the angular velocity vector $\boldsymbol{\omega}^i = \begin{bmatrix}\omega_1^i & \omega_2^i & \omega_3^i\end{bmatrix}^T$ as

$$\dot{\mathbf{r}}^i = \dot{\mathbf{R}}^i + \boldsymbol{\omega}^i \times \mathbf{u}^i \tag{3.24}$$

Alternatively, one can substitute the second equation in Eq. 23 into velocity equation $\dot{\mathbf{r}}^i = \dot{\mathbf{R}}^i + \dot{\mathbf{A}}^i\overline{\mathbf{u}}^i$ and follow a procedure similar to the one previously outlined in this section to obtain an alternate expression for the absolute velocity vector in terms of the angular velocity vector $\overline{\boldsymbol{\omega}}^i = \begin{bmatrix}\overline{\omega}_1^i & \overline{\omega}_2^i & \overline{\omega}_3^i\end{bmatrix}^T$ defined in the body coordinate system as

$$\dot{\mathbf{r}}^i = \dot{\mathbf{R}}^i + \mathbf{A}^i\left(\overline{\boldsymbol{\omega}}^i \times \overline{\mathbf{u}}^i\right) \tag{3.25}$$

where $\boldsymbol{\omega}^i = \mathbf{A}^i\overline{\boldsymbol{\omega}}^i$.

Absolute Acceleration Vector The angular acceleration vector $\boldsymbol{\alpha}^i = \begin{bmatrix}\alpha_1^i & \alpha_2^i & \alpha_3^i\end{bmatrix}^T$ is defined as the time derivative of angular velocity vector $\boldsymbol{\omega}^i$: that is, $\boldsymbol{\alpha}^i = \dot{\boldsymbol{\omega}}^i$. Differentiating the velocity equation $\dot{\mathbf{r}}^i = \dot{\mathbf{R}}^i + \boldsymbol{\omega}^i \times \mathbf{u}^i$ with respect to time, and using the definition of the angular acceleration vector and the identity $\dot{\mathbf{u}}^i = \dot{\mathbf{A}}^i\overline{\mathbf{u}}^i = \boldsymbol{\omega}^i \times \mathbf{u}^i$, one can show that the absolute acceleration vector of the arbitrary point on the body can be written as

$$\ddot{\mathbf{r}}^i = \ddot{\mathbf{R}}^i + \boldsymbol{\alpha}^i \times \mathbf{u}^i + \boldsymbol{\omega}^i \times \left(\boldsymbol{\omega}^i \times \mathbf{u}^i\right) \tag{3.26}$$

In this equation, $\ddot{\mathbf{R}}^i$ is the absolute acceleration vector of the body reference point. The term $\boldsymbol{\alpha}^i \times \mathbf{u}^i$ on the right-hand side of Eq. 26 is the *tangential component* of the acceleration, while the term $\boldsymbol{\omega}^i \times (\boldsymbol{\omega}^i \times \mathbf{u}^i)$ is the *normal component,* which is also referred to as the *centripetal acceleration.* As in the case of the absolute velocity vector, the absolute acceleration vector can be expressed in terms of the angular velocity and acceleration vectors defined in the body coordinate system as

$$\ddot{\mathbf{r}}^i = \ddot{\mathbf{R}}^i + \mathbf{A}^i\left(\bar{\boldsymbol{\alpha}}^i \times \bar{\mathbf{u}}^i\right) + \mathbf{A}^i\left(\bar{\boldsymbol{\omega}}^i \times \left(\bar{\boldsymbol{\omega}}^i \times \bar{\mathbf{u}}^i\right)\right) \tag{3.27}$$

where $\boldsymbol{\alpha}^i = \mathbf{A}^i\bar{\boldsymbol{\alpha}}^i$. The absolute acceleration vector of an arbitrary point on the body can be used to define the virtual work of inertia forces, as discussed in Chapter 6.

3.6 GENERALIZED COORDINATES

Because $\dot{\mathbf{A}}^i = \tilde{\boldsymbol{\omega}}^i\mathbf{A}^i$ and $\dot{\mathbf{A}}^i = \mathbf{A}^i\tilde{\bar{\boldsymbol{\omega}}}^i$ (Eq. 23), regardless of the orientation parameters used, vectors $\boldsymbol{\omega}^i$ and $\bar{\boldsymbol{\omega}}^i$ are always linear functions in the time derivatives of the orientation parameters $\boldsymbol{\theta}^i$. Therefore, regardless of the orientation parameters used, one can always write angular velocity vectors $\boldsymbol{\omega}^i$ and $\bar{\boldsymbol{\omega}}^i$ in terms of the time derivatives of the orientation parameters, respectively, as

$$\boldsymbol{\omega}^i = \mathbf{G}^i\dot{\boldsymbol{\theta}}^i, \qquad \bar{\boldsymbol{\omega}}^i = \bar{\mathbf{G}}^i\dot{\boldsymbol{\theta}}^i \tag{3.28}$$

where coefficient matrices $\mathbf{G}^i$ and $\bar{\mathbf{G}}^i$ are expressed in terms of orientation parameters $\boldsymbol{\theta}^i$. For example, in the case of using Euler angles $\boldsymbol{\theta}^i = \left[\psi^i \;\; \phi^i \;\; \theta^i\right]^T$ with the sequence $Z^i - X^i - Y^i$, matrices $\mathbf{G}^i$ and $\bar{\mathbf{G}}^i$ are

$$\mathbf{G}^i = \begin{bmatrix} 0 & \cos\psi^i & -\sin\psi^i\cos\phi^i \\ 0 & \sin\psi^i & \cos\psi^i\cos\phi^i \\ 1 & 0 & \sin\phi^i \end{bmatrix}, \qquad \bar{\mathbf{G}}^i = \begin{bmatrix} -\cos\phi^i\sin\theta^i & \cos\theta^i & 0 \\ \sin\phi^i & 0 & 1 \\ \cos\phi^i\cos\theta^i & \sin\theta^i & 0 \end{bmatrix} \tag{3.29}$$

Using the equations $\tilde{\boldsymbol{\omega}}^i = \dot{\mathbf{A}}^i\mathbf{A}^{i^T}$ and $\tilde{\bar{\boldsymbol{\omega}}}^i = \mathbf{A}^{i^T}\dot{\mathbf{A}}^i$ to determine the angular velocity vectors can be very cumbersome. Such an approach can be avoided by using a much simpler approach based on a physical interpretation of the columns of the two matrices $\mathbf{G}^i$ and $\bar{\mathbf{G}}^i$. While finite rotations are not commutative and cannot be added, the angular velocities can be added and can be treated as vectors. Therefore, the angular velocity vectors can be written as $\boldsymbol{\omega}^i = \mathbf{g}_1^i\,\dot{\psi}^i + \mathbf{g}_2^i\,\dot{\phi}^i + \mathbf{g}_3^i\,\dot{\theta}^i$ and $\bar{\boldsymbol{\omega}}^i = \bar{\mathbf{g}}_1^i\,\dot{\psi}^i + \bar{\mathbf{g}}_2^i\,\dot{\phi}^i + \bar{\mathbf{g}}_3^i\,\dot{\theta}^i$. In these two equations, $\mathbf{g}_k^i$ and $\bar{\mathbf{g}}_k^i$, $k = 1, 2, 3$, are unit vectors defined, respectively, in the global and body coordinate systems. Vectors $\mathbf{g}_k^i$ and $\bar{\mathbf{g}}_k^i$, $k = 1, 2, 3$, are the axes about which the Euler rotations are performed. It follows that $\mathbf{G}^i = \left[\mathbf{g}_1^i \;\; \mathbf{g}_2^i \;\; \mathbf{g}_3^i\right]$ and $\bar{\mathbf{G}}^i = \left[\bar{\mathbf{g}}_1^i \;\; \bar{\mathbf{g}}_2^i \;\; \bar{\mathbf{g}}_3^i\right]$. Therefore, instead of using the equations $\tilde{\boldsymbol{\omega}}^i = \dot{\mathbf{A}}^i\mathbf{A}^{i^T}$ and $\tilde{\bar{\boldsymbol{\omega}}}^i = \mathbf{A}^{i^T}\dot{\mathbf{A}}^i$ to define vectors $\boldsymbol{\omega}^i$ and $\bar{\boldsymbol{\omega}}^i$, respectively, which can be used to determine the two matrices $\mathbf{G}^i$ and $\bar{\mathbf{G}}^i$, these matrices can be easily determined by recognizing that the columns of $\mathbf{G}^i$ are unit vectors, defined in the global coordinate system, about which the three Euler rotations are performed; and the columns of matrix $\bar{\mathbf{G}}^i$ are the same vectors defined in the body coordinate system. Consequently, $\mathbf{G}^i = \mathbf{A}^i\bar{\mathbf{G}}^i$.

For the four Euler parameters $\boldsymbol{\theta}^i = \left[\theta_0^i \;\; \theta_1^i \;\; \theta_2^i \;\; \theta_3^i\right]^T$, matrices $\mathbf{G}^i$ and $\bar{\mathbf{G}}^i$ are 3×4 matrices defined as

$$\mathbf{G}^i = 2\mathbf{E}^i, \qquad \bar{\mathbf{G}}^i = 2\bar{\mathbf{E}}^i \tag{3.30}$$

where matrices $\mathbf{E}^i$ and $\overline{\mathbf{E}}^i$ are defined in Eq. 20. The columns of matrices $\mathbf{G}^i$ and $\overline{\mathbf{G}}^i$ should not be interpreted the same way as for Euler angles, since Euler parameters are quaternions and are not interpreted as rotations about axes of rotation as is the case with Euler angles.

Example 3.5 In this example, the columns of matrices $\mathbf{G}^i = [\mathbf{g}_1^i \ \mathbf{g}_2^i \ \mathbf{g}_3^i]$ and $\overline{\mathbf{G}}^i = [\overline{\mathbf{g}}_1^i \ \overline{\mathbf{g}}_2^i \ \overline{\mathbf{g}}_3^i]$ are determined based on physical considerations instead of using the cumbersome approach that requires the time derivatives of the Euler-angle transformation matrix. The case of the three Euler angles $\boldsymbol{\theta}^i = [\psi^i \ \phi^i \ \theta^i]^T$ with the sequence $Z^i - X^i - Y^i$ is considered in this example. The first rotation ψ^i is performed about the Z^i axis, which is defined in the global coordinate system XYZ by the unit vector $\mathbf{g}_1^i = [0 \ \ 0 \ \ 1]^T$. The second rotation ϕ^i is performed about the new body X^i axis, which is defined in the global coordinate system XYZ by the unit vector $\mathbf{g}_2^i = \mathbf{A}_\psi^i [1 \ \ 0 \ \ 0]^T = [\cos\psi^i \ \sin\psi^i \ 0]^T$, where the simple transformation matrix $\mathbf{A}_\psi^i$ is defined in Eq. 11. The last Euler rotation θ^i is performed about the body Y^i axis, which can be defined in the global coordinate system XYZ after the first two rotations ψ^i and ϕ^i by the unit vector $\mathbf{g}_3^i = \mathbf{A}_\psi^i \mathbf{A}_\phi^i [0 \ \ 1 \ \ 0]^T = [-\sin\psi^i \cos\phi^i \ \ \cos\psi^i \cos\phi^i \ \ \sin\phi^i]^T$, where matrix $\mathbf{A}_\phi^i$ is given in Eq. 11. The three vectors $\mathbf{g}_k^i, \ k = 1, 2, 3$, form the columns of matrix $\mathbf{G}^i$ in Eq. 29.

The columns $\overline{\mathbf{g}}_k^i, \ k = 1, 2, 3$, of matrix $\overline{\mathbf{G}}^i$ can be obtained in a similar manner, with the understanding that these vectors are unit vectors along the axes of the Euler rotations defined in the body coordinate system. Vector $\overline{\mathbf{g}}_1^i$, about which the first rotation ψ^i is performed, is defined in the body coordinate system by the unit vector $\overline{\mathbf{g}}_1^i = \mathbf{A}_\theta^{iT} \mathbf{A}_\phi^{iT} [0 \ \ 0 \ \ 1]^T = [-\cos\phi^i \sin\theta^i \ \ \sin\phi^i \ \ \cos\phi^i \cos\theta^i]^T$. Vector $\overline{\mathbf{g}}_2^i$, about which the second rotation ϕ^i is performed, is defined in the body coordinate system by the unit vector $\overline{\mathbf{g}}_2^i = \mathbf{A}_\theta^{iT} [1 \ \ 0 \ \ 0]^T = [\cos\theta^i \ \ 0 \ \ \sin\theta^i]^T$. Vector $\overline{\mathbf{g}}_3^i$, about which the third rotation θ^i is performed, is simply defined in the body coordinate system by the unit vector $\overline{\mathbf{g}}_3^i = [0 \ \ 1 \ \ 0]^T$. The three vectors $\overline{\mathbf{g}}_k^i, \ k = 1, 2, 3$, form the columns of matrix $\overline{\mathbf{G}}^i$ in Eq. 29.

Example 3.6 In this example, the columns of the matrices $\mathbf{G}^i = [\mathbf{g}_1^i \ \ \mathbf{g}_2^i \ \ \mathbf{g}_3^i]$ and $\overline{\mathbf{G}}^i = [\overline{\mathbf{g}}_1^i \ \ \overline{\mathbf{g}}_2^i \ \ \overline{\mathbf{g}}_3^i]$ are determined again using a sequence different from that used in the preceding example. As in the preceding example, these columns are determined based on physical considerations instead of using the cumbersome approach that requires the time derivatives of the Euler-angle transformation matrix. In this example, the case of the three Euler angles $\boldsymbol{\theta}^i = [\phi^i \ \ \theta^i \ \ \psi^i]^T$ with the sequence $Z^i - X^i - Z^i$ is considered. This sequence, which was used by Euler to study gyroscopic motion, was considered in Example 2, where the simple rotation matrices used to define the Euler angle transformation matrix are provided.

As discussed in Example 2, the first rotation ϕ^i is performed about the Z^i axis, which is defined in the global coordinate system XYZ by the unit vector $\mathbf{g}_1^i = [0 \ \ 0 \ \ 1]^T$. The second rotation θ^i is performed about the new body X^i axis, which is defined in the global coordinate system XYZ by the unit vector $\mathbf{g}_2^i = \mathbf{A}_\phi^i[1 \;\; 0 \;\; 0]^T = [\cos\phi^i \;\; \sin\phi^i \;\; 0]^T$, where the simple transformation matrix $\mathbf{A}_\phi^i$ is defined in Example 2. The last Euler rotation ψ^i is performed about the body Z^i axis, which can be defined in the global coordinate system XYZ after the first two rotations ϕ^i and θ^i by the unit vector $\mathbf{g}_3^i = \mathbf{A}_\phi^i\mathbf{A}_\theta^i[0 \;\; 0 \;\; 1]^T = [\sin\theta^i\sin\phi^i \;\; -\sin\theta^i\cos\phi^i \;\; \cos\theta^i]^T$, where matrix $\mathbf{A}_\theta^i$ is given in Example 2. The three vectors $\mathbf{g}_k^i$, $k = 1, 2, 3$, form the columns of matrix $\mathbf{G}^i$: that is,

$$\mathbf{G}^i = \begin{bmatrix} 0 & \cos\phi^i & \sin\theta^i\sin\phi^i \\ 0 & \sin\phi^i & -\sin\theta^i\cos\phi^i \\ 1 & 0 & \cos\theta^i \end{bmatrix}$$

The columns $\bar{\mathbf{g}}_k^i$, $k = 1, 2, 3$, of matrix $\bar{\mathbf{G}}^i$ can be obtained in a similar manner, with the understanding that these vectors are unit vectors along the axes of the Euler rotations defined in the body coordinate system. Vector $\bar{\mathbf{g}}_1^i$, about which the first rotation ϕ^i is performed, is defined in the body coordinate system by the unit vector $\bar{\mathbf{g}}_1^i = \mathbf{A}_\theta^{iT}\mathbf{A}_\psi^{iT}[0 \;\; 0 \;\; 1]^T = [\sin\theta^i\sin\psi^i \;\; \sin\theta^i\cos\psi^i \;\; \cos\theta^i]^T$, where the simple rotation matrix $\mathbf{A}_\psi^i$ is defined in Example 2. Vector $\bar{\mathbf{g}}_2^i$, about which the second rotation θ^i is performed, is defined in the body coordinate system by the unit vector $\bar{\mathbf{g}}_2^i = \mathbf{A}_\psi^{iT}[1 \;\; 0 \;\; 0]^T = [\cos\psi^i \;\; -\sin\psi^i \;\; 0]^T$. Vector $\bar{\mathbf{g}}_3^i$, about which the third rotation ψ^i is performed, is simply defined in the body coordinate system by the unit vector $\bar{\mathbf{g}}_3^i = [0 \;\; 0 \;\; 1]^T$. The three vectors $\bar{\mathbf{g}}_k^i$, $k = 1, 2, 3$, form the columns of matrix $\bar{\mathbf{G}}^i$, which can be written as

$$\bar{\mathbf{G}}^i = \begin{bmatrix} \sin\theta^i\sin\psi^i & \cos\psi^i & 0 \\ \sin\theta^i\cos\psi^i & -\sin\psi^i & 0 \\ \cos\theta^i & 0 & 1 \end{bmatrix}$$

Euler Parameter Identities Because $\boldsymbol{\omega}^i = \mathbf{G}^i\dot{\boldsymbol{\theta}}^i$, the angular acceleration vector can be written as $\boldsymbol{\alpha}^i = \dot{\boldsymbol{\omega}}^i = \mathbf{G}^i\ddot{\boldsymbol{\theta}}^i + \dot{\mathbf{G}}^i\dot{\boldsymbol{\theta}}^i$. Therefore, the angular acceleration vector $\boldsymbol{\alpha}^i$ has, in general, two terms: one linear in the second derivative of the orientation coordinates, and the other quadratic in the first derivative of these orientation coordinates. It is clear, however, from the identities presented in Table 1, that when using Euler parameters $\dot{\mathbf{G}}^i\dot{\boldsymbol{\theta}}^i = 2\dot{\mathbf{E}}^i\dot{\boldsymbol{\theta}}^i = \mathbf{0}$, and in this case, the angular acceleration vector reduces to $\boldsymbol{\alpha}^i = \dot{\boldsymbol{\omega}}^i = \mathbf{G}^i\ddot{\boldsymbol{\theta}}^i$. However, this is not, the case when Euler angles are used. Similarly, one has $\dot{\bar{\mathbf{G}}}^i\dot{\boldsymbol{\theta}}^i = 2\dot{\bar{\mathbf{E}}}^i\dot{\boldsymbol{\theta}}^i = \mathbf{0}$, and $\bar{\boldsymbol{\alpha}}^i = \bar{\mathbf{G}}^i\ddot{\boldsymbol{\theta}}^i$ when Euler parameters are used.

Derivatives with Respect to Orientation Parameters It was previously shown in this chapter that the absolute velocity vector can be written as $\dot{\mathbf{r}}^i = \dot{\mathbf{R}}^i + \boldsymbol{\omega}^i \times \mathbf{u}^i$ or $\dot{\mathbf{r}}^i = \dot{\mathbf{R}}^i + \mathbf{A}^i \left(\bar{\boldsymbol{\omega}}^i \times \bar{\mathbf{u}}^i\right)$. These two equations can be written in the forms $\dot{\mathbf{r}}^i = \dot{\mathbf{R}}^i - \mathbf{u}^i \times \boldsymbol{\omega}^i = \dot{\mathbf{R}}^i - \tilde{\mathbf{u}}^i\mathbf{G}^i\dot{\boldsymbol{\theta}}^i$ and $\dot{\mathbf{r}}^i = \dot{\mathbf{R}}^i - \mathbf{A}^i \left(\bar{\mathbf{u}}^i \times \bar{\boldsymbol{\omega}}^i\right) = \dot{\mathbf{R}}^i - \mathbf{A}^i \left(\tilde{\bar{\mathbf{u}}}^i\bar{\mathbf{G}}^i\dot{\boldsymbol{\theta}}^i\right)$. Using these equations, one can show that the absolute velocity and acceleration vectors of an arbitrary point on the body can be written in terms of generalized velocities $\dot{\mathbf{q}}^i$ as

$$\dot{\mathbf{r}}^i = \mathbf{L}^i\dot{\mathbf{q}}^i, \qquad \ddot{\mathbf{r}}^i = \mathbf{L}^i\ddot{\mathbf{q}}^i + \dot{\mathbf{L}}^i\dot{\mathbf{q}}^i \tag{3.31}$$

where $\mathbf{q}^i = \left[\mathbf{R}^{i^T}\ \ \boldsymbol{\theta}^{i^T}\right]^T$, and

$$\mathbf{L}^i = \left[\mathbf{I}\ \ -\tilde{\mathbf{u}}^i\mathbf{G}^i\right] = \left[\mathbf{I}\ \ -\mathbf{A}^i\tilde{\bar{\mathbf{u}}}^i\bar{\mathbf{G}}^i\right] \tag{3.32}$$

One can show that

$$\dot{\mathbf{L}}^i\dot{\mathbf{q}}^i = \left(\tilde{\boldsymbol{\omega}}^i\right)^2\mathbf{u}^i - \tilde{\mathbf{u}}^i\dot{\mathbf{G}}^i\dot{\boldsymbol{\theta}}^i = \mathbf{A}^i\left(\left(\tilde{\bar{\boldsymbol{\omega}}}^i\right)^2\bar{\mathbf{u}}^i - \tilde{\bar{\mathbf{u}}}^i\dot{\bar{\mathbf{G}}}^i\dot{\boldsymbol{\theta}}^i\right) \tag{3.33}$$

This is with the understanding that when the four Euler parameters are used, $\dot{\mathbf{G}}^i\dot{\boldsymbol{\theta}}^i = \mathbf{0}$ and $\dot{\bar{\mathbf{G}}}^i\dot{\boldsymbol{\theta}}^i = \mathbf{0}$. It is also clear from the preceding equations that

$$\left.\begin{aligned} \frac{\partial \mathbf{r}^i}{\partial \mathbf{R}^i} &= \frac{\partial \dot{\mathbf{r}}^i}{\partial \dot{\mathbf{R}}^i} = \frac{\partial \ddot{\mathbf{r}}^i}{\partial \ddot{\mathbf{R}}^i} = \mathbf{I}, \\ \frac{\partial \mathbf{r}^i}{\partial \boldsymbol{\theta}^i} &= \frac{\partial \dot{\mathbf{r}}^i}{\partial \dot{\boldsymbol{\theta}}^i} = \frac{\partial \ddot{\mathbf{r}}^i}{\partial \ddot{\boldsymbol{\theta}}^i} = -\tilde{\mathbf{u}}^i\mathbf{G}^i = -\mathbf{A}^i\tilde{\bar{\mathbf{u}}}^i\bar{\mathbf{G}}^i \end{aligned}\right\} \tag{3.34}$$

The equations presented in this section will be used in Chapter 6 to develop the dynamic equations of a railroad vehicle system by developing the expressions for the generalized inertia and applied forces. Equation 34 is also used to define the Jacobian matrix of the kinematic constraint equations that describe mechanical joints and specified motion trajectories.

3.7 KINEMATIC SINGULARITIES

Kinematic singularities are encountered whenever three parameters are used to describe a body orientation in space. Kinematic singularities can take different forms depending on the parameters used. When using the three Euler angles, singularities appear in the two matrices $\mathbf{G}^i$ and $\bar{\mathbf{G}}^i$ used to define the angular velocity vector as $\boldsymbol{\omega}^i = \mathbf{G}^i\dot{\boldsymbol{\theta}}^i$ and $\bar{\boldsymbol{\omega}}^i = \bar{\mathbf{G}}^i\dot{\boldsymbol{\theta}}^i$, respectively. At the singular configurations, one cannot solve for $\dot{\boldsymbol{\theta}}^i$ in terms of the components of the angular velocities. The configurations at which the singularities occur depend on the sequence of Euler angles used. Therefore, these singularities must be viewed as mathematical singularities due to the techniques used in the motion description; such singularities have nothing to do with the physics of the problem.

For the sequence of Euler angles $Z^i - X^i - Y^i$ used in railroad vehicle system dynamics, the singularities in the two matrices $\mathbf{G}^i$ and $\bar{\mathbf{G}}^i$ of Eq. 29 appear when $\phi^i = \pm\pi/2$. One can show that at this configuration, the Z^i and Y^i axes of rotation become parallel, and these two axes of Euler rotations are not independent. At this configuration, one cannot determine the derivatives of Euler angles from the angular velocities using the equations $\boldsymbol{\omega}^i = \mathbf{G}^i\dot{\boldsymbol{\theta}}^i$ and $\overline{\boldsymbol{\omega}}^i = \overline{\mathbf{G}}^i\dot{\boldsymbol{\theta}}^i$. Determining $\dot{\boldsymbol{\theta}}^i$ is necessary during dynamic simulations since the angular velocity vectors $\boldsymbol{\omega}^i$ and $\overline{\boldsymbol{\omega}}^i$ are not exact differentials and cannot be integrated to determine rotational coordinates. In fact, there are no known rotational coordinates such that the matrices $\mathbf{G}^i$ and $\overline{\mathbf{G}}^i$ are the identity matrices: therefore, $\boldsymbol{\omega}^i \neq \dot{\boldsymbol{\theta}}^i$ and $\overline{\boldsymbol{\omega}}^i \neq \dot{\boldsymbol{\theta}}^i$.

In railroad vehicle dynamics, the Euler-angle sequence $Z^i - X^i - Y^i$ is used to avoid these kinematic singularities in most simulations. Figure 6 shows the coordinate system of a wheelset in which the Y^i axis is selected along the wheel axle. In most railroad vehicle system problems, the yaw angle ψ^i and the roll angle ϕ^i remain small. Therefore, in these simulation scenarios, it is not expected that the roll angle ϕ^i will approach the value $\pm\pi/2$ at which the singularity occurs.

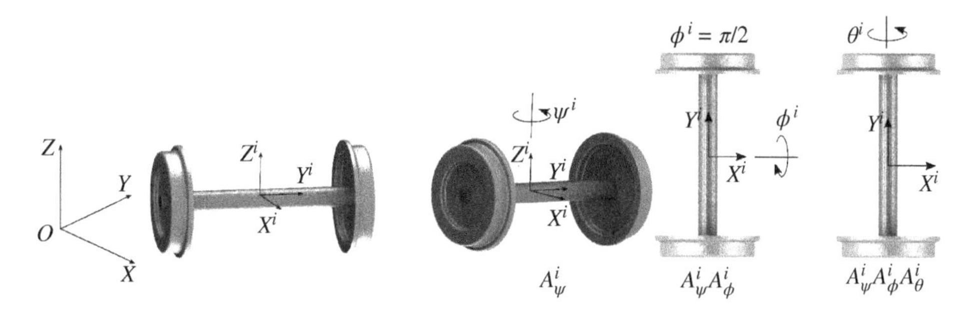

Figure 3.6 Euler-angle singularities.

The singular configurations depend on the sequence of Euler angles used, as previously mentioned. If the sequence $Z^i - X^i - Z^i$ of Examples 2 and 6 is used with the angles ϕ^i, θ^i, and ψ^i, respectively, the singular configuration occurs when $\theta^i = 0$ or $\theta^i = \pi$. At this configuration, one cannot distinguish between the first and second Z^i axes about which the Euler rotations ϕ^i and ψ^i are performed, respectively. One can show that if $\theta^i = 0$ or $\theta^i = \pi$, the two matrices $\mathbf{G}^i$ and $\overline{\mathbf{G}}^i$ obtained in Example 6, which have the determinants $|\mathbf{G}^i| = |\overline{\mathbf{G}}^i| = -\sin\theta^i$, are singular, and the derivatives of Euler angles cannot be determined from the angular velocity vectors.

No such kinematic singularities are associated with Euler parameters. When Euler parameters are used, one can still write the angular velocity vectors as $\boldsymbol{\omega}^i = \mathbf{G}^i\dot{\boldsymbol{\theta}}^i$ and $\overline{\boldsymbol{\omega}}^i = \overline{\mathbf{G}}^i\dot{\boldsymbol{\theta}}^i$, where $\mathbf{G}^i = 2\mathbf{E}^i$ and $\overline{\mathbf{G}}^i = 2\overline{\mathbf{E}}^i$. In order to solve for $\dot{\boldsymbol{\theta}}^i$ using, for example, the equation $\boldsymbol{\omega}^i = \mathbf{G}^i\dot{\boldsymbol{\theta}}^i$, one can pre-multiply by $\mathbf{G}^{i^T}$ to obtain $\mathbf{G}^{i^T}\boldsymbol{\omega}^i = \mathbf{G}^{i^T}\mathbf{G}^i\dot{\boldsymbol{\theta}}^i$. Using the identities of Euler parameters presented in Table 1, one has $\mathbf{G}^{i^T}\boldsymbol{\omega}^i = 4\left(\mathbf{I}_4 - \boldsymbol{\theta}^i\boldsymbol{\theta}^{i^T}\right)\dot{\boldsymbol{\theta}}^i$. Because $\boldsymbol{\theta}^{i^T}\dot{\boldsymbol{\theta}}^i = 0$, one has $\dot{\boldsymbol{\theta}}^i = (1/4)\,\mathbf{G}^{i^T}\boldsymbol{\omega}^i$ for any values of Euler parameters. Consequently, when Euler parameters are used, the kinematic singularities associated with Euler angles can be avoided. This is one of the main reasons Euler parameters are often used in developing general MBS algorithms.

Example 3.7

The orientation of a body i is defined by the Euler parameters $\boldsymbol{\theta}^i = \begin{bmatrix}0 & 0 & 1 & 0\end{bmatrix}^T$. At this configuration, the absolute angular velocity vector in the global coordinate system is given by $\boldsymbol{\omega}^i = \begin{bmatrix}20.0 & 0 & 5.0\end{bmatrix}^T$. Determine the time derivatives of the Euler parameters.

Solution The absolute angular velocity vector defined in the global XYZ coordinate system is written in terms of the derivatives of Euler parameters as $\boldsymbol{\omega}^i = \mathbf{G}^i \dot{\boldsymbol{\theta}}^i = 2\mathbf{E}^i \dot{\boldsymbol{\theta}}^i$. Pre-multiplying both sides of this equation by $\mathbf{G}^{i^T}$, and using the identity $\mathbf{G}^{i^T}\mathbf{G}^i = 4\left(\mathbf{I}_4 - \boldsymbol{\theta}^i\boldsymbol{\theta}^{i^T}\right)$ given in Table 1 (Nikravesh 1988; Shabana et al. 2001), one can write $\dot{\boldsymbol{\theta}}^i = (1/4)\,\mathbf{G}^{i^T}\boldsymbol{\omega}^i = (1/2)\,\mathbf{E}^{i^T}\boldsymbol{\omega}^i$, as previously explained. Therefore, one has

$$\dot{\boldsymbol{\theta}}^i = \frac{1}{2}\mathbf{E}^{i^T}\boldsymbol{\omega}^i = \frac{1}{2}\begin{bmatrix} -\theta_1^i & -\theta_2^i & -\theta_3^i \\ \theta_0^i & \theta_3^i & -\theta_2^i \\ -\theta_3^i & \theta_0^i & \theta_1^i \\ \theta_2^i & -\theta_1^i & \theta_0^i \end{bmatrix}\begin{bmatrix}\omega_1^i \\ \omega_2^i \\ \omega_3^i\end{bmatrix}$$

$$= \frac{1}{2}\begin{bmatrix} 0 & 1 & 0 \\ 0 & 0 & -1 \\ 0 & 0 & 0 \\ 1 & 0 & 0 \end{bmatrix}\begin{bmatrix}20.0 \\ 0 \\ 5.0\end{bmatrix} = \begin{bmatrix}0 \\ -2.5 \\ 0 \\ 10\end{bmatrix}$$

3.8 EULER ANGLES AND TRACK GEOMETRY

In railroad vehicle system dynamics, Euler angles can be used for two fundamentally different purposes, as previously mentioned. First, Euler angles can be used to define the orientation of a body in space, as explained in the preceding sections of this chapter. In this case, three discrete Euler angles are assigned to each body in the system, and these Euler angles are not treated as field variables since the body orientation is a characteristic of the entire body and does not depend on a particular point. Euler angles used in motion description vary with time as the bodies change their orientations due to the applied forces and moments.

For the second purpose, Euler angles are used to represent the geometry of the track and rail space curves without consideration of the track and rail motion. While each rail can be modeled as a separate body that can move independently, and can be assigned its discrete Euler angles to define its orientation, another set of field-variable Euler angles based on another sequence can be used to develop the geometry of the space curves associated with each rail, as described in the remainder of this chapter. To understand the process of defining track geometry, it is important to understand the basic sections and definitions used in practice when the track layout is designed.

Track Sections The rail track is designed using a few sections with specific geometries. In the layout of the track, it necessary to ensure smoothness during the transition between sections that have different geometries, in order to avoid undesirable impulsive forces, vibration, noise, and possibly derailments. Non-smooth transitions can lead to sudden changes in the wheel/rail contact forces, and such impulsive forces not only produce undesirable effects such as noise, severe vibrations, and ride discomfort, but can also cause derailments and serious accidents that can be deadly and result in environmental and economic damage.

In general, three different types of segments with different geometries are used in the layout of the rail track. These segments are

1. *Tangent* (straight), which has zero curvature
2. *Curve*, which has constant curvature
3. *Spiral*, which has linearly varying curvatures

The geometries of these three segments are shown in Figure 7. The spiral segment is used to connect two segments that have two different curvature values. The spiral is designed to have linearly varying curvature, with the spiral ends having the curvatures of the two segments (tangent or curve) connected by the spiral. For example, to connect a tangent section of track that has zero curvature to a curve track segment with non-zero curvature, the spiral connecting these two segments has zero curvature at the first end and the curvature of the curve section at the second end. Within the spiral, the curvature varies linearly; an example is shown in Figure 8.

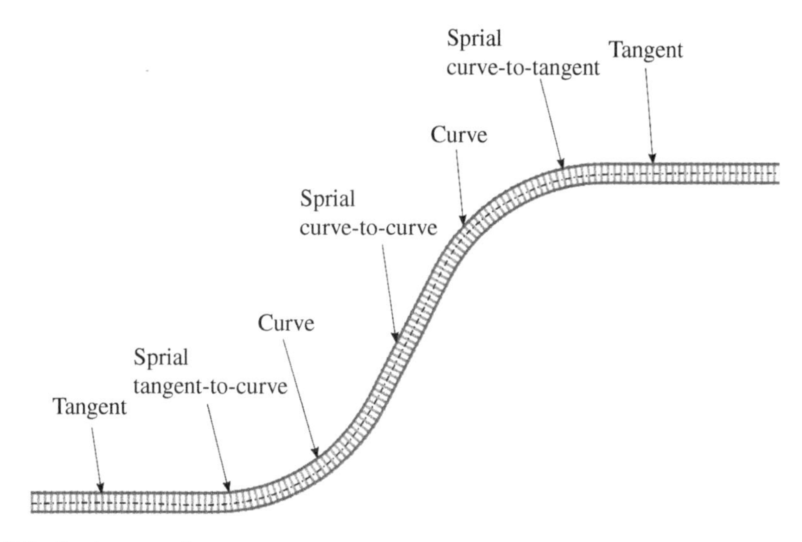

Figure 3.7 Track segments.

Track Geometry Parameters As also discussed in Chapter 1, important geometric parameters are used in the construction of the track. These are the track *gage* G, which measures the spacing between the two rails of the track and is defined as the lateral distance between two points on the right and left rails located at a distance 5/8 in. (14 mm)

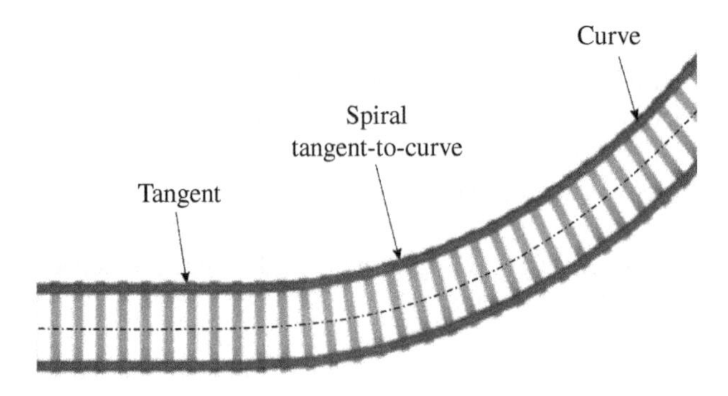

Figure 3.8 Spiral section.

from the top of the railhead, as shown in Figure 9. The gage can have different values in different countries. The standard gage value in North America varies from 56 to 57.25 in. The *superelevation h*, also shown in Figure 9, is defined as the vertical distance between the right and left rails. The *horizontal curvature* is defined as the angle ψ required to obtain a 100′-length chord of constant radius of curvature R_H in the horizontal plane, as shown in Figure 10. The *grade* is the ratio, given as a percentage, between the vertical elevation and the longitudinal distance on a rail. The *cant angle* measures the rotation of a rail about its longitudinal axis, as shown in Figure 11. *Rail irregularities* are defined using two types of deviations: *profile* and *alignment*. The profile and alignment deviations are, respectively, the vertical and lateral deviations of the rail space curve, as shown in Figure 12.

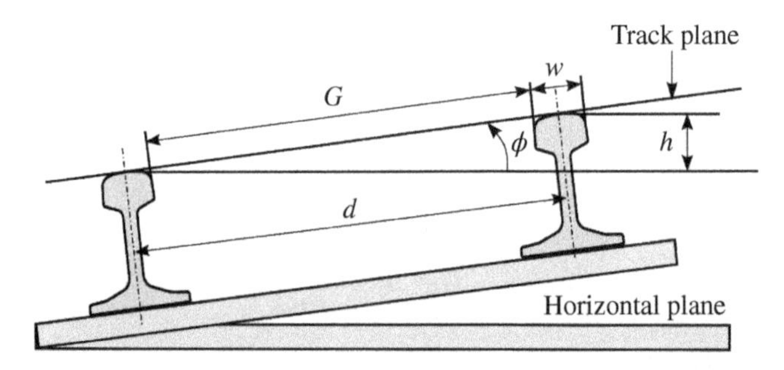

Figure 3.9 Gage and superelevation.

Track Geometry Definitions In practice, very few inputs are provided to obtain the discretized data used in computer simulations and virtual prototyping of railroad vehicle systems. These data can be provided at the points on the rail at which the geometry changes (tangent to spiral, spiral to curve, spiral to tangent, etc.). As explained in Chapter 4, in addition to the distance (arc length) at which these points are located, three pieces of information are needed at these geometry-change points in order to define the track geometry in a track preprocessor computer program. The geometry data produced by the track preprocessor at every nodal point include the distance, the Cartesian coordinates of the nodes in

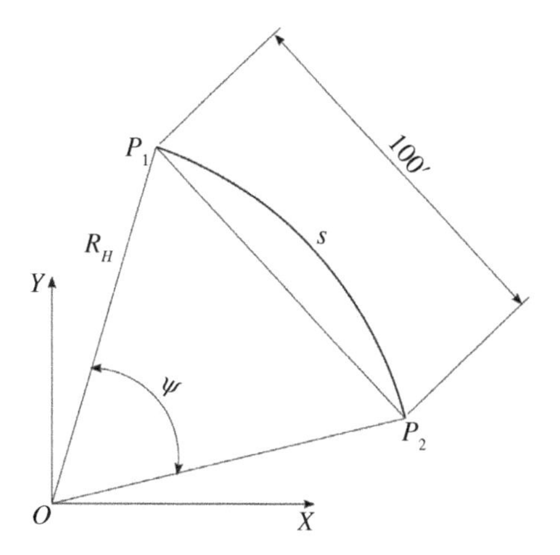

Figure 3.10 Horizontal curvature.

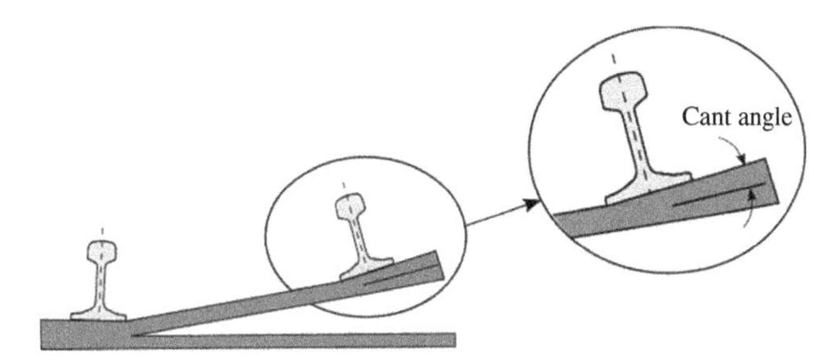

Figure 3.11 Cant angle.

a selected track or rail coordinate system, and three Euler angles that define the geometry of the rail. In order for the track preprocessor to compute the positions of and the angles at the nodes defined by the distance specified, the following three pieces of given industry inputs are provided at the points at which there is a change in the track geometry:

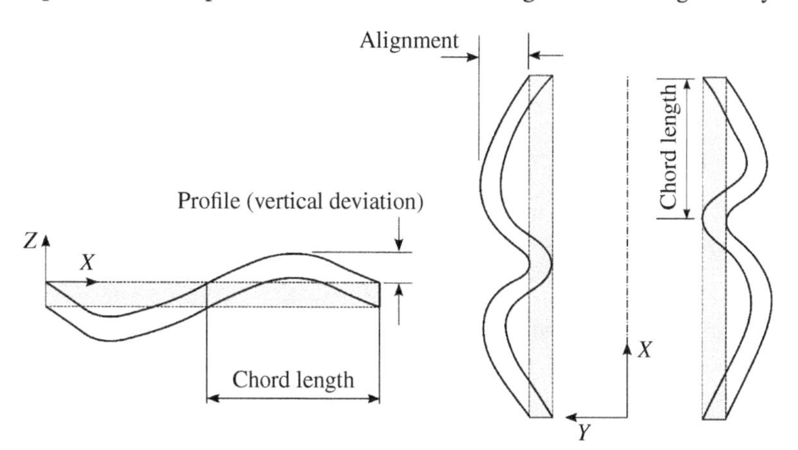

Figure 3.12 Track irregularities.

1. *Horizontal curvature* measures the curvature of the rail or track space curve when projected on a horizontal plane, as shown in Figure 10. The rail space curve is assumed to have arc length s, while the projected curve is parameterized by the arc length S. The horizontal curvature is used to define a Euler angle ψ, used as a measure of the curvature of a curve or spiral segment.
2. *Superelevation* measures the vertical elevation between the two rails. The superelevation is used to define another Euler angle called the *bank angle* ϕ.
3. *Grade* or *development* is the ratio between the vertical elevation and the longitudinal distance along the space curve. The grade is used to define a third Euler angle, called the *vertical - development (elevation) angle* θ.

Using these three inputs, the three Euler angles ψ, ϕ, and θ can be obtained as field variables that depend on the arc length of the space curve. It is important to point out that these Euler angles are not used to describe the orientation of bodies, but are used to define the geometry of space curves. These angles have a more general interpretation as the elements of *position-gradient vectors*, as will be discussed in Chapter 4. Therefore, the Euler angles considered in this section and the remaining sections of this chapter are used to define geometry and are not used in the motion description. Their use as geometry variables gives physical meaning to the basic steps used in the design of the track layout.

Track Euler Angles To develop a general approach for modeling railroad vehicle systems, the geometry of three space curves needs to be defined: the track centerline, right rail, and left rail space curves. Each of these space curves can have its own geometry. As discussed in Chapter 4, a track preprocessor computer program can be designed to generate the data for these three space curves for use in nonlinear dynamic simulations of railroad vehicle systems.

The sequence of rotation used for the three Euler angles that define the geometry of the track or rail space curves is different from the sequence previously used in this chapter to describe a body orientation. The reason for using a different sequence is to be consistent with the positive sign definitions of the curvature, superelevation, and grade used in practice. In this book, the X axis is considered the longitudinal axis, the Y axis is the lateral axis, and the Z axis is the axis perpendicular to the plane of the track. The sequence of rotations used for Euler angles that define the track geometry is a rotation ψ about the Z axis, a rotation θ about the $-Y$ axis, and a rotation ϕ about the $-X$ axis. Using this sequence of Euler angles, one can write the following simple rotation matrices:

$$\mathbf{A}_{\psi} = \begin{bmatrix} \cos\psi & -\sin\psi & 0 \\ \sin\psi & \cos\psi & 0 \\ 0 & 0 & 1 \end{bmatrix}, \quad \mathbf{A}_{\theta} = \begin{bmatrix} \cos\theta & 0 & -\sin\theta \\ 0 & 1 & 0 \\ \sin\theta & 0 & \cos\theta \end{bmatrix}, \quad \mathbf{A}_{\phi} = \begin{bmatrix} 1 & 0 & 0 \\ 0 & \cos\phi & \sin\phi \\ 0 & -\sin\phi & \cos\phi \end{bmatrix} \tag{3.35}$$

In the railroad literature, the angle ψ is called the *horizontal curvature angle*, the angle θ is called the *vertical - development (elevation) angle*, and the angle ϕ is called the *bank angle*. When the track geometry is considered, these angles are not called the yaw, pitch, and roll angles. The horizontal curvature angle ψ, elevation angle θ, and bank angle ϕ can be obtained from a simple track geometry description, as discussed in later sections of this chapter and in Chapter 4. Using the simple rotation matrices in the preceding equation, one can define the following transformation matrix:

$$\mathbf{A} = \mathbf{A}_\psi \mathbf{A}_\theta \mathbf{A}_\phi$$
$$= \begin{bmatrix} \cos\psi\cos\theta & -\sin\psi\cos\phi + \cos\psi\sin\theta\sin\phi & -\sin\psi\sin\phi - \cos\psi\sin\theta\cos\phi \\ \sin\psi\cos\theta & \cos\psi\cos\phi + \sin\psi\sin\theta\sin\phi & \cos\psi\sin\phi - \sin\psi\sin\theta\cos\phi \\ \sin\theta & -\cos\theta\sin\phi & \cos\theta\cos\phi \end{bmatrix} \tag{3.36}$$

This transformation can be used to completely define the geometry of the track and rail space curves. To this end, the Euler angles ψ, θ, and ϕ must be expressed as field variables, as explained in a later section of this chapter. The first column of this matrix is assumed to be the unit tangent to the space curve parameterized by the arc length s. In this case, one can write $\psi = \psi(s)$, $\theta = \theta(s)$, and $\phi = \phi(s)$.

3.9 ANGLE REPRESENTATION OF THE CURVE GEOMETRY

When Euler angles are used as time-dependent motion coordinates to describe a body orientation, these angles are assumed to be independent coordinates in the case of unconstrained motion. In curve geometry, on the other hand, the Frenet frame that completely defines the curve geometry is a function of only one parameter, which can be the curve arc length. That is, the Frenet frame, which is defined by three orthogonal unit vectors, can be completely defined using the arc length parameter. Therefore, if Euler angles employed in developing the transformation of Eq. 36 are used to define the space curve geometry, these Euler angles must be related, and one should be able to write them as functions of the arc length. By so doing, the columns of the transformation matrix in Eq. 36 become the same as the columns of the transformation matrix that defines the Frenet frame, with the exception that the matrix of Eq. 36 is expressed in angles that have the physical meaning used in practice by the rail industry in the construction of a track layout.

A track or rail space curve $\mathbf{r}$ can be defined by considering the first column in the transformation matrix of Eq. 36 as the unit tangent vector. Assuming that curve $\mathbf{r}$ is given in its parametric form in terms of the curve arc length parameter s, one can use the first column of the transformation matrix of Eq. 36 to write a segment of the space curve as

$$d\mathbf{r} = \left(\frac{d\mathbf{r}}{ds}\right) ds = \begin{bmatrix} \cos\psi\cos\theta \\ \sin\psi\cos\theta \\ \sin\theta \end{bmatrix} ds \tag{3.37}$$

The curvature vector to the curve $\mathbf{r}$ is obtained by differentiating the unit tangent vector $\mathbf{t}(s) = d\mathbf{r}/ds = \begin{bmatrix} \cos\psi\cos\theta & \sin\psi\cos\theta & \sin\theta \end{bmatrix}^T$ with respect to the arc length s as

$$\mathbf{r}''(s) = \frac{d\mathbf{t}}{ds} = \begin{bmatrix} -\psi'\sin\psi\cos\theta - \theta'\cos\psi\sin\theta \\ \psi'\cos\psi\cos\theta - \theta'\sin\psi\sin\theta \\ \theta'\cos\theta \end{bmatrix} \tag{3.38}$$

In this equation, $\psi' = \partial\psi/\partial s$ and $\theta' = \partial\theta/\partial s$. The curvature, which is the magnitude of this curvature vector, is defined as

$$\kappa = \left|\mathbf{r}''(s)\right| = \left|\frac{d\mathbf{t}}{ds}\right| = \sqrt{(\psi'\cos\theta)^2 + (\theta')^2} \tag{3.39}$$

Therefore, the unit normal vector to the curve $\mathbf{r}$ defined by Eq. 37 is along the curvature vector and is defined as

$$\mathbf{n} = \frac{1}{\kappa}\begin{bmatrix} -\psi'\sin\psi\cos\theta - \theta'\cos\psi\sin\theta \\ \psi'\cos\psi\cos\theta - \theta'\sin\psi\sin\theta \\ \theta'\cos\theta \end{bmatrix} \tag{3.40}$$

As explained in Chapter 2, the binormal vector $\mathbf{b}$ is the cross product of the tangent vector $\mathbf{t}$ and the normal vector $\mathbf{n}$. This cross product leads to

$$\mathbf{b} = \mathbf{t}\times\mathbf{n} = \frac{1}{\kappa}\begin{bmatrix} -\psi'\cos\psi\sin\theta\cos\theta + \theta'\sin\psi \\ -\psi'\sin\psi\sin\theta\cos\theta - \theta'\cos\psi \\ \psi'\cos^2\theta \end{bmatrix} \tag{3.41}$$

The Frenet frame $\mathbf{A}_f$ is formed by the three orthogonal unit vectors $\mathbf{t}$, $\mathbf{n}$, and $\mathbf{b}$: that is,

$$\mathbf{A}_f = \begin{bmatrix}\mathbf{t} & \mathbf{n} & \mathbf{b}\end{bmatrix} \tag{3.42}$$

As discussed in Chapter 2, performing additional differentiations leads to the Serret–Frenet formulas defined as

$$\left.\begin{aligned} \mathbf{t}' &= \kappa\mathbf{n} \\ \mathbf{n}' &= -\kappa\mathbf{t} + \tau\mathbf{b} \\ \mathbf{b}' &= -\tau\mathbf{n} \end{aligned}\right\} \tag{3.43}$$

where τ is the curve *torsion*, which is defined using the equation $\mathbf{b}'(s) = -\tau(s)\mathbf{n}(s)$.

In general, the two orthogonal matrices of Eqs. 36 and 42 share the first column. The other two columns are not, in general, the same, and they differ by a simple planar rotation about the tangent vector. In order for the two orthogonal matrices $\mathbf{A}$ and $\mathbf{A}_f$ of Eqs. 36 and 42, respectively, to be the same, the Euler angles must be related because a curve geometry is defined by two geometric invariants only. This allows using matrix $\mathbf{A}$ of Eq. 36 to define the curve geometry using angles that can be given a physical meaning.

3.10 EULER ANGLES AS FIELD VARIABLES

By imposing conditions on the Euler angles, these angles can have a physical meaning and can be directly related to definitions used by the rail industry, such as the horizontal curvature, superelevation, and grade. Without these conditions, the three Euler angles ψ, θ, and ϕ cannot be effectively used to define the geometry of the track.

In order for the two orthogonal matrices $\mathbf{A}$ and $\mathbf{A}_f$ of Eqs. 36 and 42, respectively, to be the same, the Euler angles must be related. As previously explained, because the matrices $\mathbf{A}$ and $\mathbf{A}_f$ share the first column, the other two columns in the two matrices lie in the same plane but differ by a simple rotation about the longitudinal tangent $\mathbf{t}$. By imposing conditions on the angle ϕ, one can show that the second column in matrix **A** becomes the unit curvature vector **n**, which is the second column of matrix $\mathbf{A}_f$. To this end, the angle ϕ is defined using the equations (Rathod and Shabana 2006a,b)

$$\left.\begin{aligned}\sin\phi &= \frac{-\theta'}{\sqrt{(\psi'\cos\theta)^2+(\theta')^2}} = \frac{-\theta'}{\kappa}\\ \cos\phi &= \frac{\psi'\cos\theta}{\sqrt{(\psi'\cos\theta)^2+(\theta')^2}} = \frac{\psi'\cos\theta}{\kappa}\end{aligned}\right\} \tag{3.44}$$

By using this equation in Eq. 40, the normal vector **n** can be written as

$$\mathbf{n} = \frac{1}{\kappa}\begin{bmatrix}-\psi'\sin\psi\cos\theta-\theta'\cos\psi\sin\theta\\ \psi'\cos\psi\cos\theta-\theta'\sin\psi\sin\theta\\ \theta'\cos\theta\end{bmatrix} = \begin{bmatrix}-\sin\psi\cos\phi+\cos\psi\sin\theta\sin\phi\\ \cos\psi\cos\phi+\sin\psi\sin\theta\sin\phi\\ -\cos\theta\sin\phi\end{bmatrix} \tag{3.45}$$

which is the same as the second column of matrix **A**. In Eq. 45, the normal vector has a form that is not a function of the derivatives of the angles. The conditions of Eq. 44, which allowed eliminating the angle derivatives, implies that

$$\phi = \tan^{-1}\left(\frac{-\theta'}{\psi'\cos\theta}\right) \tag{3.46}$$

Similarly, by using Eq. 44 or, equivalently, Eq. 46, the binormal vector of Eq. 41 can be written as

$$\mathbf{b} = \frac{1}{\kappa}\begin{bmatrix}-\psi'\cos\psi\sin\theta\cos\theta+\theta'\sin\psi\\ -\psi'\sin\psi\sin\theta\cos\theta-\theta'\cos\psi\\ \psi'\cos^2\theta\end{bmatrix} = \begin{bmatrix}-\cos\psi\sin\theta\cos\phi-\sin\psi\sin\phi\\ -\sin\psi\sin\theta\cos\phi+\cos\psi\sin\phi\\ \cos\theta\cos\phi\end{bmatrix} \tag{3.47}$$

which is the same as the third column of matrix **A** of Eq. 36. Again, the angle derivatives have been eliminated from one of the vectors that define the Frenet frame.

By eliminating the angle derivatives in the normal and binormal vectors, the second equation in the Serret–Frenet equations (Eq. 43), $\mathbf{n}' = -\kappa\mathbf{t}+\tau\mathbf{b}$, leads, upon using the third element in vectors $\mathbf{n}'$, $\mathbf{t}$ and $\mathbf{b}$, to the condition $\theta'\sin\theta\sin\phi-\phi'\cos\theta\cos\phi = -\kappa\sin\theta+\tau\cos\theta\cos\phi$. This equation can be simplified using Eq. 44 as $\phi'\cos\theta\cos\phi = \kappa\sin\theta-\kappa\sin\theta\sin^2\phi-\tau\cos\theta\cos\phi$, or

$$\phi'\cos\theta = \kappa\sin\theta\cos\phi-\tau\cos\theta \tag{3.48}$$

Therefore, using Eqs. 44 and 48, one has the following three differential equations that can be used to define Euler angles as field variables:

$$\left.\begin{aligned}\psi'\cos\theta &= \kappa\cos\phi\\ \theta' &= -\kappa\sin\phi\\ \phi'\cos\theta &= \kappa\sin\theta\cos\phi-\tau\cos\theta\end{aligned}\right\} \tag{3.49}$$

The derivatives of the angles can then be written as

$$\left.\begin{aligned}\psi' &= \kappa(\cos\phi/\cos\theta)\\ \theta' &= -\kappa\sin\phi\\ \phi' &= \kappa\tan\theta\cos\phi-\tau\end{aligned}\right\} \tag{3.50}$$

These equations can be used to define the angles in terms of the curve curvature κ and the curve torsion τ, which uniquely define the curve geometry. It is clear from this equation that a singularity exists when the vertical-development angle θ approaches $\pm(\pi/2)$. Such a high value of the grade, which corresponds to the vertical rail, is not used in practice. The elevation and superelevation angles θ and ϕ, respectively, do not assume very high values in practical applications. If both of these angles remain very small, one has $\psi' \approx \kappa$ and $\phi' \approx -\tau$: that is, the angle ψ can be directly related to the curvature of the curve κ, and the angle ϕ can be directly related to the torsion of the curve τ. For a more general curve defined by the curvature $\kappa(s)$ and torsion $\tau(s)$, the derivative of the angles in Eq. 50 can be integrated to determine the angles as functions of the arc length parameter. To this end, one can use Eq. 50 to write

$$\left.\begin{aligned} \psi(s) &= \psi_o + \int_{s_o}^{s} \kappa\left(\cos\phi/\cos\theta\right) ds \\ \theta(s) &= \theta_0 - \int_{s_o}^{s} \kappa \sin\phi ds \\ \phi(s) &= \phi_o + \int_{s_o}^{s} \left(\kappa \tan\theta \cos\phi - \tau\right) ds \end{aligned}\right\} \tag{3.51}$$

where ψ_o, θ_o, and ϕ_o are specified values of the angles at s_0. In general, obtaining analytical closed-form solutions for the integrals in Eq. 51 can be difficult, and therefore, numerical integration methods can be employed to evaluate these integrals. Equation 51 shows that Euler angles can all be written in terms of the curve arc length s since the curvature κ and torsion τ depend on the arc length parameter. Therefore, Euler angles are no longer considered independent parameters and can be used to uniquely define the curve geometry. Previously, it was shown that the curve curvature κ can be written in terms of Euler angles and their derivatives as $\kappa(s) = \sqrt{(\psi'\cos\theta)^2 + (\theta')^2}$ (Eq. 39). Using Eq. 50, the curve torsion τ can also be written in terms of Euler angles and their derivatives as

$$\begin{aligned} \tau(s) &= \kappa \tan\theta \cos\phi - \phi' \\ &= \sqrt{(\psi'\cos\theta)^2 + (\theta')^2}(\tan\theta\cos\phi) - \phi' \end{aligned} \tag{3.52}$$

That is, Euler angles uniquely define the geometric curve properties. It is also clear that once the angles are determined, the curve can be defined in its parametric form by integrating Eq. 37 as

$$\mathbf{r} = \int_{s_o}^{s} \left(\frac{d\mathbf{r}}{ds}\right) ds = \int_{s_o}^{s} \begin{bmatrix} \cos\psi(s)\cos\theta(s) \\ \sin\psi(s)\cos\theta(s) \\ \sin\theta(s) \end{bmatrix} ds \tag{3.53}$$

In the layout of the track, Euler angles are expressed in terms of given specific industry inputs, as discussed in the following section. It is to be noted that the bank angle ϕ that enters into the definition of the curve geometry is different from the bank angle that defines the track super-elevation and the orientation of the track coordinate systems that follow the motion of the vehicle components in railroad vehicle algorithms (Ling and Shabana, 2020).

3.11 EULER-ANGLE DESCRIPTION OF THE TRACK GEOMETRY

To understand the mathematical foundations of some of the terminology used in track descriptions, some basic geometry concepts are discussed in this section, starting with the curvature vector in Eq. 38. This vector can be written as the sum of two vectors as

$$\mathbf{r}''(s) = \frac{d\mathbf{t}}{ds} = \begin{bmatrix} -\psi' \sin\psi\cos\theta - \theta'\cos\psi\sin\theta \\ \psi'\cos\psi\cos\theta - \theta'\sin\psi\sin\theta \\ \theta'\cos\theta \end{bmatrix} = \psi'\cos\theta \begin{bmatrix} -\sin\psi \\ \cos\psi \\ 0 \end{bmatrix} + \theta' \begin{bmatrix} -\cos\psi\sin\theta \\ -\sin\psi\sin\theta \\ \cos\theta \end{bmatrix} \tag{3.54}$$

This equation shows that the curvature vector has two components $\psi'\cos\theta$ and θ' along the two orthogonal unit vectors $\mathbf{a}_1 = \begin{bmatrix} -\sin\psi & \cos\psi & 0 \end{bmatrix}^T$ and $\mathbf{a}_2 = [-\cos\psi\sin\theta \ -\sin\psi\sin\theta\cos\theta]^T$, respectively. These two vectors are also orthogonal to the longitudinal unit tangent to the curve $\mathbf{t} = \begin{bmatrix} \cos\psi\cos\theta & \sin\psi\cos\theta & \sin\theta \end{bmatrix}^T$. Therefore, the three vectors $\mathbf{t}$, $\mathbf{a}_1$, and $\mathbf{a}_2$ represent an orthogonal triad, with vector $\mathbf{a}_1$ lying in a plane parallel to the horizontal plane and vector $\mathbf{a}_2$ perpendicular to the plane formed by the two vectors $\mathbf{t}$ and $\mathbf{a}_1$. In railroad literature, the curvature component θ' is called the *vertical curvature* C_V. Therefore, the vertical curvature is defined as

$$C_V = \theta' = \frac{d\theta}{ds} \tag{3.55}$$

That is, the vertical curvature is defined to be the derivative of the elevation angle θ with respect to the curve arc length parameter s.

Horizontal Projection and Curvature In the mathematical description of a railroad track, the track and rail space curves are first projected on the horizontal plane. The horizontal projection of the curve segment defined by Eq. 37 leads to

$$d\mathbf{r}_H = \begin{bmatrix} \cos\psi\cos\theta & \sin\psi\cos\theta & 0 \end{bmatrix}^T ds \tag{3.56}$$

By factoring out cosθ, one can write $d\mathbf{r}_H = \begin{bmatrix} \cos\psi & \sin\psi & 0 \end{bmatrix}^T \cos\theta ds$. This equation defines the unit tangent vector $\mathbf{t}_H$ to and the arc length dS of the projected planar curve shown in Figure 10, respectively, as

$$\left.\begin{aligned} \mathbf{t}_H &= \begin{bmatrix} \cos\psi & \sin\psi & 0 \end{bmatrix}^T \\ dS &= \cos\theta ds \end{aligned}\right\} \tag{3.57}$$

Using the relationship between the space curve arc length s and the arc length S of the projected planar curve, one has $\partial\psi/\partial S = (\partial\psi/\partial s)(\partial s/\partial S)$, where $\partial s/\partial S = 1/\cos\theta$. The curvature vector of the projected planar curve that is parameterized by the arc length S is defined as

$$\mathbf{c}_H = \frac{\partial \mathbf{t}_H}{\partial S} = \begin{bmatrix} -\sin\psi \\ \cos\psi \\ 0 \end{bmatrix} \frac{\partial\psi}{\partial S} \tag{3.58}$$

This equation defines the *horizontal curvature* C_H of the projected planar curve, which is the magnitude of the curvature vector $\mathbf{c}_H$, as $C_H = |\mathbf{c}_H| = \partial\psi/\partial S$, or

$$d\psi = C_H dS = \frac{dS}{R_H} \tag{3.59}$$

In this equation, R_H is the radius of curvature of the projected planar curve shown in Figure 10. It is clear from the preceding equation that the angle ψ can be determined by integration of the horizontal curvature C_H. In this case, one can write

$$\psi = \psi(S) = \psi_o + \int_{S_o}^{S} \left(C_H(S)\right) dS \tag{3.60}$$

where ψ_o is the value of ψ at S_o. The horizontal curvature C_H, as discussed in Chapter 4, is one of the standard data used in practice for the track description. In the preceding equation, for convenience, the horizontal curvature angle ψ is written in terms of the projected-curve arc length S instead of the actual curve arc length s. Recall from Eq. 57 that S and s are related by $dS = \cos\theta ds$.

In the design of railroad tracks, the *chord definition* of curvature is used. The chord definition of curvature is based on the number of degrees encompassed by a 100-ft line segment that has endpoints located on the arc of the track curve. An approximate method for measuring the curvature is to stretch a string that has length 62 ft between two points on the inside face of the outer railhead and determine the number of inches between the center point of the string and the rail. This number corresponds to the degree of curvature: 1 in. equals 1° and 2 in. equals 2°. The sharpest curve that can be handled by a single four-axle diesel locomotive is 40°. When the locomotive is in a train consist, the limit decreases to 20°. Uneven terrain, such as mountains, further limits the curvature to 5–10°. Curves of 1° or 2° are the most commonly used (McGonigal, 2006).

Elevation and Grade The vertical-development angle θ can be determined from the *vertical curvature* C_V defined using the *grade*, which is a given industry input that measures the ratio between the elevation and the longitudinal distance between two points. It is always preferable to use a tangent track when possible, to avoid more energy consumption, wheel and rail wear, and reduced speeds. This desired tangent-track option, however, is not always possible because of uneven ground terrain and the need to avoid obstacles. Consequently, track curvatures and elevations are difficult to avoid.

The grade is defined in North America in terms of the number of feet the track rises per 100 ft of horizontal distance. For example, if a track rises 2 ft over a distance of 200 ft, the grade is considered to be 1%. In other parts of the world, the grade is measured in terms of the horizontal distance required to achieve a one-foot rise. In general, grades do not exceed 1%, and grades higher than 2.2% are not common. For every 1% increase in the grade, the resistance force is approximately four times the force required when a locomotive travels on a tangent track (McGonigal 2006).

The vertical curvature was defined by Eq. 55 in terms of the derivative of the vertical-development angle θ as $C_V = \theta' = d\theta/ds$. Therefore, the data at two points defined by the arc length values s_1 and s_2 can be used to define the vertical curvature, which is assumed constant for a given track segment, as

$$C_V = \frac{\theta_2 - \theta_1}{s_2 - s_1} \tag{3.61}$$

where θ_1 and θ_2 are the vertical-development angles at the two points defined by the arc length values s_1 and s_2, respectively. Using the definition of C_V in the preceding equation, the vertical-development angle can be written in a differential form as $d\theta = C_V ds$. This equation, upon integration between the two points, defines the vertical-development angle as

$$\theta = \theta_1 + C_V \left(s - s_1\right) \tag{3.62}$$

This definition of the elevation angle θ can be used with the relationship $dS = \cos\theta ds$, if desired, to write $d\psi$ in Eq. 59 in terms of the space curve arc length s.

Bank Angle and Superelevation The bank angle ϕ can be obtained using the superelevation, which is a given industry input that measures the elevation of one rail with respect to the other as a result of a *bank-angle* rotation ϕ about the longitudinal axis. The bank angle is assumed to vary linearly as a function of the projected arc length S as

$$\begin{aligned}\phi = \phi(S) &= \left(1 - \xi_S\right)\phi_1 + \xi_S\phi_2 \\ &= \frac{\phi_2\left(S - S_1\right) - \phi_1\left(S - S_2\right)}{\left(S_2 - S_1\right)}\end{aligned} \tag{3.63}$$

where $\xi_S = S/L_S$, $L_S = S_2 - S_1$, and ϕ_1 and ϕ_2 are the bank angles at the two end nodes of the segments defined, respectively, by the projected arc length coordinates S_1 and S_2.

Equations 60, 62, and 63, which define the three Euler angles $\psi = \psi(S)$, $\theta = \theta(s)$, and $\phi = \phi(S)$ as field variables, also define the geometry of the space curve. As previously mentioned, for convenience, the angles ψ and ϕ are written in terms of the projected arc length S, while the angle θ is written in terms of the actual arc length s of the space curve. Recall that the sequence $Z, -Y, -X$ is selected when Euler angles are used to describe the track geometry in order to have positive definitions for the horizontal curvature, grade, and superelevation. The curvature is considered positive if it is the result of a positive rotation ψ about the Z axis; the vertical development is considered positive if it is the result of a rotation θ about the $-Y$ axis; and the superelevation is considered positive if it is the result of a rotation ϕ about the $-X$ axis. This convention is consistent with the measurements made in practice. These three Euler angles, obtained as field variables, can be used to define the space curve in a parametric form as described in the remainder of this section. It is important, however, to point out that the linearly interpolated bank angle ϕ of Eq. 63 is used to define the orientation of the track (body trajectory) frames that follow the vehicle components in railroad vehicle algorithms. This angle does not enter into the definition of the curve geometry which can be completely defined by two geometric parameters only (Ling and Shabana, 2020).

Curve Position Coordinates In addition to determining the angles at arbitrary points on the space curve from the given industry input, the position coordinates of these arbitrary points are also required. As previously discussed, the space curve segment is defined by considering the first column of the transformation matrix of Eq. 36 as the unit vector tangent to the curve. This allows writing the curve equation in the differential form $d\mathbf{r} = \left[\cos\psi\cos\theta \;\; \sin\psi\cos\theta \;\; \sin\theta\right]^T ds$. Using the second equation in Eq. 57, which defines the relationship between s and S, $dS = \cos\theta ds$, the equation for $d\mathbf{r}$ can be written as

$$d\mathbf{r} = \begin{bmatrix} \cos\psi dS \\ \sin\psi dS \\ \sin\theta ds \end{bmatrix} \tag{3.64}$$

This equation can be used, upon integration, to define the position coordinates of an arbitrary point on the curve $\mathbf{r} = \begin{bmatrix} r_1 & r_2 & r_3 \end{bmatrix}^T$. This also defines the curve in its parametric form as

$$\mathbf{r} = \begin{bmatrix} r_{o1} \\ r_{o2} \\ r_{o3} \end{bmatrix} + \begin{bmatrix} \int_{S_o}^{S_1} \cos\psi dS \\ \int_{S_o}^{S_1} \sin\psi dS \\ \int_{s_o}^{s_1} \sin\theta ds \end{bmatrix} = \begin{bmatrix} r_1(S) \\ r_2(S) \\ r_3(s) \end{bmatrix} \tag{3.65}$$

where $\mathbf{r}_o = \begin{bmatrix} r_{o1} & r_{o2} & r_{o3} \end{bmatrix}^T$ is vector $\mathbf{r}$ at the start of the segment. This equation shows that, by using Euler angles as field variables, the coordinates of the points on the space curve can be determined. Euler angles, as field variables, are assumed to be known from the horizontal curvature C_H, grade θ, and superelevation h, which are given industry inputs as discussed in Chapter 4. The locations of the points on the space curve and the Euler angles that define the curve geometry can be determined for the tangent, curve, and spiral rail sections. These segment types have different geometries because the tangent section has zero curvature, the curve section has constant curvature, and the spiral section is designed to have linearly varying horizontal curvature, as discussed in Chapter 4. It is also important to note that Eq. 65 requires using only two angles because the curve geometry is completely defined in terms of two geometric invariants only (Ling and Shabana, 2020).

3.12 GEOMETRIC MOTION CONSTRAINTS

Railroad vehicle systems consist of large number of components that have independent motion relative to each other. Example of these railroad components are the rails, car bodies, frames, and wheelsets. For example, the wheelsets, which are mounted on the frame using a suspension system, experience large relative rotations with respect to all other vehicle components. Mechanical joints and force elements are used to connect the rail vehicle components in order to restrict the motion amplitudes and ensure the safe operation of the rail vehicles. The joints and force elements are also selected in a manner that allows obtaining the desired design and performance characteristics of the system. Using the motion description introduced at the beginning of this chapter, each vehicle component i with distributed inertia can have the set of generalized coordinates $\mathbf{q}^i = \begin{bmatrix} \mathbf{R}^{i^T} & \boldsymbol{\theta}^{i^T} \end{bmatrix}^T$, $i = 1, 2, \ldots, n_b$, where $\mathbf{R}^i$ is the vector that defines the global position of the body reference point O^i in the global coordinate system; θ^i is the set of orientation parameters that can be selected to be Euler angles or Euler parameters, as previously discussed in this chapter; and n_b is the total number of bodies in the system. Therefore, the system generalized coordinates can be written as

$$\begin{aligned} \mathbf{q} &= \begin{bmatrix} \mathbf{q}^{1^T} & \mathbf{q}^{2^T} & \ldots & \mathbf{q}^{n_b{}^T} \end{bmatrix}^T \\ &= \begin{bmatrix} \mathbf{R}^{1^T} & \boldsymbol{\theta}^{1^T} & \mathbf{R}^{2^T} & \boldsymbol{\theta}^{2^T} & \ldots & \mathbf{R}^{n_b{}^T} & \boldsymbol{\theta}^{n_b{}^T} \end{bmatrix}^T \end{aligned} \tag{3.66}$$

If Euler angles are used, the number of system coordinates is $6 \times n_b$, while if Euler parameters are used, the number of the system coordinates is $7 \times n_b$.

Degrees of Freedom Due to mechanical joints and specified motion trajectories, the system coordinates in Eq. 66 are not totally independent. The joints and specified motion trajectories can be formulated mathematically using a set of nonlinear algebraic *constraint equations* that represent relationships between the system coordinates. Because of these kinematic relationships, *dependent coordinates* can be determined using other coordinates, called *independent coordinates.* If the constraint equations are independent, the number of dependent coordinates is equal to the number of the algebraic constraint equations n_c. Therefore, if the total number of system coordinates in Eq. 66 is n, the total number of independent coordinates – also called the system *degrees of freedom* – is $n_d = n - n_c$. Knowledge of the system degrees of freedom is important in understanding a rail vehicle dynamic behavior and stability. This knowledge is also important in the control of any mechanical system if under- or over-actuation is to be avoided. For example, to fully control a mechanical system such as a robotic manipulator, the number of actuators and motors must be equal to the number of system degrees of freedom, to avoid under- or over-actuation. In railroad vehicle systems, not all of the degrees of freedom are controlled, due to the complexity of the system and also because of the need to provide the vehicle with the freedom to negotiate unexpected disturbances at a reasonable operating cost. The simple formula $n_d = n - n_c$, which can be used to give an estimate of the number of degrees of freedom n_d, is called the *mobility criterion*. This criterion defines the number of degrees of freedom as the total number of coordinates minus the number of algebraic constraint equations that relate these coordinates.

Nonlinear Algebraic Constraint Equations Most mechanical joints such as spherical, revolute, prismatic, cylindrical, and universal joints can be formulated mathematically using nonlinear algebraic constraint equations in terms of the Cartesian coordinates of Eq. 66. These algebraic constraint equations can be written in a vector form as

$$\mathbf{C}(\mathbf{q}, t) = \begin{bmatrix} C_1(\mathbf{q}, t) & C_2(\mathbf{q}, t) & \dots & C_{n_c}(\mathbf{q}, t) \end{bmatrix}^T = \mathbf{0} \tag{3.67}$$

where n_c is the total number of constraint functions. Geometric constraints that can be written at the position level, like the ones given in the preceding equation, are called *holonomic constraint equations.* An example of these holonomic constraints is the *spherical (ball) joint* constraint equations, which eliminate the relative translations and allow only three relative rotations between the two bodies connected by this joint. Figure 13 shows two bodies i and j connected by a spherical (ball) joint. The joint attachment points on bodies i and j are, respectively, denoted as P^i and P^j. The three nonlinear algebraic scalar constraint equations that describe the spherical joint can be written using the kinematic equations previously presented in this chapter as

$$\begin{aligned}\mathbf{C}\left(\mathbf{q}^i, \mathbf{q}^j\right) &= \begin{bmatrix} C_1 & C_2 & C_3 \end{bmatrix}^T = \mathbf{r}_P^i - \mathbf{r}_P^j \\ &= \left(\mathbf{R}^i + \mathbf{A}^i \bar{\mathbf{u}}_P^i\right) - \left(\mathbf{R}^j + \mathbf{A}^j \bar{\mathbf{u}}_P^j\right) = \mathbf{0}\end{aligned} \tag{3.68}$$

These three scalar equations eliminate all the relative translational degrees of freedom between the two bodies.

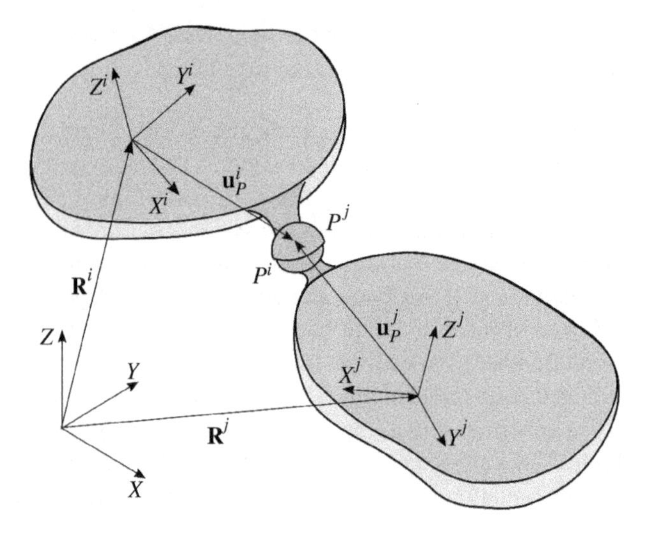

Figure 3.13 Spherical joint.

Other joints such as the *cylindrical joint* can also be systematically formulated in terms of the Cartesian coordinates in Eq. 66. As shown in Figure 14, the cylindrical joint allows one relative translation and one relative rotation between the two bodies connected by the joint along the joint axis. Therefore, this joint eliminates four degrees of freedom: two translations in the plane perpendicular to the joint axis and two rotations about two axes perpendicular to the joint axis. Therefore, the cylindrical joint has two degrees of freedom. If $\mathbf{v}^i = \mathbf{A}^i\overline{\mathbf{v}}^i$ and $\mathbf{v}^j = \mathbf{A}^j\overline{\mathbf{v}}^j$ are two vectors on bodies i and j, respectively, defined along the joint axis; and P^i and P^j are two points on the two bodies located on the joint axis, as shown in Figure 14, the relative rotations between the two bodies about two axes perpendicular to the joint axis can be prevented by imposing the mathematical conditions $\mathbf{v}_1^i \cdot \mathbf{v}^j = \left(\mathbf{A}^i\overline{\mathbf{v}}_1^i\right)^T \left(\mathbf{A}^j\overline{\mathbf{v}}^j\right) = 0$ and $\mathbf{v}_2^i \cdot \mathbf{v}^j = \left(\mathbf{A}^i\overline{\mathbf{v}}_2^i\right)^T \left(\mathbf{A}^j\overline{\mathbf{v}}^j\right) = 0$, where $\mathbf{v}_1^i$ and $\mathbf{v}_2^i$ are two vectors orthogonal to $\mathbf{v}^i$; $\overline{\mathbf{v}}^i, \overline{\mathbf{v}}_1^i$, and $\overline{\mathbf{v}}_2^i$ are constant vectors defined in the coordinate system of body i; and $\overline{\mathbf{v}}^j$ is a constant vector defined with respect to the coordinate system of body j. The relative translation between the two bodies in a plane perpendicular to the joint axis can be prevented using the two scalar equations $\mathbf{v}_1^i \cdot \mathbf{r}_P^{ij} = 0$ and $\mathbf{v}_2^i \cdot \mathbf{r}_P^{ij} = 0$, where $\mathbf{r}_P^{ij} = \mathbf{r}_P^i - \mathbf{r}_P^j = \left(\mathbf{R}^i + \mathbf{A}^i\overline{\mathbf{u}}_P^i\right) - \left(\mathbf{R}^j + \mathbf{A}^j\overline{\mathbf{u}}_P^j\right)$. One can, therefore, write the four constraint equations of the cylindrical joint as

$$\begin{aligned} \mathbf{C}\left(\mathbf{q}^i, \mathbf{q}^j\right) &= \begin{bmatrix} C_1 & C_2 & C_3 & C_4 \end{bmatrix}^T \\ &= \begin{bmatrix} \left(\mathbf{v}_1^{i^T}\mathbf{v}^j\right) & \left(\mathbf{v}_2^{i^T}\mathbf{v}^j\right) & \left(\mathbf{v}_1^{i^T}\mathbf{r}_P^{ij}\right) & \left(\mathbf{v}_2^{i^T}\mathbf{r}_P^{ij}\right) \end{bmatrix}^T = \mathbf{0} \end{aligned} \tag{3.69}$$

These four constraint equations define the two-degree-of-freedom cylindrical joint.

The *revolute joint*, also called the *pin joint*, which allows one relative rotation between two bodies about the joint axis, can be obtained from the cylindrical joint constraint equations by adding an additional constraint equation that eliminates the relative translation between the two bodies. This condition can be defined by requiring that the distance between the two points P^i and P^j to remain constant: that is, $\mathbf{r}_P^{ij^T}\mathbf{r}_P^{ij} = c$, where c is a constant that defines the distance between the two points P^i and P^j in the initial configuration. Therefore, the five

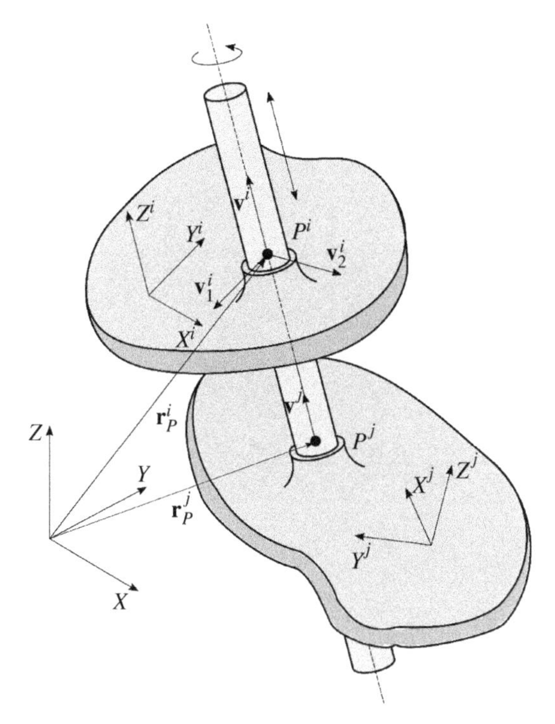

Figure 3.14 Cylindrical joint.

constraint equations that define the single-degree-of-freedom revolute joint can be written as

$$\mathbf{C}\left(\mathbf{q}^i,\ \mathbf{q}^j\right) = \begin{bmatrix} C_1 & C_2 & C_3 & C_4 & C_5 \end{bmatrix}^T$$
$$= \left[\left(\mathbf{v}_1^{iT}\mathbf{v}^j\right)\ \left(\mathbf{v}_2^{iT}\mathbf{v}^j\right)\ \left(\mathbf{v}_1^{iT}\mathbf{r}_P^{ij}\right)\ \left(\mathbf{v}_2^{iT}\mathbf{r}_P^{ij}\right)\ \left(\mathbf{r}_P^{ijT}\mathbf{r}_P^{ij} - c\right)\right]^T = \mathbf{0} \tag{3.70}$$

An alternate formulation of the revolute joint uses the spherical joint constraints, which eliminate the relative translations between the two bodies with two additional constraints that eliminate two relative rotations between the two bodies about two axes perpendicular to the joint axis. This alternate formulation of the revolute joint can be written as

$$\mathbf{C}\left(\mathbf{q}^i,\ \mathbf{q}^j\right) = \begin{bmatrix} \mathbf{C}_{1-3}^T & C_4 & C_5 \end{bmatrix}^T$$
$$= \left[\left(\mathbf{r}_P^i - \mathbf{r}_P^j\right)^T\ \ \mathbf{v}_1^{iT}\mathbf{v}^j\ \ \mathbf{v}_2^{iT}\mathbf{v}^j\right]^T = \mathbf{0} \tag{3.71}$$

Equations 70 and 71 lead to the same physical constraints on the motion of the two bodies connected by the revolute joint.

The constraint equations for the *prismatic joint*, also called the *translational joint*, allow one translational degree of freedom along the joint axis. These constraint equations can also be obtained by adding to the cylindrical joint constraint equations one algebraic equation that eliminates the relative rotation between the two bodies connected by the joint. As shown in Figure 15, one can define the two orthogonal vectors $\mathbf{h}^i = \mathbf{A}^i\overline{\mathbf{H}}^i$ and $\mathbf{h}^j = \mathbf{A}^j\overline{\mathbf{h}}^j$ on bodies i and j, respectively, perpendicular to the joint axis, where $\overline{\mathbf{H}}^i$ and $\overline{\mathbf{h}}^j$ are constant

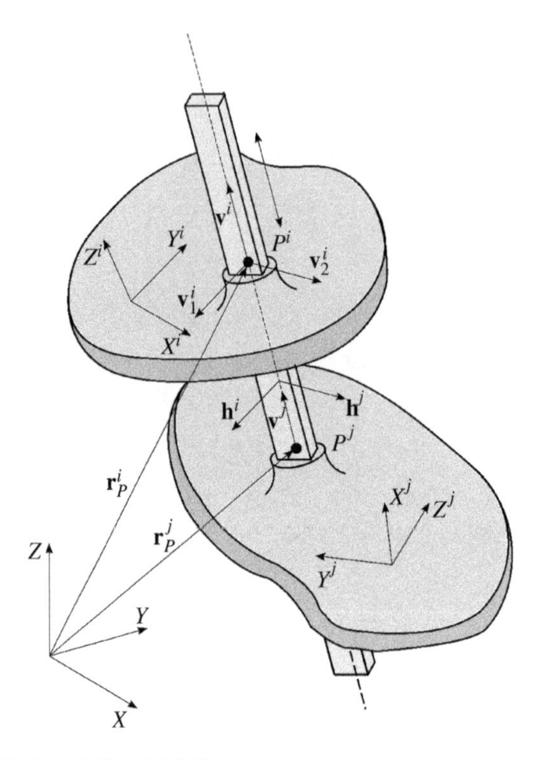

Figure 3.15 Prismatic (translational joint).

vectors defined, respectively, in the coordinate systems of bodies i and j. In order to prevent the relative rotations between the two bodies, one can introduce the condition $\mathbf{h}^{i^T}\mathbf{h}^j = \left(\mathbf{A}^i\overline{\mathbf{H}}^i\right)^T\left(\mathbf{A}^j\overline{\mathbf{h}}^j\right) = 0$. Therefore, expanding on the cylindrical joint constraint equations of Eq. 69, the five constraint equations that define the single-degree-of-freedom prismatic joint can be written as

$$\begin{aligned}\mathbf{C}\left(\mathbf{q}^i, \mathbf{q}^j\right) &= \begin{bmatrix} C_1 & C_2 & C_3 & C_4 & C_5 \end{bmatrix}^T \\ &= \left[\left(\mathbf{v}_1^{i^T}\mathbf{v}^j\right) \ \left(\mathbf{v}_2^{i^T}\mathbf{v}^j\right) \ \left(\mathbf{v}_1^{i^T}\mathbf{r}_P^{ij}\right) \ \left(\mathbf{v}_2^{i^T}\mathbf{r}_P^{ij}\right) \ \left(\mathbf{h}^{i^T}\mathbf{h}^j\right)\right]^T = \mathbf{0}\end{aligned} \tag{3.72}$$

These five equations allow only one translational degree of freedom along the joint axis.

The constraint equations of other joints can be formulated in a similar manner. In addition to the joint constraints, *specified motion trajectory constraints* are often used in the dynamic simulation of railroad vehicle systems. For example, the specified motion trajectory constraints can be used to specify the forward velocity of the vehicle. Other types of geometric constraints are the constraints imposed on the coordinates in Eq. 66 when Euler parameters are used. As previously described in this chapter, if the four Euler parameters are used to define the orientation of a body, one must impose the condition $\sum_{k=0}^{3}\left(\theta_k^i\right)^2 = 1$.

Treatment of the Algebraic Constraint Equations It is clear from the examples presented in this section for the formulation of some widely used mechanical joints that the mathematical description of different constraints imposed on the motion of the system can be systematically developed in terms of the generalized coordinates used. Kinematic constraints produce constraint forces that must be distinguished from applied forces such as gravity, spring, damper, actuator, and aerodynamic forces. Applied forces do not eliminate degrees of freedom regardless of their magnitude. Constraint forces, on the other hand, produce reaction or constraint forces that do not do work, as explained in Chapter 6. As demonstrated by the joint examples presented in this section, the geometric constraint conditions can always be formulated using a set of nonlinear algebraic equations. When these algebraic constraint equations are combined with the second-order ordinary differential equations of the motion of the system, one obtains a system of *differential/algebraic equations* that need to be solved simultaneously using numerical methods.

As discussed in Chapter 6, two different MBS approaches can be used for the treatment of the resulting differential/algebraic equations. In the first approach, referred to in this book as the *augmented formulation*, the equations of motion are formulated in terms of the vector of redundant coordinates $\mathbf{q}$ of Eq. 66. Because these coordinates are not independent in the presence of the algebraic constraint equations, constraint forces appear in the equations of motion. In the augmented formulation, the algebraic constraint equations are combined with the system differential equations of motion to form a large system that has a sparse matrix structure. This system can be efficiently solved for accelerations and constraint forces using *sparse matrix techniques*. In the second approach, referred to in this book as the *embedding technique*, the nonlinear algebraic constraint equations are used to eliminate the dependent variables and write the final form of the differential equations of motion in terms of the independent accelerations. Because the dependent accelerations are eliminated and independent accelerations are not related by any algebraic equations, using this approach allows for systematically eliminating the constraint forces and obtaining a minimum number of equations equal to the number of the system degrees of freedom. The obtained system of equations, which often has a dense and highly nonlinear inertia matrix associated with the independent coordinates, can be solved to determine the independent accelerations, which can be integrated forward in time to determine the independent coordinates and velocities. The dependent variables (coordinates, velocities, and accelerations) can be determined from the independent variables using the algebraic constraint equations. This numerical procedure is consistent with the Lagrange–D'Alembert principle in which a velocity transformation matrix is formulated to write dependent variables in terms of independent variables.

3.13 TRAJECTORY COORDINATES

Using the absolute Cartesian coordinates of Eq. 66 allows for developing general MBS algorithms for railroad vehicle systems. Algorithms based on absolute Cartesian coordinates are easier to generalize for flexible body dynamics as compared to other algorithms specialized for specific applications. Nonetheless, specialized railroad vehicle system algorithms have been developed in the literature based on a set of coordinates called the *trajectory coordinates*. The trajectory coordinates are also used in specialized railroad software designed to study *longitudinal train dynamics* (LTD). LTD software is designed to efficiently solve the equations of motion of trains consisting of a large number of cars. In some LTD software, each car is assumed to have only one degree of freedom that defines the distance traveled by the car along the track.

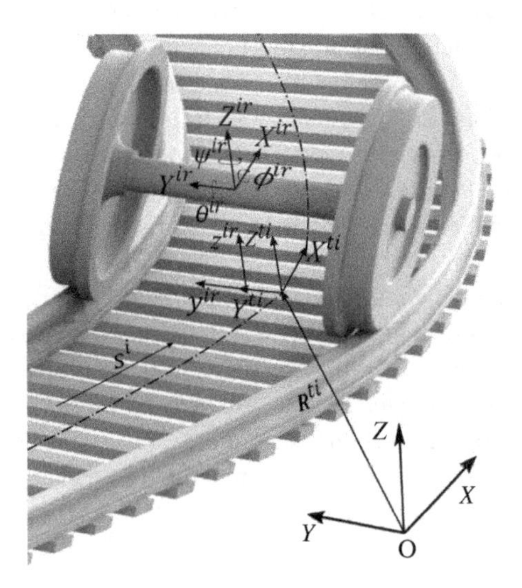

Figure 3.16 Trajectory coordinates.

As discussed in Chapter 6, trajectory coordinates can also be used to formulate the equations of motion of railroad vehicle systems, and they are related to the absolute Cartesian coordinates of Eq. 66 by nonlinear relationships. When trajectory coordinates are used, the configuration of each body like the one shown in Figure 16 can be described using six coordinates: three translational and three rotational. For each body in the system, as shown in Figure 16, three coordinate systems are used to define the body configuration. These three coordinate systems are the following:

- The global coordinate system XYZ is used for all bodies in the system to define global vectors and absolute velocities and accelerations. In this book, the global coordinate system is assumed to be fixed.
- The trajectory body coordinate system $X^{ti}Y^{ti}Z^{ti}$ follows the motion of body i. This trajectory body coordinate system is uniquely defined using the arc length parameter s^i, as previously described in this chapter. Three Euler rotations $\psi^{ti} = \psi^{ti}(s^i)$, $\theta^{ti} = \theta^{ti}(s^i)$, and $\phi^{ti} = \phi^{ti}(s^i)$, respectively, about the Z^{ti}, $-Y^{ti}$, and $-X^{ti}$ axes, respectively, are used to define this coordinate system. This sequence leads to the transformation matrix defined by Eq. 36 and denoted here as $\mathbf{A}^{ti}$. Note that the Euler angles $\psi^{ti} = \psi^{ti}(s^i)$, $\theta^{ti} = \theta^{ti}(s^i)$, and $\phi^{ti} = \phi^{ti}(s^i)$ are not generalized coordinates and can be obtained once the arc length coordinate s^i is known and the track geometry is given. If the track is fixed, the global position of the origin of the trajectory body coordinate system $X^{ti}Y^{ti}Z^{ti}$ is defined by vector $\mathbf{R}^{ti} = \mathbf{R}^{ti}(s^i(t))$.
- The body coordinate system $X^iY^iZ^i$ has an origin rigidly attached to a point on the body, as shown in Figure 16. The orientation of this coordinate system $X^iY^iZ^i$ is defined with respect to the trajectory body coordinate system $X^{ti}Y^{ti}Z^{ti}$ using three Euler angles $\boldsymbol{\theta}^{ir} = \left[\psi^{ir} \quad \phi^{ir} \quad \theta^{ir}\right]^T$, where $\psi^{ir} = \psi^{ir}(t)$ is the yaw angle resulting from rotation about the Z^i

axis, $\phi^{ir} = \phi^{ir}(t)$ is the roll angle resulting from rotation about the X^i axis, and $\theta^{ir} = \theta^{ir}(t)$ is the pitch angle resulting from rotation about the Y^i axis. Therefore, the transformation matrix that defines the orientation of the body $X^iY^iZ^i$ coordinate system with respect to the trajectory body coordinate system is given by Eq. 12.

It is important to understand the difference between the angles $\boldsymbol{\theta}^{ti} = \left[\psi^{ti}\left(s^i\right) \quad \theta^{ti}\left(s^i\right) \quad \phi^{ti}\left(s^i\right)\right]^T$ used to formulate the transformation matrix $\mathbf{A}^{ti}$ that defines the orientation of the trajectory body coordinate system $X^{ti}Y^{ti}Z^{ti}$ and the angles $\boldsymbol{\theta}^{ir} = \left[\psi^{ir} \quad \phi^{ir} \quad \theta^{ir}\right]^T$ used to formulate the transformation matrix $\mathbf{A}^{ir}$ that defines the orientation of the body coordinate system $X^iY^iZ^i$, as previously discussed in this chapter. The angles used for the coordinate system $X^{ti}Y^{ti}Z^{ti}$ describe geometry, while the angles used for the coordinate system $X^iY^iZ^i$ describe motion. The two transformation matrices $\mathbf{A}^{ti}$ and $\mathbf{A}^{ir}$ are reproduced here for convenience (Sanborn et al. 2007; Sinokrot et al. 2008; Shabana et al. 2008):

$$\mathbf{A}^{ti} = \begin{bmatrix} \cos\psi^{ti}\cos\theta^{ti} & -\sin\psi^{ti}\cos\phi^{ti} + \cos\psi^{ti}\sin\theta^{ti}\sin\phi^{ti} & -\sin\psi^{ti}\sin\phi^{ti} - \cos\psi^{ti}\sin\theta^{ti}\cos\phi^{ti} \\ \sin\psi^{ti}\cos\theta^{ti} & \cos\psi^{ti}\cos\phi^{ti} + \sin\psi^{ti}\sin\theta^{ti}\sin\phi^{ti} & \cos\psi^{ti}\sin\phi^{ti} - \sin\psi^{ti}\sin\theta^{ti}\cos\phi^{ti} \\ \sin\theta^{ti} & -\cos\theta^{ti}\sin\phi^{ti} & \cos\theta^{ti}\cos\phi^{ti} \end{bmatrix} \tag{3.73}$$

and

$$\mathbf{A}^{ir} = \begin{bmatrix} \cos\psi^{ir}\cos\theta^{ir} - \sin\psi^{ir}\sin\phi^{ir}\sin\theta^{ir} & -\sin\psi^{ir}\cos\phi^{ir} & \cos\psi^{ir}\sin\theta^{ir} + \sin\psi^{ir}\sin\phi^{ir}\cos\theta^{ir} \\ \sin\psi^{ir}\cos\theta^{ir} + \cos\psi^{ir}\sin\phi^{ir}\sin\theta^{ir} & \cos\psi^{ir}\cos\phi^{ir} & \sin\psi^{ir}\sin\theta^{ir} - \cos\psi^{ir}\sin\phi^{ir}\cos\theta^{ir} \\ -\cos\phi^{ir}\sin\theta^{ir} & \sin\phi^{ir} & \cos\phi^{ir}\cos\theta^{ir} \end{bmatrix} \tag{3.74}$$

Two sets of generalized trajectory coordinates, which are different from the absolute Cartesian coordinates previously used in this chapter, can then defined as follows:

1. The first coordinate is an arc length coordinate $s^i = s^i(t)$ that defines the distance traveled by the body along the track, and the second and third coordinates are the lateral and vertical displacements $y^{ir} = y^{ir}(t)$ and $z^{ir} = z^{ir}(t)$ relative to a trajectory body coordinate system $X^{ti}Y^{ti}Z^{ti}$ that follows the body, as shown in Figure 16. As previously described in this chapter, the trajectory body coordinate system can be uniquely defined in terms of the arc length s^i, using the definition of the track geometry. The coordinates y^{ir} and z^{ir} are relative coordinates and are not absolute even if the track is fixed. In most railroad vehicle system applications, and under normal operating conditions, the two relative coordinates y^{ir} and z^{ir} are normally small.
2. Three relative (not absolute) rotation angles define the body orientation with respect to the trajectory body coordinate system $X^{ti}Y^{ti}Z^{ti}$. The first is a yaw angle $\psi^{ir} = \psi^{ir}(t)$ obtained by a rotation about the Z^i axis of the body coordinate system, the second is a roll angle $\phi^{ir} = \phi^{ir}(t)$ obtained by a rotation about the X^i axis of the body coordinate system, and the third is a pitch angle $\theta^{ir} = \theta^{ir}(t)$ obtained by a rotation about the Y^i axis of the body coordinate system. In most railroad vehicle system applications, and under normal operating conditions, the two relative rotations ψ^{ir} and ϕ^{ir} are normally small, while the pitch angle θ^{ir} can be very large.

The equations of motion of a body can be defined using the translational coordinates s^i, y^{ir}, and z^{ir} and the relative orientation angles $\boldsymbol{\theta}^{ir} = \begin{bmatrix} \psi^{ir} & \phi^{ir} & \theta^{ir} \end{bmatrix}^T$. Knowing the arc length parameter $s^i = s^i(t)$, the location of the origin and the orientation of the trajectory body coordinate system $X^{ti}Y^{ti}Z^{ti}$ that follows the motion of the body can be uniquely defined, as previously discussed in this chapter. Therefore, the vector of the six time-dependent trajectory coordinates of body i is given by (Sanborn et al. 2007; Sinokrot et al. 2008; Shabana et al. 2008)

$$\mathbf{p}^i(t) = \begin{bmatrix} s^i & y^{ir} & z^{ir} & \psi^{ir} & \phi^{ir} & \theta^{ir} \end{bmatrix}^T \tag{3.75}$$

If the rails are assumed fixed, the global position vector of the body reference point (origin of the coordinate system $X^iY^iZ^i$) can be written as

$$\mathbf{R}^i = \mathbf{R}^{ti}\left(s^i\right) + \mathbf{A}^{ti}\left(s^i\right)\bar{\mathbf{u}}^{ir} \tag{3.76}$$

where vector $\bar{\mathbf{u}}^{ir} = \begin{bmatrix} 0 & y^{ir} & z^{ir} \end{bmatrix}^T$ is the position vector of the body reference point with respect to the origin of the trajectory body coordinate system. As previously mentioned, y^{ir} and z^{ir} are, respectively, the lateral and normal components of the position vector of the body reference point with respect to the origin of the trajectory body coordinate system. In writing Eq. 76, it is assumed that the longitudinal component in vector $\bar{\mathbf{u}}^{ir}$ is zero. Equation 76 can be used to define the global position of an arbitrary point on the body, as previously discussed in this chapter, as $\mathbf{r}^i = \mathbf{R}^i + \mathbf{A}^i\bar{\mathbf{u}}^i$, where $\mathbf{A}^i = \mathbf{A}^{ti}\mathbf{A}^{ir}$.

It is clear that using the equation $\mathbf{r}^i = \mathbf{R}^i + \mathbf{A}^i\bar{\mathbf{u}}^i$ with Eq. 76 can make the formulation of the velocity and acceleration equations as well as the equations of motion cumbersome. It is easier, for a given arbitrary body i, to develop the velocity transformation that relates the derivatives of the absolute Cartesian coordinates $\mathbf{q}^i = \begin{bmatrix} \mathbf{R}^{i^T} & \boldsymbol{\theta}^{i^T} \end{bmatrix}^T$ to the derivatives of the trajectory coordinates $\mathbf{p}^i = \begin{bmatrix} s^i & y^{ir} & z^{ir} & \psi^{ir} & \phi^{ir} & \theta^{ir} \end{bmatrix}^T$ and use this velocity transformation with the relatively simple Newton–Euler equations obtained in Chapter 6 to derive the equations of motion in terms of trajectory coordinates. To develop this velocity transformation, Eq. 76 is differentiated with respect to time to yield

$$\dot{\mathbf{R}}^i = \dot{\mathbf{R}}^{ti} + \dot{\mathbf{A}}^{ti}\bar{\mathbf{u}}^{ir} + \mathbf{A}^{ti}\dot{\bar{\mathbf{u}}}^{ir} = \mathbf{L}^i\dot{\mathbf{p}}^i \tag{3.77}$$

where $\mathbf{L}^i = \begin{bmatrix} \left(\left(\partial\mathbf{R}^{ti}/\partial s^i\right) + \left(\partial\mathbf{A}^{ti}/\partial s^i\right)\bar{\mathbf{u}}^{ir}\right) & \mathbf{a}_2^{ti} & \mathbf{a}_3^{ti} & \mathbf{0} \end{bmatrix}$ is a 3×6 matrix in which $\mathbf{a}_2^{ti}$ and $\mathbf{a}_3^{ti}$ are the second and third columns of the transformation matrix $\mathbf{A}^{ti}$ of Eq. 73, and $\mathbf{0}$ is a 3×3 null matrix resulting from the fact that the position vector of Eq. 76 does not depend on the angles $\boldsymbol{\theta}^{ir} = \begin{bmatrix} \psi^{ir} & \phi^{ir} & \theta^{ir} \end{bmatrix}^T$. By differentiating Eq. 77 with respect to time, one obtains

$$\ddot{\mathbf{R}}^i = \mathbf{L}^i\ddot{\mathbf{p}}^i + \boldsymbol{\gamma}_R^i \tag{3.78}$$

where $\boldsymbol{\gamma}_R^i$ is a vector that includes terms that are quadratic in the velocities.

Because angular velocities can be added and treated as vectors, one can write the absolute angular velocity vector of body i as $\boldsymbol{\omega}^i = \boldsymbol{\omega}^{ti} + \boldsymbol{\omega}^{ir}$, which is the sum of the absolute angular velocity $\boldsymbol{\omega}^{ti} = \mathbf{G}^{ti}\dot{\boldsymbol{\theta}}^{ti}$ of the trajectory body coordinate system and $\boldsymbol{\omega}^{ir} = \mathbf{A}^{ti}\mathbf{G}^{ir}\dot{\boldsymbol{\theta}}^{ir}$, which defines the angular velocity of the body with respect to the trajectory body coordinate system. Therefore, the absolute angular velocity $\boldsymbol{\omega}^i$ of body i can be written in terms of the time derivatives of the trajectory coordinates as

$$\boldsymbol{\omega}^i = \mathbf{G}^{ti}\dot{\boldsymbol{\theta}}^{ti} + \mathbf{A}^{ti}\mathbf{G}^{ir}\dot{\boldsymbol{\theta}}^{ir} = \mathbf{H}^i\dot{\mathbf{p}}^i \tag{3.79}$$

where matrices $\mathbf{G}^{ti}$ and $\mathbf{G}^{ir}$ are, respectively, defined as

$$\mathbf{G}^{ti} = \begin{bmatrix} 0 & \sin\psi^{ti} & -\cos\psi^{ti}\cos\theta^{ti} \\ 0 & -\cos\psi^{ti} & -\sin\psi^{ti}\cos\theta^{ti} \\ 1 & 0 & -\sin\theta^{ti} \end{bmatrix}, \quad \mathbf{G}^{ir} = \begin{bmatrix} 0 & \cos\psi^{ir} & -\sin\psi^{ir}\cos\phi^{ir} \\ 0 & \sin\psi^{ir} & \cos\psi^{ir}\cos\phi^{ir} \\ 1 & 0 & \sin\phi^{ir} \end{bmatrix} \tag{3.80}$$

and $\mathbf{H}^i = \left[\left(\mathbf{G}^{ti}\left(\partial\boldsymbol{\theta}^{ti}/\partial s^i\right)\right) \;\; \mathbf{0} \;\; \mathbf{0} \;\; \left(\mathbf{A}^{ti}\mathbf{G}^{ir}\right)\right]$ is a 3×6 matrix in which $\mathbf{0}$ is a three-dimensional zero vector. Differentiating Eq. 79 with respect to time, the absolute angular acceleration vector $\boldsymbol{\alpha}^i$ of body i can be written in terms of the second time derivatives of the trajectory coordinates as

$$\boldsymbol{\alpha}^i = \mathbf{H}^i\ddot{\mathbf{p}}^i + \boldsymbol{\gamma}_\alpha^i \tag{3.81}$$

where $\boldsymbol{\gamma}_\alpha^i$ is a vector that absorbs terms that are quadratic in the velocities (Sanborn et al. 2007; Sinokrot et al. 2008; Shabana et al. 2008). Equation 81 can also be used to write the absolute angular acceleration vector defined in the body coordinate system $X^iY^iZ^i$ as

$$\bar{\boldsymbol{\alpha}}^i = \mathbf{A}^{i^T}\boldsymbol{\alpha}^i = \bar{\mathbf{H}}^i\ddot{\mathbf{p}}^i + \bar{\boldsymbol{\gamma}}_\alpha^i \tag{3.82}$$

where $\bar{\mathbf{H}}^i = \mathbf{A}^{i^T}\mathbf{H}^i = \mathbf{A}^{ir^T}\mathbf{A}^{ti^T}\mathbf{H}^i$ and $\bar{\boldsymbol{\gamma}}_\alpha^i = \mathbf{A}^{i^T}\boldsymbol{\gamma}_\alpha^i$. Combining Eqs. 78 and 82, one obtains (Sanborn et al. 2007; Sinokrot et al. 2008)

$$\begin{bmatrix} \ddot{\mathbf{R}}^i \\ \bar{\boldsymbol{\alpha}}^i \end{bmatrix} = \begin{bmatrix} \mathbf{L}^i \\ \bar{\mathbf{H}}^i \end{bmatrix}\ddot{\mathbf{p}}^i + \begin{bmatrix} \boldsymbol{\gamma}_R^i \\ \bar{\boldsymbol{\gamma}}_\alpha^i \end{bmatrix} \tag{3.83}$$

As demonstrated in Chapter 6, the mass matrix associated with the absolute Cartesian accelerations $\left[\ddot{\mathbf{R}}^{i^T} \;\; \bar{\boldsymbol{\alpha}}^{i^T}\right]^T$ is constant in the Newton–Euler formulation of the equations of motion; therefore, it is preferred to use the angular acceleration vector $\bar{\boldsymbol{\alpha}}^i$ instead of $\boldsymbol{\alpha}^i$. Equation 83 can be used with Newton–Euler equations to obtain the equations of motion in terms of the trajectory coordinates, as discussed in Chapter 6. These equations are obtained without using any small angle assumptions; therefore, they include all the nonlinear terms that can be significant in some railroad-vehicle motion scenarios.

Chapter 4

RAILROAD GEOMETRY

Wheel and rail geometry has a significant influence on the dynamic behavior of railroad vehicle systems. Consequently, an accurate description of the geometry is necessary in order to correctly predict wheel/rail contact forces. These forces have a significant effect not only on vehicle dynamics and stability but also on the integrity of the track structure. The geometry, which enters into formulating the normal and tangential wheel/rail contact forces, can be described using the theories of curves and surfaces introduced in Chapter 2. For example, to determine the dimensions of the wheel/rail contact area, the *principal curvatures* and *principal directions* of the wheel and rail surfaces in the contact region must be evaluated. While the wheel geometry can be described using a surface of revolution, the rail geometry can be defined by extruding the rail profile in the rail longitudinal direction. More complex surface geometries can be defined using numerical approximation methods to capture details that cannot be captured using analytical techniques that are more suited for simple or idealized geometries.

The geometry of an unworn wheel can be described as a surface of revolution by rotating the profile curve about the wheel axis. While the profile can assume any shape, conical wheels are often used to improve vehicle stability and avoid derailments. Different profiles with different conicity values are used depending on the type of vehicle, speed of operation, and loading conditions. The functions that define the profiles are not simple straight-line functions, and in most practical applications, a numerical description of the profile function is required for accurate computer modeling and virtual prototyping.

Developing an accurate description of track geometry is one of the basic steps in formulating railroad-vehicle nonlinear dynamic equations of motion and the numerical solution of these equations. For the most part, tracks are constructed using three segment types with different geometries: tangent, curve, and spiral segments. A *tangent segment* is a straight section of track with zero curvature, which corresponds to a radius of curvature equal to infinity. A *curve segment* is a circular section of track that has a constant radius of curvature and constant curvature. To connect tangent and curve segments or two curves with different curvature values, a *spiral segment* is used. When a spiral segment is used to connect a tangent section and a curve section, for example, the geometry of the spiral segment is designed such that the spiral has zero curvature at the end connected to the tangent segment and the value of the curve curvature at the end connected to the curve segment. This spiral design allows for smoothly varying the curvature and ensures smooth operation of rail vehicles during the transition between tangent and curve sections and vice versa.

Mathematical Foundation of Railroad Vehicle Systems: Geometry and Mechanics,
First Edition. Ahmed A. Shabana.

When a railroad vehicle negotiates a curve or spiral section, the forces exerted on the vehicle can be significantly different from the forces that arise when the vehicle negotiates a straight segment. During curve and spiral negotiations, centrifugal forces must be taken into account in order to ensure safe operation of the rail vehicle. To avoid derailments, the magnitude of the centrifugal forces is used to put a limit on the vehicle speed; this limit is referred to as the *balance speed*. For a curve that has a constant radius of curvature, the variations in the centrifugal forces acting on the vehicle are not as significant as for a spiral, which has variable curvature. When constructing a track, the geometry of spiral sections is often designed to have a linearly varying curvature.

This chapter discusses the geometries of wheel and rail surfaces, which are fundamental in defining the kinematic and dynamic equations of railroad vehicles. The wheel and rail surface equations are defined in terms of two independent surface parameters, as explained in Chapter 2. These surface parameters, which define the surface equations in their parametric form, can be used to determine the locations of arbitrary points on wheel and rail surfaces. The parametric surface equations are necessary for developing wheel/rail contact conditions, which are used to determine online wheel/rail contact points. This chapter discusses two approaches for describing rail geometry: the *semi-analytical approach* and the *absolute nodal coordinate formulation (ANCF) interpolation approach*. The semi-analytical approach has two main disadvantages. The first disadvantage is the need to evaluate the derivatives of angles with respect to the rail longitudinal surface parameter in order to determine the tangent, normal, and curvature vectors. The second disadvantage is the low order of interpolation used to determine the position coordinates of arbitrary points on the rail with specific output from the *track preprocessor*, which is designed to generate the data required at discrete nodal points for use in nonlinear dynamic simulations of railroad vehicles. The ANCF interpolation approach is preferred because it does not have these disadvantages: it does not require differentiation of angles, and it allows for higher-order interpolation for the position coordinates of arbitrary points on the rail space curve.

4.1 WHEEL SURFACE GEOMETRY

In the case of unworn wheels, the wheel surface can be considered a surface of revolution generated by rotating the profile curve about the wheel axis, as shown in Figure 1.

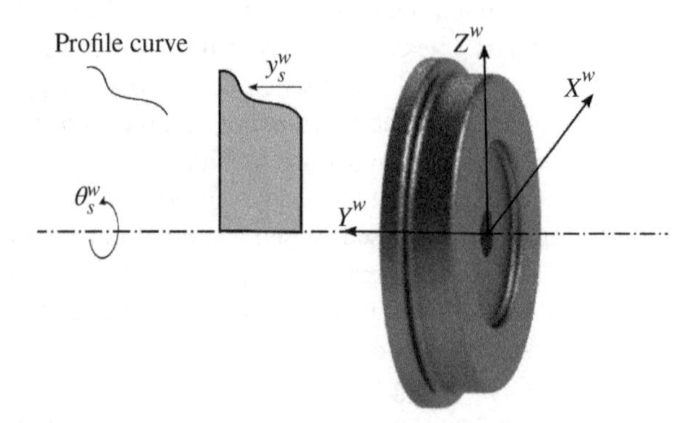

Figure 4.1 Wheel profile curve.

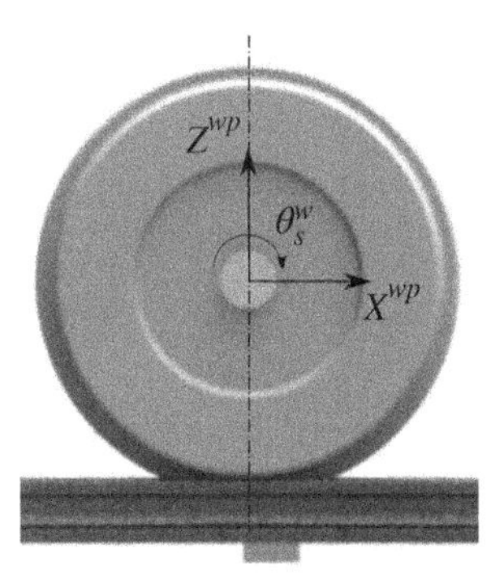

Figure 4.2 Wheel profile frame.

As explained in Chapter 2, the locations of points on a surface can be expressed in terms of two surface parameters. In this chapter, the wheel surface is assumed to have the parameterization $s_1^w = y_s^w$ and $s_2^w = \theta_s^w$, as shown in Figure 1. To develop a mathematical definition of the wheel surface, a *profile frame* $X^{wp}Y^{wp}Z^{wp}$ is first introduced, as shown in Figure 2, for the convenience of defining the profile curve. As shown in the figure, the angular surface parameter θ_s^w is measured from the Z^{wp} axis. The position of the origin of the profile frame with respect to the wheel coordinate system $X^wY^wZ^w$ is defined by the Cartesian coordinates x_o^{wp}, y_o^{wp}, and z_o^{wp}. The Y^w axis is assumed to coincide with the wheel axis of rotation. In the case of a rigid wheelset, the wheelset coordinate system $X^wY^wZ^w$ is located at the center of mass of the wheelset. The negative sign is used in the third element of the vector of Eq. 1 to associate the contact point in the initial configuration with the zero value of the radial surface parameter s_2^w. A profile function g^{wp} is used to define the wheel profile curve in the wheel profile frame. In the case of *unworn wheels*, profile function g^{wp} does not depend on the wheel angular surface parameter $s_2^w = \theta_s^w$: in this special case, one has $g^{wp} = g^{wp}(s_1^w)$. In this case of unworn wheels, the coordinates of a point on the wheel surface that comes into contact with the rail can be defined mathematically in the selected wheel coordinate system $X^wY^wZ^w$ in terms of surface parameters s_1^w and s_2^w as

$$\bar{\mathbf{u}}^w(s_1^w, s_2^w) = \bar{\mathbf{R}}^{wp} + \bar{\mathbf{u}}^{wp} = \begin{bmatrix} x_o^{wp} + g^{wp}(s_1^w)\sin s_2^w \\ y_o^{wp} + s_1^w \\ z_o^{wp} - g^{wp}(s_1^w)\cos s_2^w \end{bmatrix} \tag{4.1}$$

where $\bar{\mathbf{R}}^{wp} = [x_o^{wp} \ \ y_o^{wp} \ \ z_o^{wp}]^T$ is the vector that defines the origin of the wheel profile frame $X^{wp}Y^{wp}Z^{wp}$ with respect to the coordinate system $X^wY^wZ^w$ of the wheel or wheelset, and $\bar{\mathbf{u}}^{wp} = \left[g^{wp}(s_1^w)\sin s_2^w \ \ s_1^w \ \ -g^{wp}(s_1^w)\cos s_2^w\right]^T$ is the vector that defines the location of the point in the profile frame. For a single wheel, the coordinates x_o^{wp}, y_o^{wp}, and z_o^{wp} define the position of the origin of the profile frame $X^{wp}Y^{wp}Z^{wp}$ in the wheel coordinate system $X^wY^wZ^w$. For a rigid wheelset that has two wheels rigidly connected by an axle, the coordinate systems of the rigidly connected right and left wheels can be assumed the same and have origins located at the axle center point, as shown in Figure 3; in this case, $x_o^{wp} = z_o^{wp} = 0$, and y_o^{wp} can be selected to be half the distance between the centers of the two wheels. Using the coordinates x_o^{wp}, y_o^{wp}, and z_o^{wp} of the origin of the wheel profile frame allows the systematic generalization of the geometric description presented in this section to the case of a deformable wheel axle or independent non-rigidly connected wheels.

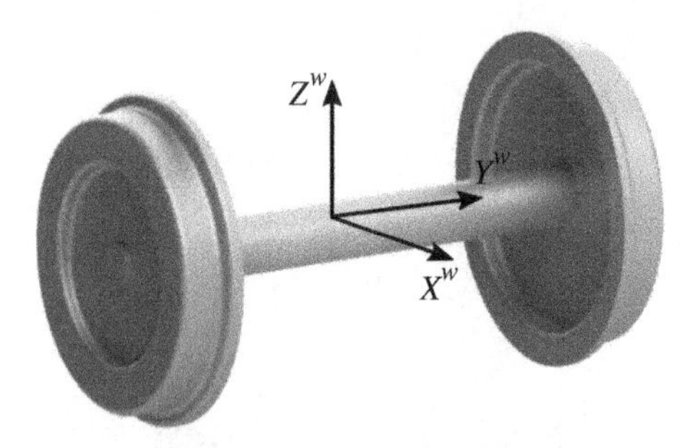

Figure 4.3 Wheelset coordinate system.

Profile Geometry and Wheel Flanges Wheel profiles are designed to have conical shapes to improve stability and steering. In practice, the conical profile is not an exact straight line, but a curve generated by combinations of arcs. Flanges are added to wheels to restrict lateral displacements and prevent derailments. Table 1 shows details of wheel sections, profile terminologies, and an example of wheel profiles that have been used in practice. The profile can be measured using a device known as a *MiniProf,* which generates $y-z$ numerical data that can be used with a cubic spline function representation to define the wheel profile function numerically: $g^{wp} = g^{wp}(s_1^w)$.

Different wheel designs with different profile conicity are used in practice (Cummings 2018). An example is the AAR 1:20 profile, which was widely used before 1990 and which represented a significant improvement over the cylindrical profile. This wheel design has a flange angle of 70°, flange fillet radius of 0.75 in., and tread taper of 1:20. The AAR 1:20 wheel design was altered to obtain the AAR-1 design that has been used since 1990. The AAR-1 wheel was initially introduced in the 1980s based on measurements of worn wheel profiles. It demonstrated better performance during curve negotiations and has improved rolling resistance at the expense of lower critical hunting speed: 49 mph compared to 70 mph for the AAR 1:20 wheel. Revised versions, AAR-1A and AAR-1B, were introduced later to improve stability characteristics and wear resistance. The AAR-1 wheel has a 75° flange angle; and because its design is based on worn-wheel measurements, it has multiple flange fillet radii. Like the AAR 1:20, AAR-1 designs have a 1:20 tread taper (Cummings 2018).

Another design based on worn-wheel geometry, which was introduced later, is the AAR-2A design, which has a 75° flange angle, a 1:20 tread taper, and multiple flange fillet radii. The AAR-2A design, which was originally called TTCI-1A and later SRI-1A, offers improved curving performance. As reported by Cummings (2018), the final version of this design reduces the flange thickness by 1/8 in. and is available with a wide flange (1.25 in.) or narrow flange (1.15). Several other profile designs were also presented in a report produced by the American Public Transportation Association (APTA Press Task Force 2007).

In general, profile dimensions are measured relative to a point on the profile called the *tread datum position*, shown in Table 1. The tread datum position defines a circle on the tread or a tread cross section, which is 70 mm from the flange back face. For example, the flange height, shown in Table 1, is defined as the difference between the wheel radius at the flange tip and the wheel radius at the tread datum position. In addition to using the tread datum position to define the dimensions of the profile, profile gauges

Table 4.1 Wheel geometry.

Wheel structure

Hub
Tire
Disc

- **Hub**, the center portion of the wheel, is used to mount the wheel on the axle.
- **Disc** is the section of the wheel that connects the hub to the tire.
- **Tire** is the section of the wheel that has the wheel profile, which contacts the rail.

Tire details

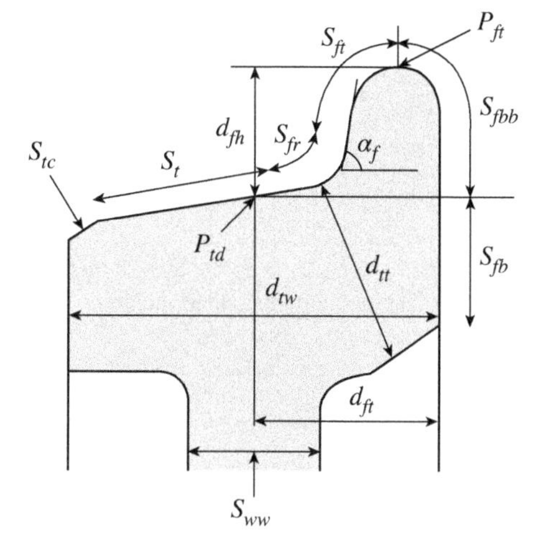

Sections

S_t is the *tread.*

S_{fr} is the *flange root.*

S_{ft} is the *flange toe.*

S_{fbb} is the *flange back blend.*

S_{fb} is the *flange back.*

S_{tc} is the *tread chamfer.*

S_{ww} is the *wheel web.*

Points

P_{ft} is the *flange tip.*

P_{td} is the *tread datum.*

Dimensions

d_{fh} is the *flange height.*

d_{tw} is the *tire/rim width.*

d_{tt} is the *throat thickness.*

d_{ft} is the *flange-back/tread-datum distance* (normally 70 mm).

Angles

α_f is the *flange angle.*

Example of basic dimensions (Riftek Sensors and Instruments 2008)

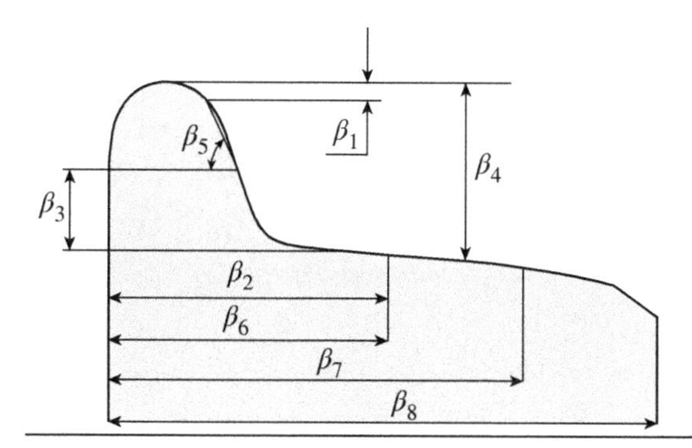

(Continued)

Table 4.1 (Continued)

Parameter	Locomotive	Rolling stock coach	Description
β_1	2 mm	5 mm	Used to calculate flange slope
β_2	70 mm	70 mm	Defines wheel rolling circle position
β_3	13 mm	18 mm	Used to calculate flange thickness
β_4	30 mm	28 mm	Used to calculate tire roll wear = height of reference profile flange
β_5	—	60°	Reference profile slope
β_6	70	70	Used to calculate rolling surface section slope
β_7	105	105	Used to calculate rolling surface section slope
β_8	140	140	Used to calculate tire/rim width

Profile example (APTA Press Task Force 2007)

APTA 120 wheel profile (based on former AAR S-621-79, 1:20 taper; dimensions in inches)

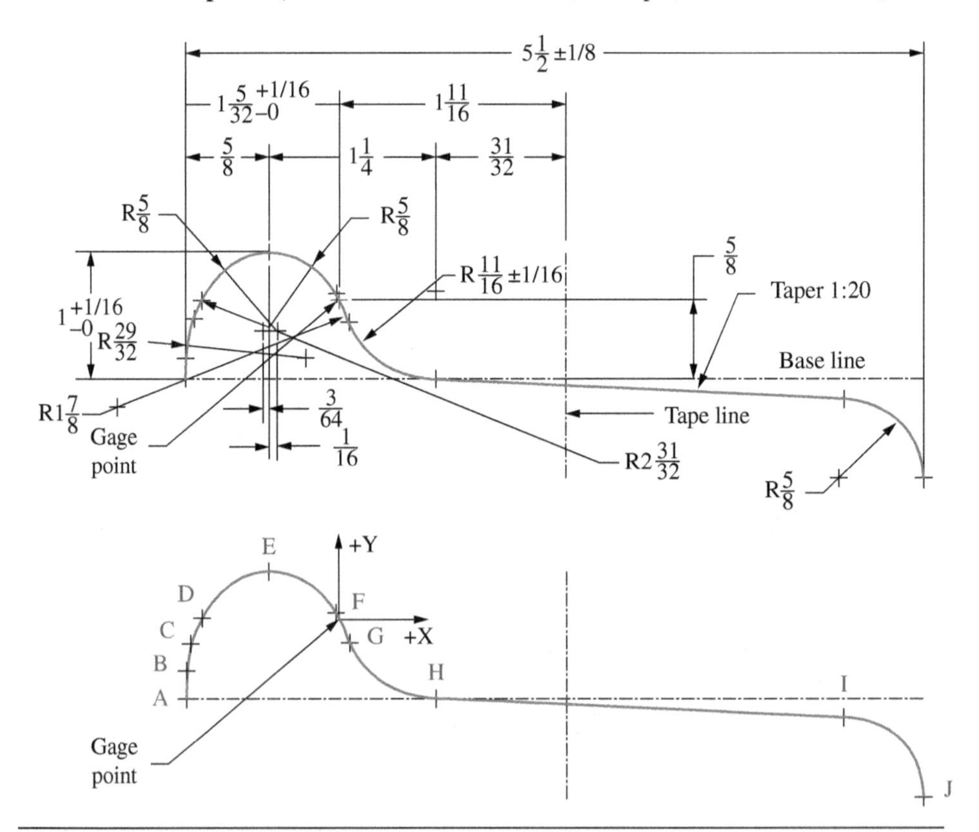

Table 4.1 (Continued)

	Node Coordinates	
Point	X	Y
A	−1.1563	−0.6250
B	−1.1562	−0.4583
C	−1.1013	−0.1476
D	−1.0438	−0.0019
E	−0.5313	0.3750
F	−0.0270	0.0465
Gage Point	0.0000	0.0000
G	0.0751	−0.1790
H	0.7188	−0.6250
I	3.7500	−0.7766
J	4.3438	−1.4008

	Segment Details		
Segement	Radius *Line*	X – Center	Y – Center
A–B	*Line* 90°		
B–C	0.9063	−0.2500	−0.4583
C–D	2.9688	1.6875	−1.1654
D–E	0.6250	−0.4688	−0.2469
E–F	0.6250	−0.5781	−0.2482
F–G	1.8750	−1.6805	−0.8376
G–H	0.6875	0.7188	0.0625
H–I	*Line* 1:20		
I–J	0.6250	3.7188	−1.4008
Beyond J	*Line* 90°		

also use this position as a reference point. Another important dimension that is used to ensure wheel strength is *throat thickness*. In determining this dimension, which depends on the size of the wheel bearings, wheel/rail forces and axle loads must be considered. The throat thickness is the distance between the root of the tread profile radius at the flange root and the inner side of the flange-side wheel rim, measured at the narrowest point (British Railway Board 1996). The minimum value for the throat thickness, which depends on the type of wheel and its use, is in the range 30–50 mm. While more examples of profiles and information on the wheel geometry and various wheel designs can be found in the literature, Table 1 shows the complexity of the wheel profile geometry, which is constructed using circular arcs. Because developing an analytical representation of the wheel profile geometry can be difficult, a numerical description based on spline functions is often used in the virtual prototyping of railroad vehicle systems.

Worn-Wheel Geometry For worn wheels, the wheel surface geometry can be determined from measurements. Point clouds can be created and used to define the profile function g^{wp} numerically at tabulated data points for different values of the surface parameters $s_1^w = y_s^w$ and $s_2^w = \theta_s^w$. With the advancement in scanning and imaging techniques and computer technology, accurate point-cloud data can be obtained and used to define the geometry of worn wheels. The profile function depends on the two parameters s_1^w and s_2^w – that is, $g^{wp} = g^{wp}(s_1^w, s_2^w)$ – and Eq. 1 can be replaced by the following more general equation:

$$\bar{\mathbf{u}}^w(s_1^w, s_2^w) = \overline{\mathbf{R}}^{wp} + \bar{\mathbf{u}}^{wp} = \begin{bmatrix} x_o^{wp} + g^{wp}(s_1^w, s_2^w)\sin s_2^w \\ y_o^{wp} + s_1^w \\ z_o^{wp} - g^{wp}(s_1^w, s_2^w)\cos s_2^w \end{bmatrix} \tag{4.2}$$

In this equation, $\overline{\mathbf{R}}^{wp} = \begin{bmatrix} x_o^{wp} & y_o^{wp} & z_o^{wp} \end{bmatrix}^T$ is the same as previously defined and is the vector that defines the origin of the wheel profile frame $X^{wp}Y^{wp}Z^{wp}$ with respect to the coordinate system $X^wY^wZ^w$ of the wheel or wheelset, and $\bar{\mathbf{u}}^{wp} = \begin{bmatrix} g^{wp}(s_1^w, s_2^w)\sin s_2^w & s_1^w & -g^{wp}(s_1^w, s_2^w)\cos s_2^w \end{bmatrix}^T$ is the vector that defines the location of an arbitrary point on the profile frame that may come into contact with the rail surface. Clearly, the case of worn wheels cannot be considered the simple case of a surface of revolution as can be assumed for unworn wheels. When numerical methods are used to describe the profile function $g^{wp}(s_1^w, s_2^w)$, the value of this function at an arbitrary point on the wheel surface can be determined using interpolations based on tabulated data points obtained from measurements and/or scanning and imaging techniques.

Geometric tolerances are defined for new wheel profiles to estimate the level of wear of wheels in service (British Railway Board, 1996). Profiles that do not meet these profile tolerances can lead to vehicle instability and derailments and can also result in significantly increased wear. Specific measures are used in practice to define the profile wear, including the *tread run-out*, which defines the change in the radius at a tread position in one complete revolution about the wheel axle. The *wear allowance* is defined by measuring the difference between the as-machined and fully worn flange height (British Railway Board 1996). In general, wear allowance is defined as the reduction allowed in the wheel radius after profiling. While wear allowance can be in the range of 6.5 mm, tighter limits may be adopted depending on wheel usage, loads, and operating speeds.

4.2 WHEEL CURVATURES AND GLOBAL VECTORS

It is clear from Eqs. 1 and 2 that only one function g^{wp} is needed to define the coordinates of arbitrary points on the wheel surface that come into contact with the rail. For unworn wheels, this function depends only on the lateral surface parameter $s_1^w = y_s^w$: that is, $g^{wp} = g^{wp}(s_1^w)$. For worn wheels, g^{wp} depends on the lateral surface parameter $s_1^w = y_s^w$ as well as the angular surface parameter $s_2^w = \theta_s^w$: that is, $g^{wp} = g^{wp}(s_1^w, s_2^w)$. As previously mentioned, the profile function g^{wp} can be defined analytically or numerically using interpolation methods, including spline functions. Such a numerical approach offers generality and allows for using measured data.

To formulate the *contact conditions* used to determine the locations of contact points on the wheel, tangent and normal vectors must be evaluated. These tangent and normal vectors are also required in order to define normal and tangential contact forces. By knowing the profile function g^{wp}, the tangent and normal vectors at an arbitrary point on the wheel surface can be determined, as discussed in Chapter 2. The two tangent vectors $\bar{\mathbf{t}}_1^w$ and $\bar{\mathbf{t}}_2^w$ define the tangent plane and also define the normal vector $\bar{\mathbf{n}}^w$ in the wheel or wheelset coordinate system. These tangent and normal vectors are defined, respectively, in the wheel or wheelset coordinate system $X^wY^wZ^w$ as

$$\bar{\mathbf{t}}_1^w = \frac{\partial \bar{\mathbf{u}}^w}{\partial s_1^w}, \qquad \bar{\mathbf{t}}_2^w = \frac{\partial \bar{\mathbf{u}}^w}{\partial s_2^w}, \qquad \bar{\mathbf{n}}^w = \bar{\mathbf{t}}_1^w \times \bar{\mathbf{t}}_2^w \tag{4.3}$$

where vector $\bar{\mathbf{u}}^w$ of Eq. 1 or 2 defines the location of an arbitrary point on the wheel surface with respect to the wheel or wheelset coordinate system. If the unit normal vector must be computed, it is defined as $\hat{\bar{\mathbf{n}}}^w = (\bar{\mathbf{t}}_1^w \times \bar{\mathbf{t}}_2^w)/|\bar{\mathbf{t}}_1^w \times \bar{\mathbf{t}}_2^w|$.

Higher Derivatives As discussed in later chapters of this book, it may be necessary to compute higher derivatives with respect to the wheel surface parameters, particularly when the contact conditions are solved numerically using the iterative Newton–Raphson procedure. In this case, one needs to evaluate the curvature vectors as well as the derivative of the normal vector. The curvature vectors at an arbitrary point on the wheel surface can be written as

$$\left.\begin{aligned}\partial^2\bar{\mathbf{u}}^w/\partial(s_1^w)^2 &= \partial\bar{\mathbf{t}}_1^w/\partial s_1^w\\ \partial^2\bar{\mathbf{u}}^w/\partial(s_2^w)^2 &= \partial\bar{\mathbf{t}}_2^w/\partial s_2^w\\ \partial^2\bar{\mathbf{u}}^w/\partial s_1^w\partial s_2^w &= \partial\bar{\mathbf{t}}_1^w/\partial s_2^w = \partial\bar{\mathbf{t}}_2^w/\partial s_1^w\end{aligned}\right\} \tag{4.4}$$

The derivatives of the normal vector with respect to the surface parameters can be determined using the derivatives of the two tangent vectors as

$$\left.\begin{aligned}\partial\bar{\mathbf{n}}^w/\partial s_1^w &= (\partial\bar{\mathbf{t}}_1^w/\partial s_1^w)\times\bar{\mathbf{t}}_2^w + \bar{\mathbf{t}}_1^w\times(\partial\bar{\mathbf{t}}_2^w/\partial s_1^w)\\ \partial\bar{\mathbf{n}}^w/\partial s_2^w &= (\partial\bar{\mathbf{t}}_1^w/\partial s_2^w)\times\bar{\mathbf{t}}_2^w + \bar{\mathbf{t}}_1^w\times(\partial\bar{\mathbf{t}}_2^w/\partial s_2^w)\end{aligned}\right\} \tag{4.5}$$

It is clear from the preceding two equations that the derivatives of the normal vector are functions of the curvature vectors. When using the unit normal vector $\hat{\bar{\mathbf{n}}}^w = (\bar{\mathbf{t}}_1^w\times\bar{\mathbf{t}}_2^w)/|\bar{\mathbf{t}}_1^w\times\bar{\mathbf{t}}_2^w|$, the derivatives can also be evaluated in closed form. For example, if $\hat{\mathbf{a}}$ is a unit vector, one has the identity $\partial\hat{\mathbf{a}}/\partial\mathbf{a} = (\mathbf{I}-\hat{\mathbf{a}}\hat{\mathbf{a}}^T)/|\mathbf{a}|$, where $\hat{\mathbf{a}}\hat{\mathbf{a}}^T$ is the outer or dyadic product. It follows that $\partial\hat{\mathbf{a}}/\partial s = (\partial\hat{\mathbf{a}}/\partial\mathbf{a})(\partial\mathbf{a}/\partial s)$: that is, $\partial\hat{\mathbf{a}}/\partial s = ((\mathbf{I}-\hat{\mathbf{a}}\hat{\mathbf{a}}^T)/|\mathbf{a}|)(\partial\mathbf{a}/\partial s)$ for any parameter s. Using this identity, one can write the derivative of the unit normal to the wheel surface as

$$\partial\hat{\bar{\mathbf{n}}}^w/\partial s_k^w = ((\mathbf{I}-\hat{\bar{\mathbf{n}}}^w\hat{\bar{\mathbf{n}}}^{w^T})/|\bar{\mathbf{n}}^w|)(\partial\bar{\mathbf{n}}^w/\partial s_k^w),\quad k=1,2 \tag{4.6}$$

In this equation, $\partial\bar{\mathbf{n}}^w/\partial s_k^w, k=1,2$, can be determined using Eq. 5. Therefore, the derivatives of the unit normal to the wheel surface can be systematically evaluated using the tangent vectors and their derivatives. The numerical approach based on interpolations allows using the equations developed in this section, as discussed in later chapters.

Table 2 shows a summary of the wheel surface position and derivative equations in the cases of unworn and worn wheels where $g^{wp} = g^{wp}(s_1^w)$ and $g^{wp} = g^{wp}(s_1^w, s_2^w)$, respectively. The expressions presented in this table are used in the numerical implementation of the formulations presented in this book. These expressions show how position and derivative functions vary as a function of the profile function g^{wp} and its derivatives.

Wheel Global Vectors Using the equation for the position vector of an arbitrary point with respect to the wheel coordinate system $X^wY^wZ^w$, the position vector $\mathbf{r}^w$ of arbitrary points can be defined in the global coordinate system XYZ, as explained in the preceding chapter, as

$$\mathbf{r}^w = \mathbf{R}^w + \mathbf{A}^w\bar{\mathbf{u}}^w = \mathbf{R}^w + \mathbf{A}^w(\bar{\mathbf{R}}^{wp}+\bar{\mathbf{u}}^{wp}) \tag{4.7}$$

where $\mathbf{R}^w = \mathbf{R}^w(t)$ is the global position vector of the origin of the wheel coordinate system $X^wY^wZ^w$, $\mathbf{A}^w$ is the transformation matrix that defines the orientation of the wheel coordinate system, and t is time. Because $\mathbf{R}^w$ and $\mathbf{A}^w$ do not depend on the wheel surface parameters, one can show that the tangent and normal vectors defined in the global coordinate system XYZ are given by

$$\left.\begin{aligned}\mathbf{t}_1^w &= \frac{\partial\mathbf{r}^w}{\partial s_1^w} = \mathbf{A}^w\left(\frac{\partial\bar{\mathbf{u}}^w}{\partial s_1^w}\right) = \mathbf{A}^w\bar{\mathbf{t}}_1^w,\\ \mathbf{t}_2^w &= \frac{\partial\mathbf{r}^w}{\partial s_2^w} = \mathbf{A}^w\left(\frac{\partial\bar{\mathbf{u}}^w}{\partial s_2^w}\right) = \mathbf{A}^w\bar{\mathbf{t}}_2^w,\\ \mathbf{n}^w &= \mathbf{t}_1^w\times\mathbf{t}_2^w = (\mathbf{A}^w\bar{\mathbf{t}}_1^w)\times(\mathbf{A}^w\bar{\mathbf{t}}_2^w) = \mathbf{A}^w(\bar{\mathbf{t}}_1^w\times\bar{\mathbf{t}}_2^w) = \mathbf{A}^w\bar{\mathbf{n}}^w\end{aligned}\right\} \tag{4.8}$$

Table 4.2 Summary of the position and derivative equations of the wheel surface.

$g^{wp} = g^{wp}(s_1^w)$	$g^{wp} = g^{wp}(s_1^w, s_2^w)$
$\bar{\mathbf{u}}^w(s_1^w, s_2^w) = \begin{bmatrix} x_o^{wp} + g^{wp} \sin s_2^w \\ y_o^{wp} + s_1^w \\ z_o^{wp} - g^{wp} \cos s_2^w \end{bmatrix}$	$\bar{\mathbf{u}}^w(s_1^w, s_2^w) = \begin{bmatrix} x_o^{wp} + g^{wp}(s_1^w, s_2^w) \sin s_2^w \\ y_o^{wp} + s_1^w \\ z_o^{wp} - g^{wp}(s_1^w, s_2^w) \cos s_2^w \end{bmatrix}$
$\bar{\mathbf{t}}_1^w = \dfrac{\partial \bar{\mathbf{u}}^w}{\partial s_1^w} = \begin{bmatrix} (\partial g^{wp}/\partial s_1^w) \sin s_2^w \\ 1 \\ -(\partial g^{wp}/\partial s_1^w) \cos s_2^w \end{bmatrix}$	$\bar{\mathbf{t}}_1^w = \dfrac{\partial \bar{\mathbf{u}}^w}{\partial s_1^w} = \begin{bmatrix} (\partial g^{wp}/\partial s_1^w) \sin s_2^w \\ 1 \\ -(\partial g^{wp}/\partial s_1^w) \cos s_2^w \end{bmatrix}$
$\bar{\mathbf{t}}_2^w = \dfrac{\partial \bar{\mathbf{u}}^w}{\partial s_2^w} = \begin{bmatrix} g^{wp} \cos s_2^w \\ 0 \\ g^{wp} \sin s_2^w \end{bmatrix}$	$\bar{\mathbf{t}}_2^w = \dfrac{\partial \bar{\mathbf{u}}^w}{\partial s_2^w} = \begin{bmatrix} (\partial g^{wp}/\partial s_2^w) \sin s_2^w + g^{wp} \cos s_2^w \\ 0 \\ -(\partial g^{wp}/\partial s_2^w) \cos s_2^w + g^{wp} \sin s_2^w \end{bmatrix}$
$\bar{\mathbf{n}}^w = \bar{\mathbf{t}}_1^w \times \bar{\mathbf{t}}_2^w = \begin{bmatrix} g^{wp} \sin s_2^w \\ -g^{wp}(\partial g^{wp}/\partial s_1^w) \\ -g^{wp} \cos s_2^w \end{bmatrix}$	$\bar{\mathbf{n}}^w = \bar{\mathbf{t}}_1^w \times \bar{\mathbf{t}}_2^w = \begin{bmatrix} -(\partial g^{wp}/\partial s_2^w) \cos s_2^w + g^{wp} \sin s_2^w \\ -g^{wp}(\partial g^{wp}/\partial s_1^w) \\ -[(\partial g^{wp}/\partial s_2^w) \sin s_2^w + g^{wp} \cos s_2^w] \end{bmatrix}$
$\hat{\bar{\mathbf{n}}}^w = \dfrac{\bar{\mathbf{n}}^w}{\lvert \bar{\mathbf{n}}^w \rvert} = \dfrac{1}{\lvert \bar{\mathbf{n}}^w \rvert} \begin{bmatrix} g^{wp} \sin s_2^w \\ -g^{wp}(\partial g^{wp}/\partial s_1^w) \\ -g^{wp} \cos s_2^w \end{bmatrix},$	$\hat{\bar{\mathbf{n}}}^w = \dfrac{\bar{\mathbf{n}}^w}{\lvert \bar{\mathbf{n}}^w \rvert} = \begin{bmatrix} -(\partial g^{wp}/\partial s_2^w) \cos s_2^w + g^{wp} \sin s_2^w \\ -g^{wp}(\partial g^{wp}/\partial s_1^w) \\ -[(\partial g^{wp}/\partial s_2^w) \sin s_2^w + g^{wp} \cos s_2^w] \end{bmatrix},$
$\lvert \bar{\mathbf{n}}^w \rvert = g^{wp} \sqrt{1 + (\partial g^{wp}/\partial s_1^w)^2}$	$\lvert \bar{\mathbf{n}}^w \rvert = \sqrt{(g^{wp})^2 + (\partial g^{wp}/\partial s_2^w)^2 + (g^{wp}(\partial g^{wp}/\partial s_1^w))^2}$
$\partial \bar{\mathbf{t}}_1^w/\partial s_1^w = \partial^2 \bar{\mathbf{u}}^w/\partial (s_1^w)^2 = \begin{bmatrix} (\partial^2 g^{wp}/\partial (s_1^w)^2) \sin s_2^w \\ 0 \\ -(\partial^2 g^{wp}/\partial (s_1^w)^2) \cos s_2^w \end{bmatrix}$	$\partial \bar{\mathbf{t}}_1^w/\partial s_1^w = \partial^2 \bar{\mathbf{u}}^w/\partial (s_1^w)^2 = \begin{bmatrix} (\partial^2 g^{wp}/\partial (s_1^w)^2) \sin s_2^w \\ 0 \\ -(\partial^2 g^{wp}/\partial (s_1^w)^2) \cos s_2^w \end{bmatrix}$
$\partial \bar{\mathbf{t}}_2^w/\partial s_2^w = \partial^2 \bar{\mathbf{u}}^w/\partial (s_2^w)^2 = \begin{bmatrix} -g^{wp} \sin s_2^w \\ 0 \\ g^{wp} \cos s_2^w \end{bmatrix}$	$\partial \bar{\mathbf{t}}_2^w/\partial s_2^w = \partial^2 \bar{\mathbf{u}}^w/\partial (s_2^w)^2 = \begin{bmatrix} (\partial^2 g^{wp}/\partial (s_2^w)^2) \sin s_2^w + 2(\partial g^{wp}/\partial s_2^w) \cos s_2^w - g^{wp} \sin s_2^w \\ 0 \\ -(\partial^2 g^{wp}/\partial (s_2^w)^2) \cos s_2^w + 2(\partial g^{wp}/\partial s_2^w) \sin s_2^w + g^{wp} \cos s_2^w \end{bmatrix}$
$\partial \bar{\mathbf{t}}_1^w/\partial s_2^w = \partial \bar{\mathbf{t}}_2^w/\partial s_1^w = \partial^2 \bar{\mathbf{u}}^w/\partial s_1^w \partial s_2^w = \begin{bmatrix} (\partial g^{wp}/\partial s_1^w) \cos s_2^w \\ 0 \\ (\partial g^{wp}/\partial s_1^w) \sin s_2^w \end{bmatrix}$	$\partial \bar{\mathbf{t}}_1^w/\partial s_2^w = \partial \bar{\mathbf{t}}_2^w/\partial s_1^w = \partial^2 \bar{\mathbf{u}}^w/\partial s_1^w \partial s_2^w = \begin{bmatrix} (\partial g^{wp}/\partial s_1^w) \cos s_2^w + (\partial^2 g^{wp}/\partial s_1^w \partial s_2^w) \sin s_2^w \\ 0 \\ (\partial g^{wp}/\partial s_1^w) \sin s_2^w - (\partial^2 g^{wp}/\partial s_1^w \partial s_2^w) \cos s_2^w \end{bmatrix}$

Table 4.2 (Continued)

$\partial\bar{\mathbf{n}}^w/\partial s_1^w = (\partial\bar{\mathbf{t}}_1^w/\partial s_1^w)\times\bar{\mathbf{t}}_2^w + \bar{\mathbf{t}}_1^w\times(\partial\bar{\mathbf{t}}_2^w/\partial s_1^w) = \begin{bmatrix} (\partial g^{wp}/\partial s_1^w)\sin s_2^w \\ -(\partial g^{wp}/\partial s_1^w)^2 - g^{wp}(\partial^2 g^{wp}/\partial(s_1^w)^2) \\ -(\partial g^{wp}/\partial s_1^w)\cos s_2^w \end{bmatrix}$	$\partial\bar{\mathbf{n}}^w/\partial s_1^w = (\partial\bar{\mathbf{t}}_1^w/\partial s_1^w)\times\bar{\mathbf{t}}_2^w + \bar{\mathbf{t}}_1^w\times(\partial\bar{\mathbf{t}}_2^w/\partial s_1^w) = \begin{bmatrix} -(\partial^2 g^{wp}/\partial s_1^w\partial s_2^w)\cos s_2^w + (\partial g^{wp}/\partial s_1^w)\sin s_2^w \\ -(\partial g^{wp}/\partial s_1^w)^2 \\ -(\partial^2 g^{wp}/\partial s_1^w\partial s_2^w)\sin s_2^w - (\partial g^{wp}/\partial s_1^w)\cos s_2^w \end{bmatrix}$
$\partial\bar{\mathbf{n}}^w/\partial s_2^w = (\partial\bar{\mathbf{t}}_1^w/\partial s_2^w)\times\bar{\mathbf{t}}_2^w + \bar{\mathbf{t}}_1^w\times(\partial\bar{\mathbf{t}}_2^w/\partial s_2^w) = \begin{bmatrix} g^{wp}\cos s_2^w \\ 0 \\ g^{wp}\sin s_2^w \end{bmatrix}$	$\partial\bar{\mathbf{n}}^w/\partial s_2^w = (\partial\bar{\mathbf{t}}_1^w/\partial s_2^w)\times\bar{\mathbf{t}}_2^w + \bar{\mathbf{t}}_1^w\times(\partial\bar{\mathbf{t}}_2^w/\partial s_2^w) = \begin{bmatrix} -(\partial^2 g^{wp}/\partial(s_2^w)^2)\cos s_2^w \\ -(\partial g^{wp}/\partial s_1^w)(\partial g^{wp}/\partial s_2^w) \\ -(\partial^2 g^{wp}/\partial(s_2^w)^2)\sin s_2^w \end{bmatrix} + \begin{bmatrix} 2(\partial g^{wp}/\partial s_2^w)\sin s_2^w + g^{wp}\cos s_2^w \\ -g^{wp}(\partial^2 g^{wp}/\partial s_1^w\partial s_2^w) \\ -2(\partial g^{wp}/\partial s_2^w)\cos s_2^w + g^{wp}\sin s_2^w \end{bmatrix}$

Similarly, the curvature vectors can be defined in the global coordinate system XYZ as

$$\left.\begin{aligned} \frac{\partial^2\mathbf{r}^w}{\partial(s_1^w)^2} &= \frac{\partial\mathbf{t}_1^w}{\partial s_1^w} = \mathbf{A}^w\left(\frac{\partial^2\bar{\mathbf{u}}^w}{\partial(s_1^w)^2}\right) = \mathbf{A}^w\left(\frac{\partial\bar{\mathbf{t}}_1^w}{\partial s_1^w}\right) \\ \frac{\partial^2\mathbf{r}^w}{\partial(s_2^w)^2} &= \frac{\partial\mathbf{t}_2^w}{\partial s_2^w} = \mathbf{A}^w\left(\frac{\partial^2\bar{\mathbf{u}}^w}{\partial(s_2^w)^2}\right) = \mathbf{A}^w\left(\frac{\partial\bar{\mathbf{t}}_2^w}{\partial s_2^w}\right) \\ \frac{\partial^2\mathbf{r}^w}{\partial s_1^w\partial s_2^w} &= \frac{\partial\mathbf{t}_1^w}{\partial s_2^w} = \frac{\partial\mathbf{t}_2^w}{\partial s_1^w} = \mathbf{A}^w\left(\frac{\partial^2\bar{\mathbf{u}}^w}{\partial s_1^w\partial s_2^w}\right) = \mathbf{A}^w\left(\frac{\partial\bar{\mathbf{t}}_1^w}{\partial s_2^w}\right) = \mathbf{A}^w\left(\frac{\partial\bar{\mathbf{t}}_2^w}{\partial s_1^w}\right) \end{aligned}\right\} \tag{4.9}$$

The derivatives with respect to the surface parameters of the wheel surface normal vector defined in the global XYZ coordinate system can be developed in a similar manner.

4.3 SEMI-ANALYTICAL APPROACH FOR RAIL GEOMETRY

This section introduces the semi-analytical approach for describing rail geometry. The ANCF interpolation approach is discussed in the following section. To allow for independent movement of the rails of a track during dynamic simulations in both approaches, the kinematics of each rail is defined separately to allow accounting for scenarios that include gage variations, vertical rail movements, different motion constraints imposed on the rails, and rail rotations. For this reason, each rail is assumed to have its own coordinate system $X^rY^rZ^r$ that is different from the track coordinate system $X^tY^tZ^t$, as shown in Figure 4. Here, superscript r stands for the right rail (rr) or left rail (lr): that is, $r = rr$, lr. As in the case of the wheel, the rail surface can be defined in a general parametric form in terms of two surface parameters s_1^r and s_2^r. This rail surface description allows for a general representation of the rail profile using analytical methods, tabulated data, or measurements. As shown in Figure 5, and without any loss of generality, the two surface parameters of the rail can be selected to be a longitudinal surface parameter $s_1^r = x_s^r$ and a lateral surface parameter $s_2^r = y_s^r$. The rail profile curve can be defined in a profile frame $X^{rp}Y^{rp}Z^{rp}$ using the profile function g^{rp}. The origin of the profile frame is assumed to be

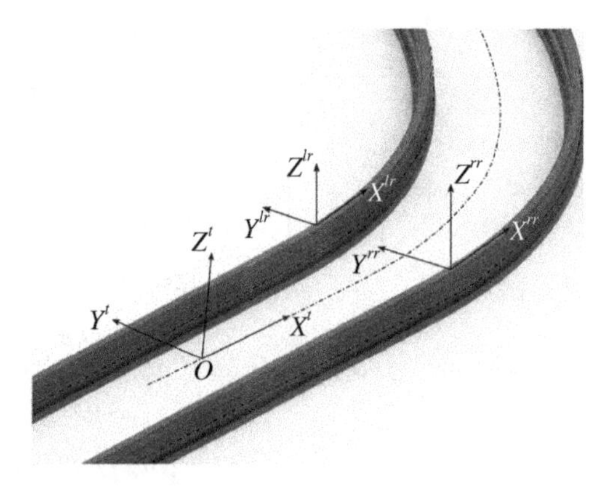

Figure 4.4 Rail coordinate system.

located on the rail space curve, as shown in Figure 5. If the profile does not change as a function of the longitudinal surface parameter, one can write $g^{rp} = g^{rp}(s_2^r)$. If the profile changes as a function of the longitudinal surface parameter, one has $g^{rp} = g^{rp}(s_1^r, s_2^r)$. These two important cases are discussed in this section.

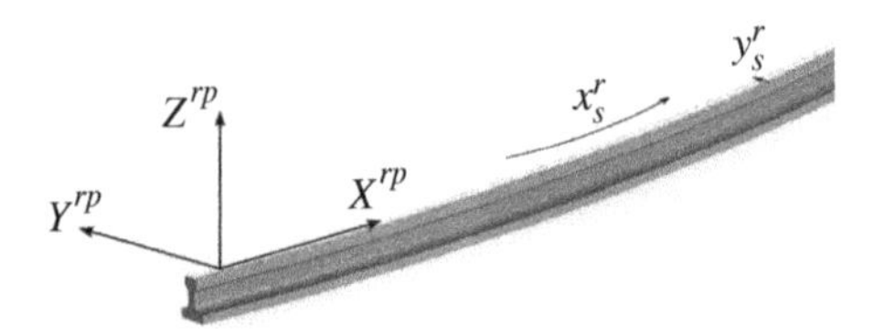

Figure 4.5 Rail surface parameters.

Constant Rail Profile Curve If the rail profile curve does not change as a function of the longitudinal surface parameter $s_1^r = x_s^r$, the rail surface can be defined by a translation of the profile function $g^{rp} = g^{rp}(s_2^r)$ in the rail longitudinal direction. In this case, the location of an arbitrary point on the rail surface can be defined in the rail coordinate system $X^rY^rZ^r$ using vector $\overline{\mathbf{u}}^r(s_1^r, s_2^r)$ as

$$\overline{\mathbf{u}}^r(s_1^r, s_2^r) = \overline{\mathbf{R}}^{rp} + \mathbf{A}^{rp}\overline{\overline{\mathbf{u}}}^{rp} \tag{4.10}$$

where $\overline{\mathbf{R}}^{rp} = \overline{\mathbf{R}}^{rp}(s_1^r) = \left[x_o^{rp}(s_1^r) \;\; y_o^{rp}(s_1^r) \;\; z_o^{rp}(s_1^r)\right]^T$ is the vector that defines the origin of the rail profile frame with respect to the rail coordinate system $X^rY^rZ^r$, $\mathbf{A}^{rp} = \mathbf{A}^{rp}(s_1^r)$ is the matrix that defines the orientation of the profile frame $X^{rp}Y^{rp}Z^{rp}$ with respect to the rail coordinate system $X^rY^rZ^r$, and $\overline{\overline{\mathbf{u}}}^{rp} = \left[0 \;\; s_2^r \;\; g^{rp}(s_2^r)\right]^T$ is the vector that defines the location of an arbitrary point in the profile frame $X^{rp}Y^{rp}Z^{rp}$. In making these definitions, it is assumed that the X^{rp} axis of the profile frame is along the longitudinal tangent of the rail space curve. As discussed in Chapter 3, matrix $\mathbf{A}^{rp} = \mathbf{A}^{rp}(s_1^r)$ in Eq. 10 depends only on the arc length parameter s_1^r of the rail space curve. This matrix, as explained in Chapter 3, can be expressed in terms of three Euler angles that are not totally independent if some specific conditions are imposed (Shabana and Rathod 2007). Euler angles are used in the design of track geometry and in the track layout, as discussed in the preceding chapter and in later sections of this chapter. Table 3 shows geometric details of rail sections, basic dimensions, and rail profile geometries.

Table 4.3 Rail geometry details.

Rail section	
Rail section 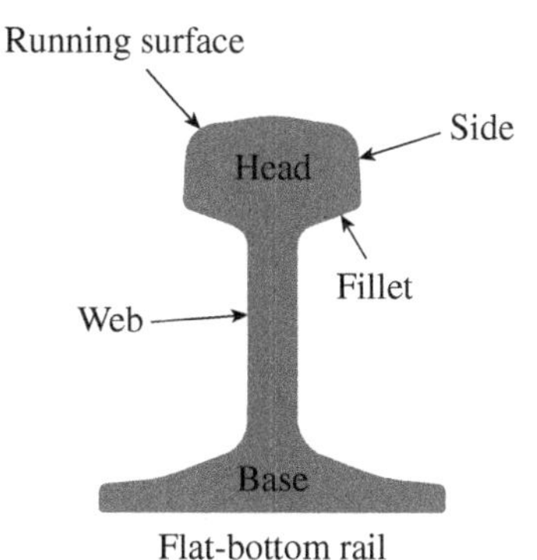	• **Head** is the portion of the rail section that contains the profile and **running surface** that comes into contact with the wheels of the railroad vehicle. • **Base** (also called **foot**) is the bottom section of the rail that rests on the track structure. • **Web** is the section that connects the head and the base.
Flat-bottom rail **Basic dimensions** 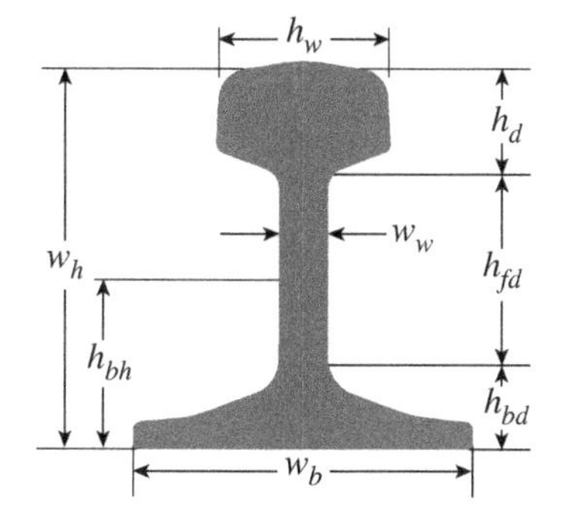	w_h is the rail section height. w_b is the base width. h_w is head width. w_w is the web thickness at the center point. h_d is the head depth. h_{fd} is the fishing. h_{bd} is the base depth. h_{bh} is the bolt-hole position.
Dimension example: 68 kg rail 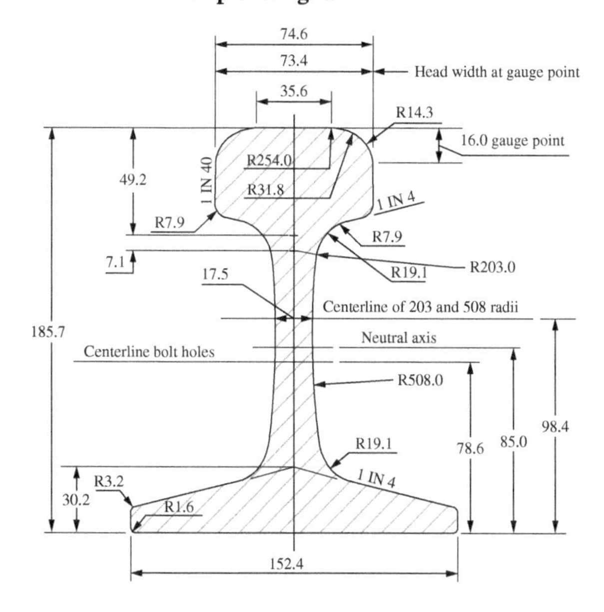	**68 kg rail profile:** head area 3117 mm^2, web area 2349 mm^2, foot area 3147 mm^2 Total area 8614 mm^2, estimated mass 67.7 kg/m, maximum rail length 25.0 m

(Continued)

Table 4.3 (Continued)

Other rail geometries

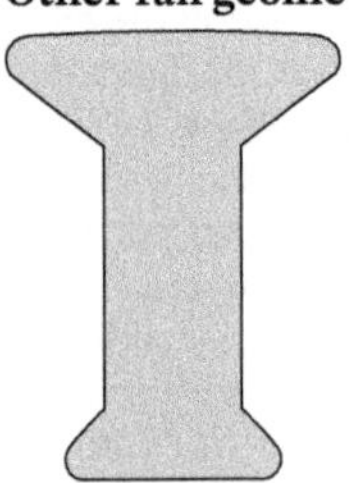

Bullhead rail was used by the British railway from the mid-nineteenth to mid-twentieth century. It has been, with very few exceptions, replaced by the flat-bottom rail.

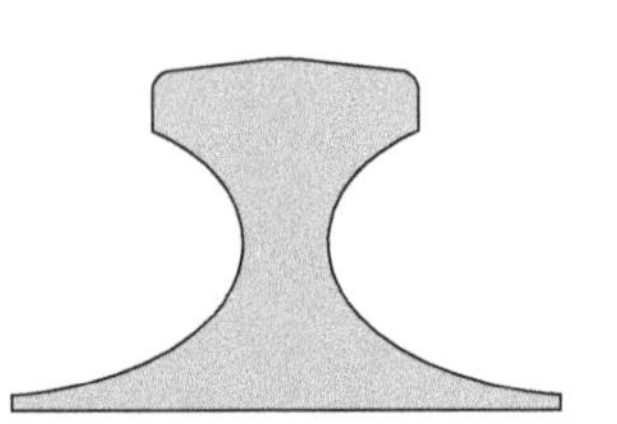

Crane rails are used in many metro areas and are available in sections that range from 104 to 175 lb per yard.

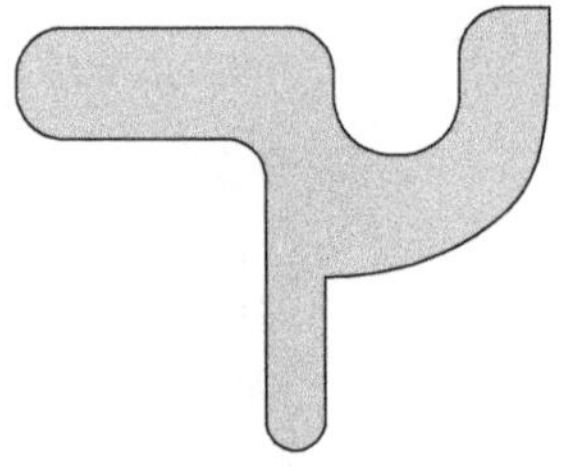

Guard rails (check rails) are placed parallel to running rails along restricted-clearance areas, such as tunnels, bridges, and/or level crossings to keep the rail wheels in alignment and thus avoid derailment and damage to the track structure.

Turnouts (Shabana et al. 2008)

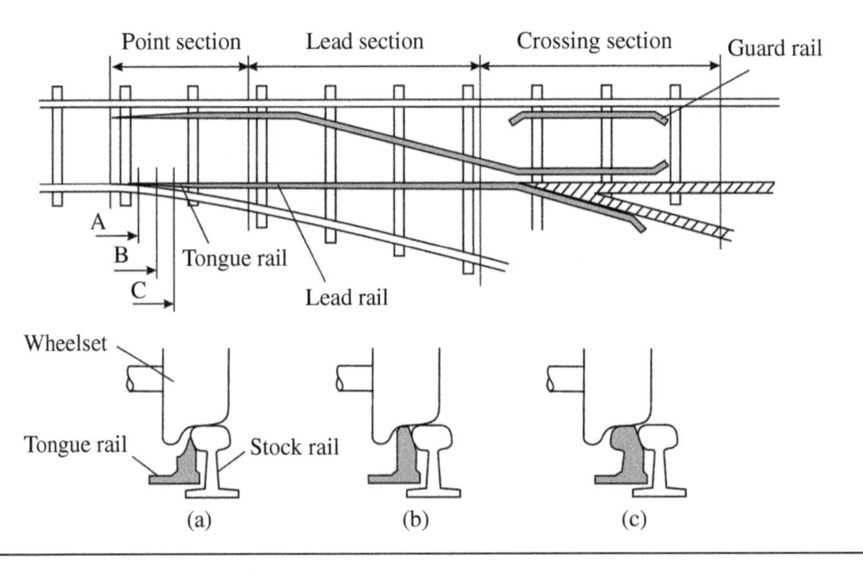

Track geometry is characterized by different types of *discontinuities* resulting from non-smooth changes in the rail geometry. These geometric discontinuities, which produce significant impulsive forces as the vehicle negotiates track sections, can be attributed to misalignments, change in soil properties, a sudden change in the track structural stiffness, or unavoidable structural designs, as is the case with switches and turnouts. Discontinuities can also be encountered as a result of bridge abutments and road crossing. A train direction of motion is changed using *turnouts* (*switches* and *crossings*), as shown in

Table 3 (Schupp et al. 2004; Kassa et al. 2006). For the most part, a turnout is constructed using a *switch panel* (point section), *crossing panel* (crossing section), and *closure panel* (lead rail) that connect the switch panel with the crossing panel, as shown in Table 3 (Shabana et al. 2008). A movable swing nose is used for high-speed trains (Kono et al. 2005). When a railroad vehicle negotiates switches starting from the *stock rail* to the *tongue rail* in the switch panel or the crossing panel, more than one wheel/rail contact point can occur, as shown in Table 3, leading to significant impulsive forces that influence vehicle dynamics and stability and may lead to derailments.

The fact that vector $\overline{\mathbf{R}}^{rp} = \left[x_o^{rp}(s_1^r) \;\; y_o^{rp}(s_1^r) \;\; z_o^{rp}(s_1^r)\right]^T$ is assumed to depend on the longitudinal surface parameter allows for modeling curved and spiral rail sections. In the simpler case of a tangent track with fixed straight rail, this vector reduces to $\overline{\mathbf{R}}^{rp} = \left[x_o^{rp}(s_1^r) \;\; y_o^{rp} \;\; z_o^{rp}\right]^T$, where y_o^{rp} and z_o^{rp} are constants and $s_1^r = x_s^r$ represents the longitudinal position of an arbitrary point on the rail surface. The more general representation of Eq. 10, in which $y_o^{rp} = y_o^{rp}(s_1^r)$ and $z_o^{rp} = z_o^{rp}(s_1^r)$, allows describing systematically curved rail sections. As in the case of wheels, the rail profile function g^{rp} can be described analytically using tabulated data or measurements. In the two latter cases, interpolation methods can be used to define the location of an arbitrary point on the rail surface. As discussed in the following section, using ANCF finite elements allows for obtaining a much simpler alternate expression to Eq. 10.

Non-Constant Rail Profile Curve For worn rails or when using track measurement data, the rail profile curve may vary as a function of the rail longitudinal surface parameter $s_1^r = x_s^r$ as well as the lateral rail surface parameter $s_2^r = y_s^r$. If the rail profile curve changes as a function of the longitudinal surface parameter $s_1^r = x_s^r$, the rail surface can be defined by the profile function $g^{rp} = g^{rp}(s_1^r, s_2^r)$ in the rail longitudinal direction. In this case, the location of an arbitrary point on the rail surface can be defined in the rail coordinate system $X^rY^rZ^r$ using vector $\overline{\mathbf{u}}^r(s_1^r, s_2^r)$ as

$$\overline{\mathbf{u}}^r(s_1^r, s_2^r) = \overline{\mathbf{R}}^{rp}(s_1^r) + \mathbf{A}^{rp}(s_1^r)\overline{\overline{\mathbf{u}}}^{rp}(s_1^r, s_2^r) \tag{4.11}$$

where $\overline{\mathbf{R}}^{rp} = \overline{\mathbf{R}}^{rp}(s_1^r) = \left[x_o^{rp}(s_1^r) \;\; y_o^{rp}(s_1^r) \;\; z_o^{rp}(s_1^r)\right]^T$ is the vector that defines the origin of the rail profile frame with respect to the rail coordinate system $X^rY^rZ^r$, $\mathbf{A}^{rp} = \mathbf{A}^{rp}(s_1^r)$ is the matrix that defines the orientation of the profile frame $X^{rp}Y^{rp}Z^{rp}$ with respect to the rail coordinate system $X^rY^rZ^r$, and $\overline{\overline{\mathbf{u}}}^{rp} = \left[0 \;\; s_2^r \;\; g^{rp}(s_1^r, s_2^r)\right]^T$ is the vector that defines the location of an arbitrary point in the profile frame $X^{rp}Y^{rp}Z^{rp}$. As with a constant rail profile curve, it is assumed that the X^{rp} axis of the profile frame is along the tangent of the rail space curve. Furthermore, the form of vector $\overline{\mathbf{R}}^{rp} = \left[x_o^{rp}(s_1^r) \;\; y_o^{rp}(s_1^r) \;\; z_o^{rp}(s_1^r)\right]^T$, as in the case of the constant rail profile curve, allows for modeling curved and spiral rail sections. For a tangent track with fixed straight rail, this vector, as previously mentioned, reduces to $\overline{\mathbf{R}}^{rp} = \left[x_o^{rp}(s_1^r) \;\; y_o^{rp} \;\; z_o^{rp}\right]^T$, where y_o^{rp} and z_o^{rp} are constants, and $s_1^r = x_s^r$ represents the longitudinal coordinate of an arbitrary point on the rail surface. The rail profile function $g^{rp}(s_1^r, s_2^r)$ can also be described analytically, using tabulated data, or using measurements. When tabulated data or measurements are used to define the rail surface, interpolation methods can be used to determine the location of an arbitrary point on the rail surface. Similar to the case of constant profile, using the ANCF interpolation approach allows for obtaining a more general, and much simpler, alternate expression to Eq. 11.

Tangent Plane and Normal Vector For rigid rails, vector $\overline{\mathbf{R}}^{rp}$ that defines the location of the origin of the profile frame $X^{rp}Y^{rp}Z^{rp}$ with respect to the rail or track coordinate system $X^rY^rZ^r$ can be, without any loss of generality, assumed constant. Consequently, the derivatives of this vector with respect to the surface parameters are equal to zero, and this vector does not appear in the definition of the rail surface tangent and normal vectors. However, this is not the case when the semi-analytical approach is used to describe the geometry of the rails since vector $\overline{\mathbf{R}}^{rp}$ that defines the location of the origin of the profile frame $X^{rp}Y^{rp}Z^{rp}$ with respect to the rail coordinate system $X^rY^rZ^r$ cannot, in general, be assumed constant due to its dependence on the rail longitudinal surface parameter $s_1^r = x_s^r$: that is, $\overline{\mathbf{R}}^{rp} = \overline{\mathbf{R}}^{rp}(s_1^r)$. Therefore, as is clear from Eqs. 10 and 11, both vector function $\overline{\mathbf{R}}^{rp}$ and scalar function g^{rp} are needed in order to define the coordinates of arbitrary points on the rail surface as well as the derivatives of these coordinates at these points. For unworn rails, scalar function g^{rp} depends only on the lateral surface parameter $s_2^r = y_s^r$: that is, $g^{rp} = g^{rp}(s_2^r)$. For worn rails, or in case of using measured track data, g^{rp} depends on the longitudinal surface parameter $s_1^r = x_s^r$ as well as the lateral surface parameter $s_2^r = y_s^r$: that is, $g^{rp} = g^{rp}(s_1^r, s_2^r)$. In both cases, profile function g^{rp} can be defined analytically or numerically using interpolation methods, including spline functions. As previously mentioned, the numerical approach offers generality and allows for using measured track data.

Formulating the wheel/rail *contact conditions* requires determining the tangent and normal vectors of the rail surface at contact points. The rail surface tangent and normal vectors are also required in order to define the normal and tangential wheel/rail contact forces. The two tangent vectors $\overline{\mathbf{t}}_1^r$ and $\overline{\mathbf{t}}_2^r$ define the rail tangent plane and also define normal vector $\overline{\mathbf{n}}^r$ in the rail coordinate system $X^rY^rZ^r$. Using the more general definition of the rail profile function $g^{rp} = g^{rp}(s_1^r, s_2^r)$, the tangent and normal vectors are defined as

$$\left.\begin{aligned}
\overline{\mathbf{t}}_1^r &= \frac{\partial \overline{\mathbf{u}}^r}{\partial s_1^r} = \frac{\partial \overline{\mathbf{R}}^{rp}}{\partial s_1^r} + \left(\frac{\partial \mathbf{A}^{rp}}{\partial s_1^r}\right)\overline{\overline{\mathbf{u}}}^{rp} + \mathbf{A}^{rp}\frac{\partial \overline{\overline{\mathbf{u}}}^{rp}}{\partial s_1^r},\\
\overline{\mathbf{t}}_2^r &= \frac{\partial \overline{\mathbf{u}}^r}{\partial s_2^r} = \frac{\partial \overline{\mathbf{R}}^{rp}}{\partial s_2^r} + \mathbf{A}^{rp}\frac{\partial \overline{\overline{\mathbf{u}}}^{rp}}{\partial s_2^r},\\
\overline{\mathbf{n}}^r &= \overline{\mathbf{t}}_1^r \times \overline{\mathbf{t}}_2^r
\end{aligned}\right\} \tag{4.12}$$

where vector $\overline{\mathbf{u}}^r$ of Eq. 10 or 11 defines the location of an arbitrary point on the rail surface with respect to the rail coordinate system $X^rY^rZ^r$, $\partial \overline{\overline{\mathbf{u}}}^{rp}/\partial s_1^r = \begin{bmatrix} 0 & 0 & \partial g^{rp}/\partial s_1^r \end{bmatrix}^T$, and $\partial \overline{\overline{\mathbf{u}}}^{rp}/\partial s_2^r = \begin{bmatrix} 0 & 1 & \partial g^{rp}/\partial s_2^r \end{bmatrix}^T$. If $\overline{\mathbf{R}}^{rp}$ is assumed to depend only on the longitudinal rail surface parameter s_1^r – that is, $\overline{\mathbf{R}}^{rp} = \overline{\mathbf{R}}^{rp}(s_1^r)$ – then in the preceding equation, one has $\partial \overline{\mathbf{R}}^{rp}/\partial s_2^r = \mathbf{0}$, and in this case, the lateral tangent vector $\overline{\mathbf{t}}_2^r$ reduces to $\overline{\mathbf{t}}_2^r = \mathbf{A}^{rp}(\partial \overline{\overline{\mathbf{u}}}^{rp}/\partial s_2^r)$. Furthermore, all Euler angles that define the transformation matrix of the profile frame $X^{rp}Y^{rp}Z^{rp}$ can be expressed in terms of the rail longitudinal surface parameters s_1^r, as discussed in the preceding chapter. If the rail unit normal vector must be computed, the unit normal to the rail surface is defined as $\widehat{\overline{\mathbf{n}}}^r = (\overline{\mathbf{t}}_1^r \times \overline{\mathbf{t}}_2^r)/|\overline{\mathbf{t}}_1^r \times \overline{\mathbf{t}}_2^r|$. The case in which the rail profile curve does not change as a function of the rail longitudinal surface parameter s_1^r can be obtained as a special case of Eq. 12 in which $\partial g^{rp}/\partial s_1^r = 0$ and, therefore, $\partial \bar{\bar{\mathbf{u}}}^{rp}/\partial s_1^r = \mathbf{0}$.

Higher Derivatives The computations of higher derivatives with respect to rail surface parameters are necessary, particularly when solving nonlinear contact conditions numerically using the iterative Newton–Raphson procedure. Determining higher-order derivatives requires evaluating the curvature vectors as well as the derivative of the vector normal to the rail surface. The curvature vectors at an arbitrary point on the rail surface can be written as

$$\left.\begin{aligned}
\frac{\partial^2 \bar{\mathbf{u}}^r}{\partial (s_1^r)^2} &= \frac{\partial \bar{\mathbf{t}}_1^r}{\partial s_1^r} = \frac{\partial^2 \bar{\mathbf{R}}^{rp}}{\partial (s_1^r)^2} + \left(\frac{\partial^2 \mathbf{A}^{rp}}{\partial (s_1^r)^2}\right)\bar{\bar{\mathbf{u}}}^{rp} + 2\left(\frac{\partial \mathbf{A}^{rp}}{\partial s_1^r}\right)\frac{\partial \bar{\bar{\mathbf{u}}}^{rp}}{\partial s_1^r} + \mathbf{A}^{rp}\frac{\partial^2 \bar{\bar{\mathbf{u}}}^{rp}}{\partial (s_1^r)^2}\\
\frac{\partial^2 \bar{\mathbf{u}}^r}{\partial (s_2^r)^2} &= \frac{\partial \bar{\mathbf{t}}_2^r}{\partial s_2^r} = \frac{\partial^2 \bar{\mathbf{R}}^{rp}}{\partial (s_2^r)^2} + \mathbf{A}^{rp}\frac{\partial \bar{\bar{\mathbf{u}}}^{rp}}{\partial (s_2^r)^2}\\
\frac{\partial^2 \bar{\mathbf{u}}^r}{\partial s_1^r \partial s_2^r} &= \frac{\partial \bar{\mathbf{t}}_1^r}{\partial s_2^r} = \frac{\partial \bar{\mathbf{t}}_2^r}{\partial s_1^r} = \frac{\partial \bar{\mathbf{R}}^{rp}}{\partial s_1^r \partial s_2^r} + \frac{\partial \mathbf{A}^{rp}}{\partial s_1^r}\frac{\partial \bar{\bar{\mathbf{u}}}^{rp}}{\partial s_2^r} + \mathbf{A}^{rp}\frac{\partial \bar{\bar{\mathbf{u}}}^{rp}}{\partial s_1^r \partial s_2^r}
\end{aligned}\right\} \tag{4.13}$$

where $\partial^2 \bar{\bar{\mathbf{u}}}^{rp}/\partial (s_1^r)^2 = \begin{bmatrix} 0 & 0 & \partial^2 g^{rp}/\partial (s_1^r)^2 \end{bmatrix}^T$, $\partial^2 \bar{\bar{\mathbf{u}}}^{rp}/\partial (s_2^r)^2 = \begin{bmatrix} 0 & 0 & \partial^2 g^{rp}/\partial (s_2^r)^2 \end{bmatrix}^T$, and $\partial^2 \bar{\bar{\mathbf{u}}}^{rp}/\partial s_1^r \partial s_2^r = \begin{bmatrix} 0 & 0 & \partial^2 g^{rp}/\partial s_1^r \partial s_2^r \end{bmatrix}^T$. The derivatives of the normal vector with respect to the rail surface parameters can be determined using the derivatives of the two tangent vectors as

$$\left.\begin{aligned}
\partial \bar{\mathbf{n}}^r/\partial s_1^r &= (\partial \bar{\mathbf{t}}_1^r/\partial s_1^r) \times \bar{\mathbf{t}}_2^r + \bar{\mathbf{t}}_1^r \times (\partial \bar{\mathbf{t}}_2^r/\partial s_1^r)\\
\partial \bar{\mathbf{n}}^r/\partial s_2^r &= (\partial \bar{\mathbf{t}}_1^r/\partial s_2^r) \times \bar{\mathbf{t}}_2^r + \bar{\mathbf{t}}_1^r \times (\partial \bar{\mathbf{t}}_2^r/\partial s_2^r)
\end{aligned}\right\} \tag{4.14}$$

These two equations show that, as in the case of a wheel, the derivatives of the normal vector are functions of the rail curvature vectors. When using the unit normal vector $\hat{\bar{\mathbf{n}}}^r = (\bar{\mathbf{t}}_1^r \times \bar{\mathbf{t}}_2^r)/|\bar{\mathbf{t}}_1^r \times \bar{\mathbf{t}}_2^r|$, the derivatives can also be evaluated in closed form, as previously discussed:

$$\partial \hat{\bar{\mathbf{n}}}^r/\partial s_k^r = ((\mathbf{I} - \hat{\bar{\mathbf{n}}}^r \hat{\bar{\mathbf{n}}}^{r^T})/|\bar{\mathbf{n}}^r|)(\partial \bar{\mathbf{n}}^r/\partial s_k^r), \quad k = 1, 2 \tag{4.15}$$

In this equation, $\partial \bar{\mathbf{n}}^r/\partial s_k^r,\ k = 1, 2$, can be determined using Eq. 14. Therefore, the derivatives of the unit normal to the rail surface can be systematically evaluated using the tangent vectors and their derivatives. While, as in the case of wheels, using a numerical approach based on interpolations allows using the rail equations developed in this section, a much simpler and more straightforward approach for describing rail geometry is to use ANCF interpolations. Using ANCF geometry significantly simplifies the rail surface equations, as discussed in the following section.

Rail Global Vectors In the formulations used in this book, the rail and/or track can have arbitrarily large rigid-body displacements, including finite rotations. Using these formulations may be necessary in some railroad scenarios that include foundation movements or the failure of some rail segments. Using the equation for the position vector of an arbitrary point with respect to the rail coordinate system $X^r Y^r Z^r$, the position vector $\mathbf{r}^r$ of arbitrary points can be defined in the global coordinate system XYZ as

$$\mathbf{r}^r = \mathbf{R}^r + \mathbf{A}^r\overline{\mathbf{u}}^r \tag{4.16}$$

where $\mathbf{R}^r = \mathbf{R}^r(t)$ is the position vector of the origin of the selected rail coordinate system $X^rY^rZ^r$ in the global coordinate system XYZ, $\mathbf{A}^r$ is the transformation matrix that defines the orientation of the rail coordinate system, t is time, and $\overline{\mathbf{u}}^r$ is defined by Eq. 11 as $\overline{\mathbf{u}}^r(s_1^r, s_2^r) = \overline{\mathbf{R}}^{rp}(s_1^r) + \mathbf{A}^{rp}(s_1^r)\overline{\overline{\mathbf{u}}}^{rp}(s_1^r, s_2^r)$. Because $\mathbf{R}^r$ and $\mathbf{A}^r$ do not depend on the rail surface parameters, one can show that the rail tangent and normal vectors defined in the global coordinate system XYZ are given by

$$\left.\begin{aligned}
\mathbf{t}_1^r &= \frac{\partial \mathbf{r}^r}{\partial s_1^r} = \mathbf{A}^r\left(\frac{\partial \overline{\mathbf{u}}^r}{\partial s_1^r}\right) = \mathbf{A}^r\overline{\mathbf{t}}_1^r,\\
\mathbf{t}_2^r &= \frac{\partial \mathbf{r}^r}{\partial s_2^r} = \mathbf{A}^r\left(\frac{\partial \overline{\mathbf{u}}^r}{\partial s_2^r}\right) = \mathbf{A}^r\overline{\mathbf{t}}_2^r,\\
\mathbf{n}^r &= \mathbf{t}_1^r \times \mathbf{t}_2^r = (\mathbf{A}^r\overline{\mathbf{t}}_1^r)\times(\mathbf{A}^r\overline{\mathbf{t}}_2^r) = \mathbf{A}^r(\overline{\mathbf{t}}_1^r \times \overline{\mathbf{t}}_2^r) = \mathbf{A}^r\overline{\mathbf{n}}^r
\end{aligned}\right\} \tag{4.17}$$

where the derivatives of $\overline{\mathbf{u}}^r$ with respect to surface parameters s_1^r and s_2^r are as previously defined in this section. Similarly, the curvature vectors can be defined in the global coordinate system XYZ as

$$\left.\begin{aligned}
\frac{\partial^2 \mathbf{r}^r}{\partial (s_1^r)^2} &= \frac{\partial \mathbf{t}_1^r}{\partial s_1^r} = \mathbf{A}^r\left(\frac{\partial^2 \overline{\mathbf{u}}^r}{\partial (s_1^r)^2}\right) = \mathbf{A}^r\left(\frac{\partial \overline{\mathbf{t}}_1^r}{\partial s_1^r}\right)\\
\frac{\partial^2 \mathbf{r}^r}{\partial (s_2^r)^2} &= \frac{\partial \mathbf{t}_2^r}{\partial s_2^r} = \mathbf{A}^r\left(\frac{\partial^2 \overline{\mathbf{u}}^r}{\partial (s_2^r)^2}\right) = \mathbf{A}^r\left(\frac{\partial \overline{\mathbf{t}}_2^r}{\partial s_2^r}\right)\\
\frac{\partial^2 \mathbf{r}^r}{\partial s_1^r \partial s_2^r} &= \frac{\partial \mathbf{t}_1^r}{\partial s_2^r} = \frac{\partial \mathbf{t}_2^r}{\partial s_1^r} = \mathbf{A}^r\left(\frac{\partial^2 \overline{\mathbf{u}}^r}{\partial s_1^r \partial s_2^r}\right) = \mathbf{A}^r\left(\frac{\partial \overline{\mathbf{t}}_1^r}{\partial s_2^r}\right) = \mathbf{A}^r\left(\frac{\partial \overline{\mathbf{t}}_2^r}{\partial s_1^r}\right)
\end{aligned}\right\} \tag{4.18}$$

The derivatives with respect to the surface parameters of the rail surface normal vector defined in the global XYZ coordinate system can be developed in a similar manner.

4.4 ANCF RAIL GEOMETRY

It is clear from the analysis presented in the preceding section that the computer implementation of rail geometry equations can be cumbersome as compared to wheel geometry, even for a rigid rail. The complexities arise from the dependence of more vectors and matrices in the rail equations on rail surface parameters. Furthermore, the rail is constructed using different section types that have different geometries, requiring the use of a more general approach. Even in the case of a rigid rail, a data file that defines the rail geometry must be developed at a preprocessing stage. It is necessary to have this data file containing the position of points as well as rotations at these points in order to perform dynamic simulations using general multibody system (MBS) algorithms. Rotations at discrete points are used to define the geometry of the rail, including curvature, grade (elevation), and superelevation. Therefore, using the semi-analytical approach described in the preceding section can significantly complicate the computer implementation of general rail geometry equations.

Furthermore, such a semi-analytical approach does not lend itself easily to cases in which the rail deforms. In such cases, using more general flexible-body computational approaches is necessary.

ANCF Geometry One effective and general method to describe the rail surface geometry in railroad vehicle system applications is to use the ANCF finite elements introduced in Chapter 2 (Shabana 2018). ANCF finite elements have been used effectively to integrate geometry and analysis in railroad vehicle system applications (Berzeri et al. 2000). These elements, which have been used to describe both rail and catenary geometries, are well-suited for describing arbitrary rail geometry with different section types such as tangent, curve, and spiral sections. ANCF elements allow for developing the solid model geometry and using the same geometry model in the analysis without any adjustments. Furthermore, these ANCF elements can be used for both rigid-rail and flexible-rail analysis. For example, ANCF fully parameterized beam elements can be used to develop the geometry of the rail space curve and superimpose on this curve geometry the profile geometry to define the rail surface, as described in this section.

The main reason ANCF elements can significantly simplify describing the rail geometry and its computer implementation is that ANCF elements use position gradients as nodal coordinates. By changing the orientations of position gradient vectors at element nodal points, curved elements can be easily constructed and used to define the rail space curve for different section types. ANCF elements are also related to computational geometry and computer-aided design (CAD) methods – such as B-splines (basis splines) and NURBS (non-uniform rational B-splines) – by a linear mapping and therefore can be used to generate the same geometries developed by these CAD methods. Additionally, ANCF elements are mechanics-based and do not require using control points, which are not material points, as is the case with B-spline and NURBS representation.

For a node k of an ANCF rail mesh, the vector of the nodal coordinates at the node is defined as

$$\mathbf{e}^{rk} = \begin{bmatrix} \mathbf{r}^{rk^T} & \mathbf{r}_x^{rk^T} & \mathbf{r}_y^{rk^T} & \mathbf{r}_z^{rk^T} \end{bmatrix}^T, \qquad k = 1, 2, \ldots, n_n^r \tag{4.19}$$

In this equation, $\mathbf{r}^{rk}$ is the position of node k of rail r, which can be defined in any coordinate system including the global, track, or rail coordinate system; $\mathbf{r}_x^{rk} = \partial \mathbf{r}^{rk}/\partial x$, $\mathbf{r}_y^{rk} = \partial \mathbf{r}^{rk}/\partial y$, and $\mathbf{r}_z^{rk} = \partial \mathbf{r}^{rk}/\partial z$ are the position vector gradients; x, y, and z are the ANCF element parameters or spatial coordinates; and n_n^r is the number of nodes in the rail ANCF mesh. For a long track, the number of nodes n_n^r can be relatively large if geometric variations are to be captured accurately.

When rail deformation is considered, the same ANCF mesh can be used as the analysis mesh without any change. The ANCF equations of motion of the rail can be formulated and solved to determine the nodal coordinates as functions of time. These coordinates can be used with the ANCF displacement field to determine the change in the geometry as well as the elastic forces that result from rail deformations. For a rigid rail, which is the focus of the discussion in this chapter, the ANCF position vector gradients remain constant and can be defined using three orthogonal unit vectors expressed in terms of the three Euler angles

$\psi^{rt}(s_1^r)$, $\theta^{rt}(s_1^r)$, and $\phi^{rt}(s_1^r)$ that define, respectively, the curvature, grade, and superelevation, where s_1^r is the rail arc length parameter. The Euler angle sequence used in the railroad literature is $Z^r, -Y^r$, and X^r. As explained in the preceding chapter, these three Euler angles completely define the rail geometry and can be calculated using specific track information. Also as discussed in Chapter 3, the three Euler angles $\psi^{rt}(s_1^r)$, $\theta^{rt}(s_1^r)$, and $\phi^{rt}(s_1^r)$ are, in general, independent unless additional conditions are enforced, as discussed in the literature (Shabana and Rathod 2007). Nonetheless, these angles can be defined in terms of the arc length parameter s_1^r because a space curve is uniquely defined using one parameter, which here is the arc length s_1^r. This arc length parameter can also be used to conveniently define the distance traveled by a wheel rolling and/or sliding on the rail. It was shown in Chapter 3 that an orthogonal transformation matrix that describes the rail geometry can be expressed in terms of the three angles ψ^{rt}, θ^{rt}, and ϕ^{rt} as

$$\mathbf{A}^{rt} = \begin{bmatrix} \cos\psi\cos\theta & -\sin\psi\cos\phi + \cos\psi\sin\theta\sin\phi & -\sin\psi\sin\phi - \cos\psi\sin\theta\cos\phi \\ \sin\psi\cos\theta & \cos\psi\cos\phi + \sin\psi\sin\theta\sin\phi & \cos\psi\sin\phi - \sin\psi\sin\theta\cos\phi \\ \sin\theta & -\cos\theta\sin\phi & \cos\theta\cos\phi \end{bmatrix}^{rt} \tag{4.20}$$

By imposing additional conditions, one can show that the second column of this matrix represents the unit curvature vector, and this matrix reduces to the transformation matrix that defines the Serret–Frenet frame (Shabana and Rathod 2007). Further discussion of these additional conditions, however, is not required for the analysis presented in this chapter.

As discussed in later sections of this chapter, the track data file produced by a track preprocessor computer program can be designed, based on a specified simple track geometry description, to provide information about rail segments (elements) defined by nodes. The information includes the positions of the rail nodes as well as the three Euler angles at these nodes. Therefore, for each node k on the rail, the track data file defines the following six-dimensional vector in a global, track, or rail coordinate system:

$$\mathbf{p}^{rk} = \begin{bmatrix} \mathbf{r}^{rk^T} & \psi^{rtk} & \theta^{rtk} & \phi^{rtk} \end{bmatrix}^T, \qquad k = 1, 2, \ldots, n_n^r \tag{4.21}$$

Using this information, the ANCF gradient vectors in Eq. 19 for a rigid fixed rail can be defined by equating the gradient vectors with the columns of the transformation matrix of Eq. 20 as

$$\left.\begin{aligned} \mathbf{r}_x^{rk} &= \begin{bmatrix} \cos\psi^{rtk}\cos\theta^{rtk} \\ \sin\psi^{rtk}\cos\theta^{rtk} \\ \sin\theta^{rtk} \end{bmatrix}, \quad \mathbf{r}_y^{rk} = \begin{bmatrix} -\sin\psi^{rtk}\cos\phi^{rtk} + \cos\psi^{rtk}\sin\theta^{rtk}\sin\phi^{rtk} \\ \cos\psi^{rtk}\cos\phi^{rtk} + \sin\psi^{rtk}\sin\theta^{rtk}\sin\phi^{rtk} \\ -\cos\theta^{rtk}\sin\phi^{rtk} \end{bmatrix}, \\ \mathbf{r}_z^{rk} &= \begin{bmatrix} -\sin\psi^{rtk}\sin\phi^{rtk} - \cos\psi^{rtk}\sin\theta^{rtk}\cos\phi^{rtk} \\ \cos\psi^{rtk}\sin\phi^{rtk} - \sin\psi^{rtk}\sin\theta^{rtk}\cos\phi^{rtk} \\ \cos\theta^{rtk}\cos\phi^{rtk} \end{bmatrix}, \qquad k = 1, 2, \ldots, n_n^r \end{aligned}\right\} \tag{4.22}$$

Using this definition of the ANCF gradient vectors, the vector of geometry coordinates of node k in Eq. 19 can be fully defined numerically. It is important to point out that in the

ANCF interpolation approach, Euler angles are interpreted as geometry parameters used to define the elements of the ANCF position-gradient vectors. This is a more accurate and general interpretation of these angles, which are treated as field variables that depend on the rail longitudinal surface parameter s_1^r and can be used to define rail segment geometries. As will be seen, using Euler angles to define the position vector gradients at the node allows using a higher-order interpolation for the position coordinates and also avoids differentiating the angles with respect to the arc length parameter s_1^r of the rail space curve.

4.5 ANCF INTERPOLATION OF RAIL GEOMETRY

Using fully parameterized ANCF elements to define rail surface geometry significantly simplifies the definitions of the position, tangent, and normal vectors as well as their derivatives. This is because Euler angles are used to define the ANCF gradient coordinates at discrete nodes, and these angles are no longer considered field variables. Consequently, they are not differentiated with respect to the arc length parameter s_1^r of the rail space curve. Furthermore, when the ANCF interpolation approach is used, using vectors $\mathbf{R}^{rp}$ and matrix $\mathbf{A}^{rp}$ of the semi-analytical approach is not necessary. In addition to avoiding angle differentiations, higher-order cubic position interpolation can be used based only on position and gradient coordinates at the two nodes of the elements; this is mainly due to the dependence of the ANCF position field on the Euler angles at the discrete nodal points. Using the ANCF position vector gradients leads to a 2-node element with 24 nodal coordinates; therefore, such an element is based on a cubic interpolation for the position coordinates, as explained in this section.

For a fully parameterized three-dimensional ANCF element, the displacement field that defines the position of an arbitrary point on an element j on the rail with respect to the rail coordinate system $X^rY^rZ^r$ can be written as

$$\bar{\mathbf{u}}^{rj}(x, y, z, t) = \mathbf{S}^{rj}(x, y, z)\mathbf{e}^{rj}(t) \tag{4.23}$$

In this equation, x, y, and z are the element volume parameters, t is time, $\bar{\mathbf{u}}^{rj}$ is the position vector of an arbitrary point on an element in the rail coordinate system, $\mathbf{S}^{rj}$ is the element shape function matrix, and $\mathbf{e}^{rj}$ is the element vector of nodal coordinates. If ANCF elements are used to describe only geometry without considering rail deformations, then $\mathbf{e}^{rj}$ is constant regardless of the rail rigid-body displacement, and the preceding equation becomes independent of time. In this case, the rigid-body motion of the rail is described using the coordinates of the rail or the track reference described by time-dependent vector $\mathbf{R}^r$ and rail orientation matrix $\mathbf{A}^r$. In this case of a rigid rail, vector $\bar{\mathbf{u}}^{rj}$ at an arbitrary point becomes independent of time. In this special and important case, the preceding equation reduces to

$$\bar{\mathbf{u}}^{rj}(x, y, z) = \mathbf{S}^{rj}(x, y, z)\mathbf{e}^{rj} \tag{4.24}$$

Since the focus in this chapter is on a numerical description of rail geometry, Eq. 24 is used in this chapter. However, Eq. 23 can be used for a flexible rail to update the geometry due to rail deformation. As discussed in Chapter 2, for a fully parameterized ANCF beam element, the shape function matrix $\mathbf{S}^{rj}$ and the vector of nodal coordinates $\mathbf{e}^{rj}$ can be defined using interpolating polynomials that are cubic along x and linear along y and z. In the development presented in this section, the gradient vectors allow for defining arbitrarily curved geometry; there is no loss of generality in assuming that within an element, $s_2^r = y$. The contact conditions can be used to determine s_1^r; and for an element or rail segment defined by two nodes k and $k-1$, the element longitudinal coordinates can be determined as $x = s_1^r - s_1^{rj(k-1)}$, where $s_1^{rj(k-1)}$ is the arc length parameter at the first node $(k-1)$ of element j. Replacing the 24 polynomial coefficients with 24 element nodal coordinates, as explained in Chapter 2, the shape function matrix $\mathbf{S}^{rj}$ can be written as

$$\mathbf{S}^{rj} = \left[\bar{s}_1\mathbf{I}\ \bar{s}_2\mathbf{I}\ \bar{s}_3\mathbf{I}\ \bar{s}_4\mathbf{I}\ \bar{s}_5\mathbf{I}\ \bar{s}_6\mathbf{I}\ \bar{s}_7\mathbf{I}\ \bar{s}_8\mathbf{I}\right] \tag{4.25}$$

where shape functions $\bar{s}_i,\ \ i = 1, 2, \ldots, 8$ are defined as (Yakoub and Shabana 2001)

$$\left.\begin{aligned} &\bar{s}_1 = 1 - 3\xi^2 + 2\xi^3, \quad \bar{s}_2 = l(\xi - 2\xi^2 + \xi^3), \quad \bar{s}_3 = l(\eta - \xi\eta), \quad \bar{s}_4 = l(\varsigma - \xi\varsigma), \\ &\bar{s}_5 = 3\xi^2 - 2\xi^3, \quad \bar{s}_6 = l(-\xi^2 + \xi^3), \quad \bar{s}_7 = l\xi\eta, \quad \bar{s}_8 = l\xi\varsigma \end{aligned}\right\} \tag{4.26}$$

where l is the length of the element. $\xi = x/l = (s_1^r - s_1^{rj(k-1)})/l$, $\eta = y/l = s_2^r/l$, and $\varsigma = z/l$. To define the rail surface, parameter z must be expressed as a function of x and y.

Surface Parameterization The surface of an element can assume an arbitrary shape by writing coordinate z as a function of the two other coordinates x and y. To this end, one can write $z = f(x, y)$, where f is a function that defines the shape of the surface. For a rail surface, for example, one has $z = f(y)$ if the rail profile does not depend on the longitudinal parameter x. Using the equation $z = f(x, y)$ implies that an element surface is defined using two parameters $s_1^r = x$ and $s_2^r = y$. That is, on the surface of the element, one has

$$\bar{\mathbf{u}}^{rj}(x, y, z) = \bar{\mathbf{u}}^{rj}(x, y, f(x, y)) = \bar{\mathbf{u}}^{rj}(x, y) = \bar{\mathbf{u}}^{rj}(s_1^r, s_2^r) = \mathbf{S}^{rj}(s_1^r, s_2^r)\mathbf{e}^{rj} \tag{4.27}$$

The function $z = f(x, y)$ can assume any form and can also be represented numerically in a tabulated form. In the special and important case $z = f(y)$, a cubic spline function can be used to describe this function based on tabulated data that can be obtained for the rail profile using MiniProf device measurements.

4.6 ANCF COMPUTATION OF TANGENTS AND NORMAL

In Eq. 27, a volume representation is converted to a surface representation using the function $z = f(x,y)$. Using this ANCF geometric approach, much simpler expressions – compared to the expressions of the semi-analytical approach previously presented in this chapter – can be developed for the tangent and normal vectors at arbitrary points on the rail surface.

Using Eq. 27, one can write $d\bar{\mathbf{u}}^{rj} = (\partial\bar{\mathbf{u}}^{rj}/\partial x)dx + (\partial\bar{\mathbf{u}}^{rj}/\partial y)dy + (\partial\bar{\mathbf{u}}^{rj}/\partial z)dz$. Using the functional relationship $z = f(x, y)$, one has $dz = (\partial f/\partial x)dx + (\partial f/\partial y)dy$. It follows that

$$\begin{aligned} d\bar{\mathbf{u}}^{rj} &= (\partial\bar{\mathbf{u}}^{rj}/\partial x)dx + (\partial\bar{\mathbf{u}}^{rj}/\partial y)dy + (\partial\bar{\mathbf{u}}^{rj}/\partial z)dz \\ &= ((\partial\bar{\mathbf{u}}^{rj}/\partial x) + (\partial\bar{\mathbf{u}}^{rj}/\partial z)(\partial f/\partial x))dx + ((\partial\bar{\mathbf{u}}^{rj}/\partial y) + (\partial\bar{\mathbf{u}}^{rj}/\partial z)(\partial f/\partial y))dy \end{aligned} \tag{4.28}$$

This equation defines two tangent vectors at an arbitrary point on the surface of an element as

$$\left.\begin{aligned}\bar{\mathbf{t}}_1^{rj} &= (\partial\bar{\mathbf{u}}^{rj}/\partial s_1^r) = (\partial\bar{\mathbf{u}}^{rj}/\partial x) + (\partial\bar{\mathbf{u}}^{rj}/\partial z)(\partial f/\partial x)\\ \bar{\mathbf{t}}_2^{rj} &= (\partial\bar{\mathbf{u}}^{rj}/\partial s_2^r) = (\partial\bar{\mathbf{u}}^{rj}/\partial y) + (\partial\bar{\mathbf{u}}^{rj}/\partial z)(\partial f/\partial y)\end{aligned}\right\} \tag{4.29}$$

It is important to recognize the difference between x and y when used with and without the relationship $z = f(x, y)$. In the case when the relationship $z = f(x, y)$ is not used, the volume parameters x, y, and z are independent, and the three tangent vectors at any point x, y, and z of an element $(\partial\bar{\mathbf{u}}^{rj}/\partial x)$, $(\partial\bar{\mathbf{u}}^{rj}/\partial y)$, and $(\partial\bar{\mathbf{u}}^{rj}/\partial z)$ are independent. However, if the surface of an element is specified by the function $z = f(x, y)$, one has only two independent parameters $s_1 = x$ and $s_2 = y$, and the interpretation of the parameters x and y in this case is different from their interpretation when z is not specified on the surface. When the function relationship $z = f(x, y)$ is used, variations are allowed only on a surface. This is clear from the definition of the tangent vectors given by Eq. 29 in which $\bar{\mathbf{t}}_1^{rj} = (\partial\bar{\mathbf{u}}^{rj}/\partial x) = (\partial\bar{\mathbf{u}}^{rj}/\partial s_1^r)$ with $x = s_1^r$ and $\bar{\mathbf{t}}_2^{rj} = (\partial\bar{\mathbf{u}}^{rj}/\partial y) = (\partial\bar{\mathbf{u}}^{rj}/\partial s_2^r)$ with $y = s_2^r$ have different directions, magnitudes, and interpretations as compared to the case when the relationship $z = f(x, y)$ is not used.

Using Eq. 29, the unit normal vector to the surface of an element can be defined as

$$\begin{aligned}\hat{\bar{\mathbf{n}}}^{rj} &= \frac{(\partial\bar{\mathbf{u}}^{rj}/\partial s_1^r)\times(\partial\bar{\mathbf{u}}^{rj}/\partial s_2^r)}{|(\partial\bar{\mathbf{u}}^{rj}/\partial s_1^r)\times(\partial\bar{\mathbf{u}}^{rj}/\partial s_2^r)|}\\ &= \frac{((\partial\bar{\mathbf{u}}^{rj}/\partial x) + (\partial\bar{\mathbf{u}}^{rj}/\partial z)(\partial f/\partial x))\times((\partial\bar{\mathbf{u}}^{rj}/\partial y) + (\partial\bar{\mathbf{u}}^{rj}/\partial z)(\partial f/\partial y))}{|((\partial\bar{\mathbf{u}}^{rj}/\partial x) + (\partial\bar{\mathbf{u}}^{rj}/\partial z)(\partial f/\partial x))\times((\partial\bar{\mathbf{u}}^{rj}/\partial y) + (\partial\bar{\mathbf{u}}^{rj}/\partial z)(\partial f/\partial y))|}\end{aligned} \tag{4.30}$$

As in the analytical approach, one can use the equation for position vector $\bar{\mathbf{u}}^{rj}$ of an arbitrary point with respect to the rail coordinate system $X^rY^rZ^r$ to determine position vector $\mathbf{r}^{rj}$ of arbitrary points on ANCF element j in the global coordinate system XYZ as

$$\mathbf{r}^{rj} = \mathbf{R}^r + \mathbf{A}^r\bar{\mathbf{u}}^{rj} \tag{4.31}$$

where $\mathbf{R}^r$ is the position vector of the origin of the selected rail coordinate system $X^rY^rZ^r$ in the global coordinate system XYZ, $\mathbf{A}^r$ is the transformation matrix that defines the orientation of the rail coordinate system, and $\bar{\mathbf{u}}^{rj}$ is defined by Eq. 27, which is much simpler than Eq. 11 given by $\bar{\mathbf{u}}^r(s_1^r, s_2^r) = \bar{\mathbf{R}}^{rp}(s_1^r) + \mathbf{A}^{rp}(s_1^r)\bar{\bar{\mathbf{u}}}^{rp}(s_1^r, s_2^r)$ in the semi-analytical approach. Because $\mathbf{R}^r$ and $\mathbf{A}^r$ do not depend on the rail surface parameters, one can write the rail tangent and normal vectors defined in the global coordinate system XYZ as

$$\left.\begin{aligned}\mathbf{t}_1^{rj} &= \frac{\partial\mathbf{r}^{rj}}{\partial s_1^r} = \mathbf{A}^r\left(\frac{\partial\bar{\mathbf{u}}^{rj}}{\partial s_1^r}\right) = \mathbf{A}^r\bar{\mathbf{t}}_1^{rj},\\ \mathbf{t}_2^{rj} &= \frac{\partial\mathbf{r}^{rj}}{\partial s_2^r} = \mathbf{A}^r\left(\frac{\partial\bar{\mathbf{u}}^{rj}}{\partial s_2^r}\right) = \mathbf{A}^r\bar{\mathbf{t}}_2^{rj},\\ \mathbf{n}^{rj} &= \mathbf{t}_1^{rj}\times\mathbf{t}_2^{rj} = (\mathbf{A}^r\bar{\mathbf{t}}_1^{rj})\times(\mathbf{A}^r\bar{\mathbf{t}}_2^{rj}) = \mathbf{A}^r(\bar{\mathbf{t}}_1^{rj}\times\bar{\mathbf{t}}_2^{rj}) = \mathbf{A}^r\bar{\mathbf{n}}^{rj}\end{aligned}\right\} \tag{4.32}$$

where the derivatives of $\bar{\mathbf{u}}^{rj}$ with respect to the surface parameters s_1^r and s_2^r are as previously defined in this section. Similarly, the curvature vectors can be defined in the global coordinate system XYZ as

$$\left.\begin{aligned}
\frac{\partial^2 \mathbf{r}^{rj}}{\partial (s_1^r)^2} &= \frac{\partial \mathbf{t}_1^{rj}}{\partial s_1^r} = \mathbf{A}^r\left(\frac{\partial^2 \bar{\mathbf{u}}^{rj}}{\partial (s_1^r)^2}\right) = \mathbf{A}^r\left(\frac{\partial \bar{\mathbf{t}}_1^{rj}}{\partial s_1^r}\right)\\
\frac{\partial^2 \mathbf{r}^{rj}}{\partial (s_2^r)^2} &= \frac{\partial \mathbf{t}_2^{rj}}{\partial s_2^r} = \mathbf{A}^r\left(\frac{\partial^2 \bar{\mathbf{u}}^{rj}}{\partial (s_2^r)^2}\right) = \mathbf{A}^r\left(\frac{\partial \bar{\mathbf{t}}_2^{rj}}{\partial s_2^r}\right)\\
\frac{\partial^2 \mathbf{r}^{rj}}{\partial s_1^r \partial s_2^r} &= \frac{\partial \mathbf{t}_1^{rj}}{\partial s_2^r} = \frac{\partial \mathbf{t}_2^{rj}}{\partial s_1^r} = \mathbf{A}^r\left(\frac{\partial^2 \bar{\mathbf{u}}^{rj}}{\partial s_1^r \partial s_2^r}\right) = \mathbf{A}^r\left(\frac{\partial \bar{\mathbf{t}}_1^{rj}}{\partial s_2^r}\right) = \mathbf{A}^r\left(\frac{\partial \bar{\mathbf{t}}_2^{rj}}{\partial s_1^r}\right)
\end{aligned}\right\} \tag{4.33}$$

The derivatives with respect to the surface parameters of the rail surface normal vector defined in the global XYZ coordinate system can be developed in a similar manner. The tangent vectors $\mathbf{t}_1^{rj} = (\partial \mathbf{r}^{rj}/\partial x) = (\partial \mathbf{r}^{rj}/\partial s_1^r)$ with $x = s_1^r$ and $\bar{\mathbf{t}}_2^{rj} = (\partial \mathbf{r}^{rj}/\partial y) = (\partial \mathbf{r}^{rj}/\partial s_2^r)$ with $y = s_2^r$ can be used to define the coefficients of the first fundamental form. The definition of the normal vector given in Eq. 32 and the second derivatives of vector $\mathbf{r}^{rj}$ with respect to $x = s_1^r$ and $y = s_2^r$ can be used to define the coefficients of the second fundamental form. These derivatives can be systematically evaluated using the element shape function matrix and vector of nodal coordinates of Eq. 27. The coefficients of the first and second fundamental forms of surfaces obtained using ANCF geometry can be used to systematically define the principal curvatures and directions of surfaces at wheel/rail contact points using the procedure discussed in Chapter 2.

Based on a simple track description, a track preprocessor computer program can be used to determine Euler angles at a number of discrete nodal points specified by the program user. Two choices can be made when discrete-node data are used in the algorithm of the main processor. The first choice is to convert discrete Euler angles to field variables by using linear interpolation within a segment and then use this angle field representation with the semi-analytical approach to define the position, tangent, normal, and curvature vectors as previously described in this chapter. This approach requires interpolating vector $\mathbf{R}^{rp}(s_1^r)$ and defining matrix $\mathbf{A}^{rp}$ of the semi-analytical approach. In the second approach, Euler angles at the discrete nodal points are used to evaluate ANCF position gradients; in this case, the angles are not treated as field variables and are not interpolated during the dynamic simulations. Furthermore, there is no need to define vector $\mathbf{R}^{rp}(s_1^r)$ and matrix $\mathbf{A}^{rp}$. This ANCF interpolation approach also allows for having a cubic interpolation for the position coordinates despite the fact that only two pieces of information are provided at the element nodes for each coordinate.

4.7 TRACK GEOMETRY EQUATIONS

As shown in the preceding section, using the ANCF interpolation approach can significantly simplify the calculations of the position, tangent, normal, and curvature vectors that are required to define rail geometry. Because position vector gradients are used as nodal coordinates, ANCF elements allow for describing complex geometry with a high degree of accuracy. Euler angles at selected nodes on a rail or track space curve can be used to define ANCF position-gradient vectors. These Euler angles completely define the rail space curve geometric properties.

This section summarizes some of the basic equations obtained in Chapter 3, which are the basis for computing track geometry. These equations are used to explain how simple track input data can be used to develop a *track file* to be used in dynamic computer simulations of railroad vehicle system models. This track file defines the position of and Euler angles at selected nodes on the track space curves. This information, which defines the track geometry, is used as an input to an MBS computer program that automatically constructs and numerically solves the nonlinear dynamic equations of complex railroad vehicle systems. In the development presented in the remaining sections of this chapter, to develop general railroad vehicle system models in which two rails can move independently, it is assumed that the track file has three different meshes that correspond to three different space curves: the *track space curve*, sometimes referred to as the *track centerline*, which is used as a reference to define vehicle locations and forward velocities; the *right rail space curve*; and the *left rail space curve*. The locations and geometric properties of these space curves are completely defined by the position of and Euler angles at the nodes of their space curves that are constructed using simple industry inputs as described in this chapter. Since the procedures described are applicable to any space curve (track, and right and left rails), the superscripts are dropped for the sake of simplicity and generality.

In practice, track geometry is defined using simple, specific data provided at the points where the geometry changes. Only three inputs are used to define each rail segment (tangent, curve, and spiral) regardless of the length of the segment (Dukkipati and Amyot 1988; Berzeri et al. 2000; Rathod and Shabana 2006a). These simple inputs, as discussed in Chapter 2, are a *projection* of the rail space curve on the horizontal plane to define a planar curve with curvature C_H, called the *horizontal curvature*; a *development* or *grade* that defines a *vertical-development angle* θ about the rail or track Y axis; and a *superelevation* that defines a *bank angle* ϕ about the longitudinal tangent to the rail or track space curve. The three variables C_H, θ, and ϕ completely and uniquely define the position coordinates and geometry of any track segment. Table 4 shows an example of the data used for defining the geometry of the track shown in Figure 6. These data are also used as input to the track preprocessor.

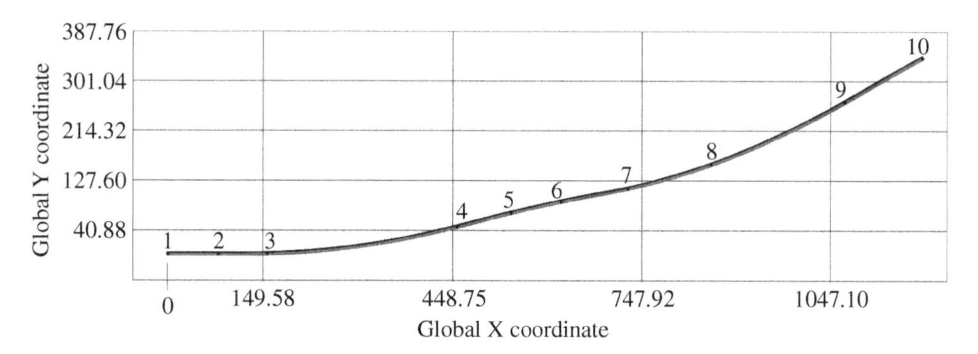

Figure 4.6 Track described in the data in Table 4.

As discussed in Chapter 3, a track or rail space curve can be defined by considering the first column in the transformation matrix of Eq. 20 as the unit tangent vector. In this case, one can write a segment of the space curve as $d\mathbf{r} = (\partial\mathbf{r}/\partial s)ds = \left[\cos\psi\cos\theta \;\; \sin\psi\cos\theta \;\; \sin\theta\right]^T ds$. The projection of this segment on the horizontal plane is

Table 4.4 Example of input data to a track preprocessor. Source: Courtesy of Shabana, A.A., Zaazaa, K.E., and Sugiyama, H.

Node #	Distance (ft)	Curvature (°)	Superelevation (in.)	Grade (%)
1	0	0	0	0
2	100	0	0	0
3	150	5	1.5	0
4	450	5	1.5	0
5	500	−3	−1	0
6	650	−3	−1	0
7	720	7	2	0
8	1020	7	2	0
9	1145	0	0	0
10	1195	0	0	0

defined by the equation $d\mathbf{r}_H = \begin{bmatrix}\cos\psi\cos\theta & \sin\psi\cos\theta & 0\end{bmatrix}^T ds$. By factoring out $\cos\theta$, one can write $d\mathbf{r}_H = \begin{bmatrix}\cos\psi & \sin\psi & 0\end{bmatrix}^T \cos\theta ds$. This equation, as discussed in Chapter 2, defines vector $\mathbf{t}_H$ tangent to and arc length dS of the projected planar curve, respectively, as

$$\mathbf{t}_H = \begin{bmatrix}\cos\psi & \sin\psi & 0\end{bmatrix}^T, \qquad dS = \cos\theta ds \tag{4.34}$$

Using the relationship between space curve arc length s and arc length S of the projected planar curve, one has $\partial\psi/\partial S = (\partial\psi/\partial s)(\partial s/\partial S)$, where $\partial s/\partial S = 1/\cos\theta$. The curvature vector of the projected planar curve is defined as

$$\mathbf{C}_H = \frac{\partial \mathbf{t}_H}{\partial S} = \begin{bmatrix}-\sin\psi \\ \cos\psi \\ 0\end{bmatrix}\frac{\partial\psi}{\partial S} \tag{4.35}$$

This equation defines the horizontal curvature of the projected planar curve, which is the magnitude of curvature vector $\mathbf{C}_H$, as $C_H = |\mathbf{C}_H| = \partial\psi/\partial S$, or

$$d\psi = C_H dS = \frac{dS}{R_H} \tag{4.36}$$

In this equation, R_H is the radius of curvature of the projected planar curve. It is clear from this equation that angle ψ can be determined by integration if the horizontal curvature C_H is given. The other two Euler angles θ and ϕ are assumed known from the grade (elevation) and superelevation (bank) data. Therefore, knowing the Euler angles, the preceding equations can be used to define angle ψ over a domain defined by endpoints S_o and S_1 as

$$\psi = \psi(S) = \psi_o + \int_{S_o}^{S_1} (C_H(S))dS \tag{4.37}$$

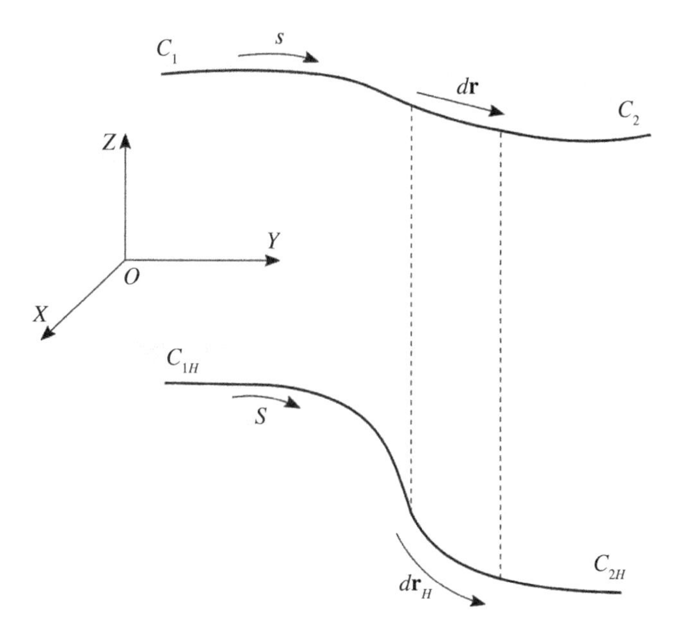

Figure 4.7 Horizontal curvature.

where ψ_o is the value of ψ at S_o. Figure 7 shows an illustration of the curve projection just discussed. In this figure, the segment $d\mathbf{r}$ of a space curve defined by endpoints C_1 and C_2 and whose arc length is s is projected on the horizontal plane to define a planar curve segment $d\mathbf{r}_H$ that has endpoints C_{H1} and C_{H2} and arc length S.

Recall that the sequence $Z, -Y, -X$ of Euler angles is selected to have positive definitions for the horizontal curvature, grade, and superelevation. Curvature is considered positive if it is the result of a positive rotation ψ about the Z axis; vertical development is considered positive if it is the result of rotation θ about the $-Y$ axis; and superelevation is considered positive if it is the result of rotation ϕ about the $-X$ axis. This convention is consistent with measurements made in practice. It is also important to point out that the equations presented in this section remain applicable when the track is not fixed. In such cases, the equations use the horizontal plane before displacements.

In addition to determining angles at arbitrary points on the space curve from track input data, one needs to determine the position coordinates of these points. Using the first column of the transformation matrix of Eq. 20 as the tangent vector to the space curve allows writing the curve equation in the differential form $d\mathbf{r} = \begin{bmatrix}\cos\psi\cos\theta & \sin\psi\cos\theta & \sin\theta\end{bmatrix}^T ds$. Using Eq. 34, which defines the relationship between s and S, the equation for $d\mathbf{r}$ can be written as

$$d\mathbf{r} = \begin{bmatrix}\cos\psi dS \\ \sin\psi dS \\ \sin\theta ds\end{bmatrix} \tag{4.38}$$

This equation, upon integration, can be used to define the curve $\mathbf{r} = \begin{bmatrix} r_1 & r_2 & r_3 \end{bmatrix}^T$ in its parametric form as

$$\mathbf{r} = \begin{bmatrix} r_{o1} \\ r_{o2} \\ r_{o3} \end{bmatrix} + \begin{bmatrix} \int_{S_o}^{S_1} \cos\psi dS \\ \int_{S_o}^{S_1} \sin\psi dS \\ \int_{s_o}^{s_1} \sin\theta ds \end{bmatrix} = \begin{bmatrix} r_1(S) \\ r_2(S) \\ r_3(s) \end{bmatrix} \tag{4.39}$$

In this equation, $\mathbf{r}_o = \begin{bmatrix} r_{o1} & r_{o2} & r_{o3} \end{bmatrix}^T$ is vector $\mathbf{r}$ at the start of the segment This equation shows clearly that the coordinates of points on the space curve can be determined using the two Euler angles, which are assumed to be known from the horizontal curvature and grade. That is, if C_H and θ are known from the track inputs, the locations of the points on the space curve and Euler angles that define the curve geometry can be completely defined for tangent, curve, and spiral rail sections. The tangent section has zero curvature, the curve section has constant curvature, and the spiral section is designed to have linearly varying horizontal curvature, as discussed in the following section.

4.8 NUMERICAL REPRESENTATION OF TRACK GEOMETRY

Using the three parameters – horizontal curvature C_H, vertical-development angle θ, and bank angle ϕ – a numerical procedure based on simple interpolation can be used to determine the position of the nodes and the three Euler angles at these nodes for the track, right rail, and left rail space curves. Using the numerical procedure is necessary since it is not possible to obtain a closed-form solution for the track equations presented in the preceding section. The nodal position and angle data, which can be determined in a preprocessor computer program prior to dynamic simulations, are required to determine the locations of wheel/rail contact points when the governing dynamic equations are numerically solved by the main processor computer program.

As discussed in Chapter 3, track consists of a few segment types. In the design of the track segments, the rail is divided into segments, and geometric variables are assumed to vary linearly within a segment. Using linear interpolation for the geometry variables allows obtaining a solution with end conditions that are provided in the description of the track. Using higher-order interpolation requires additional data that are generally not available. Linear interpolation can be used to define the position coordinates of and Euler angles at an arbitrary number of nodes specified by the user.

Rail Segment Types As discussed in Chapter 3, the track segment types include a *tangent segment* that represents a straight section with zero horizontal curvature: that is, $C_H = 0$; a curve segment with constant horizontal curvature $C_H = 1/R_H$; a *tangent-to-curve entry spiral section* with horizontal curvature $C_H = 1/R_H$ that varies linearly as a function of the projected arc length S from zero at the first point to the value of the horizontal curvature of the curve at the segment endpoint; a *curve-to-tangent exit* spiral with horizontal curvature $C_H = 1/R_H$ that varies linearly as a function of the projected arc length S from the value of the horizontal curvature of the curve at the first point to zero at the segment endpoint; and a *curve-to-curve spiral section* with horizontal curvature $C_H = 1/R_H$ that varies linearly as a function of the projected arc length S from the value of the horizontal curvature of the first curve at the first point to the value of the horizontal curvature of the second curve at the segment endpoint. The linear variation of vertical-development angle θ and bank angle ϕ within the segments is also assumed in most algorithms used to construct the track layout.

Linear Interpolation of Angles Assuming that horizontal curvature C_H varies linearly within a rail segment, the following linear interpolation can be used for C_H within the segment

$$C_H = (1 - \xi_S)C_{H1} + \xi_S C_{H2} \tag{4.40}$$

where $\xi_S = (S - S_1)/L_S$, $L_S = (S_2 - S_1)$, and C_{H1}, C_{H2}, S_1, and S_2 are the values of the horizontal curvature C_H and the projected arc length S at the first and second nodes of the segment, respectively. Using the linear interpolation of horizontal curvature C_H defined by the preceding equation and Eq. 37, which defines angle ψ, one can obtain a closed-form expression of ψ at an arbitrary point S within the segment in terms of the values of horizontal curvatures C_{H1} and C_{H2} at the two segment endpoints as

$$\psi(S) = \psi_1 + \frac{1}{2L_S}[C_{H2}(S - S_1)^2 + C_{H1}((L_S)^2 - (S - S_2)^2)] \tag{4.41}$$

In this equation, ψ_1 defines angle ψ at the first node of the segment. The preceding equation can be used to define angle ψ at the desired discrete nodal points used in the track file produced by the track preprocessor.

Vertical-development angle θ is assumed to vary with respect to arc length s according to the equation

$$d\theta = C_V ds \tag{4.42}$$

where C_V is a constant referred to in the railroad literature as the *vertical curvature*. As shown in Chapter 3, the curvature vector of a space curve can be written in terms of Euler angles and their derivatives as

$$\mathbf{c} = \begin{bmatrix} -\psi' \sin\psi \cos\theta - \theta' \cos\psi \sin\theta \\ \psi' \cos\psi \cos\theta - \theta' \sin\psi \sin\theta \\ \theta' \cos\theta \end{bmatrix} \tag{4.43}$$

where $a' = \partial a/\partial s$. This curvature vector can be written as a linear combination of two orthogonal unit vectors as

$$\mathbf{c} = \psi' \cos\theta \begin{bmatrix} -\sin\psi \\ \cos\psi \\ 0 \end{bmatrix} + \theta' \begin{bmatrix} -\cos\psi \sin\theta \\ -\sin\psi \sin\theta \\ \cos\theta \end{bmatrix} \tag{4.44}$$

The two orthogonal unit vectors in this equation are also orthogonal to the unit tangent to the space curve defined by the first column of the transformation matrix in Eq. 20. Therefore, the curvature vector has two components: $\psi' \cos\theta$ along the unit vector $[-\sin\psi \quad \cos\psi \quad 0]^T$ and $\theta' = \partial\theta/\partial s$ along the unit vector

$\begin{bmatrix}-\cos\psi\sin\theta & -\sin\psi\sin\theta & \cos\theta\end{bmatrix}^T$. The vector $\begin{bmatrix}-\cos\psi\sin\theta & -\sin\psi\sin\theta & \cos\theta\end{bmatrix}^T$ is the same as the vertical vector $\begin{bmatrix}0 & 0 & 1\end{bmatrix}^T$ before performing the two rotations ψ and θ; therefore, this vector is perpendicular to the plane of the track before the superelevation, justifying calling C_V the vertical curvature. Assuming that θ varies linearly with respect to s, the vertical curvature becomes constant within the segment and can be written as

$$C_V = \frac{d\theta}{ds} = \frac{\theta_2 - \theta_1}{s_2 - s_1} \tag{4.45}$$

where θ_1 and θ_2 are the vertical-development angles at the segment endpoints defined by the arc length parameters s_1 and s_2, respectively. It follows that vertical-development angle θ at an arbitrary point within the segment can be written as

$$\theta(s) = \theta_1 + C_V(s - s_1) \tag{4.46}$$

This equation can be used to define angle θ at the desired discrete nodal points used in the track file produced by the track preprocessor.

Similarly, bank angle ϕ can be defined using a function that is linear in the projected arc length S as

$$\begin{aligned}\phi = \phi(S) &= (1 - \xi_S)\phi_1 + \xi_S\phi_2 \\ &= \frac{\phi_2(S - S_1) - \phi_1(S - S_2)}{(S_2 - S_1)}\end{aligned} \tag{4.47}$$

where $\xi_S = S/L_S$, $L_S = S_2 - S_1$, and ϕ_1 and ϕ_2 are the bank angles at the two end nodes of the segments defined, respectively, by the projected arc length coordinates S_1 and S_2. As in the case of the other two angles ψ and θ, the preceding equation can be used to define angle ϕ at the desired discrete nodal points used in the track file produced by the track preprocessor. The distance between the nodes can be selected to ensure achieving the desired accuracy of the interpolation during computer simulations. Using a 1 ft distance between nodes in a track file is common, so the track file can be very large.

Recent investigations have demonstrated that the three Euler angles ψ, θ, and ϕ used to describe the geometry of a curve cannot all be treated as independent geometric variables. Therefore, the rail industry numerical procedure described in this section for determining the angles must be carefully evaluated for spiral geometry. The tangent and curve segments of a track have zero and constant curvature, respectively. A curve with constant curvature should, in general, be superelevated using a constant bank angle to achieve balance speed requirements. This is not the case, however, for a spiral segment that has varying curvature. Bank angle ϕ, as explained in the rail industry procedure for numerically constructing track geometry, is used with the other angles ψ and θ to determine the orientation of the track (body trajectory) frames, but such a linearly interpolated bank angle does not enter into the definition of the spiral geometry, which is completely defined using the angles ψ and θ (Shabana and Ling, 2019).

Numerical Evaluation of Position Coordinates As shown in Table 4, track geometry is described by using simple inputs in a preprocessor computer program to define the positions and Euler angles at discrete points on the rail or track space curve. Regardless of the length of a track or rail segment (tangent, curve, or spiral), only inputs at track locations where the geometry changes can be provided. That is, for a curve section of the track, which can be miles long, information is provided only at the beginning and end of the section. Linear interpolations discussed in this section are used in practice in the track layout. However, to determine the positions of the track nodes, numerical integration is necessary since closed-form solutions cannot be obtained.

Recall from Eq. 39 that position coordinates are defined in terms of angles using the integral forms $r_1(S) = r_{o1} + \int_{S_o}^{S_1} \cos\psi dS$, $r_2(S) = r_{o2} + \int_{S_o}^{S_1} \sin\psi dS$, and $r_3(s) = r_{o3} + \int_{S_o}^{S_1} \sin\theta ds$. While angle ψ is defined in terms of projected arc length S, angle θ is defined in terms of the space curve arc length s. Equation 34 shows that S and s are related by the equation $dS = \cos\theta ds$, and Eq. 42 shows that $d\theta = C_V ds$: that is, $ds = d\theta/C_V$. It follows that $dS = \cos\theta(d\theta/C_V)$: that is,

$$\frac{d\theta}{dS} = \frac{C_V}{\cos\theta} \tag{4.48}$$

which defines the derivative of vertical-development angle θ with respect to projected arc length S. The equation $dS = (\cos\theta/C_V)d\theta$, upon integration, can be used to define the projected arc length within a segment as (Berzeri et al. 2000)

$$S(s) = \begin{cases} S_1 + \frac{1}{C_V}(\sin\theta(s) - \sin\theta_1) & \text{if } C_V \neq 0 \\ S_1 + (s - s_1)\cos\theta(s) & \text{if } C_V = 0 \end{cases} \tag{4.49}$$

The definition of $S = S(s)$ from this equation can be used to evaluate the angles in Eqs. 41 and 47 as well as the coordinates of the nodal points on the space curve defined using numerical integration of the equations $r_1(S) = r_{o1} + \int_{S_o}^{S_1} \cos\psi dS$, $r_2(S) = r_{o2} + \int_{S_o}^{S_1} \sin\psi dS$, and $r_3(s) = r_{o3} + \int_{S_o}^{S_1} \sin\theta ds$.

Summary It is clear from the analysis presented in this section that knowing the horizontal curvature C_H, vertical-development angle θ, and bank angle ϕ at specific points where track geometry changes allows developing a numerical procedure based on interpolation to determine the angles, which can be used with numerical integration to define the position coordinates and angles at an arbitrary number of points on the rail or track space curve as a function of arc length parameter s. This is typically accomplished, as previously mentioned, in a track preprocessor that generates a track file that has, for a given space curve, a number of nodes specified by the user. For each node of a space curve, the file provides the corresponding entries s, r_1, r_2, r_3, ψ, θ, and ϕ. Additional information, such as horizontal curvature C_H at the nodes, can also be printed out by the track preprocessor. An organization of such a track preprocessor is discussed in the following section.

4.9 TRACK DATA

Before performing dynamic simulations of railroad vehicle systems, a data file that describes the geometry of the track space curve and the right and left rail space curves is often be created in a preprocessor computer program. In this track data file, the track and rail space curve geometries are described using the positions of and Euler angles at discrete nodes determined using the equations presented in the preceding section. During dynamic simulations, wheel/rail contact conditions are used to solve for surface parameters, as explained in Chapter 6. These surface parameters define points on the rails that come into contact with the wheels. Using the value of the rail longitudinal surface parameters s at the contact point, the segment of rail within which contact occurs can be identified. The positions and Euler angles at the two ends of this segment can be used with interpolation methods to determine the tangent, normal, and curvature vectors previously defined in this chapter. By using ANCF finite elements, interpolation of the angles is avoided, and the position vectors that define the tangent, normal, and curvature vectors are defined using cubic interpolations.

Track Preprocessor Using a track preprocessor to develop a track data file prior to dynamic simulations is recommended for the computational efficiency of the solution algorithm. The input to the track preprocessor, as shown by the data in Table 4, is simple and defines horizontal curvature C_H; the grade defining vertical-development angle θ; and the superelevation, which defines bank angle ϕ. As described in the preceding section, these three inputs can be used to completely define the track nodal positions and Euler angles at an arbitrary number of points on the rail space curve specified by the user. More nodal points lead to more accurate interpolations; therefore, the track file for long tracks can be relatively large, as previously mentioned. It is also important to note that the three inputs (horizontal curvature, grade, and superelevation) need to be provided only at track points where the geometry changes, as previously mentioned. Therefore, regardless of the number of nodal points selected by analysts, the input file to the track preprocessor contains a relatively small amount of data.

The track preprocessor can be designed to read a set of simple standard input data that are sufficient for creating discrete-point information used in nonlinear dynamic simulations of a vehicle/track system. It is clear from the analysis presented in the preceding section that the track preprocessor is required to perform the following specific functions (Berzeri et al. 2000):

1. Read the input data that define the geometry of the track segments along the track centerline. These input data define horizontal curvature C_H, the grade, and the superelevation only at points where the track geometry changes. The grade and superelevation are used to define vertical-development angle θ and bank angle ϕ, respectively, at the specified points.
2. Use horizontal curvature C_H, vertical-development angle θ, and bank angle ϕ with linear interpolation to create field variables $C_H(S)$, $\theta(s)$, and $\phi(S)$. Using horizontal curvature C_H, angle $\psi(S)$ can be defined using Eq. 41. Therefore, the values of the three Euler angles ψ, θ, and ϕ can be determined at the discrete nodal points selected by the user using linear interpolation in the track preprocessor.
3. Use Euler angles $\psi(S)$ and $\theta(s)$ and the methods of numerical integration to compute coordinates r_1, r_2, and r_3 of the nodes selected by the user on the track centerline, as discussed in the preceding section. The nodal positions can be defined with respect to the track body coordinate system in order to develop more general algorithms for simulation scenarios in which the track is not fixed.
4. Use track space curve data to define the right and left rail space curves in terms of discrete points. The right and left rail space curves can have different position coordinates and

different Euler angles at their nodes to allow for an independent description of the rail geometries. The different nodes of the different space curves can be due to superelevation, track irregularities, curvature, etc. There is no loss of generality in assuming that the track and rail body coordinate systems initially coincide, since the dynamic simulation algorithm can allow these coordinate systems to move independently.

5. Produce an output file that can be read by the main processor and used in dynamic simulations. This output file can define the track and rail geometry mesh at discrete nodal points. For node k on space curve j, the track preprocessor can provide, at minimum, the information $\begin{bmatrix} s^{jk} & r_1^{jk} & r_2^{jk} & r_3^{jk} & \psi^{jk} & \theta^{jk} & \phi^{jk} \end{bmatrix}$, where $j = t, rr, lr$, and t, rr, and lr refer, respectively, to the track, right rail, and left rail space curves. Additional information, such as the horizontal curvature, can also be included in the output file. ANCF interpolation can be used during dynamic simulations to obtain the tangent, normal, and curvature vectors at points within the segments, as previously discussed in this chapter.

Input Data to the Track Preprocessor As presented in Table 4, the input data to the track preprocessor that generates the track file meshes of the space curves are simple entries. These entries are provided at the points that define track segments. For a tangent section, which can be many miles long, only entries at the beginning and end of the section need to be provided because the section geometry remains constant. The same is for a curve, which has a constant curvature. The input at the beginning and end of a segment can be used to completely define the geometry of the segment. In addition to location s at the beginning and end of a segment, three specific geometry inputs are provided: the horizontal curvature, superelevation, and grade. This information is provided in Table 4, which shows an example of the data used as input to the track preprocessor: the node number, distance at which the node is located, horizontal curvature, superelevation, and grade at points where the geometry changes. In North America, the distance at which the node is located along the track centerline from a given origin is measured in feet. The horizontal curvature is defined as the number of degrees traversed by a 100 ft track: that is, the number of degrees for a 100 ft-chord arc length that encompasses angle ψ, as shown in Figure 8. As discussed before, the horizontal curvature can be used to determine angle ψ. For a tangent track, the horizontal curvature is zero. The fourth column in Table 4 provides values of superelevation h at the nodes in inches. These values are used to determine bank angle ϕ. The last column in Table 4 shows the grade, which is used to define the climb between two points and vertical-development angle θ. A grade value of 5.89% that defines a track rise of 413 ft over a distance of 7012 ft is considered large. Some railroad tracks in North America have a 4.7% grade value, which is also considered large since the grade of rail track rarely exceeds 1%. Very steep grades are not normally used in track design, to avoid wasting energy in braking scenarios.

The values given in Table 4, which correspond to the track space curve shown in Figure 6, can be used as examples to explain how simple track inputs relate to segment geometry. Based on the definitions given in the preceding sections, the first segment is a tangent, the second segment is a tangent-to-curve entry spiral, the third segment is a curve, the fourth segment is a curve-to-curve spiral, the fifth segment is a curve, the sixth segment is again a curve-to-curve spiral, the seventh segment is a curve, the eighth segment is a curve-to-tangent exit spiral, and the ninth segment is a tangent.

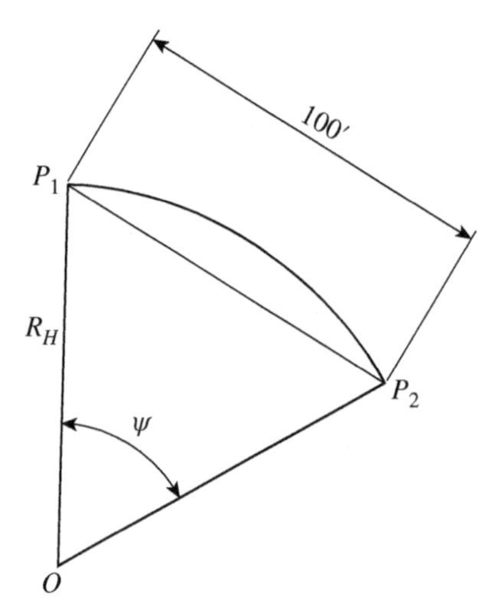

Figure 4.8 Horizontal curvature.

Computing Angles and Coordinates Given entries like those in Table 4, the Euler angles at a node can be determined and used to define nodal positions as previously described. The value of the horizontal curvature is often given using the following equation:

$$C_H = \frac{\sin(\psi/2)}{50'} \tag{4.50}$$

It is clear from this equation that Euler angle ψ is given using the table entries as $\psi = 2\sin^{-1}(50C_H)$. The values of superelevation h at the track nodes given in the fourth column in Table 4 and the value of gage G can be used to define bank angle ϕ as

$$\sin\phi = \frac{h}{d} \tag{4.51}$$

where d, shown in Figure 9, can be determined using the gage values. Finally, the grade, which describes the track climb and is defined in the fifth column of Table 4, can be used to calculate vertical-development angle θ at the nodes. Because the grade is the number of units the track rises vertically divided by the number of units traveled horizontally, the grade can be used to define $\sin\theta$.

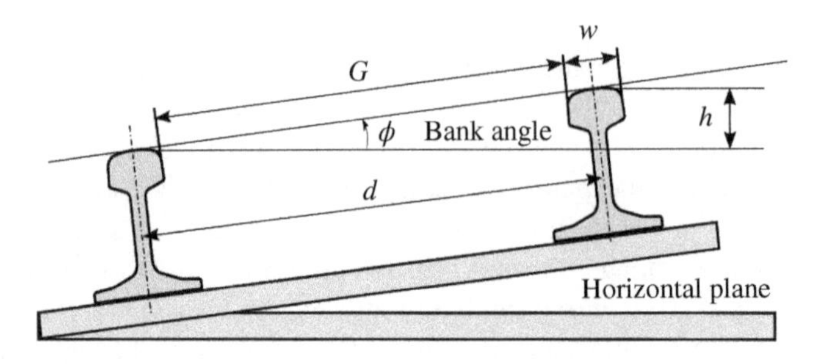

Figure 4.9 Bank angle.

As previously explained in this chapter, once the values of Euler angles ψ, θ, and ϕ are determined using the input to the track preprocessor at a relatively small number of points, these angles can be determined at an arbitrary number of points specified by the user using the interpolation described previously in this chapter. Interpolation of angles as functions of the arc length can also be used to determine the position coordinates of nodes specified by the user using Eq. 39, which cannot be evaluated in a closed form. Therefore, these position coordinates are evaluated using numerical integration methods, including the trapezoidal and Simpson rules, and Gaussian quadrature (Atkinson 1978).

Lengths of the Right and Left Rails Figure 10 illustrates that for a curved track, the length of the track space curve used in developing the entries in Table 4 is not the same as the lengths of the space curves of the right and left rails. The differences between the lengths of these three space curves can be significant and should not be ignored when computing position coordinates or angles at the discrete points specified by the user. For this reason, the lengths of the right and left rails must be properly calculated. Using Figure 10, one can show that the change in the radius of curvature of the curve projected on the horizontal plane can be written as

$$\Delta R_H = d\cos\phi \tag{4.52}$$

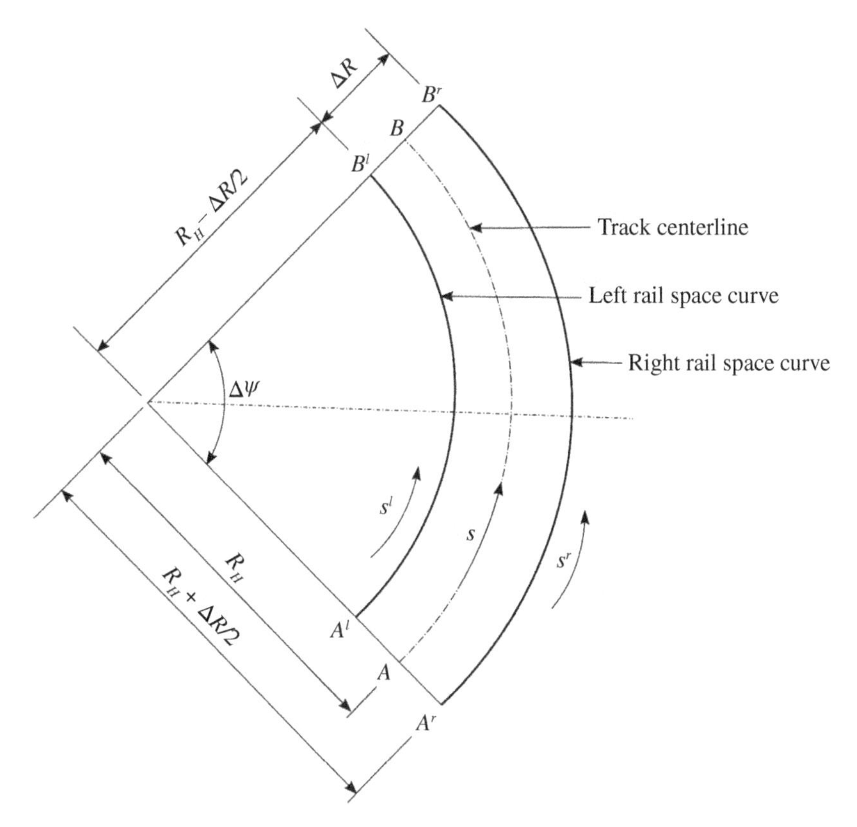

Figure 4.10 Right and left rail lengths.

where d is defined using the gage shown in Figure 9. One can then write

$$dS^j = (R_H \pm (\Delta R_H/2))d\psi$$
$$= (R_H \pm (d/2)\cos\phi)d\psi = (1 \pm C_H(d/2)\cos\phi)dS, \quad j = rr, lr \tag{4.53}$$

In this equation, superscripts rr and lr refer, respectively, to the right and left rails; the positive sign corresponds to the right rail, while the negative sign corresponds to the left rail for a positive curvature angle ψ. Using the equation $dS = (\cos\theta)ds$, which relates the projected arc length of the track space curve to the actual arc length, one can then write the projected arc length of a segment starting at a point defined by s_1 as

$$S(s) = S_1 + \int_{s_1}^{s} \cos\theta(s)ds \tag{4.54}$$

For the right and left rails, one can also write

$$ds^j = \frac{dS^j}{\cos\theta} = (1 \pm C_H(d/2)\cos\phi)ds, \quad j = rr, lr \tag{4.55}$$

This equation, which defines the difference between ds and ds^j, $j = rr$, lr, can be used to compute the change in the length Δl^j, $j = rr$, lr, of the segments of the right and left rails using the data of the track space curve as

$$\Delta l^j = \pm\frac{1}{2}\int_{s_1}^{s} C_H d\cos\phi ds, \quad j = rr, lr \tag{4.56}$$

This equation can be used to evaluate the lengths of the right and left rail segments using the equation

$$l^j = l \pm \Delta l^j, \quad j = rr, lr \tag{4.57}$$

By using this length-adjustment scheme, it is clear that the values of the position coordinates and Euler angles at the corresponding nodes of the track, right rail, and left rail space curves can vary significantly. This variation must be taken into account in order to develop an accurate algorithm for predicting wheel/rail contact forces.

Track Data File Based on the simple track description given in this section, a track preprocessor can be designed to create space curve meshes for the track centerline and right and left rail space curves. The track preprocessor produces a *track data file* that contains, at minimum, the location of the mesh track nodes and Euler angles at these nodes. As previously mentioned, for node k on space curve j, the track file can have the information given in the array $\begin{bmatrix} s^{jk} & r_1^{jk} & r_2^{jk} & r_3^{jk} & \psi^{jk} & \theta^{jk} & \phi^{jk} \end{bmatrix}$, where $j = t, rr, lr$, and t, rr and lr refer, respectively, to the track, right rail, and left rail space curves. Euler angles completely define the track geometry, as explained in this chapter. Nodal locations can be defined in the track body coordinate system or the rail body coordinate system. The latter case allows modeling motion scenarios in which the rails move independently and are treated as separate bodies during nonlinear simulations. Without any loss of generality, the track and rail body coordinate systems can be assumed to coincide initially.

Two approaches were discussed in this chapter for evaluating tangent, normal, and curvature vectors. In both approaches, the global position vector of an arbitrary point on the rail is defined using Eq. 16 as $\mathbf{r}^r = \mathbf{R}^r + \mathbf{A}^r\overline{\mathbf{u}}^r$, where the superscript r stands for rail (right or left), $\mathbf{R}^r$ is the position vector of the origin of the selected rail coordinate system $X^rY^rZ^r$ in the global coordinate system XYZ, $\mathbf{A}^r$ is the transformation matrix that defines the orientation of the rail coordinate system, and $\overline{\mathbf{u}}^r$ is the location of an arbitrary point with respect to the rail coordinate system. Vector $\mathbf{R}^r$ and matrix $\mathbf{A}^r$ do not depend on rail surface parameters. In the first approach, referred to as the *semi-analytical approach*, used to determine the tangent, normal, and curvature vectors, ANCF interpolation is not used, and the location of an arbitrary point on the rail is defined using Eq. 11 as $\overline{\mathbf{u}}^r(s_1^r, s_2^r) = \overline{\mathbf{R}}^{rp}(s_1^r) + \mathbf{A}^{rp}(s_1^r)\overline{\overline{\mathbf{u}}}^{rp}(s_1^r, s_2^r)$. Using this approach, evaluating the tangent, normal, and curvature vectors requires differentiating vector $\overline{\mathbf{R}}^{rp}(s_1^r)$ and matrix $\mathbf{A}^{rp}(s_1^r)$ with respect to rail surface parameters; this is in addition to differentiating vector $\overline{\overline{\mathbf{u}}}^{rp}(s_1^r, s_2^r)$. Using such an approach requires evaluating higher derivatives of Euler angles, including third derivatives, which are required when solving wheel/rail contact conditions. Therefore, if this semi-analytical approach is used, two additional sets of information must be provided at the nodal points for each space curve: (i) the longitudinal tangent and its first and second derivatives with respect to the curve arc length; and (ii) the first, second, and third derivative of Euler angles with respect to the curve arc length. Clearly, using the semi-analytical approach can lead to more extensive track-preprocessor and main-solver implementations that require a very large track file and many more arithmetic operations, which can have an adverse effect on the computational efficiency of the algorithm. The accuracy of the algorithm is also compromised when using the semi-analytical approach because only low-order interpolation for position coordinates can be used during nonlinear dynamic simulations.

Implementations can be significantly simplified, and the size of the track file can be reduced, by using ANCF interpolation as previously discussed in this chapter. In this case, only the data defined by the vector $\begin{bmatrix} s^{jk} & r_1^{jk} & r_2^{jk} & r_3^{jk} & \psi^{jk} & \theta^{jk} & \phi^{jk} \end{bmatrix}$ needs to be included in the track data file. This is mainly due to the fact that for ANCF interpolation, vector $\overline{\mathbf{u}}^r$ is simply defined by Eq. 24 as $\overline{\mathbf{u}}^r = \mathbf{S}^r\mathbf{e}^r$, where in this case, coordinate vector $\mathbf{e}^r$ includes position gradients evaluated using the values of Euler angles at discrete nodal points. In this case, differentiation to obtain the tangent and normal vectors is carried out by differentiating shape function matrix $\mathbf{S}^r$ with respect to the surface parameters. Therefore, when using ANCF interpolation, there is no need to differentiate Euler angles with respect to the surface parameters of the rails; this can lead to significant simplification of the implementation in both the track preprocessor and the main solver and a significant reduction in the size of the track data file produced by the track preprocessor.

Example 4.1

For the track described in Table 4, determine the location and orientation of the profile frame at the nodes of the track centerline, assuming that the track has a gage value equal to 1.42 m. Neglect the effect of the head width (Shabana et al. 2008).

(Continued)

Solution For an arbitrary node i, one can write

$$\phi^i = \sin^{-1}\left(\frac{0.0254h^i}{G}\right) \quad \text{rad}, \quad i = 1, 2, \ldots, 10$$

where h^i is the given superelevation in inches at node i and G is the gage. Angle ψ^i can be written as

$$\psi^i = \frac{\pi c_r^i}{180} \quad \text{rad}, \quad i = 1, 2, \ldots, 10$$

where c_r^i is the given curvature in degrees at the node. Angle θ^i can be defined as

$$\theta^i = \sin^{-1}\left(\frac{G_R^i}{100}\right) \quad \text{rad}, \quad i = 1, 2, \ldots, 10$$

where G_R^i is the given grade percentage. The horizontal and vertical curvatures are given, respectively, by

$$C_H^i = \frac{\sin(\psi^i/2)}{50 \times 0.3048} \ 1/\text{m}, \quad C_{v\ i} = \frac{\theta_i - \theta_{i-1}}{s_i - s_{i-1}} \ 1/\text{m}, \quad i = 1, 2, \ldots, 10$$

Since the grade is equal to zero for all nodes, one has

$$S^i = S^i + (s^i - s^{i-1})\cos\theta^i, \quad \psi^i = \psi^{i-1} + \frac{C_H^i + C_H^{i-1}}{2}(S^i - S^{i-1}), \quad i = 1, 2, \ldots, 10$$

By integrating Eq. 55 numerically, the locations and orientations of the profile frame at each node can be determined and are summarized in the following table (Shabana et al. 2008):

Node #	Distance s (m)	C_H (m^{-1})	θ (rad)	ϕ (rad)	ψ (rad)	x (m)	y (m)	z (m)
1	0	0.0	0.0	0.0	0.0	0.0	0.0	0.0
2	30.48	0.0	0.0	0.0	0.0	30.48	0.0	0.0
3	45.72	0.00286	0.0	0.02932	0.02181	45.72	0.11	0.0
4	137.16	0.00286	0.0	0.02932	0.28353	135.84	13.98	0.0
5	152.40	−0.00172	0.0	−0.01953	0.29225	150.43	18.39	0.0
6	198.12	−0.00172	0.0	−0.01953	0.21372	194.68	29.83	0.0
7	219.46	0.00401	0.0	0.03909	0.23813	215.52	34.40	0.0
8	310.90	0.00401	0.0	0.03909	0.60442	298.50	71.58	0.0
9	349.0	0.0	0.0	0.0	0.68073	328.70	94.79	0.0
10	364.24	0.0	0.0	0.0	0.68073	340.54	104.38	0.0

4.10 IRREGULARITIES AND MEASURED TRACK DATA

Track preprocessors are designed to allow for including track deviations and using measured track data to perform more realistic virtual tests to identify the causes of derailments. Before discussing the implementation of the analytical geometry variations in the track preprocessor and using measured track data, this section first discusses track quality and classes.

Track Classes Railroad vehicle speed limits are dictated by track quality. Higher track quality is required by federal regulators for passenger trains that travel at higher speeds in order to ensure safety and avoid deadly accidents. Having higher track quality for passenger routes is also necessary for ride comfort and to control the noise level. For high-speed trains, track quality must be even higher to avoid deadly accidents. Track conditions can rapidly deteriorate, particularly for freight trains, which have high axle loads. Therefore, tracks that are owned by railroad companies in North America are regularly monitored to improve their condition. The track maintenance cost for a single railroad company in North America is measured in billions of dollars annually. The Federal Railroad Administration (FRA) in the United States sets a limit on maximum train speed based on the quality of the track sections. The FRA track-quality classification, shown in Table 5, is based on the maximum allowable deviations. The data presented in this table, which show both speed and gage limits, are for tangent segments of the track. Train speeds can be significantly reduced when a train negotiates a curve or approaches a *rail crossing*.

Table 4.5 Speed and gage limits for different track classes.

Track class	Freight trains	Passenger trains	Minimum gage (in)	Maximum gage (in)
Class 1	16 km/h ≈ 10 mile/h	24 km/h ≈ 15 mile/h	56	58
Class 2	40 km/h ≈ 25 mile/h	48 km/h ≈ 30 mile/h	56	57.75
Class 3	64 km/h ≈ 40 mile/h	97 km/h ≈ 60 mile/h	56	57.75
Class 4	97 km/h ≈ 60 mile/h	129 km/h ≈ 80 mile/h	56	57.50
Class 5	129 km/h ≈ 80 mile/h	145 km/h ≈ 90 mile/h	56	57.50
Class 6	177 km/h ≈ 110 mile/h	177 km/h ≈ 110 mile/h	56	57.25
Class 7	201 km/h ≈ 125 mile/h	201 km/h ≈ 125 mile/h	56	57.25
Class 8	257 km/h ≈ 160 mile/h	257 km/h ≈ 160 mile/h	56	57.25
Class 9	354 km/h ≈ 220 mile/h	354 km/h ≈ 220 mile/h	56.25	57.25

The data in Table 5 demonstrate that, in general, lower classes that correspond to lower track quality are used for freight trains, which generally operate at lower speeds compared to passenger trains. Nonetheless, passenger trains, in some cases, can run on lower-class tracks, and freight trains are allowed in some cases to use higher-class tracks such as classes 6–9 if certain conditions are met (US Dept. of Transportation, Federal Railroad Administration, Office of Safety 2004, 2005). While Table 5 shows the gage limits for different classes, the actual distance between the right and left rails can vary from original gage values due to track irregularities, as discussed in this section.

Because track conditions can vary based on usage, *measured track data* are often used in computer simulations of railroad vehicle systems. The measured data account for the effect of all irregularities and can be used to better evaluate the response of a vehicle under more realistic operating conditions. In addition to using measured data, standard analytical *track deviations* are often used in computer simulations to evaluate vehicle stability using virtual prototyping simulation tests. Recorded track measurements can be used to develop analytical geometry variation functions (Hamid et al. 1983). These analytical track deviations, which take different shapes and can be used to represent both vertical and lateral irregularities, can be systematically integrated with track data to test different scenarios and develop guidelines for avoiding derailments. For example, new vehicle designs are virtually analyzed and evaluated using simulation test scenarios that include standard deviations. These tests are designed to excite certain modes that can influence vehicle stability and cause derailments. In the remainder of this section, track deviations and measured track data are discussed.

Analytical Geometry-Variation Functions Rail tracks are constructed using specific geometries including tangent, curve, and spiral. Due to dynamic loads resulting from wheel/rail interaction forces as well as environmental conditions that may include significant temperature variations, the track geometry can change from its original design. Changes in track geometry, referred to as *track deviations*, can include local geometry variations, gage widening, rail buckling, cant change, etc. These geometric variations, which are considered in vehicle design and when developing operation guidelines, can lead to serious accidents or derailments; therefore, their effect needs to be evaluated.

Analytical track deviations that define specific smooth track geometry variations can be systematically implemented in a track preprocessor. Recorded track measurements can also be used to define other functions using interpolation schemes. Track deviations that describe vertical or lateral geometry variations are known in a closed form. Vertical geometry variations are called *profile deviations*, and lateral geometry variations are called *alignment deviations*. These two types of deviations are illustrated in Figure 11. Analytical profile and alignment deviations, which can have arbitrary forms, are described using functions as shown in Table 6. These functions, which include *cusp*, *bump*, *jog*, *plateau*, *trough*, *sinusoid*, and *damped sinusoid*, are defined based on common irregularities observed in measured track data (Hamid et al, 1983).

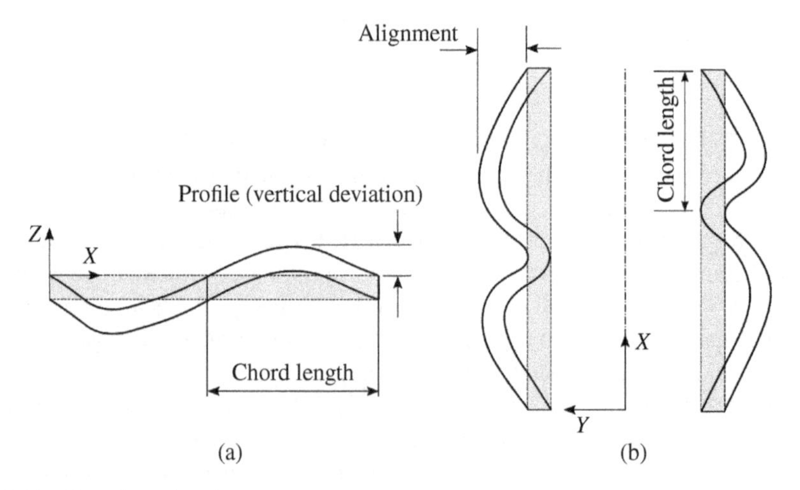

Figure 4.11 Track irregularities.

Table 4.6 Examples of analytical track deviations. Source: Courtesy of Hamid, A., Rasmussen, K., Baluja, M., and Yang, T-L.

Deviation type	Deviation shape	Function $f_d(s^r)$	Occurrence
Cusp	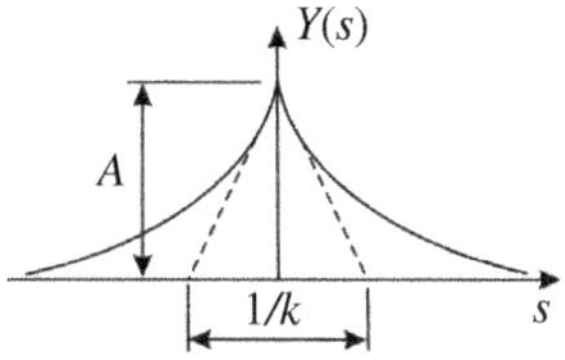	$y = Ae^{-k\|s\|}$	Joints, turnouts, interlocking, sun kinks, buffer rails, insulated joints in continuous welded rail (CWR), splice bar joint in CWR, piers at bridges
Bump	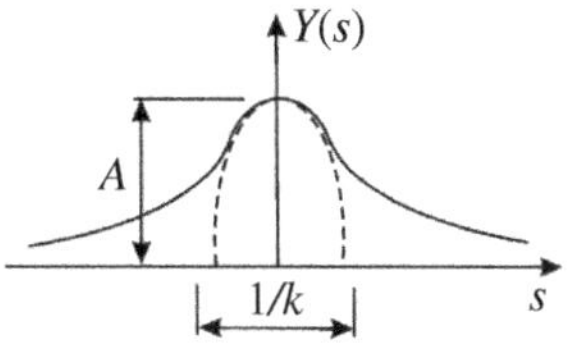	$y = Ae^{-(1/2)(ks)^2}$	Soft spots, washouts, mud spots, fouled ballast, joists, spirals, grade crossings, bridges, overpasses, loose bolts, turnouts, interlinking
Jog	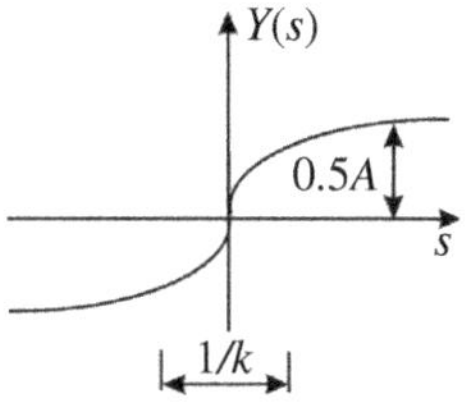	$y = \dfrac{Aks}{\sqrt{(1 + 4k^2s^2)}}$	Spirals, bridges, crossings, interlocking, fill-cut transitions
Plateau	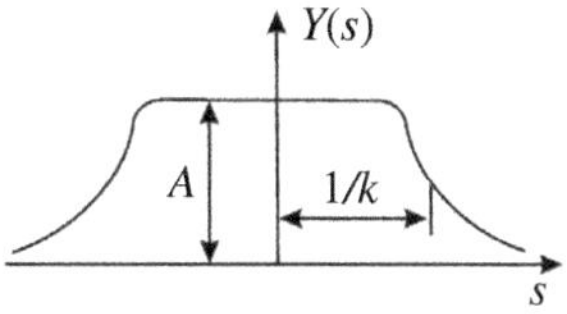	$y = \sqrt{\left(\dfrac{A^2}{1 + (ks)^8}\right)}$	Bridges, grade crossings, areas of spot maintenance
Trough	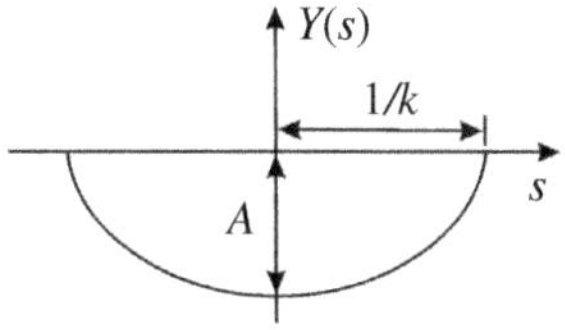	$y = Ak\sqrt{[(1/k)^2 - s^2]}$	Soft spots, soft and unstable subgrades, spirals
Sinusoid	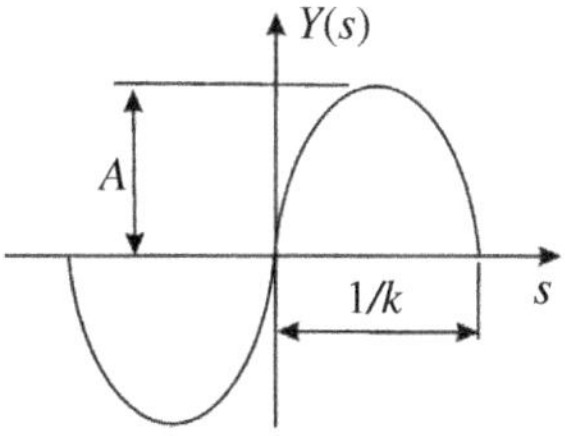	$y = A \sin \pi ks$	Spirals, soft spots, bridges
Damped sinusoid	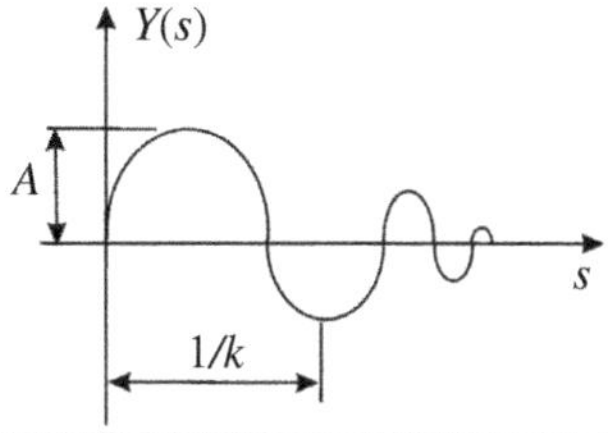	$y = Ae^{-ks} \cos \pi ks$	Spirals, turnouts, localized soft spots

The functions presented in Table 6 can be used as vertical (profile) or lateral (alignment) deviations. Some of the analytical deviations shown in Table 6 are defined using exponential functions that assume a large value as arc length s increases. These functions can be used to cover a large section of a deflected rail that assumes the shape of this particular deviation. Hamid et al. (1983) provided a range for the values of parameters A and k used in Table 6; these values are presented in Table 7. Table 6 also shows a summary of the possible occurrences for each deviation. As discussed in the literature, a single cusp occurs as a profile deviation due to the joint between two welded rail sections. Bump deviations, which are smoother and typically cover longer rail segments, can be alignment or profile deviations and can be found on both the right and left rails simultaneously. Bump deviations may be considered profile irregularities in the case of bridges or alignment irregularities in the case of spirals, where they occur at a certain distance from the start of the irregularities. A jog can occur as the result of track stiffness variations, as in an interface between soft track and a bridge. A plateau is generally caused by variations in track stiffness or wear of the high rail. A cusp/plateau combination, which can be found before a spiral section of track in some cases, can be dangerous because it leads to a sudden change in the wheel load. Trough irregularities can be the result of poor drainage or localized soft subgrades. Sinusoid irregularities occur for bridges or reverse curves connected without a tangent segment to separate them. Damped sinusoidal irregularities, which usually occur on a single rail, are the result of significant changes in track stiffness due to switches, grade crossings, and curves (Shabana et al. 2008).

Track irregularities, which can have adverse geometric effects, are taken into consideration when developing operation guidelines. For example, Table 5 shows the gage limits for different track classes. Nonetheless, lateral deviations can alter the distance between the right and left rails; and in some cases, as the result of changes in the load across the ties, the track cross-level changes. This cross-level change, called *warp* (twist), is defined more accurately as the rate of change in the cross-level. To ensure safety, the gage change for track classes 6 and higher, which are used for higher-speed trains, is limited to a maximum value of 0.5 in. within 31 ft. Due to safety requirements, limits are also imposed on the maximum allowable vertical and lateral deviations for track classes used for higher-speed trains. Tables 8 and 9 show the maximum allowable alignment and profile deviations for track classes 6 and higher, respectively. In these tables, h_1, h_2, and h_3 are, respectively, the maximum allowable deviations from the designed geometry at the middle of 31-, 62-, and 124-ft chords.

Updating Track Preprocessor Data A track preprocessor can be designed to account systematically for deviation functions that are used to modify the rail space curve geometry regardless of the method used in the main solver (semi-analytical approach or ANCF interpolation) to describe the track geometry. In the track preprocessor, the function $f_d(s^r)$ is used to define the geometry of the space curve in the region covered by a deviation. For a profile deviation, the space curve over the domain of the deviation function is updated by vector function $\mathbf{f}_d = \begin{bmatrix} 0 & 0 & f_d(s^r) \end{bmatrix}^T$; and for an alignment deviation, the space curve over the domain of the deviation function is updated by vector function $\mathbf{f}_d = \begin{bmatrix} 0 & f_d(s^r) & 0 \end{bmatrix}^T$. Vector function $\mathbf{f}_d$ can be used to define the nodal coordinates r_1^{jk}, r_2^{jk}, and r_3^{jk} in the track-preprocessor output

Table 4.7 Parameters of track deviations. Source: Courtesy of Hamid, A., Rasmussen, K., Baluja, M., and Yang, T-L.

	Parameter range							
	Gage		Alignment		Cross level		Profile	
Deviation	A (in)	k (ft^{-1})	A (in)	k (ft^{-1})	A (in)	k (ft^{-1})	A (in)	k (ft^{-1})
Cusp	0.8–1.4	0.016–0.061	0.5–0.3	0.011–0.103	0.9–3.0	0.031–0.095	0.9–3.0	0.016–0.095
Bump	0.8–1.4	0.031–0.040	0.5–2.8	0.009–0.083	1.0–3.0	0.017–0.031	0.5–4.0	0.013–0.065
Jog	–	–	0.5–3.3	0.006–0.025	1.6–2.8	0.020–0.05	0.5–5.0	0.008–0.045
Plateau	0.8–1.3	0.029–0.08	1.2–1.6	0.025–0.027	0.6–1.0	0.026–0.04	0.9–3.0	0.009–0.033
Trough	–	–	1.4–2.2	0.013–0.029	–	–	0.7–2.0	0.020–0.025
Sinusoid	–	–	0.8–1.2	0.033–0.020	–	–	1.0–1.5	0.020–0.025
Damped sinusoid	0.5–1.0	–	1.0–2.2	0.013–0.015	0.9–1.2	0.051–0.061	–	–

array $\left[s^{jk}\ r_1^{jk}\ r_2^{jk}\ r_3^{jk}\ \psi^{jk}\ \theta^{jk}\ \phi^{jk}\right]$ for a given node k of space curve j, $j = rr, lr$. Euler angles ψ^{jk} and θ^{jk} can also be adjusted due to lateral and vertical deviations, respectively. The adjustment can be made easily by updating the angles, since the change in rotation due to the deviation is assumed to be about a single axis. The change in the angle can be calculated using the equation $\psi^{jk} = \psi^{jk} + \Delta\psi^{jk} = \psi^{jk} + \tan^{-1}(df(s^r)/ds^r)$ for lateral (alignment) deviation and $\theta^{jk} = \theta^{jk} + \Delta\theta^{jk} = \theta^{jk} + \tan^{-1}(df(s^r)/ds^r)$ for vertical (profile) deviation.

Table 4.8 Alignment limits. Source: Courtesy of Shabana, A.A., Zaazaa, K.E., and Sugiyama, H.

	Single deviation			Three or more non-overlapping deviations		
Track class	h_1 **(in.)**	h_2 **(in.)**	h_3 **(in.)**	h_1 **(in.)**	h_2 **(in.)**	h_3 **(in.)**
6	0.5	0.75	1.5	0.375	0.5	1.0
7	0.5	0.5	1.25	0.375	0.375	0.875
8	0.5	0.5	0.75	0.375	0.375	0.5
9	0.5	0.5	0.75	0.375	0.375	0.5

Table 4.9 Profile limits. Source: Courtesy of Shabana, A.A., Zaazaa, K.E., and Sugiyama, H.

	Single deviation			Three or more non-overlapping deviations		
Track class	h_1 **(in.)**	h_2 **(in.)**	h_3 **(in.)**	h_1 **(in.)**	h_2 **(in.)**	h_3 **(in.)**
6	1.0	1.0	1.75	0.75	0.75	1.25
7	1.0	1.0	1.50	0.75	0.75	1.0
8	0.75	1.0	1.25	0.50	0.75	0.875
9	0.5	0.75	1.25	0.375	0.50	0.875

Measured Track Data In virtual prototyping, analysis, and design of railroad vehicle systems, it is important to examine vehicle nonlinear dynamics and stability when using actual track data obtained from field measurements. This is particularly important during accident investigations and when developing operation guidelines. The track preprocessor must be designed to allow for reading and using the measured track data when constructing the track geometry data used by the main solver. In accident investigations, in particular, it is important to use the measured track data when performing dynamic simulations to develop more realistic and credible computer models that can identify the causes of accidents. While measured track data can be provided in various formats, these data must include the coordinates of points along the track; and the gage, curvature, and superelevation at these points. In addition to these data, information about the right and left rail profile and alignment deviations is needed. Clearly, measured track data have the same information required by the track preprocessor to define the track geometry, except that for measured track data, information is required at a much larger number of points in order to capture all the irregularities recorded by the measurements. Given measured data, a simple computer procedure can be developed to define standard inputs to the track preprocessor based on the data. Using the measured track input data, which provide the same information as for analytical track models, the track preprocessor can be designed to produce the required nodal data defined by the vector $\left[s^{jk} \quad r_1^{jk} \quad r_2^{jk} \quad r_3^{jk} \quad \psi^{jk} \quad \theta^{jk} \quad \phi^{jk}\right]$, as previously discussed.

When measured data are used, the curvature, grade, and superelevation are provided at measurement points. Because the value of the gage varies and is defined at measurement points, one can assume a mean gage value and use the measured gage value at the measurement points to define alignment deviations that account for gage variations. Gage variation involves both the right and left rails, so alignment deviations designed to represent the gage variation can be equally divided between the two rails. Another important issue is using *smoothing techniques* to filter out noise from measured track data, which is often provided as raw data. For the gage and superelevation, filtering can be accomplished using a moving average window with a width that depends on the data format.

4.11 COMPARISON OF THE SEMI-ANALYTICAL AND ANCF APPROACHES

As explained in this chapter, a track preprocessor employs linear interpolation to calculate Euler angles that describe the shape of the rail and the track space curve. Using numerical integration, the position coordinates at nodes selected by the user can be determined. Each node k of space curve j in the output of the track preprocessor is defined using its arc length location, position coordinates, and Euler angles given in the node array $\left[s^{jk} \quad r_1^{jk} \quad r_2^{jk} \quad r_3^{jk} \quad \psi^{jk} \quad \theta^{jk} \quad \phi^{jk}\right]$. During dynamic simulations, contact conditions are used to define the location of points of contact between the wheels and rails. These contact points can be any points inside the rail segments defined by the track preprocessor. Therefore, it is necessary to use an accurate interpolation scheme to determine the locations of the wheel/rail contact points between two nodes of a rail segment when the surface parameters at these points become known from solving the wheel/rail contact conditions.

This chapter discussed two geometric approaches for determining the locations of points in a rail segment: *semi-analytical* and *ANCF interpolation*. In both approaches, the global position of a point in a rail segment is defined using Eq. 16, given by $\mathbf{r}^r = \mathbf{R}^r + \mathbf{A}^r\bar{\mathbf{u}}^r$, where superscript r stands here for rail (right or left), $\mathbf{R}^r$ is the position vector of the origin of the selected rail coordinate system $X^rY^rZ^r$ in the global coordinate system XYZ, $\mathbf{A}^r$ is the transformation matrix that defines the orientation of the rail coordinate system, and $\bar{\mathbf{u}}^r = \bar{\mathbf{u}}^r(s_1^r, s_2^r)$ is the location of an arbitrary point with respect to the rail coordinate system. As previously mentioned, vector $\mathbf{R}^r$ and matrix $\mathbf{A}^r$ do not depend on the rail surface parameters; therefore, they do not enter into the differentiation with respect to the surface parameters when the tangent, normal, and curvature vectors are evaluated. The semi-analytical and ANCF interpolation approaches differ in the way vector $\bar{\mathbf{u}}^r = \bar{\mathbf{u}}^r(s_1^r, s_2^r)$ is defined. The definitions of $\bar{\mathbf{u}}^r = \bar{\mathbf{u}}^r(s_1^r, s_2^r)$ in the semi-analytical approach have two major computational disadvantages: the derivatives of Euler angles with respect to rail surface parameters are required; and the degree of interpolation used for the position coordinates is limited to linear interpolation when the nodal information given in vector $\begin{bmatrix} s^{jk} & r_1^{jk} & r_2^{jk} & r_3^{jk} & \psi^{jk} & \theta^{jk} & \phi^{jk} \end{bmatrix}$ is used, as explained in this section. These major computational disadvantages are avoided by using ANCF interpolation, which does not require differentiation of Euler angles and also allows using a higher interpolation order (third-order) for the position coordinates based on nodal information given in the array $\begin{bmatrix} s^{jk} & r_1^{jk} & r_2^{jk} & r_3^{jk} & \psi^{jk} & \theta^{jk} & \phi^{jk} \end{bmatrix}$.

Semi-Analytical Approach In the *semi-analytical approach* used to determine the tangent, normal, and curvature vectors during dynamic simulations based on data produced by a track preprocessor, ANCF interpolation is not used, and the location of an arbitrary point on the rail is defined using Eq. 11, given as $\overline{\mathbf{u}}^j(s_1^j, s_2^j) = \overline{\mathbf{R}}^{jp}(s_1^j) + \mathbf{A}^{jp}(s_1^j)\overline{\overline{\mathbf{u}}}^{jp}(s_1^j, s_2^j)$, where j can be used for the right or left rail. It is clear from this equation that evaluating the tangent, normal, and curvature vectors requires differentiating vector $\overline{\mathbf{R}}^{jp}(s_1^j)$ and matrix $\mathbf{A}^{jp}(s_1^j)$ with respect to the rail surface parameters, in addition to differentiating vector $\overline{\overline{\mathbf{u}}}^{jp}(s_1^j, s_2^j)$. In general, one can write $\overline{\overline{\mathbf{u}}}^{jp}(s_1^j, s_2^j) = \begin{bmatrix} 0 & s_2^j & f(s_1^j, s_2^j) \end{bmatrix}^T$. As previously shown in this chapter, using this approach leads to the following definitions of the tangent and normal vectors, respectively: $\mathbf{t}_1^j = \partial \mathbf{r}^j/\partial s_1^j = \mathbf{A}^j(\partial \overline{\mathbf{u}}^j/\partial s_1^j)$, $\mathbf{t}_2^j = \partial \mathbf{r}^j/\partial s_2^j = \mathbf{A}^j(\partial \overline{\mathbf{u}}^j/\partial s_2^j)$, and $\mathbf{n}^j = \mathbf{t}_1^j \times \mathbf{t}_2^j$, which require differentiating both $\mathbf{R}^{jp}(s_1^j)$ and matrix $\mathbf{A}^{jp}(s_1^j)$ with respect to the rail surface parameters. Calculating the curvature vectors, as previously discussed, requires evaluating higher derivatives. To determine the position coordinates of a point inside a rail segment defined by nodes k and $(k+1)$ using the data $\begin{bmatrix} s^{jk} & r_1^{jk} & r_2^{jk} & r_3^{jk} & \psi^{jk} & \theta^{jk} & \phi^{jk} \end{bmatrix}$ for node k of rail j, only linear interpolation can be used for the components of vector $\overline{\mathbf{u}}^j(s_1^j, s_2^j)$ since for a coordinate $\overline{u}_l^j$, $l = 1, 2, 3$ of vector $\overline{\mathbf{u}}^j$, only two coordinates r_l^{jk} and $r_l^{j(k+1)}$ are available. Therefore, with the information available from the track preprocessor, one is limited to the linear interpolation $r_l^j = (1-\xi)r_l^{jk} + \xi r_l^{j(k+1)}$, where $\xi = s_1^j/l^{k,(k+1)}$ and $l^{k,(k+1)}$ is the length of the segment defined by the two nodes k and $k+1$. Using linear interpolation, which is not accurate for curved rails, fails to capture gradient and curvature continuity and can lead to incorrect force predictions if a very fine mesh is not used. A similar comment applies to the interpolation of Euler angles ψ^j, θ^j, and ϕ^j, which can only be interpolated linearly because only two coordinates α^{jk} and $\alpha^{j(k+1)}$, $\alpha = \psi, \theta, \phi$, are available at the segment ends for each of the angles. Therefore, one is limited to linear interpolation $\alpha^j = (1-\xi)\alpha^{jk} + \xi\alpha^{j(k+1)}$. Furthermore, due to the non-commutativity of finite rotations, and because interpolations imply addition, using the semi-analytical approach is not recommended.

The fact that evaluating the tangent, normal, and curvature vectors requires differentiation of Euler angles with respect to the arc length parameter when the semi-analytical approach is used is the result of the dependence of the geometric description on matrix $\mathbf{A}^{jp}(s_1^j)$. The derivative of matrix $\mathbf{A}^{jp}(s_1^j)$ with respect to the rail longitudinal arc length s_1^j can be written as

$$\partial \mathbf{A}^{jp}/\partial s_1^j = (\partial \mathbf{A}^{jp}/\partial \psi^j)(\partial \psi^j/\partial s_1^j) + (\partial \mathbf{A}^{jp}/\partial \theta^j)(\partial \theta^j/\partial s_1^j) + (\partial \mathbf{A}^{jp}/\partial \phi^j)(\partial \phi^j/\partial s_1^j) \tag{4.58}$$

Equation 37 shows that for space curve j, $\psi^j(S^j) = \psi^{jk} + \int_{S^{jk}}^{S^{j(k+1)}} (C_H^j(S^j))dS^j$, which can be used with the second equation in Eq. 34 to evaluate $\partial\psi^j/\partial s^j$ as

$$\partial\psi^j/\partial s^j = (\partial\psi^j/\partial S^j)(\partial S^j/\partial s^j) = C_H^j \cos\theta^j \tag{4.59}$$

In this equation, C_H^j at an arbitrary point can be evaluated using the linear interpolation of Eq. 40 given by $C_H^j = (1-\xi_S)C_H^{jk} + \xi_S C_H^{j(k+1)}$. Similarly, Eq. 42 can be used to evaluate $\partial\theta^j/\partial s^j$ as $\partial\theta^j/\partial s^j = C_V^j$, where vertical curvature C_V^j is assumed to be constant determined using Eq. 45. Equation 47 defines the linear interpolation of angle ϕ as $\phi^j(S^j) = (1-\xi_S)\phi^{jk} + \xi_S\phi^{j(k+1)}$, which shows that

$$\partial\phi^j/\partial s^j = (\partial\phi^j/\partial S^j)(\partial S^j/\partial s^j) = \frac{1}{L_S^{k,k+1}}(\phi^{j(k+1)} - \phi^{jk})\cos\theta^j \tag{4.60}$$

where $\xi_S = S^j/L_S^{k,k+1}$ and $L_S^{k,k+1} = S^{j(k+1)} - S^{jk}$. Following a similar procedure, higher angle derivatives, which are required to evaluate the curvature vectors, can be developed using the track preprocessor output data defined at the nodes by entries in the array $\left[s^{jk}\ r_1^{jk}\ r_2^{jk}\ r_3^{jk}\ \psi^{jk}\ \theta^{jk}\ \phi^{jk}\right]$. The angles and their derivatives are presented in Table 10. Interpolating the angles, and evaluating and interpolating their derivatives, can be avoided by using the second approach based on ANCF interpolation.

Table 4.10 Euler angles and their derivatives.

Angle	First derivative	Second derivative	Third derivative
$\psi^j(S^j) = \psi^{jk} + \int_{S^{jk}}^{S^{j(k+1)}} (C_H^j(S^j))dS^j$	$\frac{\partial\psi^j}{\partial s^j} = C_H^j \cos\theta^j$	$\frac{\partial^2\psi^j}{\partial(s^j)^2} = \frac{\partial C_H^j}{\partial S^j}\cos^2\theta^j - C_H^j C_V^j \sin\theta^j$	$\frac{\partial^3\psi^j}{\partial(s^j)^3} = -2\frac{\partial C_H^j}{\partial S^j} C_V^j \sin\theta^j \cos\theta^j - C_H^j (C_V^j)^2 \cos\theta^j$
$\theta^j(s^j) = \theta^{jk} + C_V^j(s^j - s^{jk})$	$\frac{\partial\theta^j}{\partial s^j} = C_V^j$	$\frac{\partial^2\theta^j}{(\partial s^j)^2} = 0$	$\frac{\partial^3\theta^j}{(\partial s^j)^3} = 0$
$\phi^j(S^j) = (1-\xi_S)\phi^{jk} + \xi_S\phi^{j(k+1)}$	$\frac{\partial\phi^j}{\partial s^j} = \frac{\Delta\phi}{L_S^{k,k+1}}\cos\theta^j$, $\Delta\phi = \phi^{j(k+1)} - \phi^{jk}$	$\frac{\partial^2\phi^j}{(\partial s^j)^2} = -\frac{\Delta\phi}{L_S^{k,k+1}} C_V^j \sin\theta^j$, $\Delta\phi = \phi^{j(k+1)} - \phi^{jk}$	$\frac{\partial^3\phi^j}{(\partial s^j)^3} = -\frac{\Delta\phi}{L_S^{k,k+1}}(C_V^j)^2\cos\theta^j$, $\Delta\phi = \phi^{j(k+1)} - \phi^{jk}$

ANCF Interpolation Approach With ANCF interpolation during dynamic simulations performed by the main processor, a simple procedure can be used to obtain the local rail geometry required for wheel/rail contact analysis. In this ANCF procedure, the derivatives of angles are not required, and higher-order interpolation can be used for position coordinates. Recall that the primary function of Euler angles in a track preprocessor is to define geometry. Therefore, these angles function primarily as elements of the position vector gradients that can be used to shape a rail segment. By changing the orientation of the position vector gradients, different geometry can be created. The main idea behind ANCF interpolation is to use angles to develop the elements of position vector gradients at nodal points. During dynamic simulations, wheel/rail contact conditions are used to solve for wheel and rail surface parameters at potential contact points. Using the value of a rail longitudinal surface parameter, the rail segment within which the potential contact point lies can be determined. The two nodes of rail segments can be considered the two nodes of an ANCF beam element. One can, therefore, use the elements of the track-preprocessor output array $\left[s^{jk}\;\; r_1^{jk}\;\; r_2^{jk}\;\; r_3^{jk}\;\; \psi^{jk}\;\; \theta^{jk}\;\; \phi^{jk}\right]$ to define, for each node of an ANCF element, 12 coordinates: 3 position coordinates and 9 gradient coordinates. The gradient coordinates can be defined using the columns of the transformation matrix of Eq. 20. Therefore, for node k, one can define the following vector of nodal coordinates of the ANCF element:

$$\mathbf{e}^{jk} = \begin{bmatrix} e_1^{jk} \\ e_2^{jk} \\ e_3^{jk} \\ e_4^{jk} \\ e_5^{jk} \\ e_6^{jk} \\ e_7^{jk} \\ e_8^{jk} \\ e_9^{jk} \\ e_{10}^{jk} \\ e_{11}^{jk} \\ e_{12}^{jk} \end{bmatrix} = \begin{bmatrix} r_1^{jk} \\ r_2^{jk} \\ r_3^{jk} \\ r_{x1}^{jk} \\ r_{x2}^{jk} \\ r_{x3}^{jk} \\ r_{y1}^{jk} \\ r_{y2}^{jk} \\ r_{y3}^{jk} \\ r_{z1}^{jk} \\ r_{z2}^{jk} \\ r_{z3}^{jk} \end{bmatrix} = \begin{bmatrix} r_1^{jk} \\ r_2^{jk} \\ r_3^{jk} \\ \cos\psi^{jk}\cos\theta^{jk} \\ \sin\psi^{jk}\cos\theta^{jk} \\ \sin\theta^{jk} \\ -\sin\psi^{jk}\cos\phi^{jk} + \cos\psi^{jk}\sin\theta^{jk}\sin\phi^{jk} \\ \cos\psi^{jk}\cos\phi^{jk} + \sin\psi^{jk}\sin\theta^{jk}\sin\phi^{jk} \\ -\cos\theta^{jk}\sin\phi^{jk} \\ -\sin\psi^{jk}\sin\phi^{jk} - \cos\psi^{jk}\sin\theta^{jk}\cos\phi^{jk} \\ \cos\psi^{jk}\sin\phi^{jk} - \sin\psi^{jk}\sin\theta^{jk}\cos\phi^{jk} \\ \cos\theta^{jk}\cos\phi^{jk} \end{bmatrix} \tag{4.61}$$

In this equation, the gradient vectors at node k are $\mathbf{r}_x^{jk} = \left[r_{x1}^{jk}\;\; r_{x2}^{jk}\;\; r_{x3}^{jk}\right]^T$, $\mathbf{r}_y^{jk} = \left[r_{y1}^{jk}\;\; r_{y2}^{jk}\;\; r_{y3}^{jk}\right]^T$, and $\mathbf{r}_z^{jk} = \left[r_{z1}^{jk}\;\; r_{z2}^{jk}\;\; r_{z3}^{jk}\right]^T$. Having 12 coordinates at the nodes of the ANCF element allows using cubic interpolation for the position coordinates using the ANCF displacement field of Eq. 24, given for rail segment l by the equation $\overline{\mathbf{u}}^{jl}(x,y,z) = \mathbf{S}^{jl}(x,y,z)\mathbf{e}^{jl}$, where vector $\mathbf{e}^{jl}$ has 24 elements that have the coordinates of the 2 nodes determined using the track preprocessor output data. The tangent, normal, and curvature vectors can be determined by simply differentiating the element displacement field. As previously mentioned, on the surface of the rail, element spatial coordinate z is written in terms of the other spatial coordinates x and y. Using this ANCF approach eliminates the need for the derivatives of the angles; instead, the angles at the nodes are used to determine the position gradients that define the element geometry. Furthermore, using this ANCF approach allows for using a cubic interpolation of the position coordinates instead of using less accurate linear interpolation. This approach for the **i**ntegration of **c**omputer-**a**ided **d**esign and **a**nalysis (I-CAD-A) was adopted in 2000 for developing a general procedure for the virtual prototyping of railroad vehicle systems. Such an approach can be systematically generalized for the analysis of flexible rails, as has been demonstrated in the literature.

Chapter 5

CONTACT PROBLEM

One of the most fundamental problems in the study of railroad vehicle system dynamics and stability is formulating wheel/rail contact forces. These forces must be accurately evaluated in order to obtain credible results that shed light on system behavior. When a wheel is pressed against the rail, a contact region, referred to as a *contact area* or *contact patch*, is formed. The shape of the region of contact between two solids depends on many factors that include the material properties, contact pressure, solid geometries in the contact area, etc. If the contact region covers a finite area that cannot be approximated by a point or a line, one has the case of *conformal contact*. In the case of wheel/rail contact, the dimensions of the contact area are small compared to the dimensions of the wheel and rail; therefore, a localized concentrated contact is often assumed when wheel/rail contact forces are evaluated. This assumption of *non-conformal contact* can be justified because of the shapes of the wheel and rail surfaces in the contact area.

Creep Forces Wheel/rail contact forces are the result of a combination of relative *sliding* and *rolling*. As the vehicle attains a certain speed and due to the effect of friction forces, the motion of the wheel with respect to the rail becomes predominantly rolling with a small amount of slipping that gives rise to *tangential creep forces* as well as *spin moments* that have a significant effect on vehicle dynamics and stability. In the case of traction and braking, for example, the relative velocity between the wheel and rail at the contact point increases, causing significant sliding, a case known as *full saturation* in which the tangential forces can be approximated using the *Coulomb's law of friction*. Below a certain relative velocity value, slipping as a result of the creep phenomenon produces creep forces that can be expressed in terms of normalized velocities called *creepages*. The creep phenomenon, attributed to the elasticity of the two solids in the contact area, is the source of tangential creep forces, which are functions of creepages. When two solids come into contact and are subjected to external pressure, some points on the contact surfaces may slip while other points may stick; therefore, the contact region consists of slip and adhesion areas. The small relative slip and spin between the two solids can be the result of the difference between the deformations in the contact region, and they lead to creep forces and spin moments, respectively. Different linear and nonlinear wheel/rail tangential contact force and spin moment models expressed

Mathematical Foundation of Railroad Vehicle Systems: Geometry and Mechanics,
First Edition. Ahmed A. Shabana.

in terms of velocity creepages are used in the railroad vehicle system literature, as discussed in this chapter.

Constraint and Elastic Contact Formulations Predicting the locations of the contact points online is necessary for the generality of the wheel/rail dynamic algorithm and the accuracy of predicting the wheel/rail interaction forces and moments. Fundamentally different approaches are used in formulating the wheel/rail dynamic interaction. In some of these approaches, referred to as *constraint contact formulations* (CCFs), the wheel is assumed to remain in contact with the rail: that is, wheel/rail separation is not allowed. In some other formulations, referred to in this book as *elastic contact formulations* (ECFs), wheel/rail separation is allowed. Both formulations can be used to determine the normal contact force, which is the wheel/rail interaction force in the direction normal to the surfaces of contact. To develop efficient computational algorithms for predicting wheel/rail contact point locations and forces, it is often assumed that wheel/rail contact is *non-conformal*: that is, the contact is assumed to cover a very small region such that using a point contact can be justified. Once the normal contact force is determined, the tangential creep forces can be computed using linear or nonlinear creep force models, as discussed in this chapter.

Hertz Contact Theory As previously mentioned, in the case of non-conformal contact, the wheel/rail contact area is assumed small in comparison with the wheel and rail dimensions. When the constraint contact formulation is used, the normal contact force is determined as a reaction force. In the elastic contact formulation, on the other hand, the normal contact force is determined using a compliant force model with assumed stiffness and damping coefficients. Experimental observations have shown that the wheel/rail contact area can be approximated using an elliptical shape. The dimensions of the *contact ellipse*, which enter into formulating the wheel/rail creep forces, can be determined using Hertz contact theory (Hertz 1882). The contact ellipse dimensions, wheel and rail material properties, creepages, and normal contact force predicted using the constraint or elastic contact formulation can be used to formulate the tangential creep force and spin moment. In Hertz contact theory, the geometry of two surfaces in contact is used to determine the principal curvatures, which are used to determine the dimensions of the contact ellipse.

Maglev Trains In this chapter, Hertz contact theory, wheel/rail contact formulations, and creep force models are discussed. While most of this chapter is devoted to formulating the wheel/rail contact problem, discussions of the forces used in *magnetically levitated* (maglev) trains are also presented. In the case of maglev trains, in which there is no contact between the vehicle and the guideway, vehicles are lifted using magnetic forces. This allows for a significant increase in train speed due to the elimination of restrictions that are the result of wheel/rail and pantograph/catenary contacts. Nonetheless, maglev train technology is still being researched and developed and for the most part, with a few exceptions, remains in an experimental phase. For this reason, passenger maglev trains are not widely used for passenger transportation, and their use for freight transportation remains in doubt for the foreseeable future.

5.1 WHEEL/RAIL CONTACT MECHANISM

Efficient dynamic simulations and virtual prototyping of complex railroad vehicle systems require making simplifying assumptions in the formulation and computer implementation of wheel/rail contact models. The simplifying assumptions often made are reasonable, are based on acceptable scientific explanation, and lead to accurate prediction of wheel/rail interaction forces. In this section, some of these assumptions are discussed and the basic steps used for computing wheel/rail interaction forces are presented.

Conformal and Non-Conformal Contact Wheel/rail dynamic interaction is one of the most important problems in modeling the nonlinear dynamic behavior of railroad vehicle systems. As mentioned in the introduction of this chapter, different models can be used to describe the interaction between two solids in contact. The contact between two solids can be concentrated in a small area, such as in the case of two stiff spheres pressed against each other. If the dimensions of the contact region are small compared to the dimensions of the two solids in contact, the case of concentrated contact, referred to as *non-conformal*, can be assumed. In the case of non-conformal contact, the interaction between the two solids in the contact region can be described using a concentrated force vector and possibly a concentrated moment vector that is defined at the center of the contact area. The calculations of the force and moment vectors can still take into account the dimensions of the contact region, effect of normal pressure, material properties, and surface geometry of the two solids in contact. Therefore, using the assumptions of non-conformal contact does not imply that the dimensions of the contact region are totally ignored in computing wheel/rail contact forces. Using this assumption can be viewed as an approximation in which distributed contact forces are replaced by concentrated force and moment vectors. An example of this non-conformal contact approach is Hertz theory (1882), in which the dimensions of the contact region are taken into consideration. Hertz theory has been widely used to solve contact problems in a wide class of engineering applications, and its assumptions have been tested and accepted by the scientific community in a large number of applications in which these simplifying assumptions can be justified. In addition to numerical verification, experimentation has also been conducted to validate the results of Hertz contact theory.

If, on the other hand, the contact region cannot be assumed small in comparison with the dimensions of the two solids in contact, as in the case of a block pressed against a metal sheet or a cylinder pressed against a solid plate, then the assumptions of non-conformal contact cannot be justified. In these situations, one has the case of *conformal contact*. In the case of conformal contact, interaction between the two solids in the contact region cannot be accurately described using force and moment vectors concentrated at a point. Accurately predicting the forces of interaction between the two solids requires using a continuum-based approach, such as the finite element (FE) method. The FE approach allows for the discretization of the contact region to create point meshes that make it possible to develop a general procedure for detecting the contact points between the two solids. The FE approach is also better suited for modeling *semi-conformal contact* in which only one dimension of the contact region cannot be assumed small compared to the dimensions of the two solids in contact.

Clearly, using the continuum-based FE approach to model the contact between two solids is more general and employs fewer simplifying assumptions. However, using this approach can be computationally prohibitive, and in some applications, such an FE approach may not lead to significant improvements in model accuracy, particularly in most railroad vehicle system applications. When a wheel is pressed against a rail in most railroad applications, the contact area is very small, and experimental observations have shown that the contact area can be approximated by an elliptical shape that has very small dimensions compared to the dimensions of the wheel and rail. Furthermore, numerical experimentation has shown that using assumptions of non-conformal contact with partially nonlinear tangential creep contact formulations leads to accurate results when compared with formulations based on fully nonlinear approaches. For this reason, assumptions of non-conformal contact can be justified in most cases and have been widely used in railroad vehicle dynamics algorithms. Nonetheless, there are cases, particularly near-flange contact or worn wheels and rails, where contact is conformal. In these cases, using non-conformal contact assumptions may not lead to accurately predicting wheel/rail interaction forces. Efficient and accurate conformal contact formulations for wheel/rail interaction forces are still being researched in order to test the feasibility of using these approaches in the analysis of complex railroad vehicle systems.

Slip, Adhesion, and Coulomb's Friction When two solids come into contact, some particles on the surfaces of the two solids slide relative to each other while others stick, forming slip and adhesion regions. Therefore, the contact area can be covered by slip and adhesion areas, which are difficult to determine precisely. Due to this difficulty, several theories have been proposed, and different shapes and areas of slip and adhesion regions have been assumed. Part of the slip region can be due to relative deformation between particles of the two solids in the contact area. Pure sliding between two rigid surfaces is characterized by jump discontinuity in tangential friction force as the transition from zero to non-zero relative velocity occurs. Furthermore, tangential friction force, according to Coulomb's theory of dry friction (Coulomb 1785), assumes a constant value of μF_n in the case of sliding, where μ is the coefficient of friction and F_n is the normal contact force. This tangential friction force, which has a direction opposite to the direction of the relative velocity between the two bodies, has a value that is independent of the relative velocity between the two bodies at the contact point when Coulomb's theory of dry friction is used. While the accuracy of Coulomb's friction law has been subjected to extensive debate for many years, it is still widely used in many engineering applications to describe the friction between solids in the case of finite relative velocities. This law, however, should not be used when the relative motion between two solids is predominantly due to deformations, as in the case of the creep phenomenon.

If a local deformation is assumed in the small contact area, the change in tangential force from a zero value at zero relative velocity to saturation value μF_n at nonzero velocity can be described using a smooth transition, and such a change is not characterized by the force jump discontinuity that characterizes the force-velocity relationship in the case of contact between two rigid surfaces. Some of the particles on the two surfaces deform with respect to each other, leading to the creep phenomenon, and this gives rise to small slip that is not accounted for in Coulomb's theory of dry friction. As the relative motion between the two solids evolves with time, the relative velocity between the two solids can be attributed to a combination of small slip and rolling motion. Before the tangential friction force reaches the saturation limit μF_n, the relative velocity at the contact points is primarily due to the creep phenomenon, which is a result of the relative deformation between particles on the surfaces of the two solids. At these small values of the relative velocity in the contact area, Coulomb's friction law of the sliding between two rigid surfaces does not apply. To compute the tangential forces resulting from normal contact pressure, normalized velocity coefficients called *creepages* are evaluated. Three normalized relative velocities have been found to have a significant effect on wheel/rail interaction forces: *longitudinal creepage* ζ_x, *lateral creepage* ζ_y, and *spin creepage* ζ_s. The equations of these normalized velocity creepages, which enter into formulating the tangential creep force and spin moment, are presented later in this chapter.

Steps in the Wheel/Rail Contact Analysis To determine wheel/rail contact forces, several basic steps must be followed. The geometry and mechanics concepts discussed in the preceding chapters play a fundamental role in all of these steps for developing a general three-dimensional wheel/rail contact model. These steps, which need to be followed in order and which demonstrate the importance of integrating geometry and mechanics concepts, can be summarized as follows: (i) solve for the system configuration by numerical integration of the nonlinear dynamic equations of motion to determine the system coordinates and velocities; (ii) determine the location of the contact point on the wheel and rail surfaces online using the coordinates of the wheel and rail obtained using numerical integration of the equations of motion; (iii) compute the normal wheel/rail contact forces; (iv) determine the geometry of the wheel and rail surfaces at the contact point; (v) calculate the tangential creep or friction forces at the contact point; and (vi) use the normal and tangential contact forces to determine the generalized contact forces associated with the generalized coordinates of the wheel and rail. These six basic steps are described in more detail as follows:

1. *Solve for the system configuration.* The system configuration at the beginning of the simulation is assumed to be known from the initial coordinates and velocities. At any other instant of time, the system coordinates and velocities are assumed to be known from the numerical integration of the nonlinear equations of motion, developed in Chapter 6. Therefore, for a given wheel w and rail r, coordinates $\mathbf{q}^w = \left[\mathbf{R}^{w^T} \; \boldsymbol{\theta}^{w^T}\right]^T$ and $\mathbf{q}^r = \left[\mathbf{R}^{r^T} \; \boldsymbol{\theta}^{r^T}\right]^T$ are assumed to be known when searching for the wheel/rail contact

points, where $\mathbf{R}^k$ and θ^k, $k = w, r$, are, respectively, the vector that defines the global position of the origin of the body coordinate system and the set of rotation parameters that define the body orientation in the global coordinate system. Similarly, vectors $\dot{\mathbf{q}}^w = \left[\dot{\mathbf{R}}^{w^T} \ \dot{\boldsymbol{\theta}}^{w^T}\right]^T$ and $\dot{\mathbf{q}}^r = \left[\dot{\mathbf{R}}^{r^T} \ \dot{\boldsymbol{\theta}}^{r^T}\right]^T$ are assumed to be known from numerical integration of the system equations of motion.

2. *Determine the contact point*. At this step, it is assumed that the configurations of the wheel and rail are already determined from numerical integration of the system equations of motion. That is, vectors of coordinates $\mathbf{q}^w$ and $\mathbf{q}^r$ of the wheel and rail, respectively, as well as their time derivatives $\dot{\mathbf{q}}^w$ and $\dot{\mathbf{q}}^r$, are assumed known. Using these known wheel and rail configurations, two methods can be used to define the location of the contact points on the wheel and rail online. These are the constraint and elastic contact formulations, as previously mentioned in this chapter. As will be shown, the geometry and mechanics concepts discussed in the preceding chapters represent the foundation for the analysis performed in this step to determine the locations of the wheel/rail contact points. The constraint contact formulation is described in Section 5.3, while the elastic contact formulation is described in Section 5.4. Both formulations are based on a general three-dimensional analysis and allow for predicting the location of the contact points online. In both formulations, a set of algebraic equations is solved for the surface parameters that define the coordinates of the contact points on the wheel and rail surfaces. In both approaches, the solution for the surface parameters requires using an iterative Newton–Raphson algorithm because of the nonlinearity of the algebraic equations used. The basic difference between the two contact formulations, however, is in the number of degrees of freedom used. In the constraint contact formulation, wheel/rail separations are not allowed, and therefore, the wheel has five degrees of freedom with respect to the rail. In the elastic contact formulation, on the other hand, wheel/rail separations are allowed, and therefore, the wheel has six degrees of freedom with respect to the rail. In the constraint contact formulation, the wheel is not allowed to move with respect to the rail in a direction normal to the contact surfaces at the contact point, while this relative motion is allowed in the elastic contact formulation. Some other and fundamentally different approaches, such as *lookup tables*, are used in the literature to determine the locations of the wheel/rail contact points. By using such simplified approaches, however, it is difficult to account for the three-dimensional nature of the wheel/rail contact problem using a manageable set of precomputed data. Furthermore, using lookup tables requires using interpolation methods that have orders that depend on the number of data entries at each table point. A high degree of accuracy may require using very large data files, particularly if a lower order of interpolation between the table points is the option to be used. For three-dimensional contact analysis, the interpolation can be more demanding computationally. Using a three-dimensional approach for predicting the location of the contact points is necessary in many scenarios that include wheel climb at a large angle of attack.
3. *Compute the normal contact force*. Having determined the location of the contact points on the wheel and rail, the constraint or elastic contact formulation can be used to determine the normal contact force. In the constraint contact formulation, the normal contact force is determined as a constraint (reaction) force because, in this CCF approach, one degree of freedom is eliminated. As discussed later in this chapter, the local coordinates of the contact point can be used to formulate the contact constraint conditions using a set

of nonlinear algebraic equations. Chapter 6 explains how the contact constraints can be used to determine the normal contact force systematically in terms of the constraint Jacobian matrix and vector of *Lagrange multipliers*. If, on the other hand, the elastic contact formulation is used, the wheel/rail penetration at the contact point is determined from the known system configuration. The penetration and its time derivative are used to formulate a compliant force model with assumed stiffness and damping coefficients. In the ECF approach, wheel/rail separations are allowed; therefore, if there is no penetration, the normal contact force is assumed to be zero.

4. *Determine the surface geometry at the contact point*. When two solids come into contact, the dimensions of the contact area depend on the geometric properties of the surfaces of the two solids in the contact region, their material properties, and the normal contact force. Determining the dimensions of the contact area is necessary in order to be able to evaluate the tangential creep forces. In Hertz contact theory, widely used in railroad vehicle dynamics, the contact area is assumed to be elliptical. To determine the dimensions of the contact ellipse in Hertz theory, the principal curvatures and principal directions of the wheel and rail surfaces at the contact point must be determined using the approach discussed in Chapter 2. To determine the principal curvature and principal directions, the geometry of the wheel and rail surfaces must be defined in terms of the surface parameters and used to evaluate the coefficients of the first and second fundamental forms of the wheel and rail surfaces. These coefficients are used to determine the principal curvatures and directions, as discussed in Chapter 2. It is explained in this chapter how the principal curvatures enter into the definition of the dimensions of the contact ellipse when Hertz theory is used.
5. *Calculate the tangential creep forces*. Tangential creep forces, which play a fundamental role in railroad vehicle dynamics and stability, can be computed using the normal contact force, dimensions of the contact ellipse, wheel and rail material properties, and relative velocities between the wheel and rail at the contact point. If the coordinates of the contact points on the wheel and rail are known in their respective body coordinate systems and defined by vectors $\overline{\mathbf{u}}_P^k,\ \ k = w, r$, the global position vectors of the contact points can be written as $\mathbf{r}_P^k = \mathbf{R}^k + \mathbf{A}^k\overline{\mathbf{u}}_P^k,\ \ k = w, r$, where $\mathbf{A}^k$, $k = w, r$ is the transformation matrix that defines the orientation of body k, as discussed in Chapter 3. Using this definition of the global position vector, the absolute velocity vector of the contact points on the wheel and rail can be obtained, as discussed in Chapter 3, as $\mathbf{v}_P^k = \dot{\mathbf{R}}^k + \boldsymbol{\omega}^k \times \left(\mathbf{A}^k\overline{\mathbf{u}}_P^k\right)$, $k = w, r$, where $\boldsymbol{\omega}^k$, $k = w, r$, is the absolute angular velocity vector of body k. Using these definitions of the absolute velocity vector and absolute angular velocity vector, the relative velocity between the wheel and rail in the tangential plane as well as the relative angular velocity in the direction of the normal to the surfaces at the contact point can be determined. These relative velocities are used to define the *creepages* that enter into the definition of the creep force models. Linear and nonlinear creep force models have been used in the literature. Some of these models are discussed in this chapter.
6. *Evaluate the generalized forces*. As discussed in Chapter 6, the equations of motion of the wheel and rail can be formulated using Newton–Euler equations or the generalized form of those equations. Newton–Euler equations are written in terms of angular acceleration vectors, while their generalized form is written in terms of the second time derivatives of the orientation parameters. Having determined the normal contact forces, tangential creep forces, and spin moments at the contact points, either form of

the Newton–Euler equations can be used to formulate the nonlinear system equations of motion. For railroad vehicle systems subjected to motion constraints, it is more convenient to use the generalized form of the Newton–Euler equations. In this case, the generalized contact forces associated with the generalized coordinates of the wheel and rail must be evaluated before numerically integrating the accelerations, as explained in Chapter 6. In general, the generalized wheel/rail contact forces are highly nonlinear functions of the wheel and rail coordinates. For this reason, a fully nonlinear approach that does not employ small-rotation assumptions needs to be used in order to be able to predict the dynamics of railroad vehicle systems accurately.

Summary of the Wheel and Rail Surface Equations The contact force and constraint formulations discussed in this chapter, which require an understanding of the geometry and mechanics concepts discussed in the preceding chapters, will be used to formulate the nonlinear dynamic equations of complex railroad vehicle systems developed in Chapter 6. For the three-dimensional analysis procedure developed in this book, the wheel surface is parameterized using surface parameters s_1^w and s_2^w, while the surface parameters of the rail are s_1^r and s_2^r. Therefore, the vector of the wheel and rail surface parameters can be written as $\mathbf{s} = \begin{bmatrix} s_1^w & s_2^w & s_1^r & s_2^r \end{bmatrix}^T$, where superscripts w and r refer, respectively, to the wheel and rail. The locations of points on the wheel and rail in the respective body coordinate system can be written using these surface parameters, respectively, as

$$\left.\begin{aligned} \overline{\mathbf{u}}^w\left(s_1^w, s_2^w\right) &= \begin{bmatrix} x^w\left(s_1^w, s_2^w\right) & y^w\left(s_1^w, s_2^w\right) & z^w\left(s_1^w, s_2^w\right) \end{bmatrix}^T, \\ \overline{\mathbf{u}}^r\left(s_1^r, s_2^r\right) &= \begin{bmatrix} x^r\left(s_1^r, s_2^r\right) & y^r\left(s_1^r, s_2^r\right) & z^r\left(s_1^r, s_2^r\right) \end{bmatrix}^T \end{aligned}\right\} \tag{5.1}$$

The tangents and normal to the surfaces at the contact point are defined in the body coordinate system as

$$\overline{\mathbf{t}}_1^k = \frac{\partial \overline{\mathbf{u}}^k}{\partial s_1^k}, \qquad \overline{\mathbf{t}}_2^k = \frac{\partial \overline{\mathbf{u}}^k}{\partial s_2^k}, \qquad \overline{\mathbf{n}}^k = \overline{\mathbf{t}}_1^k \times \overline{\mathbf{t}}_2^k, \qquad k = w, r \tag{5.2}$$

Equations 1 and 2 are used in the following sections to define the wheel/rail contact conditions. It is important, however, to recall from the analysis presented in Chapter 4 that the preceding two equations are highly nonlinear functions of the surface parameters. For example, in the case of the wheel, vector $\overline{\mathbf{u}}^w$, which is defined in the wheel coordinate system $X^wY^wZ^w$ can be written as $\overline{\mathbf{u}}^w\left(s_1^w, s_2^w\right) = \overline{\mathbf{R}}^{wp} + \overline{\mathbf{u}}^{wp}$, where $\overline{\mathbf{R}}^{wp} = \begin{bmatrix} x_o^{wp} & y_o^{wp} & z_o^{wp} \end{bmatrix}^T$ is the vector that defines the origin of the wheel profile frame $X^{wp}Y^{wp}Z^{wp}$ with respect to the origin of the coordinate system $X^wY^wZ^w$ of the wheel or wheelset, $\overline{\mathbf{u}}^{wp} = \begin{bmatrix} g^{wp}\left(s_1^w\right) \sin s_2^w & s_1^w & -g^{wp}\left(s_1^w\right) \cos s_2^w \end{bmatrix}^T$ is the vector that defines the location of the arbitrary point in the profile frame, and g^{wp} is the profile function, which can be a highly nonlinear function in the wheel lateral surface parameter s_1^w. Therefore, vector $\overline{\mathbf{u}}^w$ is a highly nonlinear function of the wheel surface parameters, as discussed in more detail in Chapter 4. Similarly, in the case of the rail, vector $\overline{\mathbf{u}}^r$ is a highly nonlinear function in the rail surface parameters. This vector was defined in Chapter 4 in its most general form as $\overline{\mathbf{u}}^r\left(s_1^r, s_2^r\right) = \overline{\mathbf{R}}^{rp}\left(s_1^r\right) + \mathbf{A}^{rp}\left(s_1^r\right)\overline{\mathbf{u}}^{rp}\left(s_1^r, s_2^r\right)$, where $\overline{\mathbf{R}}^{rp} = \overline{\mathbf{R}}^{rp}\left(s_1^r\right) = \begin{bmatrix} x_o^{rp}\left(s_1^r\right) & y_o^{rp}\left(s_1^r\right) & z_o^{rp} \end{bmatrix}^T$ is the vector that defines the origin of the rail profile frame $X^{rp}Y^{rp}Z^{rp}$ in the rail coordinate system $X^rY^rZ^r$, $\mathbf{A}^{rp} = \mathbf{A}^{rp}\left(s_1^r\right)$ is the matrix that defines the orientation of the profile frame

$X^{rp}Y^{rp}Z^{rp}$ in the rail coordinate system $X^rY^rZ^r$, $\bar{\bar{\mathbf{u}}}^{rp} = \begin{bmatrix} 0 & s_2^r & g^{rp}\left(s_1^r, s_2^r\right) \end{bmatrix}^T$ is the vector that defines the location of the arbitrary point in the profile frame $X^{rp}Y^{rp}Z^{rp}$, and g^{rp} is the rail profile function. As discussed in Chapter 4, the use of the ANCF interpolation leads to significant simplifications of the rail kinematic equations.

5.2 CONSTRAINT CONTACT FORMULATION (CCF)

In the wheel/rail constraint contact formulation, the wheel is assumed to remain in contact with the rail, and therefore, separation is not allowed. In this contact formulation, the freedom of the wheel to translate with respect to the rail along the direction normal to the surface at the contact point is eliminated by enforcing a set of nonlinear algebraic constraint equations at the position, velocity, and acceleration levels. In general, one constraint equation is sufficient to eliminate one degree of freedom. However, in the case of wheel/rail contact in which the contact point is not fixed and slides on the surfaces, the four surface parameters $\mathbf{s} = \begin{bmatrix} s_1^w & s_2^w & s_1^r & s_2^r \end{bmatrix}^T$ are introduced in order to be able to determine the location of contact points on the wheel and rail surfaces online. Since four additional parameters are used, one must enforce five constraint equations; four of them can be used to solve for surface parameters $\mathbf{s} = \begin{bmatrix} s_1^w & s_2^w & s_1^r & s_2^r \end{bmatrix}^T$ that define the location of the contact points given by Eq. 1. For a wheel/rail contact, the five nonlinear algebraic constraint equations used in the non-conformal constraint contact formulation can be written as

$$\mathbf{C}\left(\mathbf{q}^w, \mathbf{q}^r, \mathbf{s}^w, \mathbf{s}^r\right) = \begin{bmatrix} \mathbf{r}_P^{wr} \\ \mathbf{t}_1^{w^T}\mathbf{n}^r \\ \mathbf{t}_2^{w^T}\mathbf{n}^r \end{bmatrix} = \mathbf{0} \tag{5.3}$$

These equations, in which $\mathbf{r}_P^{wr} = \mathbf{r}_P^w - \mathbf{r}_P^r$ and $\mathbf{r}_P^k = \mathbf{R}^k + \mathbf{A}^k\bar{\mathbf{u}}_P^k$, $k = w, r$, is the global position vector of contact point P on body k, ensure that there are no wheel/rail separations and the tangent planes to the wheel and rail surfaces at contact point P are parallel. The first three scalar equations $\mathbf{r}_P^{wr} = \mathbf{r}_P^w - \mathbf{r}_P^r = \mathbf{0}$ ensure that there is no separation, while the last two equations $\mathbf{t}_1^w \cdot \mathbf{n}^r = 0$ and $\mathbf{t}_2^w \cdot \mathbf{n}^r = 0$ ensure the parallelism of the tangent planes for the wheel and rail surfaces at the contact point: a necessary non-conformal contact condition. Figure 1, which shows a wheel and a rail in contact at point P, illustrates the parallelism of the two tangent planes at the contact point.

To better explain the meaning of the constraint conditions of Eq. 3, these equations can be rewritten using another coordinate system as

$$\mathbf{C}\left(\mathbf{q}^w, \mathbf{q}^r, \mathbf{s}^w, \mathbf{s}^r\right) = \begin{bmatrix} \left(\mathbf{t}_1^{r^T}\mathbf{r}_P^{wr}\right) & \left(\mathbf{t}_2^{r^T}\mathbf{r}_P^{wr}\right) & \left(\mathbf{n}^{r^T}\mathbf{r}_P^{wr}\right) & \left(\mathbf{t}_1^{w^T}\mathbf{n}^r\right) & \left(\mathbf{t}_2^{w^T}\mathbf{n}^r\right) \end{bmatrix}^T = \mathbf{0} \tag{5.4}$$

While Eqs. 3 and 4 are equivalent, the first three scalar equations in Eq. 4 show more clearly that the longitudinal, lateral, and normal relative position of the contact point on the wheel and rail must remain equal to zero. As discussed at the end of the preceding section, the vectors that appear in Eqs. 3 and 4 are highly nonlinear functions of the surface parameters; therefore, the constraint conditions in these two equations are highly nonlinear functions in these surface parameters as well as the wheel and rail generalized coordinates. For this reason, an analytical solution for these constraint equations cannot be obtained

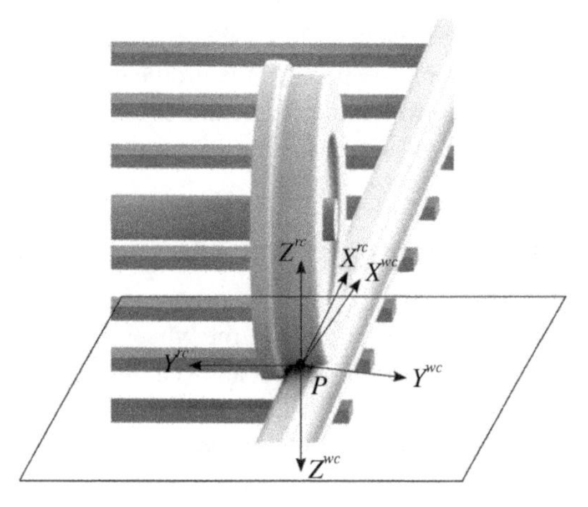

Figure 5.1 Contact conditions.

in practical applications, and the equations must be solved numerically using an iterative procedure. For example, to solve for the four surface parameters $\mathbf{s} = \begin{bmatrix} s_1^w & s_2^w & s_1^r & s_2^r \end{bmatrix}^T$, one can choose the first two equations and last two equations in Eq. 4 to form the vector of constraint equations $\mathbf{C}^d(\mathbf{q}^w, \mathbf{q}^r, \mathbf{s}^w, \mathbf{s}^r) = \begin{bmatrix} \mathbf{t}_1^{r^T} \mathbf{r}_P^{wr} & \mathbf{t}_2^{r^T} \mathbf{r}_P^{wr} & \mathbf{t}_1^{w^T} \mathbf{n}^r & \mathbf{t}_2^{w^T} \mathbf{n}^r \end{bmatrix}^T = \mathbf{0}$. If the generalized coordinates of the wheel and rail $\mathbf{q}^w$ and $\mathbf{q}^r$, respectively, are assumed to be known from the numerical integration, these four nonlinear algebraic equations can be solved using an iterative Newton–Raphson algorithm to determine the four surface parameters that ensure there is no longitudinal and lateral shift between the contact points on the wheel and rail and also ensure that the tangent planes to the wheel and rail surfaces at the contact point are parallel. The remaining equation $\mathbf{n}^{r^T} \mathbf{r}_P^{wr} = 0$ imposes a constraint on the generalized coordinates to ensure that there is no relative displacement between the wheel and rail along the normal to the contact surfaces. In this CCF approach, the constraint conditions must be satisfied at the position, velocity, and acceleration levels, and the normal contact force is considered a reaction force. The first and second derivatives of the constraint equations with respect to time must be evaluated and used with the equations of motion to solve for the system accelerations and constraint forces. Chapter 6 discusses computational algorithms for solving the constraint conditions of Eq. 4 with the dynamic equations of motion to determine the surface parameters, system coordinates, and constraintforces.

5.3 ELASTIC CONTACT FORMULATION (ECF)

Another formulation that can be used to determine the locations of the wheel/rail contact points online is the elastic contact formulation. Unlike the constraint contact formulation, the elastic contact formulation does not eliminate any degrees of freedom and allows for wheel/rail separations. Therefore, a rigid wheel is assumed to have six degrees of freedom with respect to the rail. This is despite the fact that a set of algebraic equations similar to the ones used in the constraint contact formulation are used in the elastic contact formulation. In the ECF approach, the algebraic equations used to solve for the surface parameters do not impose any constraints on the motion of the wheel or wheelset, and no constraint (reaction) forces are associated with these equations.

Using the algebraic contact conditions to determine the locations of the contact points on the wheel and rail is not the only approach that can be used. Other alternate approaches that can be used include *nodal search methods* and *lookup tables* (Shabana et al. 2008). In the nodal search approach, point meshes are created on the wheel and rail surfaces, and the distances between these points are checked to determine whether contact occurs. In the lookup table approach, precomputed wheel and rail point data are tabulated and used with an interpolation scheme to determine whether there is contact. For surfaces that have a certain degree of smoothness, there is no clear advantage of using the nodal search and lookup table approaches. Furthermore, implementing these two approaches can be cumbersome, particularly in the case of three-dimensional contacts that cannot be ignored in important motion scenarios, including wheel climbs and derailments. Therefore, the focus in this section is on using algebraic equations to determine the locations of the contact points. This approach has proven to be robust and allows for the implementation of efficient algorithms for the wheel/rail contact analysis.

In all the elastic contact formulations, such as using algebraic equations, nodal search, and lookup tables, distances between points on the wheel and rail are checked along the normal to the contact surfaces to determine whether penetration δ occurs. If there is penetration, penetration δ and its time derivative are computed and used to compute the normal contact forces using a compliant force model with assumed stiffness and damping coefficients.

In the elastic contact formulation discussed in this section, four algebraic equations are solved for each wheel/rail contact to determine the four surface parameters $\mathbf{s} = \begin{bmatrix} s_1^w & s_2^w & s_1^r & s_2^r \end{bmatrix}^T$ that define the locations of contact points $\overline{\mathbf{u}}_p^w\left(s_1^w, s_2^w\right)$ and $\overline{\mathbf{u}}_p^r\left(s_1^r, s_2^r\right)$ on the wheel and rail surfaces, respectively. It is assumed that the wheel and rail configurations are known from the integration of the system equations of motion. Because the contact algebraic equations are nonlinear functions of the surface parameters, an iterative Newton–Raphson algorithm is used to solve the contact conditions for the four surface parameters. The four equations used in the ECF approach are the same as the first two and last two equations of Eq. 4 used in the CCF approach. These equations are rewritten as

$$\mathbf{g}\left(\mathbf{s}^w, \mathbf{s}^r\right) = \begin{bmatrix} g_1\left(\mathbf{s}^w, \mathbf{s}^r\right) \\ g_2\left(\mathbf{s}^w, \mathbf{s}^r\right) \\ g_3\left(\mathbf{s}^w, \mathbf{s}^r\right) \\ g_4\left(\mathbf{s}^w, \mathbf{s}^r\right) \end{bmatrix} = \begin{bmatrix} \mathbf{t}_1^{r^T} \mathbf{r}_P^{wr} \\ \mathbf{t}_2^{r^T} \mathbf{r}_P^{wr} \\ \mathbf{t}_1^{w^T} \mathbf{n}^r \\ \mathbf{t}_2^{w^T} \mathbf{n}^r \end{bmatrix} = \mathbf{0} \tag{5.5}$$

These four nonlinear algebraic equations can be solved for surface parameters $\mathbf{s}^w = \begin{bmatrix} s_1^w & s_2^w \end{bmatrix}^T$ and $\mathbf{s}^r = \begin{bmatrix} s_1^r & s_2^r \end{bmatrix}^T$ using an iterative Newton–Raphson algorithm. Toward this end, the set of algebraic equations $(\partial \mathbf{g}/\partial \mathbf{s})\Delta \mathbf{s} = -\mathbf{g}$ are iteratively solved for each contact, where $\Delta \mathbf{s} = \begin{bmatrix} \Delta \mathbf{s}^{w^T} & \Delta \mathbf{s}^{r^T} \end{bmatrix}^T = \begin{bmatrix} \Delta s_1^w & \Delta s_2^w & \Delta s_1^r & \Delta s_2^r \end{bmatrix}^T$ is the vector of Newton differences and

$$\frac{\partial \mathbf{g}}{\partial \mathbf{s}} = \begin{bmatrix} \frac{\partial \mathbf{g}}{\partial \mathbf{s}^w} & \frac{\partial \mathbf{g}}{\partial \mathbf{s}^r} \end{bmatrix} = \begin{bmatrix} \frac{\partial \mathbf{g}}{\partial s_1^w} & \frac{\partial \mathbf{g}}{\partial s_2^w} & \frac{\partial \mathbf{g}}{\partial s_1^r} & \frac{\partial \mathbf{g}}{\partial s_2^r} \end{bmatrix}$$

$$= \begin{bmatrix} \frac{\partial g_1}{\partial s_1^w} & \frac{\partial g_1}{\partial s_2^w} & \frac{\partial g_1}{\partial s_1^r} & \frac{\partial g_1}{\partial s_2^r} \\ \frac{\partial g_2}{\partial s_1^w} & \frac{\partial g_2}{\partial s_2^w} & \frac{\partial g_2}{\partial s_1^r} & \frac{\partial g_2}{\partial s_2^r} \\ \frac{\partial g_3}{\partial s_1^w} & \frac{\partial g_3}{\partial s_2^w} & \frac{\partial g_3}{\partial s_1^r} & \frac{\partial g_3}{\partial s_2^r} \\ \frac{\partial g_4}{\partial s_1^w} & \frac{\partial g_4}{\partial s_2^w} & \frac{\partial g_4}{\partial s_1^r} & \frac{\partial g_4}{\partial s_2^r} \end{bmatrix} \tag{5.6}$$

This 4×4 Jacobian matrix can be written more explicitly as

$$\frac{\partial \mathbf{g}}{\partial \mathbf{s}} = \begin{bmatrix} \mathbf{t}_1^{r^T}\mathbf{t}_1^w & \mathbf{t}_1^{r^T}\mathbf{t}_2^w & \left(\left(\partial \mathbf{t}_1^r/\partial s_1^r\right)^T \mathbf{r}^{wr} - \mathbf{t}_1^{r^T}\mathbf{t}_1^r\right) & \left(\left(\partial \mathbf{t}_1^r/\partial s_2^r\right)^T \mathbf{r}^{wr} - \mathbf{t}_1^{r^T}\mathbf{t}_2^r\right) \\ \mathbf{t}_2^{r^T}\mathbf{t}_1^w & \mathbf{t}_2^{r^T}\mathbf{t}_2^w & \left(\left(\partial \mathbf{t}_2^r/\partial s_1^r\right)^T \mathbf{r}^{wr} - \mathbf{t}_2^{r^T}\mathbf{t}_1^r\right) & \left(\left(\partial \mathbf{t}_2^r/\partial s_2^r\right)^T \mathbf{r}^{wr} - \mathbf{t}_2^{r^T}\mathbf{t}_2^r\right) \\ \left(\partial \mathbf{t}_1^w/\partial s_1^w\right)^T \mathbf{n}^r & \left(\partial \mathbf{t}_1^w/\partial s_2^w\right)^T \mathbf{n}^r & \left(\partial \mathbf{n}^r/\partial s_1^r\right)^T \mathbf{t}_1^w & \left(\partial \mathbf{n}^r/\partial s_2^r\right)^T \mathbf{t}_1^w \\ \left(\partial \mathbf{t}_2^w/\partial s_1^w\right)^T \mathbf{n}^r & \left(\partial \mathbf{t}_2^w/\partial s_2^w\right)^T \mathbf{n}^r & \left(\partial \mathbf{n}^r/\partial s_1^r\right)^T \mathbf{t}_2^w & \left(\partial \mathbf{n}^r/\partial s_2^r\right)^T \mathbf{t}_2^w \end{bmatrix} \tag{5.7}$$

In the case of non-conformal contact, this Jacobian matrix is nonsingular; therefore, the system of equations $(\partial \mathbf{g}/\partial \mathbf{s})\Delta \mathbf{s} = -\mathbf{g}$ can be solved for the Newton differences, which can be used to update the surface parameters. The iterative process continues until convergence is achieved. Implementing this method showed that using the values of the surface parameters from the previous time step leads to a good initial guess for the Newton–Raphson iterations and fast convergence. It is important to emphasize again that while the four nonlinear ECF algebraic equations used to solve for the surface parameters are the same as four of the algebraic equations used in the constraint contact formulation, the ECF approach cannot be considered a constraint formulation because the ECF algebraic equations are not imposed at the velocity and acceleration levels and do not give rise to constraint forces. Furthermore, the ECF procedure allows for wheel/rail penetrations and separations.

The converged solution obtained for surface parameters $\mathbf{s} = \begin{bmatrix} s_1^w & s_2^w & s_1^r & s_2^r \end{bmatrix}^T$ can be used to define the locations of potential contact points $\overline{\mathbf{u}}_P^w\left(s_1^w, s_2^w\right)$ and $\overline{\mathbf{u}}_P^r\left(s_1^r, s_2^r\right)$ on the wheel and rail surfaces, respectively. These local position vectors can be used to define the global position vectors of the potential contact points as $\mathbf{r}_P^k = \mathbf{R}^k + \mathbf{A}^k\overline{\mathbf{u}}_P^k,\ k = w, r$, which can be used in turn to define vector $\mathbf{r}_P^{wr} = \mathbf{r}_P^w - \mathbf{r}_P^r$. The distance between the two potential contact points on the wheel and rail surfaces along the normal can be evaluated using an equation similar to the third equation of Eq. 4. Toward this end, the wheel/rail penetration can be computed as

$$\delta = \mathbf{r}_P^{wr^T}\hat{\mathbf{n}}^r = \left(\mathbf{r}_P^w - \mathbf{r}_P^r\right)^T \hat{\mathbf{n}}^r \tag{5.8}$$

where $\hat{\mathbf{n}}^r = \left(\mathbf{t}_1^r \times \mathbf{t}_2^r\right) / \left|\mathbf{t}_1^r \times \mathbf{t}_2^r\right|$ is the unit normal to the rail surface at the potential contact point. The sign of the penetration in the preceding equation can be used to define whether there is wheel/rail contact or separation.

5.4 NORMAL CONTACT FORCES

The fundamental differences between the CCF and ECF approaches will become clearer from the development of the governing dynamic equations and computational procedures presented in Chapter 6. While a CCF algorithm does not allow for wheel/rail separations and is not well-suited for the analysis of derailments, such an algorithm can shed more light on stability and critical speed issues without making assumptions in the force model. For this reason, implementing the constraint contact formulation is recommended for the study of motion scenarios that require precise calculations of some parameters under normal operating conditions and for investigating phenomena without making assumptions related to the nature of the wheel/rail interaction. The ECF implementation, on the other hand, is necessary for the analysis of motion scenarios in which the wheels are separated from the rails. These scenarios include derailments, and for this reason, the ECF implementation is required for credible accident investigations.

In the constraint contact formulation, the non-conformal wheel/rail contact conditions presented in this chapter can be used to identify a dependent coordinate and a constraint force for each contact. As shown in Chapter 6, the constraint reaction force, which defines the wheel/rail normal contact force, can be systematically determined using the Jacobian matrix of one of the contact constraint equations, which is taken to be the third equation in Eq. 4, $\mathbf{r}_P^{wr^T}\mathbf{n}^r = \left(\mathbf{r}_P^w - \mathbf{r}_P^r\right)^T\mathbf{n}^r = 0$. Therefore, after the surface parameters and generalized coordinates are determined from a position analysis, the constraint equations can be formulated and used to define the constraint Jacobian matrix, which is used in the Lagrangian approach to define the wheel/rail normal contact force F_n. This normal contact force can be used to compute the tangential creep forces and spin moment, as discussed later in this chapter.

In the case of the elastic contact formulation, on the other hand, the normal contact force can be calculated using a compliant force model if the wheel and rail surfaces penetrate. This compliant force model is based on assumed stiffness and damping coefficients. The penetration can be determined from the equation $\delta = \mathbf{r}_P^{wr^T}\hat{\mathbf{n}}^r = \left(\mathbf{r}_P^w - \mathbf{r}_P^r\right)^T\hat{\mathbf{n}}^r$ (Eq. 8), and the relative velocity along the normal to the rail surface can be written as $v_{rn} = \dot{\mathbf{r}}_P^{wr^T}\hat{\mathbf{n}}^r = \left(\dot{\mathbf{r}}_P^w - \dot{\mathbf{r}}_P^r\right)^T\hat{\mathbf{n}}^r$ with the understanding that the unit normal to the rail surface is assumed fixed when evaluating the normal component of the relative velocity. In general, $d\left(\mathbf{r}_P^{wr^T}\hat{\mathbf{n}}^r\right)/dt = \dot{\mathbf{r}}_P^{wr^T}\hat{\mathbf{n}}^r + \mathbf{r}_P^{wr^T}\dot{\hat{\mathbf{n}}}^r$; therefore, $v_{rn} = \dot{\mathbf{r}}_P^{wr^T}\hat{\mathbf{n}}^r$ is the component of the relative velocity along the normal to the rail surface at the contact point and is not exactly the time derivative of δ, unless an assumption is made that the normal to the rail surface is fixed because the generalized coordinates and surface parameters, as well as their time derivatives, are assumed known. It is also clear that vector $\dot{\hat{\mathbf{n}}}^r$ is perpendicular to vector $\hat{\mathbf{n}}^r$; therefore, the velocity component $\mathbf{r}_P^{wr^T}\dot{\hat{\mathbf{n}}}^r$ is not along a direction normal to the contact surface. For this reason, the equation $\dot{\delta} = v_{rn} = \dot{\mathbf{r}}_P^{wr^T}\hat{\mathbf{n}}^r$ is often used to measure the time rate of the penetration along the normal to the contact surfaces. Using the expressions for δ and v_{rn}, the normal contact force F_n can be defined as the sum of an elastic force F_{ns} and a damping force F_{nd} as (Shabana et al. 2004):

$$F_n = F_{ns} + F_{nd} = -K(\delta)^{1.5} - Cv_{rn}\left|\delta\right| \tag{5.9}$$

where $F_{ns} = -K\delta^{1.5}$ is the elastic contact force used in Hertz theory, K is the Hertzian stiffness coefficient that depends on the material properties and the geometry of the two surfaces in contact, $F_{nd} = -Cv_{rn}|\delta|$ is the component of the damping force, and C is an assumed damping coefficient. The absolute value of penetration $|\delta|$ is included in the damping force to ensure that the normal contact force assumes a zero value when penetration is zero.

5.5 CONTACT SURFACE GEOMETRY

One computational method that can be used to predict the locations of the contact point to study wheel/rail interaction forces is to create point meshes on the wheel and rail surfaces. Using the generalized coordinates of the wheel and rail, which are assumed to be known from the numerical integration of the equations of motion, the distances between these mesh points along a direction normal to the surface can be computed. The surface discretization can be performed using the FE method or a set of arbitrary points selected by the analyst. In this case, the forces can be calculated directly without the need to specify a contact area or have a certain degree of smoothness of the two surfaces in contact. In the case of non-smooth surfaces, an approximation of the normal to the contact surfaces can be made if the differential geometry smoothness requirements cannot be applied. While this approach is more general and allows for modeling both conformal and non-conformal contacts, its use for describing the contact surface geometry can lead to inefficient implementation of computational algorithms designed for solving railroad vehicle system problems. For this reason, a different semi-analytical geometry approach has been used in the railroad vehicle dynamics literature. This approach, which requires computing the principal curvatures and principal directions of the wheel and rail surfaces at the contact point, has proven to capture the nonlinear dynamics of the wheel/rail contact accurately. It also allows for defining the dimensions of the wheel/rail contact area, which is assumed to be elliptical based on experimental observations (Hertz 1882; Johnson 1985). In Hertz contact theory, non-conformal contact is assumed. This assumption is justified in most railroad vehicle system applications due to the shapes of the solid wheel and rail. Because the dimensions of the wheel/rail contact ellipse are small in comparison with the wheel and rail dimensions and the relative radii of curvatures, the assumption of *half-space* is often used (Johnson 1985). This assumption allows for treating contact stresses separately from the stress distributions in the wheel and rail due to other forces. Hertz contact theory, widely used in railroad vehicle dynamics, is based on static considerations, does not account for dynamic and friction effects, and employs the assumptions of small deformations and linear elasticity.

Surface Parametric Forms Figure 2 shows two solids in contact. The surfaces of these two solids, denoted as i and j, can be represented in parametric forms as previously mentioned in this chapter and as explained in more detail in Chapter 2. If two different coordinate systems are used to define the two surfaces, one can write the equations of the two

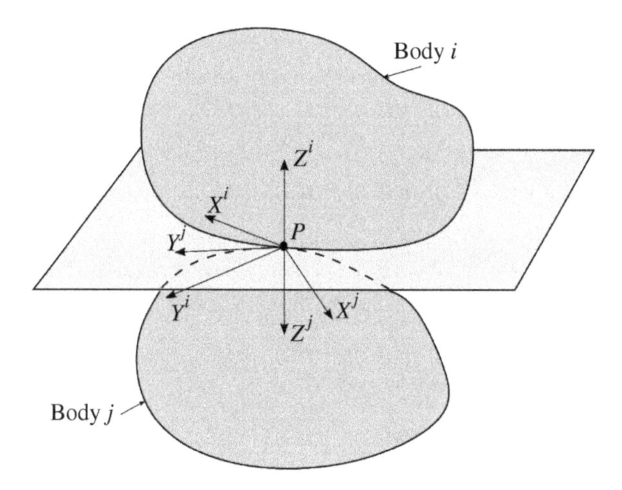

Figure 5.2 Contact between two solids.

surfaces, respectively, as

$$\mathbf{r}^i = \begin{bmatrix} r_1^i \\ r_2^i \\ r_3^i \end{bmatrix} = \begin{bmatrix} x^i \\ y^i \\ a_1^i(x^i)^2 + a_2^i(y^i)^2 + a_3^i(x^i y^i) + \cdots \end{bmatrix} \tag{5.10}$$

and

$$\mathbf{r}^j = \begin{bmatrix} r_1^j \\ r_2^j \\ r_3^j \end{bmatrix} = \begin{bmatrix} x^j \\ y^j \\ a_1^j(x^j)^2 + a_2^j(y^j)^2 + a_3^j(x^j y^j) + \cdots \end{bmatrix} \tag{5.11}$$

In these equations, the surface parameters are assumed to be Cartesian coordinates: that is, $\mathbf{s}^k = \begin{bmatrix} s_1^k & s_2^k \end{bmatrix}^T = \begin{bmatrix} x^k & y^k \end{bmatrix}^T$, $k = i, j$. A polynomial with coefficients a_l^k, $l = 1, 2, \ldots, n_p^k$ is used to define the shapes of the two surfaces, where the number of coefficients n_p^k, $k = i, j$, can be used to define the order of the polynomial. Because static and small deformation assumptions are made in Hertz theory – that is, no rigid body motion of the surfaces with respect to their coordinate systems – the absence of the linear terms of the polynomials of the preceding equations can be justified. The two curve equations can also be defined in the same coordinate system using one set of parameters and two different sets of coefficients that define the different geometries of the two surfaces. Using this representation, and assuming that the third axis of the coordinate system is normal to the contact surfaces as shown in Figure 2, one can define the gap between the two surfaces in the neighborhood of the origin of the coordinate system as $h^{ij} = r_3^i - r_3^j$, which can be written using the parametric equations of the two surfaces defined in the same coordinate system as

$$h^{ij} = a_1(x)^2 + a_2(y)^2 + a_3(xy) + \cdots \tag{5.12}$$

where a_l, $l = 1, 2, \ldots, n_p$, are the resulting polynomial coefficients and n_p is the number of coefficients that can be used to define the polynomial order. The gap $h^{ij} = h^{ij}(x, y)$ is used to define the area of contact between the two solids.

Surface Curvatures Because the contact area is assumed to be small, a quadratic interpolation is sufficient for the surface representation of the two solids in the contact region. Therefore, Eqs. 10 and 11 can be written, respectively, as $\mathbf{r}^k = \left[x^k \ y^k \ \left(a_1^k(x^k)^2 + a_2^k(y^k)^2 + a_3^k(x^k y^k)\right)\right]^T$, $k = i, j$. Furthermore, the coordinate system can be selected such that coefficient a_3^k is equal to zero. This leads to the surface equation $\mathbf{r}^k = \left[x^k \ y^k \ \left(a_1^k(x^k)^2 + a_2^k(y^k)^2\right)\right]^T$, $k = i, j$. Using the coefficients of the first and second fundamental forms of surfaces presented in Chapter 2, one can determine the principal curvatures κ_1^k and κ_2^k, which define the maximum and minimum curvatures. For a surface described by the parametric equation $\mathbf{r}^k = \left[x^k \ y^k \ \left(a_1^k(x^k)^2 + a_2^k(y^k)^2\right)\right]^T$, one can show that this equation at the origin yields

$$\left.\begin{aligned} \partial^2 r_3^k/\partial(x^k)^2 &= 2a_1^k = \kappa_1^k = 1/R_1^k \\ \partial^2 r_3^k/\partial(y^k)^2 &= 2a_2^k = \kappa_2^k = 1/R_2^k, \quad k = i,j \end{aligned}\right\} \tag{5.13}$$

where R_1^k and R_2^k are the *principal radii of curvature* at the origin, which must be given the appropriate signs depending on the signs of the principal curvatures, which can be positive or negative. It is clear from Figure 2, for example, that one surface has positive or negative curvatures, and the other surface has curvatures with opposite signs. Therefore, the surface equation can be written in terms of the principal radii of curvatures for the two solids as

$$\mathbf{r}^k = \begin{bmatrix} r_1^k \\ r_2^k \\ r_3^k \end{bmatrix} = \begin{bmatrix} x^k \\ y^k \\ \dfrac{(x^k)^2}{2R_1^k} + \dfrac{(y^k)^2}{2R_2^k} \end{bmatrix}, \quad k = i,j \tag{5.14}$$

Regardless of the coordinate system used, the principal curvatures and principal radii of curvature are unique. Similarly, the gap function of Eq. 12 can be represented using a quadratic function and can be written, by selecting a coordinate system such that $a_3 = 0$, as

$$h^{ij} = a_1(x)^2 + a_2(y)^2 = \frac{(x)^2}{2R_1} + \frac{(y)^2}{2R_2} \tag{5.15}$$

where R_1 and R_2 are the *relative principal radii of curvature* at the origin.

Mathematical Derivation Because the axes of the coordinate lines $x^i y^i$ and $x^j y^j$ are not, in general, parallel, their orientations can differ by an angle α^{ij}, as shown in Figure 3. The orientation of coordinates $x^i y^i$ and $x^j y^j$ also differ from coordinates xy by angles α^i and α^j, respectively. The convention used in Figure 3 for measuring the angles differs slightly from the one used in the literature (Johnson 1985). Due to this difference, the formulas obtained in this section can have sign differences when compared with the formulas presented in some publications. Using the coordinate systems and angles shown in Figure 3, one can write the following coordinate transformation:

$$\begin{bmatrix} x^k \\ y^k \end{bmatrix} = \begin{bmatrix} \cos\alpha^k & \sin\alpha^k \\ -\sin\alpha^k & \cos\alpha^k \end{bmatrix} \begin{bmatrix} x \\ y \end{bmatrix}, \quad k = i,j \tag{5.16}$$

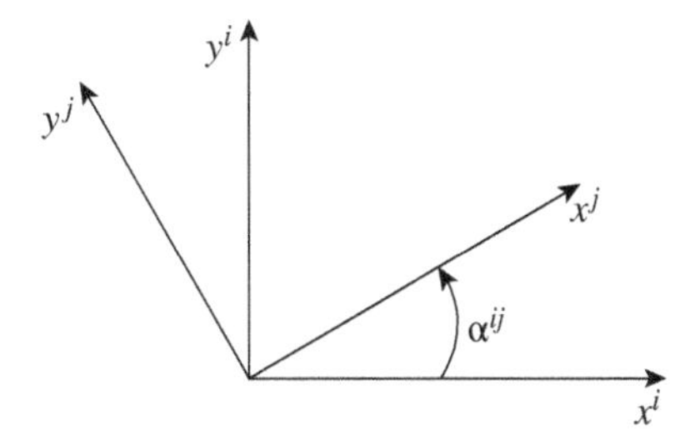

Figure 5.3 Relative orientation of the coordinate system.

It follows that $(x^k)^2 = (x)^2\cos^2\alpha^k + 2xy\sin 2\alpha^k + (y)^2\sin^2\alpha^k$ and $(y^k)^2 = (x)^2\sin^2\alpha^k - 2xy \sin 2\alpha^k + (y)^2\cos^2\alpha^k$. Therefore, if the quadratic form of Eq. 12 is used, one has

$$\begin{aligned}
h^{ij}(x,y) &= r_3^i - r_3^j = a_1^i\left(x^i\right)^2 + a_2^i\left(y^i\right)^2 - a_1^j\left(x^j\right)^2 - a_2^j\left(y^j\right)^2 \\
&= a_1^i\left((x)^2\cos^2\alpha^i + xy\sin 2\alpha^i + (y)^2\sin^2\alpha^i\right) + a_2^i\left((x)^2\sin^2\alpha^i - xy\sin 2\alpha^i + (y)^2\cos^2\alpha^i\right)^2 \\
&\quad - a_1^j\left((x)^2\cos^2\alpha^j + xy\sin 2\alpha^j + (y)^2\sin^2\alpha^j\right) - a_2^j\left((x)^2\sin^2\alpha^j - xy\sin 2\alpha^j + (y)^2\cos^2\alpha^j\right)^2
\end{aligned} \tag{5.17}$$

This equation can be written as

$$\begin{aligned}
h^{ij}(x,y) = &\left(a_1^i\cos^2\alpha^i + a_2^i\sin^2\alpha^i - a_1^j\cos^2\alpha^j - a_2^j\sin^2\alpha^j\right)(x)^2 \\
&+ \left(a_1^i\sin^2\alpha^i + a_2^i\cos^2\alpha^i - a_1^j\sin^2\alpha^j - a_2^j\cos^2\alpha^j\right)(y)^2 \\
&+ \left(\left(a_1^i - a_2^i\right)\sin 2\alpha^i - \left(a_1^j - a_2^j\right)\sin 2\alpha^j\right)(xy)
\end{aligned} \tag{5.18}$$

Using this equation, coefficients a_1, a_2, and a_3 of Eq. 12 in the case of a quadratic polynomial can be written as

$$\left.\begin{aligned}
a_1 &= a_1^i\cos^2\alpha^i + a_2^i\sin^2\alpha^i - a_1^j\cos^2\alpha^j - a_2^j\sin^2\alpha^j \\
a_2 &= a_1^i\sin^2\alpha^i + a_2^i\cos^2\alpha^i - a_1^j\sin^2\alpha^j - a_2^j\cos^2\alpha^j \\
a_3 &= \left(a_1^i - a_2^i\right)\sin 2\alpha^i - \left(a_1^j - a_2^j\right)\sin 2\alpha^j
\end{aligned}\right\} \tag{5.19}$$

Using Eq. 13, coefficient a_3 can be written as

$$a_3 = \frac{1}{2}\left(\frac{1}{R_1^i} - \frac{1}{R_2^i}\right)\sin 2\alpha^i - \frac{1}{2}\left(\frac{1}{R_1^j} - \frac{1}{R_2^j}\right)\sin 2\alpha^j \tag{5.20}$$

The condition that $a_3 = 0$, which is necessary to write Eq. 15, is then given by

$$\left(\frac{1}{R_1^i} - \frac{1}{R_2^i}\right)\sin 2\alpha^i = \left(\frac{1}{R_1^j} - \frac{1}{R_2^j}\right)\sin 2\alpha^j \tag{5.21}$$

It is also clear from Eq. 19 that

$$\begin{aligned}
a_2 - a_1 &= -a_1^i\left(\cos^2\alpha^i - \sin^2\alpha^i\right) - a_2^i\left(\sin^2\alpha^i - \cos^2\alpha^i\right) \\
&\quad + a_1^j\left(\cos^2\alpha^j - \sin^2\alpha^j\right) + a_2^j\left(\sin^2\alpha^j - \cos^2\alpha^j\right) \\
&= -\left(a_1^i - a_2^i\right)\cos 2\alpha^i + \left(a_1^j - a_2^j\right)\cos 2\alpha^j
\end{aligned} \tag{5.22}$$

This equation can be written using Eqs. 13 and 21 as

$$\begin{aligned} a_2 - a_1 &= -\frac{1}{2}\left(\frac{1}{R_1^i} - \frac{1}{R_2^i}\right)\cos 2\alpha^i + \frac{1}{2}\left(\frac{1}{R_1^j} - \frac{1}{R_2^j}\right)\cos 2\alpha^j \\ &= \frac{1}{2}\sqrt{\left(\frac{1}{R_1^i} - \frac{1}{R_2^i}\right)^2 + \left(\frac{1}{R_1^j} - \frac{1}{R_2^j}\right)^2 + 2\left(\frac{1}{R_1^i} - \frac{1}{R_2^i}\right)\left(\frac{1}{R_1^j} - \frac{1}{R_2^j}\right)\cos 2\alpha^{ij}} \end{aligned} \tag{5.23}$$

Using Eq. 19 and similar manipulations, one can show that $a_1 + a_2$ can be written as

$$a_1 + a_2 = \frac{1}{2}\left(\frac{1}{R_1^i} + \frac{1}{R_2^i}\right) - \frac{1}{2}\left(\frac{1}{R_1^j} + \frac{1}{R_2^j}\right) \tag{5.24}$$

Example 5.1 The parametric equation of a surface can be written in terms of two parameters x and y as $\mathbf{r} = \begin{bmatrix} x & y & f(x,y) \end{bmatrix}^T$. The tangents to the surface at x and y can be written as $\mathbf{r}_x = \partial \mathbf{r}/\partial x = \begin{bmatrix} 1 & 0 & \partial f/\partial x \end{bmatrix}^T$ and $\mathbf{r}_y = \partial \mathbf{r}/\partial y = \begin{bmatrix} 0 & 1 & \partial f/\partial y \end{bmatrix}^T$. The normal to the surface is defined as $\mathbf{n} = \mathbf{r}_x \times \mathbf{r}_y = \begin{bmatrix} -f_x & -f_y & 1 \end{bmatrix}^T$, where $f_x = \partial f/\partial x$ and $f_y = \partial f/\partial y$. Using the tangent vectors, the coefficients of the first fundamental form can be written as

$$E = \mathbf{r}_x^T \mathbf{r}_x = 1 + (f_x)^2, \quad F = \mathbf{r}_x^T \mathbf{r}_y = f_x f_y, \quad G = \mathbf{r}_y^T \mathbf{r}_y = 1 + (f_y)^2$$

The curvature vectors are defined as $\mathbf{r}_{xx} = \partial^2 \mathbf{r}/\partial x^2 = \begin{bmatrix} 0 & 0 & f_{xx} \end{bmatrix}^T$, $\mathbf{r}_{yy} = \partial^2 \mathbf{r}/\partial y^2 = \begin{bmatrix} 0 & 0 & f_{yy} \end{bmatrix}^T$, and $\mathbf{r}_{xy} = \partial^2 \mathbf{r}/\partial x \partial y = \begin{bmatrix} 0 & 0 & f_{xy} \end{bmatrix}^T$, where $f_{xx} = \partial^2 f/\partial x^2$, $f_{yy} = \partial^2 f/\partial y^2$, and $f_{xy} = \partial^2 f/\partial x \partial y$. Using the curvature vectors, the coefficients of the second fundamental form can be written as

$$L = \mathbf{r}_{xx}^T \hat{\mathbf{n}} = \frac{f_{xx}}{|\mathbf{n}|}, \quad M = \mathbf{r}_{xy}^T \hat{\mathbf{n}} = \frac{f_{xy}}{|\mathbf{n}|}, \quad N = \mathbf{r}_{yy}^T \hat{\mathbf{n}} = \frac{f_{yy}}{|\mathbf{n}|}$$

where $\hat{\mathbf{n}} = \mathbf{n}/|\mathbf{n}|$ is a unit vector along the normal to the surface, and $|\mathbf{n}| = \sqrt{(f_x)^2 + (f_y)^2 + 1}$.

If function f is defined using the quadratic polynomial $f = a_1(x)^2 + a_2(y)^2$, one has

$$f_x = 2a_1 x, \quad f_y = 2a_2 y, \quad f_{xx} = 2a_1, \quad f_{yy} = 2a_2, \quad f_{xy} = 0,$$

and $|\mathbf{n}| = \sqrt{1 + (f_x)^2 + (f_y)^2} = \sqrt{1 + (2a_1 x)^2 + (2a_2 y)^2}$. Using these derivatives, the curvature vectors can be written as

$$\mathbf{r}_{xx} = \begin{bmatrix} 0 & 0 & 2a_1 \end{bmatrix}^T, \quad \mathbf{r}_{yy} = \begin{bmatrix} 0 & 0 & 2a_2 \end{bmatrix}^T, \quad \mathbf{r}_{xy} = \mathbf{0}$$

The coefficients of the first and second fundamental forms can be written in this case as

$$\begin{aligned} &E = 1 + (2a_1 x)^2, \quad F = 4a_1 a_2 xy, \quad G = 1 + (2a_2 y)^2 \\ &L = 2a_1/|\mathbf{n}|, \quad M = 0, \quad N = 2a_2/|\mathbf{n}| \end{aligned}$$

As discussed in Chapter 2, the principal curvatures can be defined using the characteristic equation

$$\begin{vmatrix} L-\kappa_n E & M-\kappa_n F \\ M-\kappa_n F & N-\kappa_n G \end{vmatrix} =$$

$$\left(EG-(F)^2\right)\left(\kappa_n\right)^2-(EN+GL-2FM)\,\kappa_n+LN-(M)^2=0$$

where κ_n is the normal curvature. This characteristic equation has the following two roots:

$$\kappa_{1,2}=\frac{-\overline{b}\pm\sqrt{\left(\overline{b}\right)^2-4\overline{a}\,\overline{c}}}{2\overline{a}}$$

where in this example

$$\overline{a}=EG-(F)^2=1+\left(2a_1x\right)^2+\left(2a_2y\right)^2=(|\mathbf{n}|)^2$$

$$\overline{b}=-\ (EN+GL-2FM)=-\left[2a_2\left(1+\left(2a_1x\right)^2\right)+2a_1\left(1+\left(2a_2y\right)^2\right)\right]/\,|\mathbf{n}|$$

$$\overline{c}=LN-(M)^2=4a_1a_2/(|\mathbf{n}|)^2$$

At the origin, $x=0$ and $y=0$. In this case, the characteristic equation reduces to $(\kappa_n)^2-2(a_1+a_2)\kappa_n+4a_1a_2=0$, which can be written as $(\kappa_n-2a_1)(\kappa_n-2a_2)=0$. The roots of this equation define the principal curvatures as $\kappa_1=2a_1$ and $\kappa_2=2a_2$.

Contact Ellipse The analysis presented in this section shows that the gap h^{ij} between the two solids has the geometry of an ellipse. The ratio between the ellipse axes is defined as $\sqrt{(a_1/a_2)}$. The actual size of the contact ellipse depends on the load applied to the two solids. Due to the application of the load, the two surfaces can be displaced vertically with respect to each other. Let u^i and u^j be, respectively, the displacements of the solids i and j, as shown in Figure 4. In this figure, δ^i and δ^j are the deformation of the two points P^i and

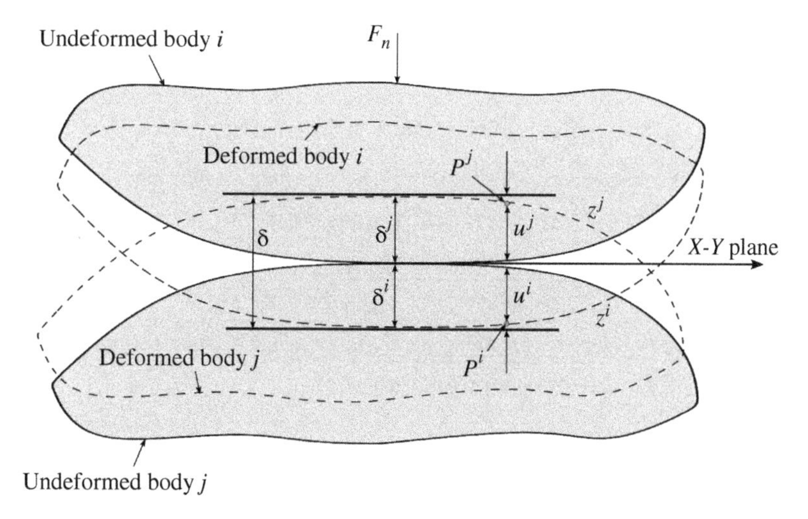

Figure 5.4 Relative displacement between the solids in contact.

P^j on two solids i and j, respectively. If two points P^i and P^j are assumed to coincide, one can write the total deformation as $\delta = \delta^i + \delta^j = u^i + u^j + h^{ij}$. Using the definition of the gap function h^{ij} given by Eq. 15, this equation can be written as

$$u^i + u^j = \delta - a_1(x)^2 - a_2(y)^2 \tag{5.25}$$

This equation is used in the static linear-elastic force analysis in Hertz theory.

5.6 CONTACT ELLIPSE AND NORMAL CONTACT FORCE

In both the constraint and elastic contact formulations, the dimensions of the contact ellipse are determined in order to evaluate the tangential creep forces and spin moment. In addition to the dimensions of the contact ellipse, which depend on the contact pressure, a compliant normal force model in the case of the elastic contact formulation can be developed based on Hertz theory, which provides an expression for the stiffness coefficient that can be used to formulate the elastic force in such a compliant force model. As previously mentioned, the assumptions of half-space are used because the contact region is assumed small in comparison to the dimensions of the two solids. Furthermore, the displacement and normal stresses as the result of the contact pressure are assumed negligible outside the contact area, and the shear stresses τ_{xz} and τ_{yz} on the contact surfaces are assumed to be zero (Johnson 1985; Goldsmith 1960).

Using these assumptions, the elastic component F_{ns} of the applied normal force F_n can be considered equal to the restoring force resulting from the normal component of pressure p in the contact region: that is, $F_{ns} = \iint p dxdy$. In Hertz theory, pressure p is assumed to be a quadratic function of x and y, which can be written in the form

$$p = p_0\sqrt{1 - \left(\frac{x}{a}\right)^2 - \left(\frac{y}{b}\right)^2} \tag{5.26}$$

where p_0 is a constant pressure value, and a and b are the lengths of the semi axes of the contact ellipse. Following the derivation presented in Appendix A, contact pressure p leads to displacement u given by (Love 1944; Johnson 1985; Goldsmith 1960)

$$u = \frac{1 - \nu^2}{\pi E}\left(L_e - M_e(x)^2 - N_e(y)^2\right) \tag{5.27}$$

In this equation, E is the modulus of elasticity, ν is the Poisson ratio, and

$$\left.\begin{aligned} M_e &= \frac{\pi p_0 ab}{2}\int_0^\infty \frac{dw}{\sqrt{\left((a)^2 + w\right)^3\left((b)^2 + w\right)w}} = \frac{\pi p_0 b}{e^2(a)^2}\left(K_e - E_e\right) \\ N_e &= \frac{\pi p_0 ab}{2}\int_0^\infty \frac{dw}{\sqrt{\left((a)^2 + w\right)\left((b)^2 + w\right)^3 w}} = \frac{\pi p_0 b}{e^2(a)^2}\left[\left(\frac{a}{b}\right)^2 E_e - K_e\right] \\ L_e &= \frac{\pi p_0 ab}{2}\int_0^\infty \frac{dw}{\sqrt{\left((a)^2 + w\right)\left((b)^2 + w\right)w}} = \pi p_0 b K_e \end{aligned}\right\} \tag{5.28}$$

where E_e and K_e are the elliptical integrals of argument $e = \sqrt{1 - (b/a)^2}$, $b < a$. These integrals are given in Appendix A. Equation 27 can be used to define the total displacement

$u^i + u^j$ as

$$u^i + u^j = \left(L_e - M_e(x)^2 - N_e(y)^2\right)/\pi E^{ij} \tag{5.29}$$

where

$$\frac{1}{E^{ij}} = \frac{1-\left(\nu^i\right)^2}{E^i} + \frac{1-\left(\nu^j\right)^2}{E^j} \tag{5.30}$$

Using this equation and Eq. 25, one has

$$\delta - a_1(x)^2 - a_2(y)^2 = \frac{L_e - M_e(x)^2 - N_e(y)^2}{\pi E^{ij}} \tag{5.31}$$

It is clear from this equation and the definitions of Eq. 28 that

$$\left.\begin{aligned} \delta &= \frac{L_e}{\pi E^{ij}} = \frac{p_0 b}{E^{ij}} K_e \\ a_1 &= \frac{M_e}{\pi E^{ij}} = \frac{p_0 b}{E^{ij}(e)^2(a)^2}\left(K_e - E_e\right) \\ a_2 &= \frac{N_e}{\pi E^{ij}} = \frac{p_0 b}{E^{ij}(e)^2(a)^2}\left[\left(\frac{a}{b}\right)^2 E_e - K_e\right] \end{aligned}\right\} \tag{5.32}$$

It follows that

$$\left.\begin{aligned} \frac{a_2}{a_1} &= \left(\frac{R_1}{R_2}\right) = \frac{(a/b)^2 E_e - K_e}{K_e - E_e} \\ \sqrt{a_1 a_2} &= \tfrac{1}{2}\sqrt{\frac{1}{R_1 R_2}} = \frac{p_0}{E^{ij}}\frac{b}{(a)^2(e)^2}\sqrt{\left((a/b)^2 E_e - K_e\right)\left(K_e - E_e\right)} \end{aligned}\right\} \tag{5.33}$$

Because the pressure distribution is assumed semi-ellipsoidal, the elastic component of the normal force F_{ns} can be written as

$$F_{ns} = \iint p\,dx\,dy = \frac{2}{3} p_0 \pi a b \tag{5.34}$$

This equation and Eq. 26 lead to (Hertz 1882; Love 1944; Goldsmith 1960)

$$p = \frac{3F_{ns}}{2\pi a b}\sqrt{1 - \left(\frac{x}{a}\right)^2 + \left(\frac{y}{b}\right)^2} \tag{5.35}$$

In Hertz contact theory, friction or damping is not considered, and in this case, the total force F_n is equal to the elastic force F_{ns}: that is $F_n = F_{ns}$. In this case, the contact ellipse semi axes a and b are defined as

$$\left.\begin{aligned} a &= m\left(3\pi F_n\left(K_1 + K_2\right)/4K_3\right)^{1/3} \\ b &= n\left(3\pi F_n\left(K_1 + K_2\right)/4K_3\right)^{1/3} \end{aligned}\right\} \tag{5.36}$$

where K_1 and K_2 are constants that depend on the material properties of the two solids, while K_3 is a constant that depends on the solid geometries. These constants are given as

$$\left.\begin{aligned} &K_1 = \frac{1-\left(\nu^i\right)^2}{\pi E^i}, \qquad K_2 = \frac{1-\left(\nu^j\right)^2}{\pi E^j}, \\ &K_3 = \left(a_1 + a_2\right) = \frac{1}{2}\left(\frac{1}{R_1^i} + \frac{1}{R_2^i}\right) - \frac{1}{2}\left(\frac{1}{R_1^j} + \frac{1}{R_2^j}\right) \end{aligned}\right\} \tag{5.37}$$

Table 5.1 Coefficients m and n in terms of angle θ defined in degrees. Source: Courtesy of Hertz, H.

θ	m	n	θ	m	n	θ	m	n
0.5	61.4	0.1018	10	6.604	0.3112	60	1.486	0.717
1	36.89	0.1314	20	3.813	1.4123	65	1.378	0.759
1.5	27.48	0.1522	30	2.731	0.493	70	1.284	0.802
2	22.26	0.1691	35	2.397	0.530	75	1.202	0.846
3	16.5	0.1964	40	2.136	0.567	80	1.128	0.893
4	13.31	0.2188	45	1.926	0.604	85	1.061	0.944
6	9.79	0.2552	50	1.754	0.641	90	1.0	1.0
8	7.86	0.285	55	1.611	0.678			

Table 1 shows the coefficients m and n in Eq. 36 as functions of the angle θ ($0° \leq \theta \leq 90°$) (Hertz 1882), where θ is defined as

$$\theta = \cos^{-1}\left(K_4/K_3\right) \tag{5.38}$$

and

$$K_4 = \left(a_2 - a_1\right) = \frac{1}{2}\sqrt{\left(\frac{1}{R_2^i} - \frac{1}{R_1^i}\right)^2 + \left(\frac{1}{R_2^j} - \frac{1}{R_1^j}\right)^2 + 2\left(\frac{1}{R_2^i} - \frac{1}{R_1^i}\right)\left(\frac{1}{R_2^j} - \frac{1}{R_1^j}\right)\cos 2\alpha^{ij}} \tag{5.39}$$

As previously explained in this chapter, in the constraint and elastic contact formulations, a set of algebraic equations is solved at the position analysis step to determine the surface parameters of a potential contact point. Using these surface parameters, the principal curvatures and principal directions can be determined and used to compute the constants defined in this section. An interpolation scheme can be employed to determine coefficients m and n using the data presented in Table 1. These coefficients and constants are substituted in Eq. 36 to determine the dimensions of the contact ellipse. An alternative to using interpolation based on the data given in Table 1 is to obtain closed-form expressions for constants m and n in terms of angle θ. The closed-form expressions given here for coefficients m and n were proposed by Berzeri (Shabana et al. 2001)

$$\left.\begin{aligned} m &= A_m \tan(\theta - \pi/2) + \frac{B_m}{(\theta)^{C_m}} + D_m, \\ n &= \frac{1}{A_n \tan(\theta - \pi/2) + 1} + B_n(\theta)^{C_n} + D_n \sin\theta \end{aligned}\right\} \tag{5.40}$$

Table 5.2 Coefficients in the closed-form expressions of m and n. Source: Courtesy of Shabana, A.A., Zaazaa, K.E., and Sugiyama, H.

Coefficient	Value	Coefficient	Value
A_m	−1.086 419 052 477	A_n	−0.773 444 080 706
B_m	−0.106 496 432 832	B_n	0.256 695 354 565
C_m	1.350 000 000 000	C_n	0.200 000 000 00
D_m	1.057 885 958 251	D_n	−0.280 958 376 499

where, in this equation, the value of θ is given in radians and the coefficients A_l, B_l, C_l, and D_l, $l = m, n$, are defined in Table 2 (Shabana et al. 2001). The preceding equation results in a good approximation of coefficients m and n and captures the asymptotic behavior of coefficient m as θ approaches zero.

Compliant Force Model Using the first equation in Eq. 32 and the second equation in Eq. 36, a compliant force model based on Hertz theory (Eq. 34) can be defined as

$$F_{ns} = K(\delta)^{1.5} = \frac{4c_{ns}}{3\left(K_1 + K_2\right)\sqrt{a_1 + a_2}}(\delta)^{1.5} \tag{5.41}$$

where $K = 4c_{ns}/\left(3\left(K_1 + K_2\right)\sqrt{a_1 + a_2}\right)$ is a stiffness coefficient, and c_{ns} is a constant given in Table 3 (Goldsmith 1960). The compliant force F_{ns} can be used to define the elastic component of the normal force F_n used in the elastic contact formulation.

Table 5.3 Coefficient c_{ns}. Source: Courtesy of Goldsmith, W.

a_1/a_2	1.0	0.7041	0.4903	0.3333	0.2174	0.1325	0.0718	0.0311	0.00765
c_{ns}	0.3215	0.3180	0.3322	0.3505	0.3819	0.4300	0.5132	0.6662	1.1450

Example 5.2

In this example, a wheelset that consists of two wheels connected by an axle is assumed to travel on a tangent track. At a given instant of time, the right wheel is assumed to be in contact with a point on the rail that has a transverse radius of curvature of 0.26 m. The wheel is assumed to have a conical profile with a rolling radius of 0.46 m at the instant of time considered. The normal load applied to the right wheel is assumed to be 50 000 N.

(Continued)

The wheel and rail are assumed to be made of steel with a modulus of elasticity equal to 2.1×10^{11} Pa and a Poisson ratio of 0.28. If the yaw angle of the wheelset is assumed to be zero, determine the dimensions of the semi axes of the contact ellipse and the maximum contact pressure.

Solution To determine the dimensions of the semi axes of the contact ellipse, the constants K_1, K_2, and K_3 of Eq. 37 must be determined first. The superscripts w and r are used in this example to refer to the wheel and rail, respectively. Because the wheel tread is assumed to be conical, $R_1^w = \infty$; and because the rolling radius is given, $R_2^w = 0.46$ m. For the tangent track assumed in this example, one has $R_1^r = \infty$ and $R_2^r = -0.26$ m. A negative sign is used for R_2^r, to be consistent with the notations used in this book. Therefore, Eq. 37 leads to

$$K_1 = \frac{1-(\nu^w)^2}{\pi E^w} = \frac{1-(0.28)^2}{\pi\left(2.1\times 10^{11}\right)} = 1.396\times 10^{-12}\ \text{m}^2/\text{N},$$

$$K_2 = \frac{1-(\nu^r)^2}{\pi E^r} = \frac{1-(0.28)^2}{\pi\left(2.1\times 10^{11}\right)} = 1.396\times 10^{-12}\ \text{m}^2/\text{N},$$

$$K_3 = \frac{1}{2}\left(\frac{1}{R_1^w}+\frac{1}{R_2^w}\right) - \frac{1}{2}\left(\frac{1}{R_1^r}+\frac{1}{R_2^r}\right) = \frac{1}{2}\left[0+\frac{1}{0.46}-0-\frac{1}{(-0.26)}\right]$$

$$= 3.01\ \text{m}^{-1}$$

The coefficient K_4 can be evaluated using Eq. 39. Because in this example, the yaw angle is assumed to be zero, the angle α^{ij} in Eq. 39 is zero; therefore,

$$K_4 = \frac{1}{2}\sqrt{\left(\frac{1}{R_2^w}-\frac{1}{R_1^w}\right)^2+\left(\frac{1}{R_2^r}-\frac{1}{R_1^r}\right)^2+2\left(\frac{1}{R_2^w}-\frac{1}{R_1^w}\right)\left(\frac{1}{R_2^r}-\frac{1}{R_1^r}\right)\cos 2\alpha^{ij}}$$

$$= \frac{1}{2}\sqrt{\left(\frac{1}{0.46}-0\right)^2+\left(\frac{1}{(-0.26)}-0\right)^2+2\left(\frac{1}{0.46}-0\right)\left(\frac{1}{(-0.26)}-0\right)\cos(0)}$$

$$= 0.8361\ \text{m}^{-1}$$

Using constants K_3 and K_4, the angle θ of Eq. 38 can be evaluated as

$$\theta = \cos^{-1}\left(K_4/K_3\right) = \cos^{-1}(0.8361/3.01) = 1.2893\ \text{rad}$$

Using the data of Table 1, one can show that m and n can be approximated as $m = 1.23$ and $n = 0.83$. The closed-form expressions of Eq. 40 can also be used to determine the values of m and n instead of the data of Table 1. Using the values of m and n and the coefficients K_1, K_2, and K_3; the dimensions of the semi axes of the contact ellipse can be evaluated as

$$a = m\left[3\pi F_n\left(K_1+K_2\right)/4K_3\right]^{1/3} = 5.8805\times 10^{-3}\ \text{m}$$

$$b = n\left[3\pi F_n\left(K_1+K_2\right)/4K_3\right]^{1/3} = 3.9681\times 10^{-3}\ \text{m}$$

where in this equation, F_n is assumed to be equal to the normal force given in this example: that is, $F_n = 50\,000$ N. Recall that in Hertz' theory, $F_n = F_{ns}$ because this theory does not account for the energy dissipation in the contact area. Because $(1/R_1^w) - (1/R_1^r) < (1/R_2^w) - (1/R_2^r)$, a and b are, respectively, the longitudinal and transverse dimensions of the semi axes of the contact ellipse. Using the dimensions of the contact ellipse, the contact ellipse area A_{ce} can be evaluated as

$$A_{ce} = \pi ab = 7.331 \times 10^{-5}\ \text{m}^2$$

The contact pressure at the origin, which represents the maximum contact pressure, can be evaluated using Eq. 35 as

$$p_0 = \frac{3}{2\pi ab} F_n = \frac{3}{2A_{ce}} F_n = 102.3092\ \text{Mpa}$$

This simple example demonstrates the small dimensions of the contact ellipse and the large value of the contact pressure for the given 50 000 normal force, material properties, and assumed geometries of the wheel and rail surfaces.

5.7 CREEPAGE DEFINITIONS

The relative motion between two solids in contact can be a combination of rolling and sliding. In the case of pure rolling, there is no relative sliding between the two points in the contact region; therefore, there are no friction forces as a result of the relative displacement between the two solids in the contact area. In the case of pure sliding, referred to as *full saturation*, the dominant tangential contact force is the friction force, which is often evaluated using Coulomb's dry friction law in which the tangential friction force is written in terms of the normal force multiplied by the coefficient of friction as μF_n, where μ is the coefficient of friction and F_n is the normal contact force. Predominantly sliding motion in the case of railroad vehicle dynamics can occur during traction and braking scenarios. Under normal operating conditions, the relative motion between the wheel and rail is predominantly rolling with some points on the wheel and rail sticking together or slipping relative to each other, giving rise to *creep forces*. The small relative slip is the result of the relative deformation displacement between points on the wheel and rail. Furthermore, the creep phenomenon can result in a relative rotation about the normal to the contact surfaces. This relative rotation, called *spin*, is produced by the relative deformation of the two surfaces in the contact area and results in a creep spin moment that can be important in some scenarios, including flange contacts. Both the slip and spin as a result of the creep phenomenon have been found to play a significant role in the dynamics and stability of railroad vehicle systems; therefore, in general, their effect cannot be ignored.

In railroad vehicle dynamics, creep forces are developed in terms of velocity parameters called creepages (Carter 1926). Three components of the relative velocities are found to be significant in the study of railroad vehicle system dynamics: the longitudinal and lateral creepages ζ_x and ζ_y, respectively, and the spin creepage ζ_s. The longitudinal and lateral

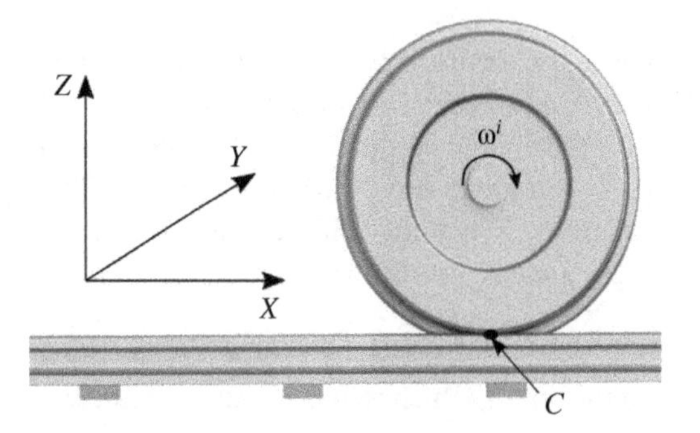

Figure 5.5 Tangential creep and friction forces.

creepages ζ_x and ζ_y, respectively, are dimensionless velocities defined by dividing the relative velocity between the wheel and rail at the contact point by the forward velocity of the wheel center relative to the rail; this forward relative velocity is denoted as V. The spin creepage ζ_s is defined by dividing the component of the relative angular velocity between the wheel and rail along the normal to the contact surface by the relative velocity V: that is, spin creepage is not dimensionless. The longitudinal and lateral creepages lead to tangential creep forces, while the spin creepage leads to a spin moment. In railroad vehicle system dynamics, slip is considered the result of the creep phenomenon if the relative velocity between the wheel and rail, shown in Figure 5, is smaller than a certain value v_s. If the relative velocity is higher than v_s, the case of relative sliding is considered, and tangential dry friction forces, defined by the equation μF_n, are used instead of creep forces. The velocity creepages are defined as

$$\zeta_x = \frac{\left(\mathbf{v}_P^w - \mathbf{v}_P^r\right)^T \hat{\mathbf{t}}_1^r}{V}, \quad \zeta_y = \frac{\left(\mathbf{v}_P^w - \mathbf{v}_P^r\right)^T \hat{\mathbf{t}}_2^r}{V}, \quad \zeta_s = \frac{(\boldsymbol{\omega}^w - \boldsymbol{\omega}^r)^T \hat{\mathbf{n}}^r}{V} \tag{5.42}$$

In this equation, $\mathbf{v}_P^w$ and $\mathbf{v}_P^r$ are, respectively, the absolute velocities of the contact points on the wheel and rail; $\hat{\mathbf{t}}_1^r, \hat{\mathbf{t}}_2^r$, and $\hat{\mathbf{n}}^r$ are, respectively, the unit longitudinal and lateral tangents and unit normal to the surface at the contact point; and $\boldsymbol{\omega}^w$ and $\boldsymbol{\omega}^r$ are the absolute angular velocities of the wheel and rail, respectively. It is clear that in the case of pure rolling about the wheel lateral axis only, all the creepage components are equal to zero. Nonetheless, because the case of pure rolling cannot be sustained for a long duration due to the variations in wheel displacements, creepages do not normally vanish, giving rise to creep forces and moments.

Using one of the contact formulations discussed in this chapter, the global positions of the contact points on the wheel and rail can be determined as $\mathbf{r}_P^k = \mathbf{R}^k + \mathbf{A}^k \overline{\mathbf{u}}_P^k$, $k = w, r$, as previously discussed. Using this equation, the absolute velocity vectors at the contact points on the wheel and rail can be defined as $\mathbf{v}_P^k = \dot{\mathbf{r}}_P^k = \dot{\mathbf{R}}^k + \boldsymbol{\omega}^k \times \left(\mathbf{A}^k \overline{\mathbf{u}}_P^k\right), \ k = w, r$; and the unit tangent and normal vectors to the rail surface can be defined at the contact point, respectively, as

$$\left.\begin{aligned} \hat{\mathbf{t}}_1^r &= \mathbf{A}^r \left(\partial \overline{\mathbf{u}}_P^r / \partial s_1^r\right) / \left|\partial \overline{\mathbf{u}}_P^r / \partial s_1^r\right|, \\ \hat{\mathbf{t}}_2^r &= \mathbf{A}^r \left(\partial \overline{\mathbf{u}}_P^r / \partial s_2^r\right) / \left|\partial \overline{\mathbf{u}}_P^r / \partial s_2^r\right|, \\ \hat{\mathbf{n}}^r &= \left(\mathbf{t}_1^r \times \mathbf{t}_2^r\right) / \left|\mathbf{t}_1^r \times \mathbf{t}_2^r\right| \end{aligned}\right\} \tag{5.43}$$

The absolute velocity of the center of the wheel with respect to the rail $\mathbf{v}^{wr} = \dot{\mathbf{R}}^w - \mathbf{v}_P^r$ can be used to define the forward velocity V as $V = \mathbf{v}^{wr^T}\hat{\mathbf{t}}_1^r$. Because the generalized coordinates and velocities are assumed to be known from the numerical integration of the equations of motion and because the locations of the contact points on the wheels and rails can be determined using the constraint or elastic contact formulation, the absolute velocities can be computed during the dynamic simulations and used to determine the velocity V and creepages ζ_x, ζ_y, and ζ_s.

Example 5.3

A wheelset w is assumed to travel on a tangent track with a forward velocity equal to 15 m/s. At a given instant of time, the yaw and roll angles of the wheelset coordinate system are assumed to be equal to zero: that is, $\psi^w = \phi^w = 0$. The pitch angle at this instant of time is assumed to be equal to multiple of 2π: that is, $\theta^w = 2k\pi$ for a given integer k. In the initial configuration, the local position vector of the contact point on the right wheel defined in the wheelset coordinate system $X^wY^wZ^w$ is assumed to be $\bar{\mathbf{u}}_P^w = \begin{bmatrix} 0 & -l_a/2 & -r_P \end{bmatrix}^T = \begin{bmatrix} 0 & -0.73 & -0.46 \end{bmatrix}^T$, where l_a is the distance between the two wheel centers, and r_P is the rolling radius of the contact of the right wheel. The angle between the lateral tangent to the right rail $\mathbf{t}_2^r$ at the contact point and the wheelset lateral axis Y^w is defined by contact angle $\delta_c = 0.027$ rad. Assuming component ω_z^w of the absolute angular velocity $\boldsymbol{\omega}^w = \begin{bmatrix} \omega_x^w & \omega_y^w & \omega_z^w \end{bmatrix}^T$ of the wheelset is equal to zero – that is, $\omega_z^w = 0$ – determine the values of the creepages for the right wheel contact at the given configuration.

Solution The longitudinal, lateral, and spin creepages are defined by Eq. 42, respectively, as

$$\zeta_x = \frac{\left(\mathbf{v}_P^w - \mathbf{v}_P^r\right)^T \hat{\mathbf{t}}_1^r}{V}, \quad \zeta_y = \frac{\left(\mathbf{v}_P^w - \mathbf{v}_P^r\right)^T \hat{\mathbf{t}}_2^r}{V}, \quad \zeta_s = \frac{(\boldsymbol{\omega}^w - \boldsymbol{\omega}^r)^T \hat{\mathbf{n}}^r}{V}$$

where $\mathbf{v}_P^w$ and $\mathbf{v}_P^r$ are, respectively, the absolute velocities of the contact points on the wheel and rail; $\hat{\mathbf{t}}_1^r$, $\hat{\mathbf{t}}_2^r$, and $\hat{\mathbf{n}}^r$ are, respectively, the unit longitudinal and lateral tangents and unit normal to the surface at the contact point; V is the wheel forward velocity; and $\boldsymbol{\omega}^w$ and $\boldsymbol{\omega}^r$ are the absolute angular velocity vectors of the wheel and rail, respectively. In the case of fixed tangent rail, one has $\mathbf{v}_P^r = \mathbf{0}$ and $\boldsymbol{\omega}^r = \mathbf{0}$; therefore, the definitions of the creepages reduce to

$$\zeta_x = \frac{\mathbf{v}_P^{wT} \hat{\mathbf{t}}_1^r}{V}, \quad \zeta_y = \frac{\mathbf{v}_P^{wT} \hat{\mathbf{t}}_2^r}{V}, \quad \zeta_s = \frac{\boldsymbol{\omega}^{w^T} \hat{\mathbf{n}}^r}{V}$$

The absolute velocity vector of the contact point on the right wheel can be written as $\mathbf{v}_P^w = \dot{\mathbf{r}}_P^w = \dot{\mathbf{R}}^w + \boldsymbol{\omega}^w \times \left(\mathbf{A}^w \bar{\mathbf{u}}_P^w\right)$, where $\mathbf{R}^w$ is the global position vector of the origin of the wheelset coordinate system, $\boldsymbol{\omega}^w = \begin{bmatrix} \omega_x^w & \omega_y^w & \omega_z^w \end{bmatrix}^T$ is the absolute angular velocity vector of the wheelset, $\bar{\mathbf{u}}_P^w$ is the local position vector of the contact point, and $\mathbf{A}^w$ is the orthogonal transformation matrix that defines the orientation of the wheelset coordinate system. This transformation matrix, as discussed in Chapter 3, can be formulated using the three Euler angles ψ^w, ϕ^w, and θ^w, which are the result of

(Continued)

rotations about the three axes Z^w, X^w, and Y^w, respectively. For the values of Euler angles given in this example, $\psi^w = 0$, $\phi^w = 0$, and $\theta^w = 2k\pi$, where k is an integer, the transformation matrix $\mathbf{A}^w$ can be written as

$$\mathbf{A}^w = \begin{bmatrix} \cos\psi^w\cos\theta^w - \sin\psi^w\sin\varphi^w\sin\theta^w & -\sin\psi^w\cos\varphi^w & \cos\psi^w\sin\theta^w + \sin\psi^w\sin\varphi^w\cos\theta^w \\ \sin\psi^w\cos\theta^w + \cos\psi^w\sin\varphi^w\sin\theta^w & \cos\psi^w\cos\varphi^w & \sin\psi^w\sin\theta^w - \cos\psi^w\sin\varphi^w\cos\theta^w \\ -\cos\varphi^w\sin\theta^w & \sin\varphi^w & \cos\varphi^w\cos\theta^w \end{bmatrix}$$

$$= \begin{bmatrix} 1 & 0 & 0 \\ 0 & 1 & 0 \\ 0 & 0 & 1 \end{bmatrix}$$

Vector $\mathbf{A}^w\bar{\mathbf{u}}_P^w$ that defines the global position vector of the contact point, where in this example $\bar{\mathbf{u}}_P^w = \begin{bmatrix} 0 & -0.73 & -0.46 \end{bmatrix}^T$, can be written as $\mathbf{A}^w\bar{\mathbf{u}}_P^w = \begin{bmatrix} 0 & -0.73 & -0.46 \end{bmatrix}^T$. The absolute velocity vector of the origin (reference point) of the wheelset is defined as $\dot{\mathbf{R}}^w = \begin{bmatrix} 15 & 0 & 0 \end{bmatrix}^T$ m/s. Because the absolute velocity vector of the contact point in the case of pure rolling is equal to zero – that is, $\mathbf{v}_P^w = \mathbf{0}$ – this vector can be written in terms of the components of the angular velocity vector $\boldsymbol{\omega}^w$ as

$$\mathbf{v}_P^w = \dot{\mathbf{r}}_P^w = \dot{\mathbf{R}}^w + \boldsymbol{\omega}^w \times \left(\mathbf{A}^w\bar{\mathbf{u}}_P^w\right)$$

$$= \begin{bmatrix} 15 \\ 0 \\ 0 \end{bmatrix} + \begin{bmatrix} -0.46\omega_y^w + 0.73\omega_z^w \\ 0.46\omega_x^w \\ -0.73\omega_x^w \end{bmatrix} = \begin{bmatrix} 0 \\ 0 \\ 0 \end{bmatrix}$$

Because the yaw angular velocity ω_z^w is assumed in this example to be zero, the nonholonomic pure rolling contact condition defines $\omega_x^w = 0$ and $\omega_y^w = 15/0.46 = 32.6086$ rad/s; therefore, the absolute angular velocity vector of the wheelset can be written as

$$\boldsymbol{\omega}^w = \begin{bmatrix} 0 & 32.6086 & 0 \end{bmatrix}^T$$

Using this angular velocity vector, the absolute velocity vector of the geometric center of the right wheel can be written as $\dot{\mathbf{r}}_C^w = \dot{\mathbf{R}}^w + \boldsymbol{\omega}^w \times \left(\mathbf{A}^w\bar{\mathbf{u}}_C^w\right)$, where $\bar{\mathbf{u}}_C^w = \begin{bmatrix} 0 & -l_a/2 & 0 \end{bmatrix}^T = \begin{bmatrix} 0 & -0.73 & 0 \end{bmatrix}^T$ is the local position vector of the geometric center of the right wheel defined in the wheelset coordinate system $X^wY^wZ^w$. Therefore, one has

$$\dot{\mathbf{r}}_C^w = \dot{\mathbf{R}}^w + \boldsymbol{\omega}^w \times \left(\mathbf{A}^w\bar{\mathbf{u}}_C^w\right) = \begin{bmatrix} 15 \\ 0 \\ 0 \end{bmatrix} + \begin{bmatrix} 0 \\ 0 \\ 0 \end{bmatrix} = \begin{bmatrix} 15 \\ 0 \\ 0 \end{bmatrix}$$

Using the value of the contact angle $\delta_c = 0.027$ rad and the wheelset transformation matrix $\mathbf{A}^w = \mathbf{I}$, where $\mathbf{I}$ is the 3×3 identity matrix, the unit tangent and normal vectors are defined in the XYZ global coordinate system as

$$\begin{bmatrix} \hat{\mathbf{t}}_1^r & \hat{\mathbf{t}}_2^r & \hat{\mathbf{n}}^r \end{bmatrix} = \mathbf{A}^w \begin{bmatrix} 1 & 0 & 0 \\ 0 & \cos\delta_c & \sin\delta_c \\ 0 & -\sin\delta_c & \cos\delta_c \end{bmatrix} = \begin{bmatrix} 1 & 0 & 0 \\ 0 & 0.9996 & 0.027 \\ 0 & -0.027 & 0.9996 \end{bmatrix}$$

Using vector $\dot{\mathbf{r}}_C^w$ previously evaluated in this example and the longitudinal tangent to the rail $\hat{\mathbf{t}}_1^r$, the forward velocity of the geometric center of the right wheel with respect to the rail can be written as $V = \dot{\mathbf{r}}_C^{w^T} \hat{\mathbf{t}}_1^r = 15$ m/s. Recall that due to the condition of pure rolling $\mathbf{v}_P^w = \mathbf{0}$, the creepages can then be evaluated as

$$\zeta_x = \frac{\mathbf{v}_P^{wT} \hat{\mathbf{t}}_1^r}{V} = 0, \quad \zeta_y = \frac{\mathbf{v}_P^{wT} \hat{\mathbf{t}}_2^r}{V} = 0, \quad \zeta_s = \frac{\boldsymbol{\omega}^{w^T} \hat{\mathbf{n}}^r}{V} = 0.0587$$

The results of this example demonstrate that while the longitudinal and lateral creepages are equal to zero, the spin creepage is not equal to zero even in the case of pure rolling. Some of the creep force formulations discussed in this chapter include coupling between the tangential creep force and spin moment; therefore, a nonzero value of the spin creepage can give rise to lateral tangential creep force as well as spin moment when these formulations are used even in the case of zero longitudinal and lateral creepages.

5.8 CREEP FORCE FORMULATIONS

The constraint and elastic contact formulations can be used to determine the locations of contact points on the wheel and rail by solving a set of nonlinear algebraic equations for the wheel and rail surface parameters. The surface parameters can be used to determine the locations of the contact points on the wheel and rail surfaces, as previously explained. Knowing the locations of the contact points, the geometric properties of the surfaces, such as principal curvatures and directions at these points, can be determined. The normal contact forces and creepages can also be computed using the coordinates of the contact points and the generalized velocities. Knowing the normal contact forces, the geometry of the two surfaces in contact, and the material properties of the wheel and rail, the dimensions of the contact ellipse can be evaluated using Hertz theory. Computing tangential creep forces at a wheel/rail contact point requires knowledge of the normal contact force, dimensions of the contact ellipse, normalized creepages, and material properties.

Several creep force formulations have been developed and used to model wheel/rail interaction forces. These formulations have different degrees of generality and are based on different simplifying assumptions. Understanding the assumptions used in some of these creep force models is necessary for their proper use in railroad vehicle system applications. For example, some of the creep force formulations are based on three-dimensional theory, while others are based on one- or two-dimensional theory. Some formulations account for spin creepage and spin moment, while others neglect the effect of spin moment. In most formulations used in railroad vehicle system dynamics, the creep force model is assumed to depend on the relative velocities; therefore, the influence of inertia on creep forces is not taken into account in formulating these force models. The inertia effect was found to be significant at speeds that exceed 500 km/h (Kalker 1986). This speed is much higher than the operating speeds of trains currently being used, and therefore, using a quasi-static creep force model is justified. Furthermore, accounting for the inertia effect in the small contact region can significantly complicate the creep force formulations, which are governed in this case by second-order ordinary differential equations. It is also doubtful that such an inertia effect in most railroad vehicle motion scenarios will have a significant effect on overall train dynamics.

More comprehensive creep force models can be developed using the general three-dimensional theory of elasticity and the computational FE method. Nonetheless, as more accuracy is demanded, the computational cost rises; therefore, a trade-off must be made in order to efficiently solve and perform computer simulations of complex railroad vehicle systems that consist of a large number of components and joints. An example of a creep force formulation is the two-dimensional *simplified theory of rolling contact* in which the relative displacement between the wheel and the rail at the contact point is defined in a coordinate system $X^cY^cZ^c$ in which the components of the traction force are defined. If $\mathbf{A}^c$ is the orthogonal transformation matrix that defines the orientation of this coordinate system with respect to the global coordinate system, and if the components of tangential force F_{tx} and F_{ty} are defined along the X^c and Y^c axes of the coordinate system $X^cY^cZ^c$, respectively, the displacement-force relationship of the simplified theory of rolling contact is assumed to take the simple form $\mathbf{A}^{c^T}\mathbf{r}_P^{wr} = \mathbf{A}^{c^T}\left(\mathbf{r}_P^w - \mathbf{r}_P^r\right) = \left[d_xF_{tx} \quad d_yF_{ty} \quad 0\right]^T$ (Kalker 1973), where d_x and d_y are, respectively, longitudinal and lateral *flexibility coefficients* that depend on the material properties and geometry of the wheel/rail contact surfaces, $\mathbf{r}_P^{wr} = \mathbf{r}_P^w - \mathbf{r}_P^r$, and $\mathbf{r}_P^k = \mathbf{R}^k + \mathbf{A}^k\overline{\mathbf{u}}_P^k$, $k = w, r$ is the global position vector of the contact point on body k. The normal contact force, which depends on the relative displacement between the wheel and rail along the normal to the contact surfaces, can still be determined using one of the formulations previously discussed in this chapter.

With the advances made in computational methods and computer technology, many of the simplified wheel/rail creep force formulations are no longer being used, because more sophisticated railroad vehicle algorithms are being developed. Nonetheless, knowledge of the history of developing such creep force models is important for understanding the assumptions originally made for the analysis of complex wheel/rail interaction forces and for explaining the need to relax some of these assumptions as more accuracy is demanded.

Carter's Theory One of the early creep force theories, based on an analytical relationship between longitudinal tangential force and longitudinal creepage, was proposed by Carter (1926). Carter's theory makes the following assumptions: (i) simplified wheel and rail surface profile geometries are considered; (ii) the wheel is represented by a cylinder with a radius that is much larger than the dimensions of the contact area; (iii) the rail is represented by a flat plate; and (iv) the shape of the contact area is assumed to be a rectangular strip with small dimensions compared to the wheel and rail dimensions such that the assumptions of an infinite elastic medium with a localized pressure distribution can be made. Carter's creep force model that defines the relationship between longitudinal tangential force and longitudinal creepage is given by (Kalker 1991)

$$F_{tl} = \begin{cases} \mu F_{nl}\left(-k_c\zeta_x + 0.25\left(k_c\right)^2\zeta_x\left|\zeta_x\right|\right) & \text{if } k_c\left|\zeta_x\right| \le 2 \\ -\mu F_{nl}\,sign\left(\zeta_x\right) & \text{if } k_c\left|\zeta_x\right| > 2 \end{cases} \tag{5.44}$$

where F_{tl} is the tangential force per unit lateral length in the longitudinal direction; μ is the coefficient of friction; F_{nl} is the total normal force per unit lateral length; ζ_x is the longitudinal creepage defined in Carter's theory as $2(V_t - V_c)/(V_t + V_c)$, where subscripts t and c refer, respectively, to tangential forward and circumferential velocities; and $k_c = 4r^w/\mu a$ is *Carter's creepage coefficient*, where r^w is the wheel rolling radius and a is the length of the contact ellipse semi axis in the rolling direction. In Carter's theory, the equivalent length l_t of the contact area in the transverse direction of the rail can be written as $l_t = (4b/3)$, where b is the length of the contact ellipse semi axis in the lateral direction. The preceding equation defines a nonlinear relationship between tangential creep force and the longitudinal creepage. The difference between the results obtained using this nonlinear relationship and the results of the linear force-creepage relationship is shown in Figure 6.

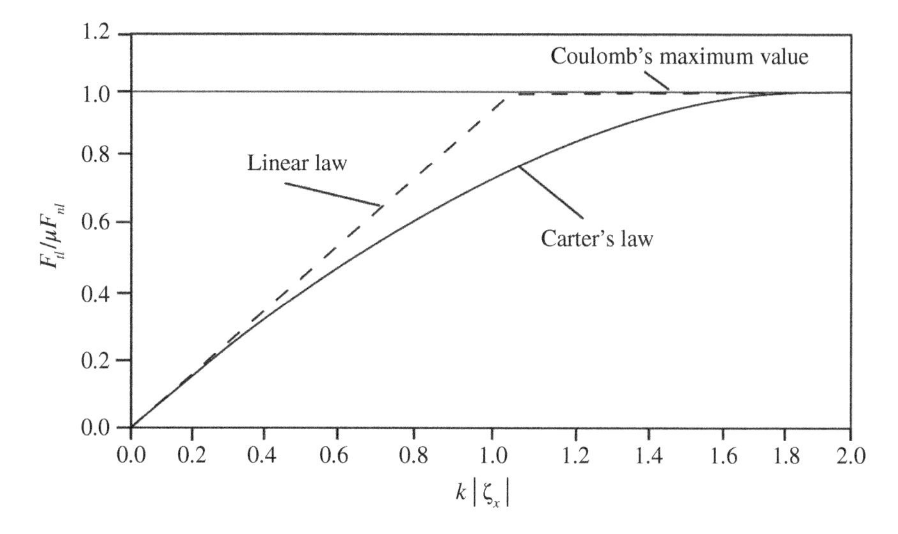

Figure 5.6 Carter's creep force model.

It is clear that Carter's theory makes several simplifying assumptions, including neglecting the effect of the lateral and spin creepages ζ_y and ζ_s, simplified geometries of the wheel and rail, a simplified expression for the forward wheel velocity, and the restriction to one-dimensional analysis. Therefore, such a creep contact theory is not suited for the analysis of complex railroad vehicle systems that can have general motion and in which wheel/rail interaction creep forces cannot be predicted using a one-dimensional theory.

Johnson and Vermeulen's Theory Carter's theory was generalized by Johnson to the three-dimensional analysis of two spheres without consideration of spin creepage (Johnson 1958a,b) and was further extended by Vermeulen and Johnson to smooth half-space without consideration of spin creepage (Vermeulen and Johnson 1964). In Johnson and Vermeulen's theory, the contact area is assumed elliptical with uneven distribution of the slip and stick regions. Furthermore, the adhesion area is also assumed elliptical, as shown in Figure 7. As illustrated in this figure, the elliptical adhesion area and the contact ellipse are assumed to

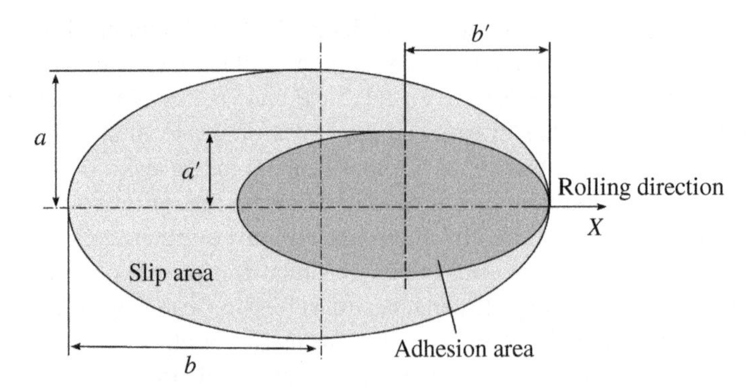

Figure 5.7 Contact area in Johnson and Vermeulen's theory.

share the leading edge. In Johnson and Vermeulen's contact model, tangential force vector $\mathbf{F}_t$ is defined as

$$\mathbf{F}_t = \begin{bmatrix} F_{tx} \\ F_{ty} \end{bmatrix} = \begin{cases} \mu F_n \left[\left(1 - (l_\zeta/3)\right)^3 - 1 \right] \left(\left(\frac{\zeta_{xn}}{l_\zeta} \right) \mathbf{i} + \left(\frac{\zeta_{yn}}{l_\zeta} \right) \mathbf{j} \right) & \text{for } \left| l_\zeta \right| \le 3, \\ -\mu F_n \left(\left(\frac{\zeta_{xn}}{l_\zeta} \right) \mathbf{i} + \left(\frac{\zeta_{yn}}{l_\zeta} \right) \mathbf{j} \right) & \text{for } \left| l_\zeta \right| > 3, \end{cases} \tag{5.45}$$

where $\zeta_{xn} = \pi abG\zeta_x/\mu F_n h_x$ is a normalized longitudinal creepage; $\zeta_{yn} = \pi abG\zeta_y/\mu F_n h_y$ is a normalized lateral creepage; $l_\zeta = \sqrt{(\zeta_{xn})^2 + (\zeta_{yn})^2}$; $\mathbf{i} = [1 \quad 0]^T$; $\mathbf{j} = [0 \quad 1]^T$; F_n is the normal contact force; G is the modulus of rigidity; ζ_x and ζ_y are, respectively, the longitudinal and lateral creepages; a and b are, respectively, the lengths of the semi axes of the contact ellipse in the rolling and lateral directions; μ is the coefficient of friction; and

$$\left. \begin{aligned} h_x &= \begin{cases} B_e - \nu\left(D_e - C_e\right) & \text{for } a \le b, e = \sqrt{1 - (a/b)^2} \\ (b/a)\left[D_e - \nu\left(D_e - C_e\right)\right] & \text{for } a \ge b, e = \sqrt{1 - (b/a)^2} \end{cases} \\ h_y &= \begin{cases} B_e - \nu(a/b)^2 C_e & \text{for } a \le b, e = \sqrt{1 - (a/b)^2} \\ (b/a)\left[D_e - \nu C_e\right] & \text{for } a \ge b, e = \sqrt{1 - (b/a)^2} \end{cases} \end{aligned} \right\} \tag{5.46}$$

In this equation, ν is Poisson's ratio, and B_e, D_e and C_e are the elliptical integrals of argument e. The differences in the results predicted by Johnson and Vermeulen's creep force model and the experimental measurements conducted were attributed to the assumption of an elliptical adhesion region. Furthermore, because Johnson and Vermeulen's theory does not account for spin creepage, which has been found to be significant, such theory is rarely used in the development of the new computational railroad vehicle system algorithms.

Kalker's Linear Theory In Kalker's linear theory, creepages are assumed to be small to justify using a linear force-creepage relationship (Kalker 1967). Because of this assumption, the contact area is assumed to be predominantly an adhesion area. In formulating the linear theory for wheel/rail contact, Kalker introduced the constants $G = (G^w + G^r)/2G^w G^r$ and $\nu = G(\nu^w G^r + \nu^r G^w)/2G^w G^r$, where G^k and ν^k, $k = w, r$ are, respectively, the modulus of rigidity and Poisson ratio of the wheel and rail, respectively. In Kalker's linear theory, which accounts for the effect of spin creepage, tangential creep forces and spin moments can be written, respectively, in terms of the creepages as

$$\left.\begin{aligned} F_{tx} &= -Gabd_{11}\zeta_x \\ F_{ty} &= -Gab\left(d_{22}\zeta_y + \sqrt{ab}d_{23}\zeta_s\right) \\ M_s &= -Gab\left(-\sqrt{ab}d_{23}\zeta_y + abd_{33}\zeta_s\right) \end{aligned}\right\} \tag{5.47}$$

where ζ_x, ζ_y, and ζ_s are, respectively, the longitudinal, lateral, and spin creepages; a and b are, respectively, the lengths of the semi axes of the contact ellipse in the rolling and lateral directions; and d_{11}, d_{22}, d_{23}, and d_{33} are creepage coefficients that are functions of the Poisson ratio and the a/b ratio and are given in Table 4 for different values of the a/b ratio and Poisson ratio (Kalker 1990).

Table 5.4 Coefficients in Kalker's linear theory. Source: Courtesy of Kalker, J.J.

	d_{11}			d_{22}			$d_{23} = -d_{32}$			d_{33}		
g	$\nu = 0$	0.25	0.5	$\nu = 0$	0.25	0.5	$\nu = 0$	0.25	0.5	$\nu = 0$	0.25	0.5
(a/b)												
0.1	2.51	3.31	4.85	2.51	2.52	2.53	0.334	0.473	0.731	6.42	8.28	11.7
0.2	2.59	3.37	4.81	2.59	2.63	2.66	0.483	0.603	0.809	3.46	4.27	5.66
0.3	2.68	3.44	4.8	2.68	2.75	2.81	0.607	0.715	0.889	2.49	2.96	3.72
0.4	2.78	3.53	4.82	2.78	2.88	2.98	0.720	0.823	0.977	2.02	2.32	2.77
0.5	2.88	3.62	4.83	2.88	3.01	3.14	0.827	0.929	1.07	1.74	1.93	2.22
0.6	2.98	6.72	4.91	2.98	3.14	33.1	0.930	1.03	1.18	1.56	1.68	1.86
0.7	3.09	3.81	4.97	3.09	3.28	3.48	1.03	1.14	1.29	1.43	1.50	1.60
0.8	3.19	3.91	5.05	3.19	3.41	3.65	1.13	1.25	1.40	1.34	1.37	1.42
0.9	3.29	4.01	5.12	3.29	3.54	3.82	1.23	1.36	1.51	1.27	1.27	1.27
(b/a)												
1.0	3.4	4.12	5.2	3.40	3.67	3.98	1.33	1.47	1.63	1.21	1.19	1.16
0.9	3.51	4.22	5.3	3.51	3.81	4.16	1.44	1.57	1.77	1.16	1.11	1.06
0.8	3.65	4.36	5.42	3.65	3.99	4.39	1.58	1.75	1.94	1.10	1.04	0.954
0.7	3.82	4.54	5.58	3.82	4.21	4.67	1.76	1.95	2.18	1.05	0.965	0.852
0.6	4.06	4.78	5.8	4.06	4.50	5.04	2.01	2.23	2.50	1.01	0.892	0.751
0.5	4.37	5.10	6.11	4.37	4.90	5.56	2.35	2.62	2.96	0.958	0.819	0.650
0.4	4.84	5.57	6.57	4.84	5.48	6.31	2.88	3.24	3.70	0.912	0.747	0.549
0.3	5.57	6.34	7.34	5.57	6.40	7.51	3.79	4.32	5.01	0.868	0.674	0.446
0.2	6.96	7.78	8.82	6.96	8.14	9.79	5.72	6.63	7.89	0.828	0.601	0.341
0.1	10.7	11.7	12.9	10.7	12.8	16.0	12.2	14.6	18.0	0.795	0.526	0.228

$g = 0$, $C_{11} = \pi^2/4(1-\nu)$; $C_{22} = \pi^2/4$; $C_{23} = -C_{32} = \pi\sqrt{g/3}$; and $C_{33} = \pi^2/16(1-\nu)g$.

The preceding equation shows that the lateral creep force depends on the spin creepage, and the spin moment depends on the lateral creepage. That is, the lateral creep force and spin moments are coupled in this linear theory. The creepage coefficients in the preceding equation were calculated using the program CONTACT with the assumption of small relative slip and without assuming an elliptical contact area. The results showed that the error is less than 5%, confirming the accuracy of the coefficients presented in Table 4 in the case of small creepages. The linear theory has been used in railroad vehicle dynamics literature where interpolations based on the data presented in Table 4 are used to determine the creepage coefficients for different values of the contact ellipse ratio (a/b) (Haque et al. 1979). Due to the limitations of the linear theory presented in this section, more general and nonlinear formulations that relax the assumptions of the linear theory were proposed by Kalker. Some of these formulations, which are used in computational railroad vehicle algorithms, are discussed before concluding this section.

Example 5.4

A wheel and a rail are assumed to be in contact and are made of steel with a Poisson ratio equal to 0.25 and modulus of rigidity equal to 8×10^{10} N/m^2. At a given configuration, the dimensions of the contact ellipse are found to be $a = 5.8805 \times 10^{-3}$ m and $b = 2.94025 \times 10^{-3}$ m; and the longitudinal, lateral, and spin creepages are, respectively, given as $\zeta_x = 0$, $\zeta_y = 0.008$, and $\zeta_s = 0.0587$ m^{-1}. Use Kalker's linear theory to compute tangential creep forces and spin moment.

Solution Using the data of Table 4, the creepage coefficients in Kalker's linear theory can be determined for the ratio of the semi axis of the contact ellipse $a/b = 5.8805/2.94025 = 2$ and for the Poisson ratio $\nu = 0.25$ as $d_{11} = 5.10$, $d_{22} = 4.9$, $d_{23} = 2.62$, and $d_{33} = 0.819$. Substituting these coefficients into Eq. 47, the longitudinal and lateral creep forces and the spin moment can be evaluated, respectively, as

$$F_{tx} = -Gabd_{11}\zeta_x = 0$$

$$F_{ty} = -Gab\left(d_{22}\zeta_y + \sqrt{ab}d_{23}\zeta_s\right) = -551.06 \text{ N}$$

$$M_s = -Gab\left(-\sqrt{ab}d_{23}\zeta_y + abd_{33}\zeta_s\right) = -1.193 \text{ N.m}$$

Heuristic Nonlinear Creep Force Model Johnson and Vermeulen's theory neglects the effect of spin creepage and considers only longitudinal and lateral creepages. The effect of spin creepage, however, cannot be ignored in many motion scenarios, including flange contacts, as previously mentioned. Kalker's linear theory also has its own limitations due to the assumption of small creepages. Therefore, Johnson and Vermeulen's theory and Kalker's linear theory are not suitable for developing general computational algorithms for solving the highly nonlinear wheel/rail contact problem characterized by geometric and force nonlinearities. Geometric nonlinearities are the result of the nonlinear geometry of the contact surfaces, and force nonlinearities are the result of the stick–slip phenomenon.

To address the limitations of the creep force theories mentioned earlier, some researchers proposed nonlinear contact force formulations that take into account the effect of spin creepage. An example of these formulations is the work of Shen et al. (1983), who proposed to combine features of Johnson and Vermeulen's nonlinear theory, which does not account for spin creepage, and Kalker's linear theory, which accounts for spin creepage, with the goal of developing a heuristic nonlinear theory. In the heuristic nonlinear theory, the creep forces of Kalker's linear theory are used in the evaluation of the creep force in Johnson and Vermeulen's theory to obtain a correction factor that takes into account the effect of spin creepage. This correction factor is a nonlinear function of the creepages; therefore, the resulting creep force formulation is nonlinear and takes into account the effect of spin creepage. To develop the heuristic nonlinear theory, the first two equations of Eq. 47 are used. These equations can be written as

$$\left.\begin{aligned}\left(F_{tx}\right)_K &= -Gabd_{11}\zeta_x\\ \left(F_{ty}\right)_K &= -Gab\left(d_{22}\zeta_y + \sqrt{ab}d_{23}\zeta_s\right)\end{aligned}\right\} \tag{5.48}$$

where subscript K refers to Kalker's linear theory. The magnitude of the resultant creep force in the preceding equation can be written as

$$\begin{aligned}\left(F_t\right)_K &= \sqrt{\left(F_{tx}\right)_K^2 + \left(F_{ty}\right)_K^2}\\ &= Gab\sqrt{\left(d_{11}\zeta_x\right)^2 + \left(d_{22}\zeta_y + \sqrt{ab}d_{23}\zeta_s\right)^2}\end{aligned} \tag{5.49}$$

While the magnitude of this resultant creep force cannot exceed the friction force μF_n, Johnson and Vermeulen's nonlinear theory can be used to determine a force limit $(F_t)_J$. To this end, l_ζ in Johnson and Vermeulen' theory is approximated as $l_\zeta \approx \bar{l}_\zeta = \left(F_t\right)_K/\mu\left(F_t\right)_J$. Substituting this equation into the resultant creep force of Johnson and Vermeulen's theory, one obtains (Shen et al. 1983)

$$\left(F_t\right)_J = \begin{cases}\mu F_n\left[\left(\bar{l}_\zeta\right) - \frac{1}{3}\left(\bar{l}_\zeta\right)^2 + \frac{1}{27}\left(\bar{l}_\zeta\right)^3\right], & \bar{l}_\zeta \le 3\\ \mu F_n, & \bar{l}_\zeta > 3\end{cases} \tag{5.50}$$

Using this value, one can define a correction factor $k_t = (F_t)_J/(F_t)_K$ and use it to define the tangential creep force as

$$\mathbf{F}_t = \begin{bmatrix}F_{tx}\\ F_{ty}\end{bmatrix} = k_t\begin{bmatrix}\left(F_{tx}\right)_K\\ \left(F_{ty}\right)_K\end{bmatrix} \tag{5.51}$$

This expression for tangential creep force accounts for the effect of spin creepage. As reported in the literature, while the heuristic nonlinear theory produces more realistic creep force results outside the linear range as compared to Kalker's linear theory (Shen et al. 1983), this theory produces unsatisfactory results in the case of high values of spin creepage (Kalker 1991). Nonetheless, Shen et al. (1983) demonstrated that the results of the heuristic theory are in good agreement with the results obtained using Kalker's simplified theory in some scenarios. Kalker's simplified theory is discussed later in this section.

Polach's Nonlinear Creep Force Model If the maximum stress distribution in Hertz contact theory is denoted as σ_m, the maximum tangential stress can be written as $\tau_m = \mu\sigma_m$, where μ is the friction coefficient. In Polach's nonlinear creep force model, the contact area is assumed to be elliptic, and the relative displacement between the two solids and tangential stress in the adhesion region are assumed to increase linearly from the first to the second edge of the contact area. Sliding is assumed to take place when tangential stress reaches its maximum value $\tau_m = \mu\sigma_m$. Tangential force is defined in this nonlinear model as

$$F_{Ptx} = -\frac{2\mu F_n}{\pi}\left(\frac{\chi_P}{1+(\chi_P)^2} + \tan^{-1}\chi_P\right) \tag{5.52}$$

where F_n is the normal contact force, $\chi_P = \left(G\pi ab\overline{\chi}_P/4\mu F_n\right) l_{\zeta s}$, G is the modulus of rigidity, $\overline{\chi}_P$ is a constant that depends on Kalker's coefficients d_{11} and d_{22} and is defined as $\overline{\chi}_P = \sqrt{\left(d_{11}\zeta_x/l_\zeta\right)^2 + \left(d_{22}\zeta_y/l_\zeta\right)^2}$, $l_\zeta = \sqrt{(\zeta_x)^2 + (\zeta_y)^2}$, $l_{\zeta s} = \sqrt{(\zeta_x)^2 + (\zeta_{ys})^2}$, and ζ_{ys} is a modified lateral creepage used to account for the effect of spin creepage. The modified lateral creepage ζ_{ys} is evaluated according to the following definition:

$$\zeta_{ys} = \begin{cases} \zeta_y, & \left|\zeta_y + \zeta_s a\right| \le \left|\zeta_y\right| \\ \left(\zeta_y + \zeta_s a\right), & \left|\zeta_y + \zeta_s a\right| > \left|\zeta_y\right| \end{cases} \tag{5.53}$$

where a is the dimension of the semi axis of the contact ellipse in the rolling direction. Making the assumption that the moment resulting from spin creepage and/or lateral creepage is small compared to other external and inertia moments applied to the wheel or wheelset, the lateral tangential creep force that takes into consideration the effect of spin creepage can be written as

$$F_{Pty} = -\frac{9}{16}a\mu F_n K_P\left[1 + 6.3\left(1 - e^{-a/b}\right)\right] \tag{5.54}$$

where K_P is a constant defined as $K_P = \left|\varepsilon_y\right|\left(\left((\delta)^3/3\right) - \left((\delta)^2/2\right) + 1/6\right) - (1/3)\sqrt{\left(1-\delta^2\right)^3}$, $\delta = ((c_P)^2 - 1)/((c_P)^2 + 1)$, $c_P = 8Gb\sqrt{ab}d_{23}\zeta_{ys}/3\mu F_n\left(1 + 6.3\left(1 - e^{-a/b}\right)\right)$, b is the dimension of the semi axis of the contact ellipse in the lateral direction, and d_{23} is Kalker's creep coefficient previously introduced in this chapter. Using these definitions, the tangential creep forces in Polach's nonlinear model are written as

$$F_{tx} = F_{Ptx}\left(\frac{\zeta_x}{l_{\zeta s}}\right), \qquad F_{ty} = F_{Ptx}\left(\frac{\zeta_y}{l_{\zeta s}}\right) + F_{Pty}\left(\frac{\zeta_s}{l_{\zeta s}}\right) \tag{5.55}$$

Because numerical investigations have shown that Polach's nonlinear tangential creep force model defined by the preceding equation is accurate, this model has been implemented in several computational algorithms developed for the nonlinear dynamic analysis of railroad vehicle systems (Polach 1999).

Simplified Theory of Contact To develop expressions for tangential forces using the simplified theory of contact, a coordinate system $X^cY^cZ^c$ is defined in the neighborhood of the contact point. In this coordinate system, the relative displacements between the wheel and rail can be defined using the transpose of the orthogonal transformation matrix $\mathbf{A}^c$ that defines the orientation of the coordinate system $X^cY^cZ^c$ in the global coordinate system XYZ. If the tangential creep forces at a point in the $X^cY^cZ^c$ coordinate system are given by the vector $\overline{\mathbf{F}}_t = \left[\overline{F}_{tx} \quad \overline{F}_{ty}\right]^T$, one can introduce flexibility coefficients d_x and d_y and write the relative displacements between the wheel and rail in the neighborhood of the contact point as

$$\left(\mathbf{r}_P^{wr}\right)^c = \mathbf{A}^{c^T}\mathbf{r}_P^{wr} = \mathbf{A}^{c^T}\left(\mathbf{r}_P^w - \mathbf{r}_P^r\right) = \left[d_x\overline{F}_{tx} \quad d_y\overline{F}_{ty} \quad 0\right]^T \tag{5.56}$$

If x and y are assumed to be the longitudinal and lateral surface parameters in the coordinate system $X^cY^cZ^c$ at contact point P, one can define the displacement vector $\overline{\mathbf{u}}_t^{wr} = \overline{\mathbf{u}}_t^{wr}(x, y) = \mathbf{A}^{c^T}\left(\mathbf{r}_P^{wr} - \left(\mathbf{r}_P^{wrT}\hat{\mathbf{n}}^r\right)\hat{\mathbf{n}}^r\right)$, which has a zero component along the normal to the contact surfaces. Using the longitudinal surface parameter x, the wheel forward velocity V can be written as $V = dx/dt$. In the case of steady-state rolling, the slip is assumed to be small in the contact area, and the components of the relative velocity along the normal can be neglected; therefore, one has (Kalker 1979)

$$\left(\mathbf{v}_P^{wr}\right)^c = \mathbf{A}^{c^T}\dot{\mathbf{r}}_P^{wr} = V\begin{bmatrix}\zeta_x - \zeta_s y\\ \zeta_y + \zeta_s x\\ 0\end{bmatrix} \tag{5.57}$$

This velocity vector can also be defined as

$$\left(\mathbf{v}_P^{wr}\right)^c = \frac{\partial\overline{\mathbf{u}}_t^{wr}}{\partial x}\frac{dx}{dt} = V\frac{\partial\overline{\mathbf{u}}_t^{wr}}{\partial x} \tag{5.58}$$

Using the preceding two equations, one has

$$V\begin{bmatrix}\zeta_x - \zeta_s y\\ \zeta_y + \zeta_s x\\ 0\end{bmatrix} = V\frac{\partial\overline{\mathbf{u}}_t^{wr}}{\partial x} \tag{5.59}$$

which upon using Eq. 56, leads to

$$V\begin{bmatrix}\zeta_x - \zeta_s y\\ \zeta_y + \zeta_s x\end{bmatrix} = V\begin{bmatrix}d_x\left(\partial\overline{F}_{tx}/\partial x\right)\\ d_y\left(\partial\overline{F}_{ty}/\partial x\right)\end{bmatrix} \tag{5.60}$$

Assuming that the longitudinal and lateral creepages are constant in the small contact area, integrating the preceding equation leads to the definition of tangential creep forces in the simplified theory of contact as

$$\overline{\mathbf{F}}_t = \begin{bmatrix}\overline{F}_{tx}\\ \overline{F}_{ty}\end{bmatrix} = \begin{bmatrix}\dfrac{x\left(\zeta_x - \zeta_s y\right) + c_1(y)}{d_x}\\ \dfrac{x\left(\zeta_y + \frac{1}{2}\zeta_s x\right) + c_2(y)}{d_y}\end{bmatrix} \tag{5.61}$$

where $c_1(y)$ and $c_2(y)$ are constants of integration, which can be evaluated using the condition that the tangential creep force is zero at the leading edge of the contact ellipse. Using this condition, one obtains

$$\bar{\mathbf{F}}_t = \begin{bmatrix} \bar{F}_{tx} \\ \bar{F}_{ty} \end{bmatrix} = \begin{bmatrix} \dfrac{(x - x_b)(\zeta_x - \zeta_s y)}{d_x} \\ \dfrac{\zeta_y (x - x_b) + \frac{1}{2}\zeta_s \left((x)^2 - (x_b)^2\right)}{d_y} \end{bmatrix} \tag{5.62}$$

where $x_b = a\sqrt{1 - (y/b)^2}$, and a and b are the dimensions of the semi axes of the contact ellipse in the rolling and lateral directions, respectively. Using the preceding equation, neglecting the spin moment, and integrating over the contact area, one obtains

$$\mathbf{F}_t = \begin{bmatrix} F_{tx} \\ F_{ty} \end{bmatrix} = \begin{bmatrix} \int_{-b}^{b} \int_{-x_b(y)}^{x_b(y)} \bar{F}_{tx}\, dx\, dy \\ \int_{-b}^{b} \int_{-x_b(y)}^{x_b(y)} \bar{F}_{ty}\, dx\, dy \end{bmatrix} = \begin{bmatrix} -\dfrac{8a^2 b\zeta_x}{3d_x} \\ -\dfrac{8a^2 b\zeta_y}{3d_y} - \dfrac{\pi a^3 b\zeta_s}{4d_y} \end{bmatrix} \tag{5.63}$$

To determine the flexibility coefficients d_x and d_y, the expressions for tangential creep forces previously obtained in Kalker's linear theory can be used. These expressions were given in Eq. 47 as $F_{tx} = -Gabd_{11}\zeta_x$ and $F_{ty} = -Gab\left(d_{22}\zeta_y + \sqrt{ab}d_{23}\zeta_s\right)$. By equating these tangential creep forces with the forces in Eq. 63, the flexibility coefficients d_x and d_y are defined in terms of the coefficients of the Kalker's linear theory as

$$d_x = \frac{8a}{3Gd_{11}}, \quad d_{y1} = \frac{8a}{3Gd_{22}}, \quad d_{y2} = \frac{\pi a^2}{4G\sqrt{ab}d_{23}} \tag{5.64}$$

Two values of d_y are provided in this equation, because the second equation in Eq. 63 has two coefficients associated with creepages ζ_y and ζ_s. Jacobson and Kalker (2001) suggested combining the three flexibility coefficients in one coefficient.

It is clear that by neglecting the effect of the spin moment, an analytical solution for the tangential creep forces can be obtained using the simplified theory of contact. This theory was used as the basis for developing the computer program FASTSim, which is widely used in railroad vehicle system applications (Kalker 1982). Using numerical techniques allows for using FASTSim in predicting the tangential creep forces in more general wheel/rail configurations. In FASTSim, a strip discretization of the wheel/rail contact surface is made, and such a discretization is used to evaluate the resultant tangential creep forces.

More General Wheel/Rail Contact Theories The assumptions of the simplified theory of contact can be relaxed to obtain more general nonlinear wheel/rail contact theories. This can be accomplished at the expense of increasing the computational cost. As previously mentioned, a general approach can be based on the FE discretization in which FE meshes are developed for the two solids in contact. The motion of the solid surfaces can be monitored to determine the distances between the points on the surfaces. The search for the contact points can be based on formulating algebraic equations and using these equations to minimize the distances between the two surfaces, or using a numerical approach and selected discrete points to determine the points of contact between the two solids. Regardless of the contact-search method used, the FE approach is more general since it can handle both conformal and non-conformal contacts and allows for predicting stresses at the contact points using a more general continuum mechanics approach without making simplifying assumptions. These stresses can be used to determine the normal forces at each pair of contact points. The contact stresses can be predicted using linear or nonlinear constitutive models based on the theory of elasticity. However, using the FE approach can be computationally prohibitive in the case of railroad vehicle systems where dynamic simulations over an extended time may be required, as previously mentioned in this chapter. Moreover, such a computationally intensive approach may not lead to significant improvements in the accuracy of the results in most railroad vehicle system applications. With this fact in mind, Kalker developed an approach that is more general than the simplified theory of contact. This approach is implemented in a computer program called USETAB (Kalker 1990), which is commercially available. The USETAB program is designed for a general wheel/rail contact problem; therefore, it requires the evaluation of more coefficients as compared to the simplified theory of contact. The evaluation of these coefficients is necessary in order to account for all possible wheel/rail contact configurations. USETAB produces two sets of coefficient tables: one contains the creepage coefficients that enter into formulating the tangential creep forces, while the second contains the constants used in Hertz theory. To account for all possible wheel/rail contact configurations, the Hertz constants included in the second table are produced using a large number of runs of another computer program called CON93, which was also developed by Kalker.

To use the USETAB program in a simulation of wheel/rail contact, one must provide the normal contact force, dimensions of the contact ellipse semi axes, creepages, and material properties. The normal force and creepages must be defined in a contact frame that has one of its axes along the normal to the contact surfaces at the contact point. The longitudinal, lateral, and spin creepages must also be defined in this contact frame. Based on this input, USETAB computes the longitudinal and lateral components of the tangential creep force as well as the spin moment. These force and moment components are defined in the directions of the axes of the contact frame in which the creepages are defined. The results of the USETAB computer program have been found to be accurate when compared with the results of the fully nonlinear contact theory. The difference between the creep force results predicted using USETAB and the fully nonlinear theory was found to be approximately 1.5%, while the difference between the results of FASTSim and the fully nonlinear theory was found to be 15%. Despite the fact that the USETAB algorithm is not based on a fully nonlinear theory, it has proven to give accurate results in a large number of railroad vehicle system applications without sacrificing computational efficiency.

5.9 CREEP FORCE AND WHEEL/RAIL CONTACT FORMULATIONS

It is clear that the accuracy of the tangential contact force results obtained using the creep force formulations presented in the preceding section depends on the accuracy of predicting the location of the wheel/rail contact points. The contact points can be determined online using one of the constraint and elastic contact formulations previously discussed in this chapter. In these two formulations, a set of nonlinear algebraic equations for each wheel/rail contact is solved for the four surface parameters $\mathbf{s} = \begin{bmatrix} s_1^w & s_2^w & s_1^r & s_2^r \end{bmatrix}^T$. These surface parameters, which allow for modeling general three-dimensional contact, can be used to determine the locations of the contact points on the wheel and rail surfaces, as previously explained. Using the locations of the contact points and the associated surface parameters, surface geometric properties such as the principal curvatures and principal directions can be computed and used in Hertz contact theory with the normal force and material properties of the wheel and rail to determine the dimensions of the contact ellipse.

While the constraint and elastic contact formulations previously discussed in this chapter are three-dimensional contact formulations and are recommended for computer implementation because of their generality and ability to capture spatial motion scenarios, other approaches can be used to determine the locations of the contact points. Some of these approaches employ only two surface parameters, one for the wheel profile and the second for the rail profile, and make the assumption that contact occurs in a plane defined by the lateral tangent and the normal to the surface. Other formulations are based on nodal search and do not require a certain degree of surface smoothness. However, these approaches may require some form of interpolation to determine the geometric properties that enter in the definition of the contact area. Using *lookup tables* to determine the location of the contact points on the wheel and rail using precomputed data is not discussed in this section because of the serious limitations of such an approach, the difficulties that might be encountered, and the amount of data storage required for capturing general three-dimensional wheel/rail contact scenarios. On the other hand, examples of the nodal search and planar contact approaches are briefly discussed in this section for completeness.

Nodal Search Method The *nodal search method* is an elastic contact formulation that can be used as an alternative to algebraic equations for solving for the surface parameters that define the contact point locations. In this method, point meshes are created on the wheel and rail surfaces to define the profile of the wheel and rail (Shabana et al. 2004, 2005). Using discrete mesh points, the distance between these points on the wheel and rail profiles can be computed and used to determine whether two points are in contact. While such an approach has the advantage of not requiring a certain degree of smoothness of the wheel and rail surfaces during the search process, as previously mentioned, it has the drawback that the change in the wheel and rail lateral surface parameters is not smooth since the contact is assumed to occur at discrete mesh points. As the contact point changes from one mesh point to another, there can be a jump discontinuity in the wheel/rail relative velocity at the contact point, and this velocity jump discontinuity leads to creepage discontinuity, which in turn leads to creep force discontinuity. Numerical experimentation has shown that the force discontinuity can be significant because creepage coefficients used in formulating creep forces are normally very large. The impulsive force resulting from such a discontinuity can lead to numerical problems and deterioration of the solution accuracy if interpolation schemes are not used to achieve a certain degree of smoothness. The coordinates of the contact points determined using the nodal search can be used with the absolute nodal coordinate formulation (ANCF) geometry to determine the principal curvatures and principal directions that are required for computing the dimensions of the contact ellipse.

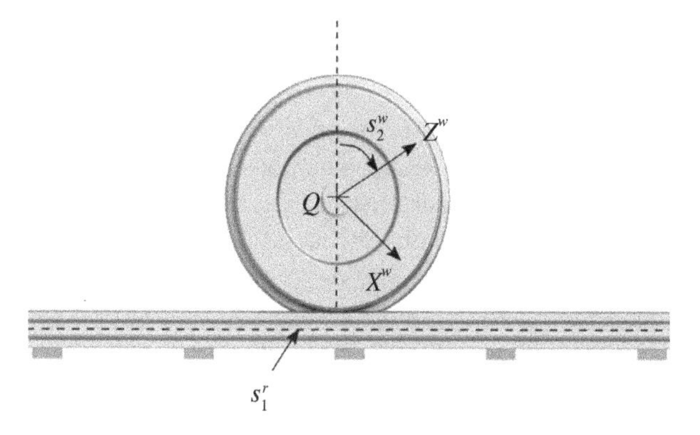

Figure 5.8 Nodal search method.

The main steps used in the nodal search for determining the locations of the contact points on the wheel and rail surfaces can be summarized as follows (Shabana et al. 2004, 2008):

1. The rail arc length s_1^r traveled by the wheel is computed and used to determine the rail cross section in which potential contact points may lie. The distance traveled by the wheel can be determined by selecting a reference point Q on the wheel, as shown in Figure 8, and solving the first-order differential equation $\dot{s}_1^r = \dot{\mathbf{r}}_Q^{wT}\hat{\mathbf{t}}_1^r$ simultaneously with the system equations of motion, where $\dot{\mathbf{r}}_Q^w$ is the absolute velocity of point Q and $\hat{\mathbf{t}}_1^r$ is a unit vector along the longitudinal tangent to the rail surface. As an alternative to using the first-order ordinary differential equation $\dot{s}_1^r = \dot{\mathbf{r}}_Q^{wT}\hat{\mathbf{t}}_1^r$, one can use the Cartesian coordinates of the wheel or wheelset and solve a set of nonlinear algebraic equations for the trajectory coordinates. By varying the rail lateral surface parameter s_2^r for the given rail longitudinal surface parameter s_1^r, the locations of the mesh points that define the rail profile can be determined, as previously explained in this chapter.
2. Wheel angular parameter s_2^w is computed and used to define the wheel diametric section in which potential contact points may lie. Knowing the component of the angular velocity of the wheel about its own axis, a simple procedure can be used to determine angular surface parameter s_2^w. Knowing angular surface parameter s_2^w, the location of any point on the wheel profile can be determined if wheel lateral surface parameter s_1^w is given. By varying wheel lateral surface parameter s_1^w for given angular surface parameter s_2^w, the locations of the mesh points that define the wheel profile can be determined, as previously explained in this chapter.
3. The distances between the discrete mesh points on the wheel and rail are computed to determine which points come in contact. If two points come into contact, rail profile parameter s_2^r and wheel profile parameter s_1^w associated with the contact points are determined. Knowing the four surface parameters $\mathbf{s} = \begin{bmatrix} s_1^w & s_2^w & s_1^r & s_2^r \end{bmatrix}^T$, the geometric properties of the surfaces can be approximated and used with the normal force and the material properties to determine the dimensions of the contact area using Hertz theory. This approach allows for using the same procedure previously described to determine the tangential creep forces when the constraint and elastic contact formulations are used to determine the contact points. Because the nodal search method is not a constraint

formulation and allows for wheel/rail separations, the normal force can be determined using a compliant force model with assumed stiffness and damping coefficients.

It is clear that one of the disadvantages of the nodal search method is the need to introduce a number of first-order ordinary differential equations equal to the number of wheel/rail contacts in order to solve for the distance traveled by each wheel. Furthermore, this method can lead to a large number of contact points for each wheel/rail contact if fine point meshes are used. On the other hand, using a coarse mesh may lead to missing potential contact points and to a higher degree of discontinuities. Therefore, when fine point meshes are used with the nodal search method, an efficient search algorithm is needed. The mesh points can be grouped in batches, and a limit can be introduced on the number of batches that have potential contact points. The number of contact points is assumed not to exceed the number of contact batches. For a given batch, one pair of points that leads to the maximum wheel/rail penetration can be selected to define the contact point. This nodal search algorithm allows for the possibility of having several wheel/rail contact points.

Planar Contact In some wheel/rail contact formulations used in existing research and commercial software, the wheel/rail contact problem is simplified. Instead of using two surface parameters for each contact surface, only one profile parameter s_1^w is used for the wheel, and one profile parameter s_2^r is used for the rail. The other two surface parameters s_2^w and s_1^r are assumed to be known from the wheel and rail configurations and do not enter into the definition of the algebraic equations used in the search algorithm. Because of this assumption, the contact between the wheel and rail is reduced to a planar curve representation instead of the three-dimensional surface representation. This simplified *planar contact approach* neglects kinematic couplings between the geometric surface parameters and leads to approximation in predicting the location of the contact points. Since one needs to solve for only two surface parameters, the planar constraint contact conditions can be formulated using three algebraic equations instead of five; two of these equations can be used to determine the two surface parameters s_1^w and s_2^r, and the third equation is used to eliminate the freedom of the wheel to translate with respect to the rail along the normal to the contact surfaces. Similar algebraic equations can be used in the elastic contact formulation, as previously discussed in this chapter.

To formulate the planar contact conditions, an *intermediate wheel coordinate system*, which does not share the wheel pitch rotation, is introduced. This coordinate system can be conveniently defined using the wheel *trajectory coordinates* introduced in Chapter 3. The six trajectory coordinates are defined by the vector $\mathbf{p}^w = \begin{bmatrix} s^w & y^{wr} & z^{wr} & \psi^{wr} & \varphi^{wr} & \theta^{wr} \end{bmatrix}^T$, where s^w is the rail arc length that defines the location of the origin of the trajectory wheel coordinate system $X^{tw}Y^{tw}Z^{tw}$; y^{wr} and z^{wr} are the coordinates of the wheel center of mass with respect to the origin of the trajectory wheel coordinate system $X^{tw}Y^{tw}Z^{tw}$; and ψ^{wr}, φ^{wr}, and θ^{wr} are the three Euler angles that define the orientation of the wheel coordinate system $X^wY^wZ^w$ with respect to the trajectory wheel coordinate system $X^{tw}Y^{tw}Z^{tw}$. As described in Chapter 3, the sequence of Euler angle rotations used is $Z^w - X^w - Y^w$. Regardless of the coordinates used (Cartesian or trajectory), the vector of trajectory coordinates $\mathbf{p}^w$ is assumed to be known from the solutions of the dynamic equations of motion since this vector can be systematically determined using absolute Cartesian coordinates.

To define the *intermediate wheel coordinate system,* which does not have the pitch rotation θ^{wr} of the wheel about its Y^w axis, two Euler angles ψ^{wr} and φ^{wr} are used to first write the transformation matrix $\mathbf{A}^{tw}$ of this coordinate system in the trajectory wheel coordinate system. Using this transformation matrix, the orthogonal transformation matrix that defines the orientation of the intermediate wheel coordinate system in the global system can be written as (Shabana and Rathod 2007; Shabana et al. 2008)

$$\mathbf{A}^{wi} = \mathbf{A}^r\mathbf{A}^{tw}\mathbf{A}_z\mathbf{A}_x = \begin{bmatrix}\mathbf{a}_1^{wi} & \mathbf{a}_2^{wi} & \mathbf{a}_3^{wi}\end{bmatrix} \tag{5.65}$$

where $\mathbf{A}^r$ is the transformation matrix that defines the orientation of the rail body coordinate system in the global coordinate system, $\mathbf{A}^{tw} = \mathbf{A}^{tw}(s^w)$ is the transformation matrix that defines the orientation of the trajectory wheel coordinate system $X^{tw}Y^{tw}Z^{tw}$ with respect to the rail body coordinate system $X^rY^rZ^r$, and

$$\mathbf{A}_z = \begin{bmatrix}\cos\psi^{wr} & -\sin\psi^{wr} & 0\\ \sin\psi^{wr} & \cos\psi^{wr} & 0\\ 0 & 0 & 1\end{bmatrix}, \quad \mathbf{A}_x = \begin{bmatrix}1 & 0 & 0\\ 0 & \cos\varphi^{wr} & -\sin\varphi^{wr}\\ 0 & \sin\varphi^{wr} & \cos\varphi^{wr}\end{bmatrix} \tag{5.66}$$

Using the preceding two equations, one can write

$$\mathbf{A}^{wi} = \mathbf{A}^r\mathbf{A}^{tw}\begin{bmatrix}\cos\psi^{wr} & -\sin\psi^{wr}\cos\varphi^{wr} & \sin\psi^{wr}\sin\varphi^{wr}\\ \sin\psi^{wr} & \cos\psi^{wr}\cos\varphi^{wr} & -\cos\psi^{wr}\sin\varphi^{wr}\\ 0 & \sin\varphi^{wr} & \cos\varphi^{wr}\end{bmatrix} \tag{5.67}$$

If absolute Cartesian coordinates are used as the generalized coordinates, transformation matrix $\mathbf{A}^{wi}$ can be obtained using the following simple matrix multiplication

$$\mathbf{A}^{wi} = \mathbf{A}^w\mathbf{A}_y^T \tag{5.68}$$

where $\mathbf{A}^w$ is the transformation matrix that defines the orientation of the wheel coordinate system $X^wY^wZ^w$ in the global coordinate system XYZ, and $\mathbf{A}_y$ is the matrix that accounts for the pitch rotation about the wheel Y^w axis and is defined as

$$\mathbf{A}_y = \begin{bmatrix}\cos\theta^{wr} & 0 & \sin\theta^{wr}\\ 0 & 1 & 0\\ -\sin\theta^{wr} & 0 & \cos\theta^{wr}\end{bmatrix} \tag{5.69}$$

The location of an arbitrary point on the rail space curve at arc length s_1^r can be written as

$$\mathbf{r}_o^r = \mathbf{R}^r + \mathbf{A}^r\bar{\mathbf{u}}_o^r\left(s_1^r\right) \tag{5.70}$$

where $\mathbf{R}^r$ is the global position vector of the origin of the rail body coordinate system $X^rY^rZ^r$ and $\bar{\mathbf{u}}_o^r\left(s_1^r\right)$ is the position vector of the arbitrary point with respect to the rail body coordinate system, as shown in Figure 9. The arc length s_1^r at which vector $\mathbf{r}_o^r$ of Eq. 70 has a zero component along the X^{wi} axis of the wheel can be determined using the equation $\mathbf{a}_1^{wi^T}\left(\mathbf{r}_o^r - \mathbf{R}^w\right) = 0$, where $\mathbf{R}^w$ is the global position vector of the origin of the wheel body coordinate system $X^wY^wZ^w$ and $\mathbf{a}_1^{wi}$, which is the first column of transformation matrix $\mathbf{A}^{wi}$, represents a unit vector along the X^{wi} axis of the intermediate wheel coordinate system. One can therefore write

$$\mathbf{a}_1^{wi^T}\left(\mathbf{R}^r + \mathbf{A}^r\bar{\mathbf{u}}_o^r - \mathbf{R}^w\right) = 0 \tag{5.71}$$

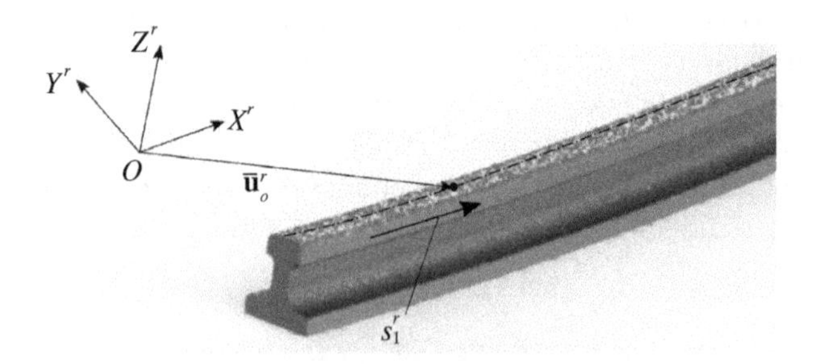

Figure 5.9 Planar contact.

In the case of arbitrary track geometry, this equation is a nonlinear function of rail arc length s_1^r. Because the wheel and rail generalized coordinates are assumed known, Eq. 71 can be solved iteratively using a Newton–Raphson algorithm to determine s_1^r. In this algorithm, the algebraic equation $\left(\mathbf{a}_1^{wi^T}\mathbf{A}^r\bar{\mathbf{t}}_1^r\right)\Delta s_1^r = -\mathbf{a}_1^{wi^T}\left(\mathbf{R}^r + \mathbf{A}^r\bar{\mathbf{u}}_o^r - \mathbf{R}^w\right)$ is iteratively solved for the Newton difference Δs_1^r, where $\bar{\mathbf{t}}_1^r = \partial\bar{\mathbf{u}}_o^r/\partial s_1^r$ is the longitudinal tangent to the rail space curve defined in the rail body coordinate system $X^rY^rZ^r$. Therefore, the Newton difference Δs_1^r can be determined using the equation $\Delta s_1^r = -\mathbf{a}_1^{wi^T}\left(\mathbf{R}^r + \mathbf{A}^r\bar{\mathbf{u}}_o^r - \mathbf{R}^w\right)/\left(\mathbf{a}_1^{wi^T}\mathbf{A}^r\bar{\mathbf{t}}_1^r\right)$. The Newton difference is used to update s_1^r until convergence is achieved. This Newton–Raphson iterative procedure is an alternate to using the first-order ordinary differential equation $\dot{s}_1^r = \dot{\mathbf{r}}_Q^{w^T}\mathbf{t}_1^r$, which was introduced previously in this section to determine s_1^r in the nodal search method. Knowing s_1^r and assuming $s_2^w = \theta^{wr}$, the following contact conditions can be applied for the wheel/rail contact

$$\left.\begin{aligned}
&\mathbf{t}_2^{r^T}\left(\mathbf{R}^r + \mathbf{A}^r\bar{\mathbf{u}}^r - \mathbf{R}^w - \mathbf{A}^{wi}\bar{\mathbf{u}}^{wi}\right) = 0\\
&\mathbf{n}^{r^T}\left(\mathbf{R}^r + \mathbf{A}^r\bar{\mathbf{u}}^r - \mathbf{R}^w - \mathbf{A}^{wi}\bar{\mathbf{u}}^{wi}\right) = 0\\
&\mathbf{n}^{r^T}\mathbf{t}_1^w = 0
\end{aligned}\right\} \tag{5.72}$$

where the vectors and matrices that appear in this equation are as previously defined. Knowing the generalized coordinates of the wheel and rail, and knowing s_2^w and s_1^r, the first and third equations in the preceding equation can be used to solve for the two profile parameters s_1^w and s_2^r. Enforcing the second equation in the preceding equation eliminates the freedom of the wheel to translate with respect to the rail along the normal to the contact surfaces. In the elastic contact formulation, the second equation can be used to determine the penetration that enters into calculating the compliant normal contact force model.

In the planar contact formulation discussed in this section, it is assumed that the surface parameters $s_2^w = \theta^{wr}$ and s_1^r are known and fixed when the preceding equations are solved iteratively to determine s_1^w and s_2^r. Consequently, the planar contact method described in this section does not ensure that the two points on the wheel and the rail coincide, and these two points may differ by a small longitudinal shift. This is mainly because Eq. 71, $\mathbf{a}_1^{wi^T}\left(\mathbf{R}^r + \mathbf{A}^r\bar{\mathbf{u}}_o^r - \mathbf{R}^w\right) = 0$, is used to determine s_1^r separately. To ensure that there is no longitudinal shift between the two points on the wheel and the rail, the following three equations can be used to iteratively solve for the three parameters s_1^w, s_1^r, and s_2^r and account for the kinematic coupling between these surface parameters (Shabana and Rathod 2007):

$$\left.\begin{aligned} &\mathbf{t}_1^{r^T}\left(\mathbf{R}^r+\mathbf{A}^r\overline{\mathbf{u}}^r-\mathbf{R}^w-\mathbf{A}^{wi}\overline{\mathbf{u}}^{wi}\right)=0\\ &\mathbf{t}_2^{r^T}\left(\mathbf{R}^r+\mathbf{A}^r\overline{\mathbf{u}}^r-\mathbf{R}^w-\mathbf{A}^{wi}\overline{\mathbf{u}}^{wi}\right)=0\\ &\mathbf{n}^{r^T}\mathbf{t}_1^w=0 \end{aligned}\right\} \tag{5.73}$$

In the constraint contact formulation, the second equation in Eq. 72, $\mathbf{n}^{r^T}\left(\mathbf{R}^r+\mathbf{A}^r\overline{\mathbf{u}}^r-\mathbf{R}^w-\mathbf{A}^{wi}\overline{\mathbf{u}}^{wi}\right)=0$, can be enforced to eliminate the freedom of the wheel to travel with respect to the rail along the normal to the contact surfaces. In this case, the normal contact force is determined as a constraint (reaction) force, as discussed in Chapter 6. In the elastic contact formulation, on the other hand, the second equation in Eq. 72 can be used to compute the penetration as $\delta=\mathbf{n}^{r^T}\left(\mathbf{R}^r+\mathbf{A}^r\overline{\mathbf{u}}^r-\mathbf{R}^w-\mathbf{A}^{wi}\overline{\mathbf{u}}^{wi}\right)$. This penetration can be used to compute the normal contact force. In both the constraint and elastic contact formulations, using Eq. 73 to solve iteratively for s_1^w, s_1^r, and s_2^r ensures that there is no longitudinal shift and the two contact points on the wheel and rail coincide.

5.10 MAGLEV FORCES

The focus of this chapter has mainly been on the analysis of the wheel/rail contact problem. While wheels and rails remain the primary means of generating the traction forces that move most freight and passenger trains, wheel/rail contact is the source of many problems, including the following: (i) train speeds are limited due to wheel/rail contact; (ii) because of wheel and rail contact, maintenance costs resulting from wear are very high – measured in billions of dollars annually for a single freight company in North America; (iii) vibration and noise levels can be very high; (iv) the suspension system and bogie structures of wheel/rail trains are not simple and require many components including axles, bearings, springs, absorbers, etc.; (v) using wheels and rails means the possibility of derailments and other accidents is high; (vi) the pantograph/catenary contact that provides the electricity required for operating high-speed passenger trains can also limit speed, be a source of wear, and lead to high maintenance costs and frequent inspections; and (vii) wheel/rail contact gives rise to creep contact forces that make it difficult to develop virtual models and computer programs that perform efficient simulations and virtual prototyping. As previously discussed in this book, wheel/rail creep forces have a direct effect on critical vehicle speeds.

Maglev and Wheel/Rail Trains To avoid the problems that arise as the result of using wheel/rail contact to move trains, *magnetically levitated* (maglev) trains have received attention over the past half-century. These trains, which do not normally rely on wheels, are floated by magnetic forces; therefore, the vehicles are not in contact with the guideway. Maglev trains were first proposed in the early 1930s but were first implemented in 2003 in Shanghai, China. They have several desirable features including higher speeds, lower maintenance costs, less vibration and noise, and light weight. But despite these features, maglev trains, while operating in some countries, are still, for the most part, at the experimental stage and are currently used mainly for short distances rather than travel between cities hundreds of miles apart. Their use for freight transportation is also still doubtful, particularly in the case of very long trains that consist of many cars with very high axle loads.

Using maglev trains for long travel distances is debated due to safety concerns and the economic advantages of air transportation, which does not require building an expensive infrastructure on the ground. Although Japan approved a maglev train line for inner-city transportation to be completed by 2027, for the foreseeable future, using maglev trains as a primary freight and passenger transportation method for long distances remains in doubt; it is likely that such maglev technology will be used, for the most part, for relatively short distances.

Nonetheless, the maglev non-contact concept eliminates many of the problems resulting from wheel/rail contact and allows for a significant increase in train speed. For example, it is expected that train speed will exceed 600 km/h, and the noise level is reduced from 75–80 dB for wheel/rail trains to 60–65 dB for maglev trains (Lee et al. 2006). Furthermore, because there is no wheel/rail contact or friction in the case of a maglev train and no need for a pantograph/catenary system, many of the wear problems that result from more concentrated contact loads can be eliminated or reduced, leading to lower maintenance costs, environmentally superior transportation, and improved ride comfort (Dukkipati 2000; Lee et al. 2006).

Maglev Train Suspensions Unlike conventional trains, which are driven by wheel/rail interaction forces, in maglev trains, a magnetic levitation force separates the vehicle from the guideway. This magnetic levitation force serves as the vehicle's suspension system. The train is driven by propulsion forces generated by a linear motor instead of the rotary motor used in conventional train systems. There are three types of maglev suspension systems: *electrodynamic suspension* (EDS), *electromagnetic suspension* (EMS), and *hybrid electromagnetic suspension* (HEMS). These three maglev suspension systems are briefly described next (Lee et al. 2006):

- *Electrodynamic suspension (EDS)*: In an EDS system, shown in Figure 10a, repulsive magnetic forces are used for levitation. Magnets are attached to the vehicle, and inducing coils or conducting sheets are placed on the guideway. As the magnets attached to the vehicle

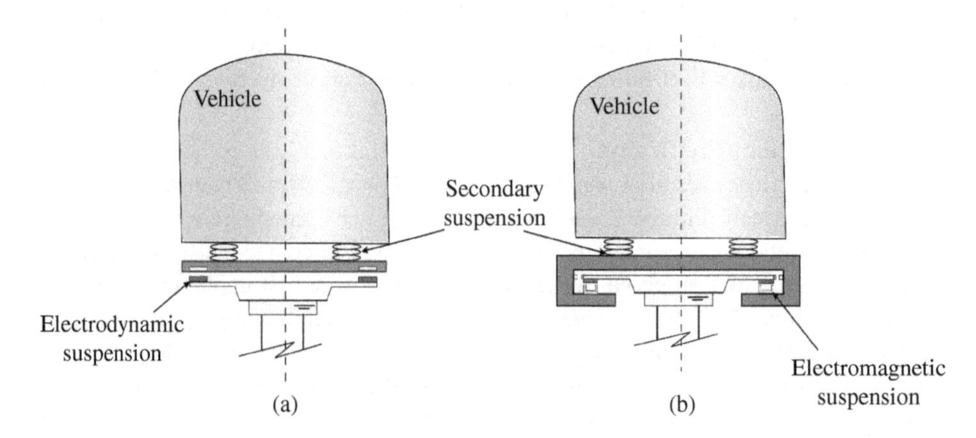

Figure 5.10 Maglev suspension.

move forward, current flows in the guideway induced coils or conducting sheets, generating the magnetic field required to produce the repulsive force that lifts the vehicle. This type of suspension is stable and does not normally require an active control system to control the air gap between the vehicle and guideway; this air gap is approximately 100 mm. Because this maglev suspension system has good stability characteristics, it is suited for high-speed operation and heavy loads. However, because the repulsive magnetic levitation forces are the result of vehicle motion above the inducing coils or conducting sheets, the vehicle must reach a certain speed (approximately 100 km/h) in order to generate the electric current required to produce the levitation forces. For this reason, when the electrodynamic suspension is used, rubber tires are used at lower speeds; therefore, this suspension system is not suitable for urban transportation that requires frequent stops and typically a low operation speed.

- *Electromagnetic suspension (EMS)*: In the EMS system, shown in Figure 10b, attraction forces instead of EDS repulsive forces produce magnetic levitation. The attraction forces are produced using electromagnets attached to the vehicle and ferromagnets placed on the guideway. The iron core of a magnetic circuit is excited by an electric current-carrying coil, and as a result, the core on the vehicle is attracted to the ferromagnetic rail. As discussed in this section, the electromagnetic suspension has stability problems because the magnetic levitation attraction force increases as the small air gap (approximately ± 10 mm) between the pole face of the electromagnet on the vehicle and the ferromagnet on the guideway decreases. Therefore, a feedback control system is required to ensure the stability of such a suspension system. Nonetheless, unlike the EDS force, the EMS levitation magnetic force does not depend on vehicle speed, and the vehicle can be lifted at low or zero speed. For this reason, this suspension system can be used for urban mass transportation that requires frequent stops and a relatively low operation speed.
- *Hybrid electromagnet suspension (HEMS)*: In the HEMS systems, permanent magnets are used with the electromagnets in some EMS systems to reduce the electric power required for magnetic levitation. For a given nearly constant air gap, permanent magnets can be designed to provide the required levitation forces without the need for electromagnets that produce a force-gap relationship that can be a source of instability. Using permanent magnets ensures that the gap remains nearly constant, reduces reliance on an active control system to control the gap, and leads to a significant reduction in the electric power required for magnetic levitation. Nonetheless, the HEMS system requires a greater variation in the amplitude of the electric current as compared to EMS because the permanent magnets have the same permeability as air (Lee et al. 2006).

Electromagnetic Suspension Forces This section formulates the magnetic levitation force produced by the EMS system in order to illustrate the force-gap relationship when such a suspension system is used in maglev trains. As previously mentioned, when EMS is used, attraction forces lift the vehicle instead of the repulsive forces produced by EDS. The force of attraction between the pole surface and the ferromagnet can be written as (Sinha 1987)

$$F_z = \frac{(\Phi_p)^2 A_p}{P_p} \tag{5.74}$$

where $A_p = l_p w_p$; l_p and w_p are, respectively, the length and width of the pole face, as shown in Figure 11; P_p is the permeability of the free space; and Φ_p is the flux density across the air gap. The flux density can be written in terms of the reluctance of the mutual flux R_M and the magneto-motive force F_m as $\Phi_p = F_m/A_p R_M$. The magneto-motive force F_m can be written in terms of the number of turns of coil n_{co} and coil electric current $i = i(t)$ as $F_m = n_{co}\, i(t)$. Therefore, the flux density Φ_p can be written as $\Phi_p = n_{co} i(t)/A_p R_M$. Substituting this equation into Eq. 74, one obtains

$$F_z = \frac{1}{P_p A_p}\left(\frac{n_{co}\, i(t)}{R_M}\right)^2 \tag{5.75}$$

The reluctance of mutual flux R_M consists of the reluctance of the air gap $2d_z/P_p l_p w_p$, electromagnet $(w_l + 2h + 2l_p)/P_e l_p w_p$, and ferromagnet $(w_l + 2l_p)/P_f h_1 w_p$, where $d_z = d_z(t)$ is the distance between the pole faces and the ferromagnet, h_1 is the thickness of the ferromagnet; w_l and h are the dimensions shown in Figure 11; and P_e and P_f are, respectively, the permeability of the electromagnet and the ferromagnet. Therefore, one can write the equation for the reluctance of mutual flux R_M as (Dukkipati 2000)

$$R_M(d_z) = \left(\frac{2d_z}{P_p l_p w_p}\right) + \left(\frac{w_l + 2h + 2l_p}{P_e l_p w_p}\right) + \left(\frac{w_l + 2l_p}{P_f h_1 w_p}\right) \tag{5.76}$$

If assumptions are made that $P_e = P_f$ and $h_1 \simeq l_p$, the preceding equation reduces to

$$R_M(d_z) \simeq \frac{2}{P_p A_p}\left(d_z(t) + c_p\right) \tag{5.77}$$

where $c_p = (P_p/P_e)(2l_p + w_l + h)$. Substituting Eq. 77 into Eq. 75 leads to

$$F_z(d_z, i) = \frac{P_p (n_{co})^2 A_p}{4}\left(\frac{i(t)}{d_z(t) + c_p}\right)^2 \le (F_z)_{\max} \tag{5.78}$$

Force $(F_z)_{\max}$, which is associated with the saturation state of the flux density Φ_p across the air gap, can be written as $(F_z)_{\max} = ((\Phi_p)_{\max} A_p)/P_p$, where $(\Phi_p)_{\max}$ is the maximum flux density defined as $(\Phi_p)_{\max} = (P_M/(P_M + P_L))\,\overline{\Phi}_p$, $P_M = 1/R_M$ is the mutual flux permeance,

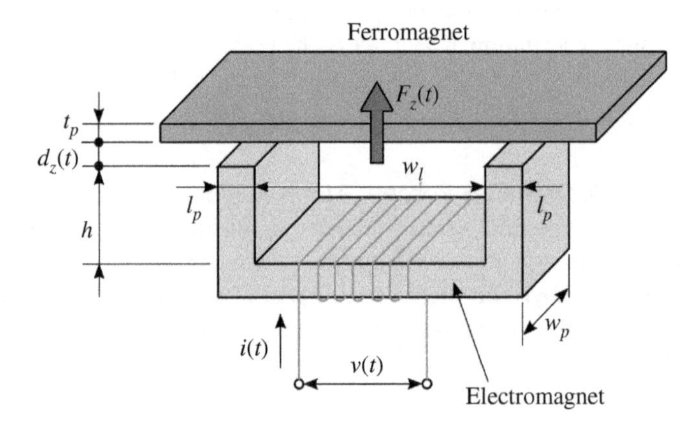

Figure 5.11 Electromagnetic suspension forces.

$P_L = P_p h((w_p/w_l) + \ln(1 + (\pi l_p/2w_l)))$ is the permeance of the leakage flux, and $\overline{\Phi}_p$ is in the range 1.5–2 Wb/m^2 (Dukkipati 2000).

The form of the magnetic levitation force of Eq. 78 sheds light on some of the dynamics and stability issues that need to be considered in the design of maglev trains. It is clear that force F_z is a quadratic function of the electric current; therefore, it increases as the electric current increases, as shown in Figure 12a. force F_z, on the other hand, is inversely proportional to $(d_z)^2$; therefore, magnetic levitation force F_z decreases as air gap d_z increases. This behavior, which is shown in Figure 12b, leads to instability. To avoid this instability, an active feedback control system or, alternatively, HEMS system may be required in order to ensure safe operation of a maglev train.

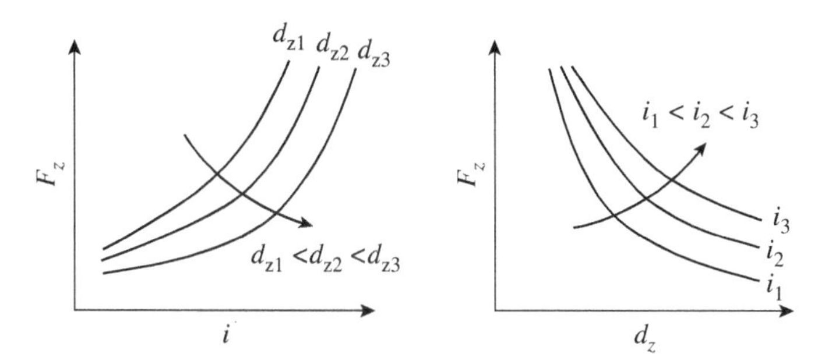

Figure 5.12 Relationship between levitation force, current, and air gap.

Having determined the magnetic levitation force F_z in the guide coordinate system, this force can be used to define the magnetic levitation force vector in any other coordinate system, including the global coordinate system, by using the proper coordinate transformation. If $\mathbf{F}_{ml}$ is the magnetic levitation force defined in the global coordinate system, one can systematically use this force vector, as described in Chapter 6, to determine the generalized forces associated with the generalized coordinates $\mathbf{R}^k$ and θ^k, $k = v, g$, where subscripts v and g refer, respectively, to the vehicle and guideway. These generalized forces can be introduced to the constrained multibody system (MBS) dynamic equations developed in the following chapter. The procedure for developing the expressions of the generalized forces and equations of motion is described in detail in Chapter 6.

Summary of Maglev Technology Concerns To place more reliance on maglev transportation, more research and technology advances are needed. Several issues and concerns need to be addressed, as follows:

- Using non-contact elements between the vehicle and guideway eliminates the options of using passive elements such as springs and dampers to control the dynamics of maglev systems. Therefore, such systems may require implementing more expensive feedback control systems to ensure safety.
- As the weight and number of rail cars increase, more electric power is needed to produce the required levitation forces to support the increased loads. Therefore, the existing maglev technology is not suited for long freight trains, which often have up to 200 cars with high axle loads.

- Changing train directions, as in the case of switching, can be difficult due to the guideway and magnetic levitation.
- The effect on passengers of the magnetic field generated by strong magnets needs to be evaluated. As the levitation force increases, strong magnets are required, and this in turn leads to a stronger magnetic field whose effect cannot be ignored.
- As a result of not using wheel/rail contact in maglev trains, traction motors must be designed to provide both propulsion and braking forces using the electromagnetic forces.

Some of these concerns are currently being researched to determine the feasibility of developing a maglev train system that can be efficiently used for mass transportation, ensures safe operation, is economically feasible, and is suited for long-distance travel. But because of the high axle load of freight trains, the focus of this research is on passenger transportation, while freight transportation will continue to rely on wheel/rail contact for the foreseeable future.

Chapter 6

EQUATIONS OF MOTION

While different approaches can be used to formulate the dynamic equations of motion of constrained dynamical systems, the *Lagrangian approach* lends itself to easily developing general algorithms for the computer-aided analysis of railroad vehicle systems. In the Lagrangian approach, the concepts of *virtual displacement*, *virtual work*, *generalized coordinates*, and *generalized forces* are fundamental. As explained in Chapter 3, the general unconstrained spatial motion of a rigid body is described using six independent coordinates. Three of these coordinates describe the translation of the body reference point, and three rotation parameters define the body orientation. The orientation coordinates can be three Euler angles or four Euler parameters, as discussed in Chapter 3. In this chapter, the Cartesian translation and orientation coordinates are referred to as the *absolute generalized coordinates*, which define, respectively, the body translation and orientation with respect to a selected global coordinate system. Another set of coordinates that is also used in formulating the dynamic equations of railroad vehicle systems is the set of *trajectory coordinates*, which includes coordinates that define the body translation and orientation with respect to a *body-track coordinate system* that follows the body motion. This body-track coordinate system is referred to in this chapter as the *trajectory body coordinate system*.

In this chapter, the generalized coordinates introduced in Chapter 3 are used to define the global position of the origin and the orientation of the body coordinate system. Because the formulations presented in this chapter are based on the *Newton–Euler equations* of motion, the body reference point, which defines the origin of the body coordinate system, is assumed to be attached to the body center of mass in order to eliminate the inertia coupling between the body translation and rotation. The virtual work principle used in the Lagrangian formulation is introduced and used with the expressions for the absolute velocity and acceleration vectors of an arbitrary point on the body, obtained in Chapter 3, to formulate the body equations of motion. This chapter also explains using *contact conditions* to formulate the wheel/rail interaction forces and discusses hunting oscillations using a constrained wheelset model that accounts for coupling between the lateral and yaw displacements.

Mathematical Foundation of Railroad Vehicle Systems: Geometry and Mechanics,
First Edition. Ahmed A. Shabana.

6.1 NEWTONIAN AND LAGRANGIAN APPROACHES

Two approaches are commonly used to formulate the constrained dynamic equations of motion of multibody systems (MBS) whose components are connected by mechanical joints: the *Newtonian approach* and the *Lagrangian approach*. The Newtonian approach, which is based on vector mechanics, requires the use of free-body diagrams by making cuts at the joints. These free-body diagrams show the joint reaction forces as well as the inertia and applied forces. This Newtonian approach is not well-suited for the development of general MBS algorithms and can be used to develop the dynamic equations of motion of relatively simple systems. For the analysis of complex systems, such as railroad vehicles, the Lagrangian approach, which has its roots in D'Alembert's principle and employs scalar quantities such as the virtual work and kinetic and potential energies, is often used to develop general-purpose MBS algorithms. When the Lagrangian approach is used, there is no need to make free-body diagrams or study the equilibrium of the bodies separately. This is mainly because the Lagrangian approach is based on the connectivity conditions. The connectivity conditions, which describe mechanical joints in the system, can be formulated using a set of nonlinear algebraic *constraint equations*. In the Lagrangian approach, the joint forces take a standard form expressed in terms of the constraint Jacobian matrix and multipliers, called *Lagrange multipliers*. The fact that algebraic constraint equations can be used to systematically define joint reaction forces allows for developing general computation procedures for the computer-aided analysis of a wide class of physics and engineering systems.

When the Lagrangian approach is used, the constraint forces can be kept in the formulation or can be systematically eliminated. Writing the equations of motion in terms of constraint forces leads to large, sparse-matrix equations of motion. The solution for the accelerations and constraint forces can be obtained by combining the equations of motion with the constraint equations at the acceleration level. This approach, in which redundant coordinates are used to formulate the dynamic equations, is called the *augmented formulation* in the MBS literature.

Alternatively, the Lagrangian approach, which is based on the *Lagrange–D'Alembert principle*, can also be used with an *embedding technique* to eliminate the constraint forces. In this case, constraint equations are used to write dependent accelerations in terms of independent accelerations using a *velocity transformation matrix*. Using this approach, the constraint forces and dependent coordinates can be systematically eliminated, leading to a number of equations of motion equal to the number of independent coordinates, which are called the *degrees of freedom* of the system. Since the independent coordinates are not related by constraint equations, the embedding technique leads to a minimum set of equations from which the constraint forces are eliminated.

In the Lagrangian approach, the concept of *generalized coordinates* is fundamental. Generalized coordinates in the MBS literature are the coordinates used to define the system configuration and formulate the *generalized forces*. Therefore, generalized coordinates do not have to be independent coordinates, as was assumed in classical mechanics literature (Greenwood 1988). The independent coordinates are called the system *degrees of freedom*. Given a constrained dynamical system, the configuration of the system can be defined using the generalized coordinates $\mathbf{q} = [q_1 \;\; q_2 \;\; \dots \;\; q_n]^T$, where n is the total number of generalized coordinates. The vector of generalized coordinates can, in general, include both independent and dependent coordinates, and therefore, one can write this vector in a partitioned form as

$$\mathbf{q} = \begin{bmatrix}\mathbf{q}_i^T & \mathbf{q}_d^T\end{bmatrix}^T \tag{6.1}$$

In this equation, $\mathbf{q}_i$ is the vector of independent coordinates, and $\mathbf{q}_d$ is the vector of dependent coordinates, with the dependency arising from motion constraints due to mechanical joints and specified motion trajectories, as explained in this chapter. If no constraints are imposed on the motion of the system, all the coordinates become independent because they are not related by algebraic equations.

6.2 VIRTUAL WORK PRINCIPLE AND CONSTRAINED DYNAMICS

The virtual work principle allows for systematic derivation of the dynamic equations of motion of complex railroad vehicle systems. In this section, the principle of virtual work is derived using a system of particles. If a rigid body is considered to consist of a large number of particles, the same principle can also be applied to rigid body systems.

Virtual Displacement and Generalized Forces The virtual work principle is based on the concept of *virtual displacement*, which refers to a change in the configuration of the system as the result of an arbitrary infinitesimal change of the coordinates, consistent with the forces and constraints imposed on the system at a given time instant t. Consider a system of n_p particles. The position vector of a particle i, denoted as $\mathbf{r}^i$, can be written in terms of the system generalized coordinates $\mathbf{q} = \begin{bmatrix}q_1 & q_2 & \dots & q_n\end{bmatrix}^T$ as

$$\mathbf{r}^i = \mathbf{r}^i(\mathbf{q}, t) = \mathbf{r}^i\left(q_1, q_2, \dots, q_n, t\right), \qquad i = 1, 2, \dots, n_p \tag{6.2}$$

At the position level, it is not always possible to write position coordinates in terms of generalized coordinates using closed-form linear relationships. The virtual displacement, on the other hand, allows writing the virtual change in position vectors in terms of the virtual change in generalized coordinates using linear equations. A virtual change in the position vector of Eq. 2 can be written as

$$\begin{aligned}\delta\mathbf{r}^i = \delta\mathbf{r}^i(\mathbf{q}) &= \frac{\partial \mathbf{r}^i}{\partial q_1}\delta q_1 + \frac{\partial \mathbf{r}^i}{\partial q_2}\delta q_2 \dots + \frac{\partial \mathbf{r}^i}{\partial q_n}\delta q_n \\ &= \sum_{j=1}^{n}\frac{\partial \mathbf{r}^i}{\partial q_j}\delta q_j, \qquad i = 1, 2, \dots, n_p\end{aligned} \tag{6.3}$$

In this expression of virtual displacement, no differentiation is made for time t because the virtual change is assumed to occur at a given time instant.

Using the definition of virtual displacement, the virtual work δW^i of a force $\mathbf{F}^i$ acting on particle i is defined as the dot product of the force vector with the virtual change in the position vector of the point of application of the force. Therefore, the virtual work of force vector $\mathbf{F}^i$ can be written as

$$\begin{aligned} \delta W^i &= \mathbf{F}^{i^T} \delta \mathbf{r}^i = \mathbf{F}^{i^T} \left(\frac{\partial \mathbf{r}^i}{\partial q_1} \delta q_1 + \frac{\partial \mathbf{r}^i}{\partial q_2} \delta q_2 \ldots + \frac{\partial \mathbf{r}^i}{\partial q_n} \delta q_n \right) \\ &= \sum_{j=1}^{n} \left(\mathbf{F}^{i^T} \frac{\partial \mathbf{r}^i}{\partial q_j} \right) \delta q_j = \sum_{j=1}^{n} Q_j \delta q_j \end{aligned} \tag{6.4}$$

where $Q_j = (\partial \mathbf{r}^i / \partial q_j)^T \mathbf{F}^i$ is called the *generalized force* associated with generalized coordinate q_j. It is clear from the preceding equation that the Cartesian space used in the Newtonian approach is replaced in the Lagrangian approach by a generalized space that has dimension n.

Generalized Inertia and Applied Forces Newton's second law is used as the starting point in developing the principle of virtual work for a system of particles. This *law of motion* states that the result of forces acting on a particle is equal to the rate of change of momentum of this particle. This statement can be written mathematically as $\mathbf{F}^i = \dot{\mathbf{P}}^i$ or, equivalently, $\mathbf{F}^i - \dot{\mathbf{P}}^i = \mathbf{0}$, where $\mathbf{P}^i = m^i \dot{\mathbf{r}}^i$ is the *linear momentum* of particle i and $\mathbf{F}^i$ is the vector of the resultant forces applied to the particle. Force vector $\mathbf{F}^i$ can be written as the sum of two vectors $\mathbf{F}^i = \mathbf{F}_e^i + \mathbf{F}_c^i$, where $\mathbf{F}_e^i$ is the vector of external forces and $\mathbf{F}_c^i$ is the vector of constraint forces. It follows that $m^i \ddot{\mathbf{r}}^i - \mathbf{F}_e^i - \mathbf{F}_c^i = \mathbf{0}$, with the assumption that the mass of the particle m^i remains constant. Multiplying the equation $m^i \ddot{\mathbf{r}}^i - \mathbf{F}_e^i - \mathbf{F}_c^i = \mathbf{0}$ by the virtual displacement $\delta \mathbf{r}^i$, one obtains

$$\left(m^i \ddot{\mathbf{r}}^i - \mathbf{F}_e^i - \mathbf{F}_c^i\right)^T \delta \mathbf{r}^i = 0, \qquad i = 1, 2, \ldots, n_p \tag{6.5}$$

This equation leads to

$$\sum_{i=1}^{n_p} \left(m^i \ddot{\mathbf{r}}^i - \mathbf{F}_e^i - \mathbf{F}_c^i\right)^T \delta \mathbf{r}^i = 0 \tag{6.6}$$

This equation, called the *Lagrange–D'Alembert* equation, can be written as $\delta W_i - \delta W_e - \delta W_c = 0$, where

$$\delta W_i = \sum_{i=1}^{n_p} \left(m^i \ddot{\mathbf{r}}^i\right)^T \delta \mathbf{r}^i, \quad \delta W_e = \sum_{i=1}^{n_p} \mathbf{F}_e^{i\,T} \delta \mathbf{r}^i, \quad \delta W_c = \sum_{i=1}^{n_p} \mathbf{F}_c^{i\,T} \delta \mathbf{r}^i \tag{6.7}$$

In this equation, δW_i is the virtual work of the inertia forces, δW_e is the virtual work of the external forces, and δW_c is the virtual work of the constraint forces. Because joint reaction forces between two particles are equal in magnitude and opposite in direction, and because the virtual change in a specified coordinate is zero, one has $\delta W_c = 0$, and the *principle of virtual work in dynamics* can be written as $\delta W_i - \delta W_e = 0$ or

$$\delta W_i = \delta W_e \tag{6.8}$$

Substituting Eqs. 3 and 7 into Eq. 8, one obtains $\sum_{j=1}^{n} \left(\sum_{i=1}^{n_p} \left(m^i \ddot{\mathbf{r}}^i - \mathbf{F}_e^i\right)^T \partial \mathbf{r}^i / \partial q_j \right) \delta q_j = 0$, which can be written as

$$\sum_{j=1}^{n} \left((Q_i)_j - (Q_e)_j \right) \delta q_j = 0 \tag{6.9}$$

where $(Q_i)_j = \sum_{i=1}^{n_p} \left(m^i \ddot{\mathbf{r}}^i\right)^T \partial \mathbf{r}^i / \partial q_j$ and $(Q_e)_j = \sum_{i=1}^{n_p} \mathbf{F}_e^{i\,T} \partial \mathbf{r}^i / \partial q_j$ are, respectively, the generalized inertia and generalized external forces associated with the generalized coordinate q_j. If the coordinates $q_1, q_2, \ldots, q_n$ are independent, Eq. 9 defines the second-order ordinary differential equations of the system as

$$(Q_i)_j = (Q_e)_j, \qquad j = 1, 2, \ldots, n \tag{6.10}$$

This equation can also be written in vector form as $\mathbf{Q}_i = \mathbf{Q}_e$, where $\mathbf{Q}_i = \left[(Q_i)_1 \ (Q_i)_2 \ \cdots \ (Q_i)_n\right]^T$ and $\mathbf{Q}_e = \left[(Q_e)_1 \ (Q_e)_2 \ \cdots \ (Q_e)_n\right]^T$.

Constrained Dynamics In railroad vehicle system dynamics, the coordinates $q_1, q_2, \ldots, q_n$ may not be independent because of mechanical joints and specified motion trajectories. Examples of mechanical joints that impose restrictions on the system motion were provided in Chapter 3. In this case of constrained dynamics, the system coordinates are related by a set of algebraic constraint equations, which can be nonlinear functions of the coordinates and can be written in a vector form as

$$\mathbf{C}(\mathbf{q}, t) = \begin{bmatrix} C_1 & C_2 & \ldots & C_{n_c} \end{bmatrix}^T = \mathbf{0} \tag{6.11}$$

where n_c is the number of constraint functions, which must be less than or equal to the number of coordinates n: that is, $n_c \leq n$. If $n_c = n$, the system is referred to as a *kinematically driven system*. For a virtual change in the coordinates, the preceding equation leads to

$$\delta \mathbf{C} = \left(\frac{\partial \mathbf{C}}{\partial \mathbf{q}}\right) \delta \mathbf{q} = \mathbf{C}_{\mathbf{q}}(\mathbf{q}, t)\, \delta \mathbf{q} = \begin{bmatrix} \partial C_1 / \partial \mathbf{q} \\ \partial C_2 / \partial \mathbf{q} \\ \vdots \\ \partial C_{n_c} / \partial \mathbf{q} \end{bmatrix} \delta \mathbf{q} = \mathbf{0} \tag{6.12}$$

In this equation, $\mathbf{C}_{\mathbf{q}} = \partial \mathbf{C} / \partial \mathbf{q}$ is the *constraint Jacobian matrix* defined as

$$\mathbf{C}_{\mathbf{q}}(\mathbf{q}, t) = \begin{bmatrix} \partial C_1 / \partial \mathbf{q} \\ \partial C_2 / \partial \mathbf{q} \\ \vdots \\ \partial C_{n_c} / \partial \mathbf{q} \end{bmatrix} = \begin{bmatrix} \partial C_1 / \partial q_1 & \partial C_1 / \partial q_2 & \ldots & \partial C_1 / \partial q_n \\ \partial C_2 / \partial q_1 & \partial C_2 / \partial q_2 & \ldots & \partial C_2 / \partial q_n \\ \vdots & \vdots & \ddots & \vdots \\ \partial C_{n_c} / \partial q_1 & \partial C_{n_c} / \partial q_2 & \ldots & \partial C_{n_c} / \partial q_n \end{bmatrix} \tag{6.13}$$

and $\partial C_k / \partial \mathbf{q}$, $k = 1, 2, \ldots n_c$, is a row vector. The number of rows in the constraint Jacobian matrix $\mathbf{C}_{\mathbf{q}}$ is equal to the number of constraint functions n_c, and the number of columns is equal to the number of coordinates n. The constraint functions must be linearly independent, and therefore, the constraint Jacobian matrix is assumed to have a full row rank. That is, one must be able to identify an $n_c \times n_c$ non-singular square matrix formed by n_c columns of the constraint Jacobian matrix $\mathbf{C}_{\mathbf{q}}$. This $n_c \times n_c$ non-singular square matrix allows, in the Lagrangian dynamics, writing the dependent variables in terms of the independent variables or degrees of freedom.

Example 6.1 As discussed in Chapter 3, a spherical (ball) joint eliminates all the degrees of freedom of the relative translation between two bodies connected by this joint. Because this joint eliminates three translational degrees of freedom, it is formulated using three algebraic constraint equations. It was shown in Chapter 3 that the constraint equations of the spherical joint between two arbitrary bodies i and j connected by the joint at points P^i and P^j, respectively, can be written as

$$\mathbf{r}_P^i - \mathbf{r}_P^j = \mathbf{R}^i + \mathbf{A}^i\bar{\mathbf{u}}_P^i - \mathbf{R}^j - \mathbf{A}^j\bar{\mathbf{u}}_P^j = \mathbf{0}$$

where $\mathbf{r}_P^k, \mathbf{R}^k, \mathbf{A}^k$ and $\bar{\mathbf{u}}_P^k,\ k = i, j$, are, respectively, the global position vector of the joint definition point P^k, the global position vector of the reference point of body k, the transformation matrix that defines the body orientation, and the local position vector of point P^k with respect to the body coordinate system. In the formulation of the spherical joint, the absolute Cartesian coordinates $\mathbf{q}^k = \begin{bmatrix}\mathbf{R}^{k^T} & \boldsymbol{\theta}^{k^T}\end{bmatrix}^T$ introduced in Chapter 3 are used.

As shown in Chapter 3, the derivatives of vector $\mathbf{r}_P^k$ with respect to the orientation parameters $\boldsymbol{\theta}^k$ of body k can always be written as $\partial\mathbf{r}_P^k/\partial\boldsymbol{\theta}^k = -\mathbf{A}^k\tilde{\bar{\mathbf{u}}}_P^k\bar{\mathbf{G}}^k$, where $\tilde{\bar{\mathbf{u}}}_P^k$ is the skew-symmetric matrix associated with vector $\bar{\mathbf{u}}_P^k$, and $\bar{\mathbf{G}}^k$ is the matrix that relates the angular velocity vector $\bar{\boldsymbol{\omega}}^k$ to the time derivatives of the orientation parameters: that is, $\bar{\boldsymbol{\omega}}^k = \bar{\mathbf{G}}^k\dot{\boldsymbol{\theta}}^k$. Therefore, the constraint Jacobian matrix of the spherical joint can be written as

$$\mathbf{C}_\mathbf{q} = \begin{bmatrix}\dfrac{\partial\mathbf{C}}{\partial\mathbf{q}^i} & \dfrac{\partial\mathbf{C}}{\partial\mathbf{q}^j}\end{bmatrix} = \begin{bmatrix}\dfrac{\partial\mathbf{C}}{\partial\mathbf{R}^i} & \dfrac{\partial\mathbf{C}}{\partial\boldsymbol{\theta}^i} & \dfrac{\partial\mathbf{C}}{\partial\mathbf{R}^j} & \dfrac{\partial\mathbf{C}}{\partial\boldsymbol{\theta}^j}\end{bmatrix}$$
$$= \begin{bmatrix}\mathbf{I} & -\mathbf{A}^i\tilde{\bar{\mathbf{u}}}_P^i\bar{\mathbf{G}}^i & -\mathbf{I} & \mathbf{A}^j\tilde{\bar{\mathbf{u}}}_P^j\bar{\mathbf{G}}^j\end{bmatrix}$$

This matrix has three rows and six columns if Euler angles are used, and it has three rows and seven columns if Euler parameters are used.

Example 6.2 As discussed in Chapter 3, a cylindrical joint is a two-degree-of-freedom joint that allows relative translation and rotation between two bodies only along the joint axis. Therefore, a cylindrical joint is formulated mathematically using four algebraic constraint equations. The four constraint equations of the cylindrical joint between two bodies i and j can be written in terms of the coordinates of the two bodies, as shown in Chapter 3, as

$$\mathbf{C}\left(\mathbf{q}^i, \mathbf{q}^j\right) = \begin{bmatrix}\mathbf{v}_1^{i^T}\mathbf{v}^j & \mathbf{v}_2^{i^T}\mathbf{v}^j & \mathbf{v}_1^{i^T}\mathbf{r}_P^{ij} & \mathbf{v}_2^{i^T}\mathbf{r}_P^{ij}\end{bmatrix}^T = \mathbf{0}$$

where $\mathbf{r}_P^{ij} = \mathbf{r}_P^i - \mathbf{r}_P^j$; $\mathbf{r}_P^i$ and $\mathbf{r}_P^j$ are, respectively, the global position vectors of two points P^i and P^j on bodies i and j defined on the joint axis; $\mathbf{v}^j = \mathbf{A}^j\bar{\mathbf{v}}^j$ is a vector on body j along the joint axis; and $\mathbf{v}_1^i = \mathbf{A}^i\bar{\mathbf{v}}_1^i$ and $\mathbf{v}_2^i = \mathbf{A}^i\bar{\mathbf{v}}_2^i$ are two vectors defined on body i perpendicular to the joint axis. In making these definitions, $\mathbf{A}^k$ is the matrix that defines the orientation of the body coordinate system, and $\bar{\mathbf{v}}^k$ is a constant vector defined in the coordinate system of body k, $k = i, j$. The constraint Jacobian matrix of the cylindrical joint can be written as

$$\mathbf{C_q} = \begin{bmatrix} \frac{\partial \mathbf{C}}{\partial \mathbf{q}^i} & \frac{\partial \mathbf{C}}{\partial \mathbf{q}^j} \end{bmatrix} = \begin{bmatrix} \frac{\partial \mathbf{C}}{\partial \mathbf{R}^i} & \frac{\partial \mathbf{C}}{\partial \boldsymbol{\theta}^i} & \frac{\partial \mathbf{C}}{\partial \mathbf{R}^j} & \frac{\partial \mathbf{C}}{\partial \boldsymbol{\theta}^j} \end{bmatrix}$$

$$= \begin{bmatrix} \mathbf{0} & -\mathbf{v}^{j^T} \mathbf{A}^i \tilde{\bar{\mathbf{v}}}_1^i \bar{\mathbf{G}}^i & \mathbf{0} & -\mathbf{v}_1^{i^T} \mathbf{A}^j \tilde{\bar{\mathbf{v}}}^j \bar{\mathbf{G}}^j \\ \mathbf{0} & -\mathbf{v}^{j^T} \mathbf{A}^i \tilde{\bar{\mathbf{v}}}_2^i \bar{\mathbf{G}}^i & \mathbf{0} & -\mathbf{v}_2^{i^T} \mathbf{A}^j \tilde{\bar{\mathbf{v}}}^j \bar{\mathbf{G}}^j \\ \mathbf{v}_1^{i^T} & -\mathbf{v}_1^{i^T} \mathbf{A}^i \tilde{\bar{\mathbf{u}}}_P^i \bar{\mathbf{G}}^i - \mathbf{r}_P^{ij^T} \mathbf{A}^i \tilde{\bar{\mathbf{v}}}_1^i \bar{\mathbf{G}}^i & -\mathbf{v}_1^{i^T} & \mathbf{v}_1^{i^T} \mathbf{A}^j \tilde{\bar{\mathbf{u}}}_P^j \bar{\mathbf{G}}^j \\ \mathbf{v}_2^{i^T} & -\mathbf{v}_2^{i^T} \mathbf{A}^i \tilde{\bar{\mathbf{u}}}_P^i \bar{\mathbf{G}}^i - \mathbf{r}_P^{ij^T} \mathbf{A}^i \tilde{\bar{\mathbf{v}}}_2^i \bar{\mathbf{G}}^i & -\mathbf{v}_2^{i^T} & \mathbf{v}_2^{i^T} \mathbf{A}^j \tilde{\bar{\mathbf{u}}}_P^j \bar{\mathbf{G}}^j \end{bmatrix}$$

This is a 4×6 matrix when Euler angles are used and a 4×7 matrix when Euler parameters are used.

Lagrange Multipliers Because $\mathbf{C_q}\delta\mathbf{q} = \mathbf{0}$, multiplying this n_c-dimensional vector by any other arbitrary n_c-dimensional vector $\boldsymbol{\lambda}$ leads to $\boldsymbol{\lambda}^T\mathbf{C_q}\delta\mathbf{q} = \mathbf{0}$. This equation can be added to Eq. 9, which can be written as $(\mathbf{Q}_i - \mathbf{Q}_e)^T\delta\mathbf{q} = 0$ to yield

$$\left(\mathbf{Q}_i - \mathbf{Q}_e + \mathbf{C}_\mathbf{q}^T\boldsymbol{\lambda}\right)^T \delta\mathbf{q} = 0 \tag{6.14}$$

Because the generalized coordinates $q_1, q_2, \ldots, q_n$ are not independent due to the constraint equations of Eq. 11, the coefficients $\left(\mathbf{Q}_i - \mathbf{Q}_e + \mathbf{C}_\mathbf{q}^T\boldsymbol{\lambda}\right)_k$ of δq_k, $k = 1, 2, \ldots, n$, in the preceding equation cannot be assumed zero unless additional conditions are imposed on the meaning of the arbitrary vector $\boldsymbol{\lambda}$. Using the coordinate partitioning of Eq. 1, one can write $\delta\mathbf{q} = \left[\delta\mathbf{q}_i^T \ \ \delta\mathbf{q}_d^T\right]^T$, where $\mathbf{q}_d$ is an n_c-dimensional vector, while $\mathbf{q}_i$ is an $(n - n_c)$-dimensional vector. According to this coordinate partitioning, Eq. 14 can be written as

$$\left(\mathbf{Q}_i - \mathbf{Q}_e + \mathbf{C}_\mathbf{q}^T\boldsymbol{\lambda}\right)_d^T \delta\mathbf{q}_d + \left(\mathbf{Q}_i - \mathbf{Q}_e + \mathbf{C}_\mathbf{q}^T\boldsymbol{\lambda}\right)_i^T \delta\mathbf{q}_i = 0 \tag{6.15}$$

In this equation, subscripts d and i refer to vector elements associated with the dependent and independent coordinates, respectively: that is, $\left(\mathbf{Q}_i - \mathbf{Q}_e + \mathbf{C}_\mathbf{q}^T\boldsymbol{\lambda}\right)_\alpha = \left(\mathbf{Q}_i\right)_\alpha - \left(\mathbf{Q}_e\right)_\alpha + \mathbf{C}_{\mathbf{q}_\alpha}^T\boldsymbol{\lambda}$, $\alpha = i, d$. Because the constraint Jacobian matrix $\mathbf{C_q}$ always has a full row rank, the dependent coordinates $\mathbf{q}_d$ can be selected such that the $n_c \times n_c$ constraint Jacobian matrix $\mathbf{C}_{\mathbf{q}_d}$ associated with the dependent coordinates is non-singular. One can then select vector $\boldsymbol{\lambda}$ to be the solution of the system of algebraic equations $\left(\mathbf{Q}_i\right)_d - \left(\mathbf{Q}_e\right)_d + \mathbf{C}_{\mathbf{q}_d}^T\boldsymbol{\lambda} = \mathbf{0}$: that is,

$$\boldsymbol{\lambda} = -\left(\mathbf{C}_{\mathbf{q}_d}^T\right)^{-1}\left(\left(\mathbf{Q}_i\right)_d - \left(\mathbf{Q}_e\right)_d\right) \tag{6.16}$$

Using this condition, Eq. 15 reduces to $\left(\mathbf{Q}_i - \mathbf{Q}_e + \mathbf{C}_\mathbf{q}^T\boldsymbol{\lambda}\right)_i^T \delta\mathbf{q}_i = 0$. Since the elements of vector $\mathbf{q}_i$ are independent, one also has $\left(\mathbf{Q}_i - \mathbf{Q}_e + \mathbf{C}_\mathbf{q}^T\boldsymbol{\lambda}\right)_i = \left(\mathbf{Q}_i\right)_i - \left(\mathbf{Q}_e\right)_i + \mathbf{C}_{\mathbf{q}_i}^T\boldsymbol{\lambda} = \mathbf{0}$, which when combined with $\left(\mathbf{Q}_i\right)_d - \left(\mathbf{Q}_e\right)_d + \mathbf{C}_{\mathbf{q}_d}^T\boldsymbol{\lambda} = \mathbf{0}$ leads to

$$\mathbf{Q}_i - \mathbf{Q}_e + \mathbf{C}_\mathbf{q}^T\boldsymbol{\lambda} = \mathbf{0} \tag{6.17}$$

This vector equation, which has n scalar equations, is the constrained dynamical equation of motion of the system formulated in terms of redundant coordinates that are not totally independent. For a given system of forces and moments, the unknowns in this equation are the n accelerations and the n_c elements of vector $\boldsymbol{\lambda}$, which are called *Lagrange multipliers*. Since no coordinates are eliminated from Eq. 17, vector $\mathbf{C}_{\mathbf{q}}^T\boldsymbol{\lambda}$ represents the *generalized constraint forces*. It is clear that generalized constraint forces are formulated using connectivity conditions without the need for making cuts to form free-body diagrams. Equation 17, however, can be viewed as the generalized form of the Newton–Euler equations obtained using free-body diagrams that include external and reaction forces. Therefore, the Cartesian-based Newton–Euler equations can be considered a special form of Eq. 17, which is more general because it is not restricted to the Cartesian-space description.

By using redundant coordinates, the number of unknowns is $n+n_c$ in Eq. 17 when the forces are given. These unknowns, which include n accelerations and n_c Lagrange multipliers, can be determined using Eqs. 11 and 17, which can be written as

$$\left.\begin{aligned} &\mathbf{Q}_i - \mathbf{Q}_e + \mathbf{C}_{\mathbf{q}}^T\boldsymbol{\lambda} = \mathbf{0} \\ &\mathbf{C}(\mathbf{q},t) = \begin{bmatrix} C_1 & C_2 & \dots & C_{n_c} \end{bmatrix}^T = \mathbf{0} \end{aligned}\right\} \tag{6.18}$$

This equation is a system of $n+n_c$ *differential/algebraic equations* (DAEs) with the differential equations representing equations of motion and the algebraic equations representing constraint equations. In MBS dynamics, two different techniques can be used to solve Eq. 18 for the n accelerations and n_c Lagrange multipliers or, equivalently, the generalized constraint forces. These two techniques, which are discussed in later sections of this chapter, are the *augmented formulation* and the *embedding technique*.

6.3 SUMMARY OF RIGID-BODY KINEMATICS

The Newton–Euler equations used in rigid body dynamics can be systematically derived using Newton's law for particles and the kinematic equations developed in Chapter 3. By assuming that a rigid body consists of a large number of particles, one can obtain an expression for the inertia forces of the rigid body. By equating the virtual work of the inertia forces to the virtual work of the applied forces and using the assumption that the body reference point is located at the body center of mass, the Newton–Euler equations can be developed as demonstrated in the following section. Because of the assumption of a centroidal body coordinate system, the Newton–Euler equations do not include inertia coupling between the rigid body translation and rotation. Before deriving the Newton–Euler equations for rigid bodies, a summary of the rigid body kinematic equations developed in Chapter 3 is first presented.

Position Equations The basic rigid-body kinematic equations presented in Chapter 3 are used in this chapter to derive the Newton–Euler equations. As explained in Chapter 3, the unconstrained spatial motion of a rigid body i in a railroad vehicle system is described using six independent coordinates: three coordinates $\mathbf{R}^i = \begin{bmatrix} R_x^i & R_y^i & R_z^i \end{bmatrix}^T$ define the translation of a selected reference point O^i on the body, and the other three coordinates $\boldsymbol{\theta}^i$ define the

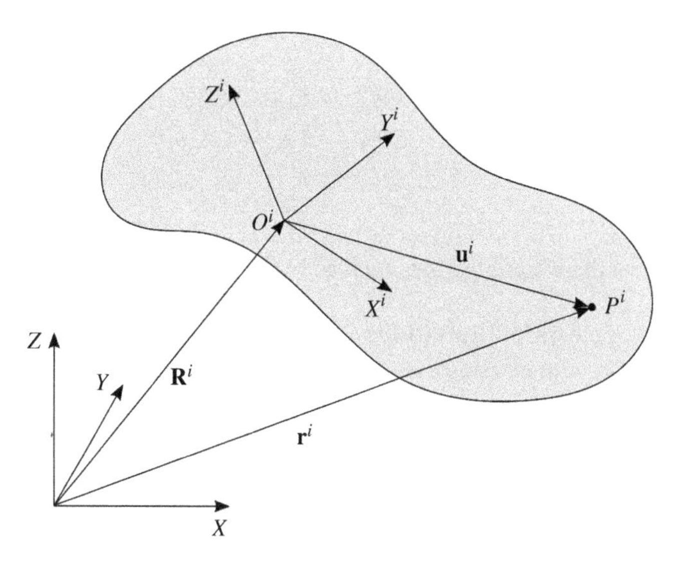

Figure 6.1 Rigid-body kinematics.

orientation of the body coordinate system $X^iY^iZ^i$ with respect to the global coordinate system XYZ. Using these coordinates, the global position of an arbitrary point on the rigid body can be written as shown in Figure 1 as $\mathbf{r}^i = \mathbf{R}^i + \mathbf{A}^i\overline{\mathbf{u}}^i$, where $\mathbf{A}^i$ is a 3×3 transformation matrix, expressed in terms of the orientation parameters $\boldsymbol{\theta}^i$, that defines the orientation of the body coordinate system $X^iY^iZ^i$ in the global coordinate system; and $\overline{\mathbf{u}}^i = \begin{bmatrix}\overline{u}_x^i & \overline{u}_y^i & \overline{u}_z^i\end{bmatrix}^T = \begin{bmatrix}x^i & y^i & z^i\end{bmatrix}^T$ is the constant position vector of the arbitrary point with respect to the reference point O^i. As previously mentioned, in the Newton–Euler formulation, reference point O^i is assumed to be the body center of mass. If Euler angles are used to define the body orientation, one has $\boldsymbol{\theta}^i = \begin{bmatrix}\psi^i & \phi^i & \theta^i\end{bmatrix}^T$. By using the Euler-angle sequence $Z^i - X^i - Y^i$, the orthogonal transformation matrix $\mathbf{A}^i$ can be written as

$$\mathbf{A}^i = \begin{bmatrix} \cos\psi^i\cos\theta^i - \sin\psi^i\sin\phi^i\sin\theta^i & -\sin\psi^i\cos\phi^i & \cos\psi^i\sin\theta^i + \sin\psi^i\sin\phi^i\cos\theta^i \\ \sin\psi^i\cos\theta^i + \cos\psi^i\sin\phi^i\sin\theta^i & \cos\psi^i\cos\phi^i & \sin\psi^i\sin\theta^i - \cos\psi^i\sin\phi^i\cos\theta^i \\ -\cos\phi^i\sin\theta^i & \sin\phi^i & \cos\phi^i\cos\theta^i \end{bmatrix} \tag{6.19}$$

Using this sequence of Euler rotations in railroad vehicle dynamics, the first rotation ψ^i about the Z^i axis is the *yaw angle*, the second rotation ϕ^i about the X^i axis is the *roll angle*, and the third rotation θ^i about the Y^i axis is the *pitch angle.*

As discussed in Chapter 3, using three independent parameters such as Euler angles leads to singular configurations (Roberson and Schwertassek 1988; Shabana 2010). For this reason, Euler parameters are often used in general MBS algorithms. In the case of Euler parameters, the orientation coordinates are defined by vector $\boldsymbol{\theta}^i = \begin{bmatrix}\theta_0^i & \theta_1^i & \theta_2^i & \theta_3^i\end{bmatrix}^T$, related by the algebraic relationship $\sum_{k=0}^{3}\left(\theta_k^i\right)^2 = 1$. Because of this algebraic relationship, the transformation matrix in terms of Euler parameters can assume different forms. One of these forms is

$$\mathbf{A}^i = \begin{bmatrix} 1-2(\theta_2^i)^2-2(\theta_3^i)^2 & 2\left(\theta_1^i\theta_2^i-\theta_0^i\theta_3^i\right) & 2\left(\theta_1^i\theta_3^i+\theta_0^i\theta_2^i\right) \\ 2\left(\theta_1^i\theta_2^i+\theta_0^i\theta_3^i\right) & 1-2(\theta_1^i)^2-2(\theta_3^i)^2 & 2\left(\theta_2^i\theta_3^i-\theta_0^i\theta_1^i\right) \\ 2\left(\theta_1^i\theta_3^i-\theta_0^i\theta_2^i\right) & 2\left(\theta_2^i\theta_3^i+\theta_0^i\theta_1^i\right) & 1-2(\theta_1^i)^2-2(\theta_2^i)^2 \end{bmatrix} \tag{6.20}$$

Euler parameters have many identities that can be used to simplify kinematic and dynamic equations. Some of these identities are provided in Chapter 3.

Velocity and Acceleration Equations The absolute velocity of an arbitrary point on a rigid body can be written using one of the two forms $\dot{\mathbf{r}}^i = \dot{\mathbf{R}}^i + \boldsymbol{\omega}^i \times \mathbf{u}^i$ and $\dot{\mathbf{r}}^i = \dot{\mathbf{R}}^i + \mathbf{A}^i\left(\overline{\boldsymbol{\omega}}^i \times \overline{\mathbf{u}}^i\right)$, where $\dot{\mathbf{R}}^i$ is the absolute velocity vector of reference point O^i, $\mathbf{u}^i = \mathbf{A}^i\overline{\mathbf{u}}^i$, $\boldsymbol{\omega}^i = \mathbf{A}^i\overline{\boldsymbol{\omega}}^i$ is the absolute angular velocity vector defined in the global coordinate system, and $\overline{\boldsymbol{\omega}}^i$ is the absolute angular velocity vector defined in the body $X^iY^iZ^i$ coordinate system. As discussed in Chapter 3, the angular velocity vectors can be written in terms of the time derivatives of the orientation coordinates as $\boldsymbol{\omega}^i = \mathbf{G}^i\dot{\boldsymbol{\theta}}^i$ and $\overline{\boldsymbol{\omega}}^i = \overline{\mathbf{G}}^i\dot{\boldsymbol{\theta}}^i$. In the case of the Euler angles $\boldsymbol{\theta}^i = \left[\psi^i \;\; \phi^i \;\; \theta^i\right]^T$ with the Z^i, X^i, Y^i rotation sequence, matrices $\mathbf{G}^i$ and $\overline{\mathbf{G}}^i$ are given, respectively, by

$$\mathbf{G}^i = \begin{bmatrix} 0 & \cos\psi^i & -\sin\psi^i\cos\phi^i \\ 0 & \sin\psi^i & \cos\psi^i\cos\phi^i \\ 1 & 0 & \sin\phi^i \end{bmatrix}, \qquad \overline{\mathbf{G}}^i = \begin{bmatrix} -\cos\phi^i\sin\theta^i & \cos\theta^i & 0 \\ \sin\phi^i & 0 & 1 \\ \cos\phi^i\cos\theta^i & \sin\theta^i & 0 \end{bmatrix} \tag{6.21}$$

When the four Euler parameters are used, the two matrices $\mathbf{G}^i$ and $\overline{\mathbf{G}}^i$ are given, respectively, by

$$\mathbf{G}^i = 2\begin{bmatrix} -\theta_1^i & \theta_0^i & -\theta_3^i & \theta_2^i \\ -\theta_2^i & \theta_3^i & \theta_0^i & -\theta_1^i \\ -\theta_3^i & -\theta_2^i & \theta_1^i & \theta_0^i \end{bmatrix}, \quad \overline{\mathbf{G}}^i = 2\begin{bmatrix} -\theta_1^i & \theta_0^i & \theta_3^i & -\theta_2^i \\ -\theta_2^i & -\theta_3^i & \theta_0^i & \theta_1^i \\ -\theta_3^i & \theta_2^i & -\theta_1^i & \theta_0^i \end{bmatrix} \tag{6.22}$$

Using the definition of matrices $\mathbf{G}^i$ and $\overline{\mathbf{G}}^i$, the velocity vector and the virtual change in the position vector of the arbitrary point can be written, respectively, as

$$\dot{\mathbf{r}}^i = \mathbf{L}^i\dot{\mathbf{q}}^i, \qquad \delta\mathbf{r}^i = \mathbf{L}^i\delta\mathbf{q}^i \tag{6.23}$$

where $\mathbf{L}^i = \left[\mathbf{I} \;\; -\tilde{\mathbf{u}}^i\mathbf{G}^i\right] = \left[\mathbf{I} \;\; -\mathbf{A}^i\tilde{\overline{\mathbf{u}}}^i\overline{\mathbf{G}}^i\right]$, and $\mathbf{I}$ is the 3×3 identity matrix. The acceleration vector can be written as

$$\ddot{\mathbf{r}}^i = \mathbf{L}^i\ddot{\mathbf{q}}^i + \dot{\mathbf{L}}^i\dot{\mathbf{q}}^i \tag{6.24}$$

Other alternate forms of the acceleration vector are $\ddot{\mathbf{r}}^i = \ddot{\mathbf{R}}^i + \boldsymbol{\alpha}^i \times \mathbf{u}^i + \boldsymbol{\omega}^i \times \left(\boldsymbol{\omega}^i \times \mathbf{u}^i\right)$ and $\ddot{\mathbf{r}}^i = \ddot{\mathbf{R}}^i + \mathbf{A}^i\left(\overline{\boldsymbol{\alpha}}^i \times \overline{\mathbf{u}}^i\right) + \mathbf{A}^i\left(\overline{\boldsymbol{\omega}}^i \times \left(\overline{\boldsymbol{\omega}}^i \times \overline{\mathbf{u}}^i\right)\right)$, where $\boldsymbol{\alpha}^i = \mathbf{A}^i\overline{\boldsymbol{\alpha}}^i$ and $\overline{\boldsymbol{\alpha}}^i$are the angular acceleration vectors defined, respectively, in the global and body coordinate systems. The form of the acceleration vector of Eq. 24 is used in the following section to derive Newton–Euler equations for rigid bodies.

6.4 INERTIA FORCES

Inertia forces play a significant role in the dynamics and stability of railroad vehicle systems. Because of inertia forces, limits on the speed of the railroad vehicle in motion scenarios must be observed and are specified in the operation and safety guidelines. In general, inertia forces can be developed using two types of forces: one is linear in accelerations, and the second is quadratic in velocities. The quadratic velocity forces give rise to centrifugal forces and gyroscopic moments that must be accounted for, particularly during curve negotiations.

The dynamics of rigid bodies are governed by the well-known Newton–Euler equations, which will be derived in this chapter after developing expressions for the inertia and applied forces. Two forms of Newton–Euler equations can be developed. In the first form, the equations are expressed in terms of the angular acceleration vector and the Cartesian definition of the moments. In the second form, the equations are expressed in terms of the second derivatives of the orientation coordinates, which are not directly associated with the actual Cartesian moments. Recall that the angular acceleration vector $\boldsymbol{\alpha}^i$ can be written in terms of the time derivatives of the orientation parameters $\ddot{\boldsymbol{\theta}}^i$ as $\boldsymbol{\alpha}^i = \mathbf{G}^i\ddot{\boldsymbol{\theta}}^i + \dot{\mathbf{G}}^i\dot{\boldsymbol{\theta}}^i$. Similarly, one can write $\overline{\boldsymbol{\alpha}}^i = \overline{\mathbf{G}}^i\ddot{\boldsymbol{\theta}}^i + \dot{\overline{\mathbf{G}}}^i\dot{\boldsymbol{\theta}}^i$. These kinematic relationships can be used to define alternate forms of the inertia forces of rigid bodies.

Mass Matrix A rigid body can be assumed to consist of an infinite number of particles. The mass of an infinitesimal volume of body i can be written as $dm^i = \rho^i dV^i$, where ρ^i is the mass density and V^i is the body volume. Therefore, the inertia force of this infinitesimal volume can be written, as shown in Figure 2, as $\left(dm^i\right)\ddot{\mathbf{r}}^i = \left(\rho^i dV^i\right)\ddot{\mathbf{r}}^i$. Therefore, the virtual work of the inertia force of body i can be written, upon using Eq. 24, as

$$\begin{aligned}\delta W_i^i &= \int_{V^i} \rho^i \ddot{\mathbf{r}}^{i^T} \delta\mathbf{r}^i dV^i \\ &= \int_{V^i} \rho^i \left(\mathbf{L}^i\ddot{\mathbf{q}}^i + \dot{\mathbf{L}}^i\dot{\mathbf{q}}^i\right)^T \delta\mathbf{r}^i dV^i \end{aligned} \tag{6.25}$$

Using the definition of $\delta\mathbf{r}^i = \mathbf{L}^i\delta\mathbf{q}^i$ of Eq. 23, one can show that the virtual work of the inertia forces can be written as

$$\delta W_i^i = \left(\mathbf{M}^i\ddot{\mathbf{q}}^i - \mathbf{Q}_v^i\right)^T \delta\mathbf{q}^i \tag{6.26}$$

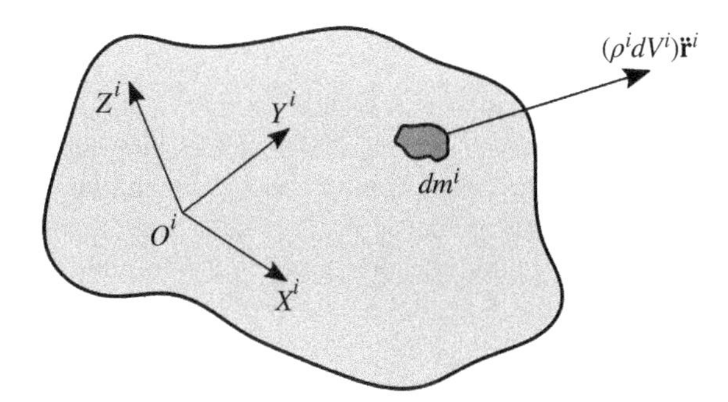

Figure 6.2 Inertia forces.

where

$$\mathbf{M}^i = \int_{V^i} \rho^i \mathbf{L}^{i^T} \mathbf{L}^i dV^i, \qquad \mathbf{Q}_v^i = -\int_{V^i} \rho^i \mathbf{L}^{i^T} \dot{\mathbf{L}}^i \dot{\mathbf{q}}^i dV^i \tag{6.27}$$

In these equations, $\mathbf{M}^i$ is the symmetric mass matrix of the body and $\mathbf{Q}_v^i$ is a vector of inertia forces that is quadratic in the velocities. Using the definition $\mathbf{L}^i = [\mathbf{I} \ \ -\mathbf{A}^i \tilde{\bar{\mathbf{u}}}^i \overline{\mathbf{G}}^i]$ in which $\bar{\mathbf{u}}^i = [x^i \ \ y^i \ \ z^i]^T$ is the position of the arbitrary point with respect to the body coordinate system, the mass matrix can be written as

$$\begin{aligned}\mathbf{M}^i &= \int_{V^i} \rho^i \mathbf{L}^{i^T} \mathbf{L}^i dV^i = \int_{V^i} \rho^i \begin{bmatrix} \mathbf{I} \\ -(\mathbf{A}^i \tilde{\bar{\mathbf{u}}}^i \overline{\mathbf{G}}^i)^T \end{bmatrix} [\mathbf{I} \ \ -\mathbf{A}^i \tilde{\bar{\mathbf{u}}}^i \overline{\mathbf{G}}^i] \, dV^i \\ &= \int_{V^i} \rho^i \begin{bmatrix} \mathbf{I} & -\mathbf{A}^i \tilde{\bar{\mathbf{u}}}^i \overline{\mathbf{G}}^i \\ -(\mathbf{A}^i \tilde{\bar{\mathbf{u}}}^i \overline{\mathbf{G}}^i)^T & \overline{\mathbf{G}}^{i^T} \tilde{\bar{\mathbf{u}}}^{i^T} \tilde{\bar{\mathbf{u}}}^i \overline{\mathbf{G}}^i \end{bmatrix} dV^i = \begin{bmatrix} \mathbf{m}_{RR}^i & \mathbf{m}_{R\theta}^i \\ \mathbf{m}_{\theta R}^i & \mathbf{m}_{\theta\theta}^i \end{bmatrix}\end{aligned} \tag{6.28}$$

where

$$\left.\begin{aligned} \mathbf{m}_{RR}^i &= \int_{V^i} \rho^i \mathbf{I} dV^i = m^i \mathbf{I}, \\ \mathbf{m}_{R\theta}^i &= \mathbf{m}_{\theta R}^{i^T} = -\mathbf{A}^i \left(\int_{V^i} \rho^i \tilde{\bar{\mathbf{u}}}^i dV^i \right) \overline{\mathbf{G}}^i, \\ \mathbf{m}_{\theta\theta}^i &= \overline{\mathbf{G}}^{i^T} \left(\int_{V^i} \rho^i \tilde{\bar{\mathbf{u}}}^{i^T} \tilde{\bar{\mathbf{u}}}^i dV^i \right) \overline{\mathbf{G}}^i = \overline{\mathbf{G}}^{i^T} \bar{\mathbf{I}}_{\theta\theta}^i \overline{\mathbf{G}}^i \end{aligned}\right\} \tag{6.29}$$

In this equation, $\mathbf{I}$ is the 3×3 identity matrix, and $\bar{\mathbf{I}}_{\theta\theta}^i = \int_{V^i} \rho^i \tilde{\bar{\mathbf{u}}}^{i^T} \tilde{\bar{\mathbf{u}}}^i dV^i$ is the constant and symmetric *inertia tensor* of the body, which can be written more explicitly as

$$\begin{aligned}\bar{\mathbf{I}}_{\theta\theta}^i &= \int_{V^i} \rho^i \tilde{\bar{\mathbf{u}}}^{i^T} \tilde{\bar{\mathbf{u}}}^i dV^i = \begin{bmatrix} i_{xx}^i & i_{xy}^i & i_{xz}^i \\ i_{xy}^i & i_{yy}^i & i_{yz}^i \\ i_{xz}^i & i_{yz}^i & i_{zz}^i \end{bmatrix} \\ &= \int_{V^i} \rho^i \begin{bmatrix} \left((y^i)^2 + (z^i)^2\right) & -x^i y^i & -x^i z^i \\ -x^i y^i & \left((x^i)^2 + (z^i)^2\right) & -y^i z^i \\ -x^i z^i & -y^i z^i & \left((x^i)^2 + (y^i)^2\right) \end{bmatrix} dV^i\end{aligned} \tag{6.30}$$

The diagonal elements i_{xx}^i, i_{yy}^i, and i_{zz}^i are called the *mass moments of inertia*, while the off-diagonal elements i_{xy}^i, i_{xz}^i, and i_{yz}^i are called the *products of inertia*.

Principal Mass Moments of Inertia It is clear from Eq. 30 that the definitions of the mass moments and products of inertia depend on the choice of the body coordinate system in which vector $\bar{\mathbf{u}}^i$ is defined. The orientation of the body coordinate system $X^iY^iZ^i$ can be selected such that the products of inertia i_{xy}^i, i_{xz}^i, and i_{yz}^i are all zeros, and the resulting mass moments of inertia include the maximum and minimum values. These mass moments of inertia are called the *principal moments of inertia,* and the axes of the body coordinate system in this case are called the *principal axes* or *principal directions*. To determine the principal moments of inertia and principal axes, one can solve the eigenvalue problem $\left[\bar{\mathbf{I}}_{\theta\theta}^i - \mu^i\mathbf{I}\right]\mathbf{d}^i = \mathbf{0}$, where μ^i is the eigenvalue that defines the principal moments of inertia and $\mathbf{d}^i$ is the eigenvector that defines the principal directions. To obtain a nontrivial solution, the matrix $\left[\bar{\mathbf{I}}_{\theta\theta}^i - \mu^i\mathbf{I}\right]$ is assumed to be singular: that is, $\left|\bar{\mathbf{I}}_{\theta\theta}^i - \mu^i\mathbf{I}\right| = 0$. The cubic polynomial resulting from the condition $\left|\bar{\mathbf{I}}_{\theta\theta}^i - \mu^i\mathbf{I}\right| = 0$ is called the *characteristic equation*. The roots of the characteristic equation define the three eigenvalues $\mu_k^i,\ k = 1, 2, 3$, which are the three principal moments of inertia. Associated with these three eigenvalues are three eigenvectors $\mathbf{d}_k^i,\ k = 1, 2, 3$, which can be determined to within an arbitrary constant using the equation $\left[\bar{\mathbf{I}}_{\theta\theta}^i - \mu_k^i\mathbf{I}\right]\mathbf{d}_k^i = \mathbf{0},\ k = 1, 2, 3$. These three eigenvectors define the three principal axes or directions. Because the inertia tensor $\bar{\mathbf{I}}_{\theta\theta}^i$ is symmetric, the eigenvalues μ_k^i are real, and the eigenvectors $\mathbf{d}_k^i$ are orthogonal vectors. Because the eigenvectors are determined to within an arbitrary constant, these eigenvectors can be scaled to be orthogonal unit vectors $\hat{\mathbf{d}}_k^i,\ k = 1, 2, 3$, which can be used to form the orthogonal matrix $\mathbf{D}^i = \left[\hat{\mathbf{d}}_1^i\ \hat{\mathbf{d}}_2^i\ \hat{\mathbf{d}}_3^i\right]$. Using this orthogonal matrix and the symmetry of the inertia tensor, one can show that $\left(\bar{\mathbf{I}}_{\theta\theta}^i\right)_{pr} = \mathbf{D}^{i^T}\bar{\mathbf{I}}_{\theta\theta}^i\mathbf{D}^i$ is a diagonal matrix whose diagonal elements are the eigenvalues $\mu_k^i = \mathbf{d}_k^{i^T}\bar{\mathbf{I}}_{\theta\theta}^i\mathbf{d}_k^i,\ k = 1, 2, 3$.

Quadratic Velocity Inertia Forces To develop an expression for the quadratic velocity inertia force vector $\mathbf{Q}_v^i = -\int_{V^i}\rho^i\mathbf{L}^{i^T}\dot{\mathbf{L}}^i\dot{\mathbf{q}}^i dV^i$ of Eq. 27, one can write $\dot{\mathbf{L}}^i = \left[\mathbf{0}\ \ -\left(\dot{\mathbf{A}}^i\tilde{\bar{\mathbf{u}}}^i\bar{\mathbf{G}}^i + \mathbf{A}^i\tilde{\bar{\mathbf{u}}}^i\dot{\bar{\mathbf{G}}}^i\right)\right]$, which can be written as $\dot{\mathbf{L}}^i = \left[\mathbf{0}\ \ -\left(\mathbf{A}^i\tilde{\bar{\boldsymbol{\omega}}}^i\tilde{\bar{\mathbf{u}}}^i\bar{\mathbf{G}}^i + \mathbf{A}^i\tilde{\bar{\mathbf{u}}}^i\dot{\bar{\mathbf{G}}}^i\right)\right]$. It follows that

$$\begin{aligned}\dot{\mathbf{L}}^i\dot{\mathbf{q}}^i &= -\left(\mathbf{A}^i\tilde{\bar{\boldsymbol{\omega}}}^i\tilde{\bar{\mathbf{u}}}^i\bar{\boldsymbol{\omega}}^i + \mathbf{A}^i\tilde{\bar{\mathbf{u}}}^i\dot{\bar{\mathbf{G}}}^i\dot{\boldsymbol{\theta}}^i\right)\\ &= \mathbf{A}^i\left(\bar{\boldsymbol{\omega}}^i \times \left(\bar{\boldsymbol{\omega}}^i \times \bar{\mathbf{u}}^i\right)\right) + \mathbf{A}^i\left(\dot{\bar{\mathbf{G}}}^i\dot{\boldsymbol{\theta}}^i\right) \times \bar{\mathbf{u}}^i\end{aligned} \tag{6.31}$$

Therefore, one can write

$$\begin{aligned}\mathbf{Q}_v^i &= -\int_{V^i}\rho^i\mathbf{L}^{i^T}\dot{\mathbf{L}}^i\dot{\mathbf{q}}^i dV^i\\ &= -\int_{V^i}\rho^i\begin{bmatrix}\mathbf{A}^i\left(\bar{\boldsymbol{\omega}}^i \times \left(\bar{\boldsymbol{\omega}}^i \times \bar{\mathbf{u}}^i\right)\right) + \mathbf{A}^i\left(\dot{\bar{\mathbf{G}}}^i\dot{\boldsymbol{\theta}}^i\right) \times \bar{\mathbf{u}}^i\\ -\bar{\mathbf{G}}^{i^T}\tilde{\bar{\mathbf{u}}}^{i^T}\left[\left(\bar{\boldsymbol{\omega}}^i \times \left(\bar{\boldsymbol{\omega}}^i \times \bar{\mathbf{u}}^i\right)\right) + \left(\dot{\bar{\mathbf{G}}}^i\dot{\boldsymbol{\theta}}^i\right) \times \bar{\mathbf{u}}^i\right]\end{bmatrix}dV^i\end{aligned} \tag{6.32}$$

It can be shown that this vector can be written as

$$\mathbf{Q}_v^i = \begin{bmatrix}\left(\mathbf{Q}_v^i\right)_R\\ \left(\mathbf{Q}_v^i\right)_\theta\end{bmatrix} = -\begin{bmatrix}\mathbf{A}^i\left(\left(\bar{\boldsymbol{\omega}}^i \times \left(\bar{\boldsymbol{\omega}}^i \times \bar{\mathbf{h}}^i\right)\right) + \left(\dot{\bar{\mathbf{G}}}^i\dot{\boldsymbol{\theta}}^i\right) \times \bar{\mathbf{h}}^i\right)\\ \bar{\mathbf{G}}^{i^T}\left(\bar{\boldsymbol{\omega}}^i \times \left(\bar{\mathbf{I}}_{\theta\theta}^i\bar{\boldsymbol{\omega}}^i\right) + \bar{\mathbf{I}}_{\theta\theta}^i\left(\dot{\bar{\mathbf{G}}}^i\dot{\boldsymbol{\theta}}^i\right)\right)\end{bmatrix} \tag{6.33}$$

In this equation, $\overline{\mathbf{h}}^i = \int_{V^i} \rho^i \overline{\mathbf{u}}^i dV^i$ is the moment of mass of the body. Vector $\mathbf{Q}_v^i$ of Eq. 33 can also be written using vectors defined in the global coordinate system as

$$\mathbf{Q}_v^i = \begin{bmatrix} \left(\mathbf{Q}_v^i\right)_R \\ \left(\mathbf{Q}_v^i\right)_\theta \end{bmatrix} = -\begin{bmatrix} \boldsymbol{\omega}^i \times \left(\boldsymbol{\omega}^i \times \mathbf{h}^i\right) + \left(\dot{\mathbf{G}}^i \dot{\boldsymbol{\theta}}^i\right) \times \mathbf{h}^i \\ \mathbf{G}^{iT} \left(\boldsymbol{\omega}^i \times \left(\mathbf{I}_{\theta\theta}^i \boldsymbol{\omega}^i\right) + \mathbf{I}_{\theta\theta}^i \left(\dot{\mathbf{G}}^i \dot{\boldsymbol{\theta}}^i\right)\right) \end{bmatrix} \tag{6.34}$$

where $\mathbf{I}_{\theta\theta}^i = \mathbf{A}^i \overline{\mathbf{I}}_{\theta\theta}^i \mathbf{A}^{iT}$, $\boldsymbol{\omega}^i = \mathbf{A}^i \overline{\boldsymbol{\omega}}^i$, $\mathbf{G}^i = \mathbf{A}^i \overline{\mathbf{G}}^i$, and $\mathbf{h}^i = \mathbf{A}^i \overline{\mathbf{h}}^i$.

Centrifugal Forces and Gyroscopic Moments In Eq. 33, vector $\left(\mathbf{Q}_v^i\right)_R$ associated with the translation coordinate of the body is the *centrifugal force vector*, while vector $\overline{\boldsymbol{\omega}}^i \times \overline{\mathbf{I}}_{\theta\theta}^i \overline{\boldsymbol{\omega}}^i$ is the *gyroscopic moment*. Centrifugal forces and gyroscopic moments, which are, respectively, inertia forces and moments, play an important role in the nonlinear dynamics and stability of railroad vehicle systems. The source of the gyroscopic moment can be better understood by examining the expression of the absolute acceleration vector $\ddot{\mathbf{r}}^i = \ddot{\mathbf{R}}^i + \boldsymbol{\alpha}^i \times \mathbf{u}^i + \boldsymbol{\omega}^i \times \left(\boldsymbol{\omega}^i \times \mathbf{u}^i\right)$. The last term in this acceleration expression is the normal component, which gives rise to the centrifugal force vector $\boldsymbol{\omega}^i \times (\boldsymbol{\omega}^i \times \mathbf{h}^i)$ in vector $\left(\mathbf{Q}_v^i\right)_R$ of Eq. 34. From Eq. 32, it is clear that the gyroscopic moment can be written in terms of $\int_{V^i} \rho^i \mathbf{u}^i \times \left(\boldsymbol{\omega}^i \times \left(\boldsymbol{\omega}^i \times \mathbf{u}^i\right)\right) dV^i$, which can be interpreted as the moment of the centrifugal force. Because centrifugal forces can assume large values during curve negotiations, a limit is imposed on the speed of the vehicle to avoid derailment. This speed, called the *balance speed*, is determined using the superelevation, which is designed to ensure that the lateral component of the centrifugal force will not exceed the lateral component of the gravity force. Gyroscopic moments also have a significant effect on vehicle dynamics and stability and must be taken into consideration to ensure stability during curve negotiations, as discussed in the literature (Shabana et al. 2013).

Newton–Euler Assumption The Newton–Euler equations for a rigid body are derived using the assumption that the body reference point O^i is selected to be the body center of mass. This assumption leads to simplifications in the form of the equations of motion, including eliminating the inertia coupling between the body translation and rotation. If the reference point is selected to be the body center of mass, the moment of mass about the center of mass (reference point) must be equal to zero. In this case, one has $\overline{\mathbf{h}}^i = \int_{V^i} \rho^i \overline{\mathbf{u}}^i dV^i = \mathbf{0}$. Using this condition, it is clear from Eq. 29 that $\mathbf{m}_{R\theta}^i = \mathbf{m}_{\theta R}^{iT} = -\mathbf{A}^i \left(\int_{V^i} \rho^i \tilde{\overline{\mathbf{u}}}^i dV^i\right) \overline{\mathbf{G}}^i = \mathbf{0}$, and the mass matrix of the body of Eq. 28 reduces to

$$\mathbf{M}^i = \begin{bmatrix} \mathbf{m}_{RR}^i & \mathbf{0} \\ \mathbf{0} & \mathbf{m}_{\theta\theta}^i \end{bmatrix} = \begin{bmatrix} m^i \mathbf{I} & \mathbf{0} \\ \mathbf{0} & \overline{\mathbf{G}}^{iT} \overline{\mathbf{I}}_{\theta\theta}^i \overline{\mathbf{G}}^i \end{bmatrix} \tag{6.35}$$

In this form of the mass matrix, there is no inertia coupling between the body translation and rotation. Furthermore, using the condition $\overline{\mathbf{h}}^i = \int_{V^i} \rho^i \overline{\mathbf{u}}^i dV^i = \mathbf{0}$, the centrifugal force vector $\left(\mathbf{Q}_v^i\right)_R$ is identically zero: that is, $\left(\mathbf{Q}_v^i\right)_R = \mathbf{0}$. In this case, vector $\mathbf{Q}_v^i$ reduces to

$$\begin{aligned} \mathbf{Q}_v^i = \begin{bmatrix} \left(\mathbf{Q}_v^i\right)_R \\ \left(\mathbf{Q}_v^i\right)_\theta \end{bmatrix} &= -\begin{bmatrix} \mathbf{0} \\ \overline{\mathbf{G}}^{iT} \left(\overline{\boldsymbol{\omega}}^i \times \left(\overline{\mathbf{I}}_{\theta\theta}^i \overline{\boldsymbol{\omega}}^i\right) + \overline{\mathbf{I}}_{\theta\theta}^i \left(\dot{\overline{\mathbf{G}}}^i \dot{\boldsymbol{\theta}}^i\right)\right) \end{bmatrix} \\ &= -\begin{bmatrix} \mathbf{0} \\ \mathbf{G}^{iT} \left(\boldsymbol{\omega}^i \times \left(\mathbf{I}_{\theta\theta}^i \boldsymbol{\omega}^i\right) + \mathbf{I}_{\theta\theta}^i \left(\dot{\mathbf{G}}^i \dot{\boldsymbol{\theta}}^i\right)\right) \end{bmatrix} \end{aligned} \tag{6.36}$$

Therefore, the virtual work of the inertia forces of Eq. 26 can be written as

$$\begin{aligned}\delta W_i^i &= \left(\mathbf{M}^i\ddot{\mathbf{q}}^i - \mathbf{Q}_v^i\right)^T\delta\mathbf{q}^i \\ &= m^i\ddot{\mathbf{R}}^{i^T}\delta\mathbf{R}^i + \left(\overline{\mathbf{G}}^{i^T}\overline{\mathbf{I}}_{\theta\theta}^i\overline{\mathbf{G}}^i\ddot{\boldsymbol{\theta}}^i + \overline{\mathbf{G}}^{i^T}\left(\overline{\boldsymbol{\omega}}^i \times \left(\overline{\mathbf{I}}_{\theta\theta}^i\overline{\boldsymbol{\omega}}^i\right) + \overline{\mathbf{I}}_{\theta\theta}^i\left(\dot{\overline{\mathbf{G}}}^i\dot{\boldsymbol{\theta}}^i\right)\right)\right)^T\delta\boldsymbol{\theta}^i \\ &= \left(\mathbf{Q}_i^i\right)_R^T\delta\mathbf{R}^i + \left(\mathbf{Q}_i^i\right)_\theta^T\delta\boldsymbol{\theta}^i\end{aligned} \tag{6.37}$$

or, alternatively, in terms of vectors defined in the global system as

$$\begin{aligned}\delta W_i^i &= \left(\mathbf{M}^i\ddot{\mathbf{q}}^i - \mathbf{Q}_v^i\right)^T\delta\mathbf{q}^i \\ &= m^i\ddot{\mathbf{R}}^{i^T}\delta\mathbf{R}^i + \left(\mathbf{G}^{i^T}\mathbf{I}_{\theta\theta}^i\mathbf{G}^i\ddot{\boldsymbol{\theta}}^i + \mathbf{G}^{i^T}\left(\boldsymbol{\omega}^i \times \left(\mathbf{I}_{\theta\theta}^i\boldsymbol{\omega}^i\right) + \mathbf{I}_{\theta\theta}^i\left(\dot{\mathbf{G}}^i\dot{\boldsymbol{\theta}}^i\right)\right)\right)^T\delta\boldsymbol{\theta}^i \\ &= \left(\mathbf{Q}_i^i\right)_R^T\delta\mathbf{R}^i + \left(\mathbf{Q}_i^i\right)_\theta^T\delta\boldsymbol{\theta}^i\end{aligned} \tag{6.38}$$

In both Eqs. 37 and 38, $\left(\mathbf{Q}_i^i\right)_R = m^i\ddot{\mathbf{R}}^i$; and in Eq. 37, $\left(\mathbf{Q}_i^i\right)_\theta = \overline{\mathbf{G}}^{i^T}\overline{\mathbf{I}}_{\theta\theta}^i\overline{\mathbf{G}}^i\ddot{\boldsymbol{\theta}}^i + \overline{\mathbf{G}}^{i^T}\left(\overline{\boldsymbol{\omega}}^i \times \left(\overline{\mathbf{I}}_{\theta\theta}^i\overline{\boldsymbol{\omega}}^i\right) + \overline{\mathbf{I}}_{\theta\theta}^i\left(\dot{\overline{\mathbf{G}}}^i\dot{\boldsymbol{\theta}}^i\right)\right)$ are, respectively, the generalized inertia forces associated with the translational and orientation coordinates of the body. The generalized inertia force associated with the orientation coordinates can also be written using Eq. 38 as $\left(\mathbf{Q}_i^i\right)_\theta = \mathbf{G}^{i^T}\mathbf{I}_{\theta\theta}^i\mathbf{G}^i\ddot{\boldsymbol{\theta}}^i + \mathbf{G}^{i^T}\left(\boldsymbol{\omega}^i \times \mathbf{I}_{\theta\theta}^i\boldsymbol{\omega}^i + \mathbf{I}_{\theta\theta}^i\left(\dot{\mathbf{G}}^i\dot{\boldsymbol{\theta}}^i\right)\right)$. These expressions for the inertia forces will be used to derive the Newton–Euler equations for rigid bodies.

6.5 APPLIED FORCES

The principle of virtual work of Eq. 8 states that the virtual work of inertia forces obtained in the preceding section is equal to the virtual work of applied forces. This section discusses the formulation of the virtual work of applied forces. It is shown that Cartesian moments are not associated with angles, and generalized forces associated with orientation parameters can be obtained from Cartesian moments using the two matrices $\mathbf{G}^i$ and $\overline{\mathbf{G}}^i$ that are used to define the angular velocity vectors in the global and body coordinate systems, respectively.

Figure 3 shows a force vector $\mathbf{F}^i$ acting at point P^i on a rigid body. The global position of point P^i is defined by vector $\mathbf{r}_P^i = \mathbf{R}^i + \mathbf{A}^i\overline{\mathbf{u}}_P^i$, where $\overline{\mathbf{u}}_P^i$ is the local position of the point of application of the force. The virtual work of this force vector is defined as the dot product of the force and the virtual change in the position vector of the point of application of the force: that is,

$$\delta W_e^i = \mathbf{F}^{i^T}\delta\mathbf{r}_P^i = \mathbf{F}^{i^T}\mathbf{L}_P^i\delta\mathbf{q}^i \tag{6.39}$$

where $\mathbf{L}_P^i = \left[\mathbf{I} \quad -\mathbf{A}^i\tilde{\overline{\mathbf{u}}}_P^i\overline{\mathbf{G}}^i\right]$. Equation 39 can then be written as

$$\delta W_e^i = \left(\mathbf{Q}_e^i\right)_R^T\delta\mathbf{R}^i + \left(\mathbf{Q}_e^i\right)_\theta^T\delta\boldsymbol{\theta}^i \tag{6.40}$$

where $\left(\mathbf{Q}_e^i\right)_R = \mathbf{F}^i$ and $\left(\mathbf{Q}_e^i\right)_\theta = -\overline{\mathbf{G}}^{i^T}\left(\tilde{\overline{\mathbf{u}}}_P^{i^T}\mathbf{A}^{i^T}\mathbf{F}^i\right)$ are, respectively, the generalized applied forces associated with the translational and orientation coordinates of the body. Equation 40 implies that force vector $\mathbf{F}^i$ acting at an arbitrary point P^i on the body is *equipollent* to

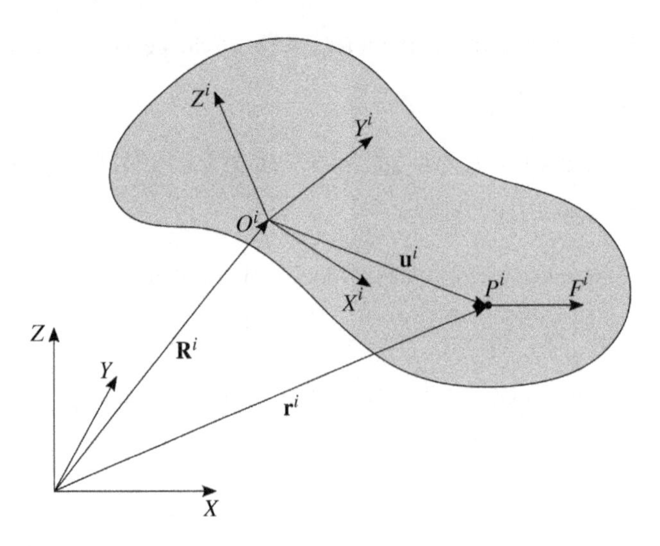

Figure 6.3 Applied force.

another system defined at point O^i, which consists of the same force vector $\mathbf{F}^i$ associated with the body translation, and a generalized force vector $\left(\mathbf{Q}_e^i\right)_\theta = -\overline{\mathbf{G}}^{i^T}\left(\tilde{\overline{\mathbf{u}}}_P^{i^T}\mathbf{A}^{i^T}\mathbf{F}^i\right)$ associated with the orientation coordinates. To understand the relationship between the generalized force vector $\left(\mathbf{Q}_e^i\right)_\theta$ associated with the orientation coordinates and the Cartesian moment, vector $\left(\mathbf{Q}_e^i\right)_\theta$ is written as

$$\left(\mathbf{Q}_e^i\right)_\theta = \overline{\mathbf{G}}^{i^T}\left(\tilde{\overline{\mathbf{u}}}_P^i\mathbf{A}^{i^T}\mathbf{F}^i\right) = \overline{\mathbf{G}}^{i^T}\left(\overline{\mathbf{u}}_P^i \times \overline{\mathbf{F}}^i\right) = \mathbf{G}^{i^T}\left(\mathbf{u}_P^i \times \mathbf{F}^i\right) \tag{6.41}$$

where $\overline{\mathbf{F}}^i = \mathbf{A}^{i^T}\mathbf{F}^i$ is the force vector defined in the body coordinate system, and $\mathbf{u}_P^i = \mathbf{A}^i\overline{\mathbf{u}}_P^i$. The preceding equation shows that the Cartesian moment of the force $\mathbf{u}_P^i \times \mathbf{F}^i = \mathbf{A}^i\left(\overline{\mathbf{u}}_P^i \times \overline{\mathbf{F}}^i\right)$ is not the moment associated with the orientation parameters $\boldsymbol{\theta}^i$. The generalized forces associated with the orientation parameters $\boldsymbol{\theta}^i$ can be obtained from the applied moments using the transpose of the matrices $\mathbf{G}^i$ and $\overline{\mathbf{G}}^i$: that is, if $\mathbf{M}_\alpha^i = \mathbf{A}^i\overline{\mathbf{M}}_\alpha^i$ is a Cartesian moment that applies to rigid body i, then the generalized forces associated with the orientation parameters are defined as

$$\left(\mathbf{Q}_e^i\right)_\theta = \mathbf{G}^{i^T}\mathbf{M}_\alpha^i = \overline{\mathbf{G}}^{i^T}\overline{\mathbf{M}}_\alpha^i \tag{6.42}$$

where $\overline{\mathbf{M}}_\alpha^i$ is the Cartesian moment defined in the body coordinate system. Equation 42 will be used in converting the generalized Newton–Euler equations obtained using the virtual work principle to the Newton–Euler equations expressed in terms of angular acceleration vectors $\boldsymbol{\alpha}^i$ and $\overline{\boldsymbol{\alpha}}^i$. In general, for any system of forces and moments acting on a rigid body, an equation similar to Eq. 40 can be developed, where $(\mathbf{Q}_e^i)_R$ and $(\mathbf{Q}_e^i)_\theta$ are vectors of the resultant generalized forces associated with the generalized translational and orientation coordinates, respectively.

6.6 NEWTON–EULER EQUATIONS

Two forms of Newton–Euler equations are presented in this section. The first is the *generalized Newton–Euler equations* obtained directly from using the principle of virtual work. This form of the Newton–Euler equations is written in terms of generalized orientation parameters. In the second form, the Newton–Euler equations are expressed in terms of the angular acceleration vector. These two alternate forms of the Newton–Euler equations show that the Cartesian moment $\mathbf{M}_\alpha^i = \mathbf{A}^i\overline{\mathbf{M}}_\alpha^i$ is associated with the angular acceleration vector and not with the second time derivative of the orientation coordinates.

Generalized Newton–Euler Equations Using the virtual work principle of Eq. 8, $\delta W_i = \delta W_e$, and using the virtual work of the inertia forces $\delta W_i^i = (\mathbf{Q}_i^i)_R^T \delta\mathbf{R}^i + (\mathbf{Q}_i^i)_\theta^T \delta\boldsymbol{\theta}^i$ of Eq. 37 and the virtual work of the applied forces $\delta W_e^i = (\mathbf{Q}_e^i)_R^T \delta\mathbf{R}^i + (\mathbf{Q}_e^i)_\theta^T \delta\boldsymbol{\theta}^i$ of Eq. 40, one obtains

$$\left((\mathbf{Q}_i^i)_R - (\mathbf{Q}_e^i)_R\right)^T \delta\mathbf{R}^i + \left((\mathbf{Q}_i^i)_\theta - (\mathbf{Q}_e^i)_\theta\right)^T \delta\boldsymbol{\theta}^i = 0 \tag{6.43}$$

In the case of unconstrained motion, the elements of vector $\mathbf{q}^i = \begin{bmatrix}\mathbf{R}^{i^T} & \boldsymbol{\theta}^{i^T}\end{bmatrix}^T$ are independent, and therefore, the coefficients of their virtual change in the preceding equation must be equal to zero: that is, $(\mathbf{Q}_i^i)_R = (\mathbf{Q}_e^i)_R$ and $(\mathbf{Q}_i^i)_\theta = (\mathbf{Q}_e^i)_\theta$. Using the definition of the inertia forces given in Eq. 37, one obtains the two equations

$$\left.\begin{aligned} m^i\ddot{\mathbf{R}}^i &= (\mathbf{Q}_e^i)_R \\ \left(\overline{\mathbf{G}}^{i^T}\overline{\mathbf{I}}_{\theta\theta}^i\overline{\mathbf{G}}^i\right)\ddot{\boldsymbol{\theta}}^i &= (\mathbf{Q}_e^i)_\theta - \overline{\mathbf{G}}^{i^T}\left(\overline{\boldsymbol{\omega}}^i \times (\overline{\mathbf{I}}_{\theta\theta}^i\overline{\boldsymbol{\omega}}^i) + \overline{\mathbf{I}}_{\theta\theta}^i\left(\dot{\overline{\mathbf{G}}}^i\dot{\boldsymbol{\theta}}^i\right)\right) \end{aligned}\right\} \tag{6.44}$$

These are the generalized Newton–Euler equations, which can be written in a matrix form as

$$\begin{bmatrix} m^i\mathbf{I} & \mathbf{0} \\ \mathbf{0} & \left(\overline{\mathbf{G}}^{i^T}\overline{\mathbf{I}}_{\theta\theta}^i\overline{\mathbf{G}}^i\right) \end{bmatrix}\begin{bmatrix}\ddot{\mathbf{R}}^i \\ \ddot{\boldsymbol{\theta}}^i\end{bmatrix} = \begin{bmatrix}(\mathbf{Q}_e^i)_R \\ (\mathbf{Q}_e^i)_\theta\end{bmatrix} + \begin{bmatrix}\mathbf{0} \\ -\overline{\mathbf{G}}^{i^T}\left(\overline{\boldsymbol{\omega}}^i \times (\overline{\mathbf{I}}_{\theta\theta}^i\overline{\boldsymbol{\omega}}^i) + \overline{\mathbf{I}}_{\theta\theta}^i\left(\dot{\overline{\mathbf{G}}}^i\dot{\boldsymbol{\theta}}^i\right)\right)\end{bmatrix} \tag{6.45}$$

This equation can be written in terms of vectors defined in the global coordinate system as

$$\begin{bmatrix} m^i\mathbf{I} & \mathbf{0} \\ \mathbf{0} & \left(\mathbf{G}^{i^T}\mathbf{I}_{\theta\theta}^i\mathbf{G}^i\right) \end{bmatrix}\begin{bmatrix}\ddot{\mathbf{R}}^i \\ \ddot{\boldsymbol{\theta}}^i\end{bmatrix} = \begin{bmatrix}(\mathbf{Q}_e^i)_R \\ (\mathbf{Q}_e^i)_\theta\end{bmatrix} + \begin{bmatrix}\mathbf{0} \\ -\mathbf{G}^{i^T}\left(\boldsymbol{\omega}^i \times (\mathbf{I}_{\theta\theta}^i\boldsymbol{\omega}^i) + \mathbf{I}_{\theta\theta}^i\left(\dot{\mathbf{G}}^i\dot{\boldsymbol{\theta}}^i\right)\right)\end{bmatrix} \tag{6.46}$$

Equations 45 and 46 can be used to obtain the Newton–Euler equations written in terms of the angular acceleration vectors $\overline{\boldsymbol{\alpha}}^i$ and $\boldsymbol{\alpha}^i$, respectively.

Newton–Euler Equations Substituting the equation $\overline{\boldsymbol{\alpha}}^i = \overline{\mathbf{G}}^i\ddot{\boldsymbol{\theta}}^i + \dot{\overline{\mathbf{G}}}^i\dot{\boldsymbol{\theta}}^i$ into Eq. 45 and using Eq. 42, one obtains

$$\begin{bmatrix} m^i\mathbf{I} & \mathbf{0} \\ \mathbf{0} & \overline{\mathbf{I}}^i_{\theta\theta} \end{bmatrix} \begin{bmatrix} \ddot{\mathbf{R}}^i \\ \overline{\boldsymbol{\alpha}}^i \end{bmatrix} = \begin{bmatrix} \left(\mathbf{Q}^i_e\right)_R \\ \overline{\mathbf{M}}^i_\alpha \end{bmatrix} + \begin{bmatrix} \mathbf{0} \\ -\overline{\boldsymbol{\omega}}^i \times \left(\overline{\mathbf{I}}^i_{\theta\theta}\overline{\boldsymbol{\omega}}^i\right) \end{bmatrix} \tag{6.47}$$

In this form of the Newton–Euler equations, the coefficient matrix is constant. Alternatively, one can use the relationship $\boldsymbol{\alpha}^i = \mathbf{G}^i\ddot{\boldsymbol{\theta}}^i + \dot{\mathbf{G}}^i\dot{\boldsymbol{\theta}}^i$ to obtain the Newton–Euler equations written in terms of vectors defined in the global coordinate system as

$$\begin{bmatrix} m^i\mathbf{I} & \mathbf{0} \\ \mathbf{0} & \mathbf{I}^i_{\theta\theta} \end{bmatrix} \begin{bmatrix} \ddot{\mathbf{R}}^i \\ \boldsymbol{\alpha}^i \end{bmatrix} = \begin{bmatrix} \left(\mathbf{Q}^i_e\right)_R \\ \mathbf{M}^i_\alpha \end{bmatrix} + \begin{bmatrix} \mathbf{0} \\ -\boldsymbol{\omega}^i \times \left(\mathbf{I}^i_{\theta\theta}\boldsymbol{\omega}^i\right) \end{bmatrix} \tag{6.48}$$

In this form of the Newton–Euler equations, the coefficient matrix is not constant since $\mathbf{I}^i_{\theta\theta} = \mathbf{A}^i\overline{\mathbf{I}}^i_{\theta\theta}\mathbf{A}^{iT}$. One can show that the Newton–Euler equations of Eq. 47 or 48 reduce to the three Newton–Euler equations used in the planar analysis, as demonstrated by the following example.

Example 6.3 In the case of planar analysis, the global Z axis and body Z^i axis coincide. In this special case, one can write the six accelerations $\ddot{\mathbf{R}}^i$ and $\overline{\boldsymbol{\alpha}}^i$ in terms of only three acceleration components $\ddot{R}^i_x$, $\ddot{R}^i_y$, and the angular acceleration $\ddot{\theta}^i_z$ about the Z axis as

$$\begin{bmatrix} \ddot{\mathbf{R}}^i \\ \overline{\boldsymbol{\alpha}}^i \end{bmatrix} = \begin{bmatrix} \ddot{R}^i_x \\ \ddot{R}^i_y \\ \ddot{R}^i_z \\ \overline{\alpha}^i_x \\ \overline{\alpha}^i_y \\ \overline{\alpha}^i_z \end{bmatrix} = \begin{bmatrix} 1 & 0 & 0 \\ 0 & 1 & 0 \\ 0 & 0 & 0 \\ 0 & 0 & 0 \\ 0 & 0 & 0 \\ 0 & 0 & 1 \end{bmatrix} \begin{bmatrix} \ddot{R}^i_x \\ \ddot{R}^i_y \\ \ddot{\theta}^i_z \end{bmatrix} = \mathbf{B}_{di}\ddot{\mathbf{q}}_i$$

where

$$\mathbf{B}_{di} = \begin{bmatrix} 1 & 0 & 0 \\ 0 & 1 & 0 \\ 0 & 0 & 0 \\ 0 & 0 & 0 \\ 0 & 0 & 0 \\ 0 & 0 & 1 \end{bmatrix}, \quad \ddot{\mathbf{q}}_i = \begin{bmatrix} \ddot{R}^i_x \\ \ddot{R}^i_y \\ \ddot{\theta}^i_z \end{bmatrix}$$

One can show that, when Euler angles are used, the acceleration transformation considered in this example is the result of imposing the three motion constraints $R^i_z = 0$, $\phi^i = 0$, and $\theta^i = 0$. In the case of planar motion, the angular velocity vector is given by $\overline{\boldsymbol{\omega}}^i = \begin{bmatrix} 0 & 0 & \dot{\theta}^i_z \end{bmatrix}^T$. Substituting for $\ddot{\mathbf{R}}^i$ and $\overline{\boldsymbol{\alpha}}^i$ into Eq. 47, and premultiplying by the transpose of matrix $\mathbf{B}_{di}$, one obtains the planar Newton–Euler equations defined as

$$\begin{bmatrix} m^i & 0 & 0 \\ 0 & m^i & 0 \\ 0 & 0 & i^i_{zz} \end{bmatrix} \begin{bmatrix} \ddot{R}^i_x \\ \ddot{R}^i_y \\ \ddot{\theta}^i_z \end{bmatrix} = \begin{bmatrix} Q^i_{ex} \\ Q^i_{ey} \\ M^i_z \end{bmatrix}$$

In this equation, Q^i_{ex} and Q^i_{ey} are the components of the resultant force acting on the body, and M^i_z is the resultant moment about the Z axis. It is clear from the previous equation that there is no gyroscopic moment in the case of planar motion since the rotation is about a fixed axis.

Euler Parameters The generalized form of the Newton–Euler equations of motion is used, for the most part, in this book. This form allows for using the same form of the constraint Jacobian matrix $\mathbf{C_q}$ for the position, velocity, and acceleration analyses. Equation 45 can be written in the case of unconstrained motion for rigid body i as

$$\mathbf{M}^i\ddot{\mathbf{q}}^i = \mathbf{Q}^i_e + \mathbf{Q}^i_v, \quad i = 1, 2, \ldots, n_b \tag{6.49}$$

where n_b is the total number of bodies in the system and

$$\left.\begin{aligned} &\mathbf{M}^i = \begin{bmatrix} m^i\mathbf{I} & \mathbf{0} \\ \mathbf{0} & \left(\overline{\mathbf{G}}^{i^T}\overline{\mathbf{I}}^i_{\theta\theta}\overline{\mathbf{G}}^i\right) \end{bmatrix}, \quad \ddot{\mathbf{q}}^i = \begin{bmatrix} \ddot{\mathbf{R}}^i \\ \ddot{\mathbf{\theta}}^i \end{bmatrix}, \quad \mathbf{Q}^i_e = \begin{bmatrix} \left(\mathbf{Q}^i_e\right)_R \\ \left(\mathbf{Q}^i_e\right)_\theta \end{bmatrix}, \\ &\mathbf{Q}^i_v = \begin{bmatrix} \mathbf{0} \\ -\overline{\mathbf{G}}^{i^T}\left(\overline{\boldsymbol{\omega}}^i \times \left(\overline{\mathbf{I}}_{\theta\theta}\,{}^i\overline{\boldsymbol{\omega}}^i\right) + \overline{\mathbf{I}}^i_{\theta\theta}\left(\dot{\overline{\mathbf{G}}}^i\dot{\mathbf{\theta}}^i\right)\right) \end{bmatrix} \end{aligned}\right\} \tag{6.50}$$

It is clear from the preceding two equations that the system inertia force vector $\mathbf{Q}_i$ of Eq. 17 or 18 can be written in terms of the inertia forces of the bodies as $\mathbf{Q}_i = \left[\mathbf{Q}^{1^T}_i \; \mathbf{Q}^{2^T}_i \; \ldots \; \mathbf{Q}^{n_b{}^T}_i\right]^T$, where

$$\mathbf{Q}^i_i = \begin{bmatrix} \left(\mathbf{Q}^i_i\right)_R \\ \left(\mathbf{Q}^i_i\right)_\theta \end{bmatrix} = \begin{bmatrix} m^i\ddot{\mathbf{R}}^i \\ \left(\overline{\mathbf{G}}^{i^T}\overline{\mathbf{I}}^i_{\theta\theta}\overline{\mathbf{G}}^i\right)\ddot{\mathbf{\theta}}^i + \overline{\mathbf{G}}^{i^T}\left(\overline{\boldsymbol{\omega}}^i \times \left(\overline{\mathbf{I}}^i_{\theta\theta}\overline{\boldsymbol{\omega}}^i\right) + \overline{\mathbf{I}}^i_{\theta\theta}\left(\dot{\overline{\mathbf{G}}}^i\dot{\mathbf{\theta}}^i\right)\right) \end{bmatrix},$$

$$i = 1, 2, \ldots, n_b \tag{6.51}$$

If Euler parameters are used as the orientation coordinates, $\dot{\overline{\mathbf{G}}}^i\dot{\mathbf{\theta}}^i = \mathbf{0}$, and the preceding equation reduces to

$$\mathbf{Q}^i_i = \begin{bmatrix} \left(\mathbf{Q}^i_i\right)_R \\ \left(\mathbf{Q}^i_i\right)_\theta \end{bmatrix} = \begin{bmatrix} m^i\ddot{\mathbf{R}}^i \\ \left(\overline{\mathbf{G}}^{i^T}\overline{\mathbf{I}}^i_{\theta\theta}\overline{\mathbf{G}}^i\right)\ddot{\mathbf{\theta}}^i + \overline{\mathbf{G}}^{i^T}\left(\overline{\boldsymbol{\omega}}^i \times \left(\overline{\mathbf{I}}^i_{\theta\theta}\overline{\boldsymbol{\omega}}^i\right)\right) \end{bmatrix}, \quad i = 1, 2, \ldots, n_b \tag{6.52}$$

The simplification resulting from using Euler parameters comes at the expense of adding an algebraic constraint equation for each body in the system, as discussed in Chapter 3. Nonetheless, using Euler parameters is recommended for developing general MBS railroad vehicle algorithms because of the singularity problem associated with any three-parameter representation. Equation 49 can be used with Eq. 17 as the basis for developing two different procedures for formulating the equations of motion of complex railroad vehicle systems subject to kinematic constraints, as discussed in the following section.

6.7 AUGMENTED FORMULATION AND EMBEDDING TECHNIQUE

It was shown previously (Eq. 17) that the equations of motion of mechanical systems such as railroad vehicle systems can be written as $\mathbf{Q}_i - \mathbf{Q}_e + \mathbf{C}_\mathbf{q}^T\boldsymbol{\lambda} = \mathbf{0}$, subject to the kinematic constraint equations $\mathbf{C}(\mathbf{q}, t) = \begin{bmatrix} C_1 & C_2 & \dots & C_{n_c} \end{bmatrix}^T = \mathbf{0}$. In this equation of motion, $-\mathbf{C}_\mathbf{q}^T\boldsymbol{\lambda}$ defines the generalized constraint forces. The vector of the inertia forces can always be written as $\mathbf{Q}_i = \mathbf{M}\ddot{\mathbf{q}} - \mathbf{Q}_v$, where $\mathbf{M}$ is the system mass matrix, $\mathbf{q}$ is the vector of the system coordinates, and $\mathbf{Q}_v$ is the vector of inertia forces that is quadratic in the velocities. Based on the development presented in the preceding section, one can write the system mass matrix $\mathbf{M}$ and the vectors $\mathbf{q}$ and $\mathbf{Q}_v$ as

$$\mathbf{M} = \begin{bmatrix} \mathbf{M}^1 & \mathbf{0} & \cdots & \mathbf{0} \\ \mathbf{0} & \mathbf{M}^2 & \cdots & \mathbf{0} \\ \vdots & \vdots & \ddots & \vdots \\ \mathbf{0} & \mathbf{0} & \cdots & \mathbf{M}^{n_b} \end{bmatrix}, \quad \mathbf{q} = \begin{bmatrix} \mathbf{q}^1 \\ \mathbf{q}^2 \\ \vdots \\ \mathbf{q}^{n_b} \end{bmatrix}, \quad \mathbf{Q}_v = \begin{bmatrix} \mathbf{Q}_v^1 \\ \mathbf{Q}_v^2 \\ \vdots \\ \mathbf{Q}_v^{n_b} \end{bmatrix} \tag{6.53}$$

Therefore, the system equations of motion $\mathbf{Q}_i - \mathbf{Q}_e + \mathbf{C}_\mathbf{q}^T\boldsymbol{\lambda} = \mathbf{0}$ can be written as $\mathbf{M}\ddot{\mathbf{q}} + \mathbf{C}_\mathbf{q}^T\boldsymbol{\lambda} = \mathbf{Q}_e + \mathbf{Q}_v$. For a given set of forces and moments that define vector $\mathbf{Q}_e$, the n scalar equations that form the matrix equation of motion $\mathbf{M}\ddot{\mathbf{q}} + \mathbf{C}_\mathbf{q}^T\boldsymbol{\lambda} = \mathbf{Q}_e + \mathbf{Q}_v$ have n unknown accelerations that form vector $\ddot{\mathbf{q}}$ plus n_c unknown Lagrange multipliers that form vector $\boldsymbol{\lambda}$. Therefore, n_c additional equations are needed in order to determine the $n + n_c$ unknowns. These additional equations are the kinematic constraint functions defined by vector $\mathbf{C}(\mathbf{q}, t) = \begin{bmatrix} C_1 & C_2 & \dots & C_{n_c} \end{bmatrix}^T = \mathbf{0}$. Differentiating this vector twice with respect to time, one obtains two types of vectors: one is linear in acceleration, and the other absorbs all other terms that are not linear in acceleration, including terms that are quadratic in velocities. Therefore, one can always define the constraint functions at the *acceleration level* as $\mathbf{C}_\mathbf{q}\ddot{\mathbf{q}} = \mathbf{Q}_d$, where $\mathbf{C}_\mathbf{q} = \partial\mathbf{C}/\partial\mathbf{q}$ is the constraint Jacobian matrix and $\mathbf{Q}_d$ is the vector that absorbs terms that are not linear in accelerations.

The two equations $\mathbf{M}\ddot{\mathbf{q}} + \mathbf{C}_\mathbf{q}^T\boldsymbol{\lambda} = \mathbf{Q}_e + \mathbf{Q}_v$ and $\mathbf{C}_\mathbf{q}\ddot{\mathbf{q}} = \mathbf{Q}_d$ represent the foundation for developing two computational procedures that are discussed in this chapter. These two equations are reproduced here for convenience:

$$\left.\begin{aligned} \mathbf{M}\ddot{\mathbf{q}} + \mathbf{C}_\mathbf{q}^T\boldsymbol{\lambda} &= \mathbf{Q}_e + \mathbf{Q}_v \\ \mathbf{C}_\mathbf{q}\ddot{\mathbf{q}} &= \mathbf{Q}_d \end{aligned}\right\} \tag{6.54}$$

In the first approach, referred to as the *augmented formulation*, redundant coordinates are used to obtain a system of equations written explicitly in terms of constraint forces. In the second approach, referred to as the *embedding technique*, constraint equations are used to systematically eliminate dependent variables, leading to a system of equations from which the constraint forces are eliminated.

Example 6.4 In Example 1, the three constraint equations of a spherical joint between two arbitrary bodies i and j at points P^i and P^j, respectively, were written as

$$\mathbf{C}\left(\mathbf{q}^i, \mathbf{q}^j\right) = \mathbf{r}_P^i - \mathbf{r}_P^j = \mathbf{R}^i + \mathbf{A}^i\bar{\mathbf{u}}_P^i - \mathbf{R}^j - \mathbf{A}^j\bar{\mathbf{u}}_P^j = \mathbf{0}$$

where $\mathbf{r}_P^k, \mathbf{R}^k, \mathbf{A}^k$ and $\bar{\mathbf{u}}_P^k,\ k = i,j$, are, respectively, the global position vector of the joint definition point P^k, the global position vector of the reference point of body k, the transformation matrix that defines the body orientation, and the local position vector of point P^k with respect to the body coordinate system. Vector $\mathbf{q}^k$ is the vector of absolute Cartesian coordinates $\mathbf{q}^k = \left[\mathbf{R}^{k^T} \quad \boldsymbol{\theta}^{k^T}\right]^T$. Differentiating the constraint equations of the spherical joint with respect to time, one obtains the constraint equations at the velocity level as

$$\dot{\mathbf{C}}\left(\mathbf{q}^i, \mathbf{q}^j\right) = \dot{\mathbf{r}}_P^i - \dot{\mathbf{r}}_P^j = \dot{\mathbf{R}}^i + \boldsymbol{\omega}^i \times \mathbf{u}_P^i - \dot{\mathbf{R}}^j - \boldsymbol{\omega}^j \times \mathbf{u}_P^j = \mathbf{0}$$

where $\mathbf{u}_P^i = \mathbf{A}^i\bar{\mathbf{u}}_P^i$, and $\boldsymbol{\omega}^i = \mathbf{G}^i\dot{\boldsymbol{\theta}}^i$ is the angular velocity vector. Differentiating the preceding equation with respect to time, one obtains the spherical-joint constraint equations at the acceleration level as

$$\begin{aligned}\ddot{\mathbf{C}}\left(\mathbf{q}^i, \mathbf{q}^j\right) &= \ddot{\mathbf{r}}_P^i - \ddot{\mathbf{r}}_P^j \\ &= \ddot{\mathbf{R}}^i + \boldsymbol{\alpha}^i \times \mathbf{u}_P^i + \boldsymbol{\omega}^i \times \left(\boldsymbol{\omega}^i \times \mathbf{u}_P^i\right) - \ddot{\mathbf{R}}^j - \boldsymbol{\alpha}^j \times \mathbf{u}_P^j - \boldsymbol{\omega}^j \times \left(\boldsymbol{\omega}^j \times \mathbf{u}_P^j\right) = \mathbf{0}\end{aligned}$$

In this equation, $\boldsymbol{\alpha}^k = \dot{\boldsymbol{\omega}}^k = \mathbf{G}^k\ddot{\boldsymbol{\theta}}^k + \dot{\mathbf{G}}^k\dot{\boldsymbol{\theta}}^k,\ k = i,j$. Therefore, the preceding equation can be written as

$$\begin{aligned}\ddot{\mathbf{C}}\left(\mathbf{q}^i, \mathbf{q}^j\right) &= \ddot{\mathbf{R}}^i - \tilde{\mathbf{u}}_P^i\boldsymbol{\alpha}^i + \boldsymbol{\omega}^i \times \left(\boldsymbol{\omega}^i \times \mathbf{u}_P^i\right) - \ddot{\mathbf{R}}^j + \tilde{\mathbf{u}}_P^j\boldsymbol{\alpha}^j - \boldsymbol{\omega}^j \times \left(\boldsymbol{\omega}^j \times \mathbf{u}_P^j\right) \\ &= \ddot{\mathbf{R}}^i - \tilde{\mathbf{u}}_P^i\left(\mathbf{G}^i\ddot{\boldsymbol{\theta}}^i + \dot{\mathbf{G}}^i\dot{\boldsymbol{\theta}}^i\right) + \boldsymbol{\omega}^i \times \left(\boldsymbol{\omega}^i \times \mathbf{u}_P^i\right) - \ddot{\mathbf{R}}^j + \tilde{\mathbf{u}}_P^j\left(\mathbf{G}^j\ddot{\boldsymbol{\theta}}^j + \dot{\mathbf{G}}^j\dot{\boldsymbol{\theta}}^j\right) - \boldsymbol{\omega}^j \\ &\quad \times \left(\boldsymbol{\omega}^j \times \mathbf{u}_P^j\right) = \mathbf{0}\end{aligned}$$

This equation can be rearranged and rewritten as

$$\begin{aligned}\ddot{\mathbf{C}}\left(\mathbf{q}^i, \mathbf{q}^j\right) &= \ddot{\mathbf{R}}^i - \tilde{\mathbf{u}}_P^i\mathbf{G}^i\ddot{\boldsymbol{\theta}}^i - \ddot{\mathbf{R}}^j + \tilde{\mathbf{u}}_P^j\mathbf{G}^j\ddot{\boldsymbol{\theta}}^i \\ &\quad + \boldsymbol{\omega}^i \times \left(\boldsymbol{\omega}^i \times \mathbf{u}_P^i\right) - \tilde{\mathbf{u}}_P^i\dot{\mathbf{G}}^i\dot{\boldsymbol{\theta}}^i - \boldsymbol{\omega}^j \times \left(\boldsymbol{\omega}^j \times \mathbf{u}_P^j\right) + \tilde{\mathbf{u}}_P^j\dot{\mathbf{G}}^j\dot{\boldsymbol{\theta}}^j = \mathbf{0}\end{aligned}$$

It is clear that this equation has two types of terms: the first is linear in accelerations, and the second absorbs all other terms that are not linear in accelerations. Therefore, the preceding equation can be written as

$$\ddot{\mathbf{C}}\left(\mathbf{q}^i, \mathbf{q}^j\right) = \ddot{\mathbf{R}}^i - \tilde{\mathbf{u}}_P^i\mathbf{G}^i\ddot{\boldsymbol{\theta}}^i - \ddot{\mathbf{R}}^j + \tilde{\mathbf{u}}_P^j\mathbf{G}^j\ddot{\boldsymbol{\theta}}^i - \mathbf{Q}_d = \mathbf{0}$$

(Continued)

where vector $\mathbf{Q}_d$ that absorbs terms that are not linear in the acceleration is recognized as

$$\mathbf{Q}_d = -\left(\boldsymbol{\omega}^i \times (\boldsymbol{\omega}^i \times \mathbf{u}_P^i) - \tilde{\mathbf{u}}_P^i \dot{\mathbf{G}}^i \dot{\boldsymbol{\theta}}^i - \boldsymbol{\omega}^j \times \left(\boldsymbol{\omega}^j \times \mathbf{u}_P^j\right) + \tilde{\mathbf{u}}_P^j \dot{\mathbf{G}}^j \dot{\boldsymbol{\theta}}^j\right)$$

For mechanical joint constraints, which are not explicit functions of time, vector $\mathbf{Q}_d$ is quadratic in velocities. It is also important to recognize that the terms that are linear in accelerations are the same as $\mathbf{C}_\mathbf{q}\ddot{\mathbf{q}}$: that is,

$$\mathbf{C}_\mathbf{q}\ddot{\mathbf{q}} = \ddot{\mathbf{R}}^i - \tilde{\mathbf{u}}_P^i \mathbf{G}^i \ddot{\boldsymbol{\theta}}^i - \ddot{\mathbf{R}}^j + \tilde{\mathbf{u}}_P^j \mathbf{G}^j \ddot{\boldsymbol{\theta}}^i$$

where $\mathbf{q} = \begin{bmatrix}\mathbf{q}^{i^T} & \mathbf{q}^{j^T}\end{bmatrix}^T, \mathbf{q}^k = \begin{bmatrix}\mathbf{R}^{k^T} & \boldsymbol{\theta}^{k^T}\end{bmatrix}^T,\ k = i, j$, and the constraint Jacobian matrix $\mathbf{C}_\mathbf{q}$ was obtained in Example 1 as

$$\begin{aligned}\mathbf{C}_\mathbf{q} &= \begin{bmatrix}\dfrac{\partial \mathbf{C}}{\partial \mathbf{q}^i} & \dfrac{\partial \mathbf{C}}{\partial \mathbf{q}^j}\end{bmatrix} = \begin{bmatrix}\dfrac{\partial \mathbf{C}}{\partial \mathbf{R}^i} & \dfrac{\partial \mathbf{C}}{\partial \boldsymbol{\theta}^i} & \dfrac{\partial \mathbf{C}}{\partial \mathbf{R}^j} & \dfrac{\partial \mathbf{C}}{\partial \boldsymbol{\theta}^j}\end{bmatrix} \\ &= \begin{bmatrix}\mathbf{I} & -\mathbf{A}^i \tilde{\overline{\mathbf{u}}}_P^i \overline{\mathbf{G}}^i & -\mathbf{I} & \mathbf{A}^j \tilde{\overline{\mathbf{u}}}_P^j \overline{\mathbf{G}}^j\end{bmatrix} \\ &= \begin{bmatrix}\mathbf{I} & -\tilde{\mathbf{u}}_P^i \mathbf{G}^i & -\mathbf{I} & \tilde{\mathbf{u}}_P^j \mathbf{G}^j\end{bmatrix}\end{aligned}$$

If Euler parameters, instead of Euler angles, are used, one has $\dot{\mathbf{G}}^i \dot{\boldsymbol{\theta}}^i = \dot{\mathbf{G}}^j \dot{\boldsymbol{\theta}}^j = \mathbf{0}$, and vector $\mathbf{Q}_d$ reduces to $\mathbf{Q}_d = -\left(\boldsymbol{\omega}^i \times (\boldsymbol{\omega}^i \times \mathbf{u}_P^i) - \boldsymbol{\omega}^j \times \left(\boldsymbol{\omega}^j \times \mathbf{u}_P^j\right)\right)$.

Augmented Formulation In the augmented formulation, the two equations in Eq. 54 are combined to form one matrix equation as

$$\begin{bmatrix}\mathbf{M} & \mathbf{C}_\mathbf{q}^T \\ \mathbf{C}_\mathbf{q} & \mathbf{0}\end{bmatrix}\begin{bmatrix}\ddot{\mathbf{q}} \\ \boldsymbol{\lambda}\end{bmatrix} = \begin{bmatrix}\mathbf{Q}_e + \mathbf{Q}_v \\ \mathbf{Q}_d\end{bmatrix} \tag{6.55}$$

For given initial coordinates and velocities, the coefficient matrix and the right-hand side of this equation can be constructed. Therefore, Eq. 55 can be solved for the acceleration vector $\ddot{\mathbf{q}}$ and the vector of Lagrange multipliers $\boldsymbol{\lambda}$. The vector of Lagrange multipliers $\boldsymbol{\lambda}$ can be used to determine the generalized constraint forces $-\mathbf{C}_\mathbf{q}^T\boldsymbol{\lambda}$, while the independent components of the acceleration vector $\ddot{\mathbf{q}}_i$ can be integrated to determine the independent coordinates and velocities. The dependent coordinates $\mathbf{q}_d$ can be determined using the constraint equations $\mathbf{C}(\mathbf{q}, t) = \begin{bmatrix}C_1 & C_2 & \dots & C_{n_c}\end{bmatrix}^T = \mathbf{0}$ with the assumption that the independent coordinates $\mathbf{q}_i$ are known from the results of the numerical integration. Therefore, the system of the n_c nonlinear constraint equations $\mathbf{C}(\mathbf{q}, t) = \begin{bmatrix}C_1 & C_2 & \dots & C_{n_c}\end{bmatrix}^T = \mathbf{0}$ becomes a function of only n_c unknown coordinates. This nonlinear system of equations can be solved using an iterative Newton–Raphson procedure. In this iterative procedure, one constructs at each iteration k the system of equations $\left(\mathbf{C}_{\mathbf{q}_d}\right)_k (\Delta\mathbf{q}_d)_k = -(\mathbf{C})_k$, where $\mathbf{C}_{\mathbf{q}_d} = \partial\mathbf{C}/\partial\mathbf{q}_d$ is the $n_c \times n_c$ constraint Jacobian matrix associated with the dependent coordinates, and $\Delta\mathbf{q}_d$ is the vector of *Newton differences* (Atkinson 1978). Because the constraint equations are assumed to be linearly independent, the constraint Jacobian matrix $\mathbf{C}_{\mathbf{q}_d} = \partial\mathbf{C}/\partial\mathbf{q}_d$ is assumed to have full rank, and therefore, a solution of the equation $\left(\mathbf{C}_{\mathbf{q}_d}\right)_k (\Delta\mathbf{q}_d)_k = -(\mathbf{C})_k$ for the Newton differences $\Delta\mathbf{q}_d$ can be obtained. At the end of each iteration, the coordinates are updated according to the equation $(\mathbf{q}_d)_{k+1} = (\mathbf{q}_d)_k + (\Delta \mathbf{q}_d)_k$. Convergence is achieved if the norm of the Newton differences or the norm of the constraint equations becomes smaller than a specified tolerance ε: that is, $|\Delta \mathbf{q}_d| \leq \varepsilon$ or $|\mathbf{C}| \leq \varepsilon$.

Having determined the coordinates using the iterative procedure and constraint equations at the *position level* $\mathbf{C}(\mathbf{q}, t) = \begin{bmatrix} C_1 & C_2 & \ldots & C_{n_c} \end{bmatrix}^T = \mathbf{0}$, the constraint equations at the *velocity level* can be defined as $\mathbf{C}_\mathbf{q}\dot{\mathbf{q}} + \mathbf{C}_t = \mathbf{0}$, where $\mathbf{C}_t = \partial \mathbf{C}/\partial t$ is the partial derivative of the constraint equations with respect to time; this vector is the zero vector if the constraint equations are not explicit functions of time. If the vector of independent velocities $\dot{\mathbf{q}}_i$ is known from the numerical integration, the constraint equations at the velocity level can be written as $\mathbf{C}_{\mathbf{q}_d}\dot{\mathbf{q}}_d + \mathbf{C}_{\mathbf{q}_i}\dot{\mathbf{q}}_i + \mathbf{C}_t = \mathbf{0}$. This equation can be arranged and written as

$$\mathbf{C}_{\mathbf{q}_d}\dot{\mathbf{q}}_d = -\left(\mathbf{C}_{\mathbf{q}_i}\dot{\mathbf{q}}_i + \mathbf{C}_t\right) \tag{6.56}$$

This is a linear system of equations in the dependent velocities $\dot{\mathbf{q}}_d$. Because the Jacobian matrix $\mathbf{C}_{\mathbf{q}_d}$ associated with the dependent velocities is non-singular, and because of the linearity of these equations, the system of Eq. 56 can be solved for the dependent velocities $\dot{\mathbf{q}}_d$ without the need to use an iterative procedure required the position analysis step. Once all the coordinates and velocities (independent and dependent) are determined, Eq. 55 can be constructed and solved for all the accelerations, which are used to advance the numerical integration.

Example 6.5 To explain some of the concepts used in the augmented formulation, a wheel/rail example is considered (Shabana et al. 2008). Figure 4 shows a wheel with radius r, mass m^w, and mass moment of inertia J^w about its center of mass. The wheel is assumed to roll without sliding on a circular rail, which is assumed to be fixed and to have a constant radius of curvature R. The external forces acting on the wheel are defined by the resultant force vector $\mathbf{F}^w = \begin{bmatrix} F_x^w & F_y^w \end{bmatrix}^T$ and the moment M^w. Because of the rolling assumption and because the rail is assumed to be fixed, there is no friction force between the wheel and rail, and the system has one degree of freedom. Using Figure 4, the tangent and normal vectors at the wheel/rail contact point can be written as

$$\mathbf{t}^r = \begin{bmatrix} -\cos\phi \\ \sin\phi \end{bmatrix}, \qquad \mathbf{n}^r = \begin{bmatrix} -\sin\phi \\ -\cos\phi \end{bmatrix}$$

The absolute velocity of the contact point on the wheel is defined as $\dot{\mathbf{r}}_c^w = \dot{\mathbf{R}}^w + \boldsymbol{\omega}^w \times \mathbf{u}_c^w$, where $\boldsymbol{\omega}^w = \begin{bmatrix} 0 & 0 & \dot{\theta}^w \end{bmatrix}^T$, $\mathbf{u}_c^w = \begin{bmatrix} -r\sin\phi & -r\cos\phi & 0 \end{bmatrix}^T$, and $\dot{\theta}^w$ is the angular velocity of the wheel. Because of the rolling and no-slipping conditions, the absolute velocity vector at the contact point is equal to zero, leading to the equation

$$\dot{\mathbf{r}}_c^w = \dot{\mathbf{R}}^w + \boldsymbol{\omega}^w \times \mathbf{u}_c^w = \begin{bmatrix} \dot{R}_x^w + r\dot{\theta}^w\cos\phi \\ \dot{R}_y^w - r\dot{\theta}^w\sin\phi \end{bmatrix} = \mathbf{0}$$

These two scalar constraint equations, which ensure that the wheel rolls without slipping, eliminate two dependent velocities; consequently, the system has only one independent velocity. The rolling conditions at the acceleration level can be written as $\ddot{\mathbf{r}}_c^w = \mathbf{0}$.

(Continued)

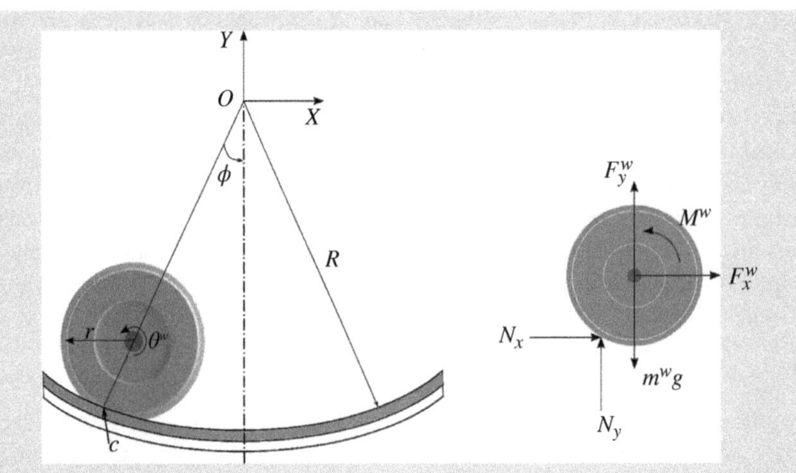

Figure 6.4 Wheel/rail example.

Using the geometry of the system shown in Figure 4, one has $R_x^w = -(R-r)\sin\phi$, $R_y^w = -(R-r)\cos\phi$, which, upon substitution in the preceding equation that defines $\dot{\mathbf{r}}_c^w$, gives the algebraic relationship between $\dot{\theta}^w$ and $\dot{\phi}$ as $r\dot{\theta}^w = (R-r)\dot{\phi}$. That is, $\dot{\phi}$ can be determined if $\dot{\theta}^w$ is known.

As previously mentioned, the Lagrangian approach does not require using free-body diagrams since the connectivity (constraint) conditions are used to define the constraint forces. To understand the relationship between Newtonian and Lagrangian mechanics, the free-body diagram shown in Figure 4, which is not required in Lagrangian dynamics, is first used. Using this free-body diagram and the Newtonian approach, one can write the three Newton–Euler equations for the planar wheel as $\mathbf{M}^w\ddot{\mathbf{q}}^w = \mathbf{Q}_e^w + \mathbf{Q}_c^w$, where $\mathbf{q}^w = \begin{bmatrix} R_x^w & R_y^w & \theta^w \end{bmatrix}^T$,

$$\mathbf{M}^w = \begin{bmatrix} m^w & 0 & 0 \\ 0 & m^w & 0 \\ 0 & 0 & J^w \end{bmatrix}, \quad \mathbf{Q}_e^w = \begin{bmatrix} F_x^w \\ F_y^w - m^w g \\ M^w \end{bmatrix}, \quad \mathbf{Q}_c^w = \begin{bmatrix} N_x \\ N_y \\ N_x r\cos\phi - N_y r\sin\phi \end{bmatrix}$$

$\mathbf{N} = \begin{bmatrix} N_x & N_y \end{bmatrix}^T$ is the vector of reaction forces at the wheel/rail contact point, and g is the gravity constant. The matrix equation of motion $\mathbf{M}^w\ddot{\mathbf{q}}^w = \mathbf{Q}_e^w + \mathbf{Q}_c^w$ has three scalar equations and the five unknowns $\ddot{R}_x^w, \ddot{R}_y^w, \ddot{\theta}^w, N_x$, and N_y. Therefore, the two constraint equations that define the rolling conditions are needed in order to have a number of equations equal to the number of unknowns. These two sets of equations can be written as

$$\left.\begin{aligned} &\mathbf{M}^w\ddot{\mathbf{q}}^w = \mathbf{Q}_e^w + \mathbf{Q}_c^w, \\ &\ddot{\mathbf{r}}_c^w = \mathbf{0} \end{aligned}\right\}$$

Therefore, when Newtonian mechanics is used, the algebraic constraint equations are also needed to solve for the unknown accelerations and constraint forces. The rolling constraint equations at the acceleration level can be written as

$$\ddot{\mathbf{C}} = \ddot{\mathbf{r}}_c^w = \mathbf{C}_\mathbf{q}\ddot{\mathbf{q}} - \mathbf{Q}_d = \begin{bmatrix} \ddot{R}_x^w + r\ddot{\theta}^w\cos\phi - r\dot{\theta}\dot{\phi}\sin\phi \\ \ddot{R}_y^w - r\ddot{\theta}^w\sin\phi - r\dot{\theta}\dot{\phi}\cos\phi \end{bmatrix} = \mathbf{0}$$

Combining this equation with the equations of motion, one obtains the augmented Lagrangian form of the equations of motion, which does not in general require using the free-body diagram as (Shabana et al. 2008)

$$\begin{bmatrix} m^w & 0 & 0 & -1 & 0 \\ 0 & m^w & 0 & 0 & -1 \\ 0 & 0 & J^w & -r\cos\phi & r\sin\phi \\ 1 & 0 & r\cos\phi & 0 & 0 \\ 0 & 1 & -r\sin\phi & 0 & 0 \end{bmatrix} \begin{bmatrix} \ddot{R}_x^w \\ \ddot{R}_y^w \\ \ddot{\theta}^w \\ N_x \\ N_y \end{bmatrix} = \begin{bmatrix} F_x^w \\ F_y^w - m^w g \\ M^w \\ r\dot{\theta}\dot{\phi}\sin\phi \\ r\dot{\theta}\dot{\phi}\cos\phi \end{bmatrix}$$

This equation can be written in the form of Eq. 55 as

$$\begin{bmatrix} \mathbf{M} & \mathbf{C}_\mathbf{q}^T \\ \mathbf{C}_\mathbf{q} & \mathbf{0} \end{bmatrix} \begin{bmatrix} \ddot{\mathbf{q}} \\ \boldsymbol{\lambda} \end{bmatrix} = \begin{bmatrix} \mathbf{Q}_e + \mathbf{Q}_v \\ \mathbf{Q}_d \end{bmatrix}$$

where, in this planar example, $\mathbf{Q}_v = \mathbf{0}, \mathbf{M} = \mathbf{M}^w, \mathbf{q} = \mathbf{q}^w$, and $\mathbf{Q}_e = \mathbf{Q}_e^w$ (previously defined in this example); and

$$\mathbf{C}_\mathbf{q} = \begin{bmatrix} 1 & 0 & r\cos\phi \\ 0 & 1 & -r\sin\phi \end{bmatrix}, \quad \mathbf{Q}_d = \begin{bmatrix} r\dot{\theta}\dot{\phi}\sin\phi \\ r\dot{\theta}\dot{\phi}\cos\phi \end{bmatrix}, \quad \boldsymbol{\lambda} = -\begin{bmatrix} N_x \\ N_y \end{bmatrix}$$

It is clear from this simple example that the vector of Lagrange multipliers $\boldsymbol{\lambda}$ is the negative of the reaction forces.

Identification of the System Degrees of Freedom Because of the complexity of railroad vehicle systems, independent and dependent coordinates are identified numerically using the $n_c \times n$ constraint Jacobian matrix $\mathbf{C}_\mathbf{q}$. This can be achieved using a Gaussian elimination process with full pivoting (Atkinson 1978; Wehage 1980). Using the Gaussian elimination procedure, matrix $\mathbf{C}_\mathbf{q}$ can be converted to a matrix in the form $[\mathbf{I}_{n_c \times n_c} \ \ \bar{\mathbf{I}}_{n_c \times n_d}]$, where $\mathbf{I}_{n_c \times n_c}$ is the $n_c \times n_c$ identity matrix, $\bar{\mathbf{I}}_{n_c \times n_d}$ is an $n_c \times n_d$ matrix that results from the Gaussian eliminate steps, and n_d is the number of independent coordinates. The coordinates associated with the columns of the square non-singular identity matrix $\mathbf{I}_{n_c \times n_c}$ are selected as the dependent coordinates $\mathbf{q}_d$, while the coordinates associated with the non-square matrix $\bar{\mathbf{I}}_{n_c \times n_d}$ are assumed to be the independent coordinates $\mathbf{q}_i$. The choice of the dependent and independent coordinates using this numerical procedure ensures that the dependent variables can always be determined using the kinematic constraint equations if the independent variables are known. In some mechanical system applications, it is sufficient to identify the set of independent coordinates (degrees of freedom) only once at the beginning of the simulation and use this set for the entire simulation time. In some other systems, particularly in the case of closed chains, the set of independent coordinates must be continuously changed in order to avoid singular configurations. While the number of degrees of freedom remains the same for a given system topology, the set of degrees of freedom is not unique. This is also clear from the fact that the constraint Jacobian matrix $\mathbf{C}_{\mathbf{q}}(\mathbf{q}, t)$ is a nonlinear function of the system coordinates; and as the system configuration changes with time, the Gaussian procedure can lead to different sets of dependent and independent coordinates.

Efficient Implementation of the Augmented Formulation In the computer implementation of the augmented formulation, partitioning the constraint Jacobian matrix as $\mathbf{C}_{\mathbf{q}} = \begin{bmatrix}\mathbf{C}_{\mathbf{q}_d} & \mathbf{C}_{\mathbf{q}_i}\end{bmatrix}$ is not necessary. Because most joints and constraints involve no more than two bodies in the system, the constraint Jacobian matrix $\mathbf{C}_{\mathbf{q}}$ is a sparse matrix that has a large number of zeros. This allows for exploiting sparse-matrix techniques to obtain efficient solutions for the position coordinates, velocities, and accelerations (Duff et al. 1986). Instead of partitioning the constraint Jacobian matrix, one can form a larger sparse matrix whose entries ensure that the Newton differences and velocities associated with the independent coordinates $\mathbf{q}_i$ remain unchanged. For example, instead of using the equation $\left(\mathbf{C}_{\mathbf{q}_d}\right)_k(\Delta\mathbf{q}_d)_k = -(\mathbf{C})_k$ in the Newton–Raphson iterations to solve for the coordinates, one can use the following equation at iteration k:

$$\begin{bmatrix}\mathbf{C}_{\mathbf{q}} \\ \mathbf{I}_d\end{bmatrix}_k (\Delta\mathbf{q})_k = \begin{bmatrix}-(\mathbf{C})_k \\ \mathbf{0}\end{bmatrix} \tag{6.57}$$

In this equation, $\mathbf{I}_d$ is an $n_d \times n$ Boolean matrix with only zero and one entries, with the ones in locations that ensure that the Newton differences associated with the independent coordinates are equal to zero: that is, $(\Delta\mathbf{q}_i)_k = \mathbf{0}$. Because the independent coordinates are assumed to be known from the numerical integration, they are not allowed to vary during the Newton–Raphson iterations. The coefficient matrix in Eq. 57 is square and sparse with a large number of zeros. The structure of this matrix does not change unless the set of independent coordinates is changed. This fact can be utilized to avoid repeated symbolic factorization and scaling at every time step.

Similarly, instead of using Eq. 56 to solve for the velocities, one can combine the two equations $\mathbf{C}_{\mathbf{q}}\dot{\mathbf{q}} = -\mathbf{C}_t$ and $\dot{\mathbf{q}}^i = \dot{\mathbf{q}}^i$ to form the following sparse-matrix equation:

$$\begin{bmatrix}\mathbf{C}_{\mathbf{q}} \\ \mathbf{I}_d\end{bmatrix}\dot{\mathbf{q}} = \begin{bmatrix}-\mathbf{C}_t \\ \dot{\mathbf{q}}_i\end{bmatrix} \tag{6.58}$$

In both the position and velocity analysis steps, the locations of the non-zero entries in the constraint Jacobian matrix $\mathbf{C}_{\mathbf{q}}$ do not change. Furthermore, if the set of independent coordinates remains the same during dynamic simulation, the locations of the non-zero entries of matrix $\mathbf{I}_d$ do not change during dynamic simulations, as previously mentioned. This allows for performing symbolic factorization of matrix $\begin{bmatrix}\mathbf{C}_{\mathbf{q}}^T & \mathbf{I}_d^T\end{bmatrix}^T$ for both the position and velocity analysis steps. This symbolic factorization can also be performed only at points in time when the set of independent coordinates is changed, to obtain an efficient solution for both Eqs. 57 and 58 and avoid the need to partition the constraint Jacobian matrix as $\mathbf{C}_{\mathbf{q}} = \begin{bmatrix}\mathbf{C}_{\mathbf{q}_d} & \mathbf{C}_{\mathbf{q}_i}\end{bmatrix}$.

Similarly, the coefficient matrix in Eq. 55 is a sparse matrix in which the locations of the non-zero entries do not change during dynamic simulations. Therefore, sparse-matrix techniques can also be used to efficiently solve for the accelerations and vector of Lagrange multipliers. It is important to emphasize again that only the independent accelerations $\ddot{\mathbf{q}}_i$ must be integrated forward in time, and vectors $\mathbf{q}_i$ and $\dot{\mathbf{q}}_i$ should not be altered outside the integrator when sophisticated integration methods for solving first-order ordinary differential equations are used (Shampine and Gordon 1975). Vectors $\mathbf{q}_i$ and $\dot{\mathbf{q}}_i$ should be used outside the integrator only to determine dependent coordinates and velocities using the constraint equations at the position and velocity levels, and also to evaluate accelerations. Therefore, the independent coordinates $\mathbf{q}_i$ and independent velocities $\dot{\mathbf{q}}_i$ should not be altered outside the integrator, and the dependent accelerations $\ddot{\mathbf{q}}_d$ should not be integrated if an explicit integration method with a well-designed error-check criterion is used (Shampine and Gordon 1975). This numerical procedure is consistent with the Lagrange–D'Alembert principle in which the equations of motion are formulated in terms of and solved for the independent accelerations, which can be integrated to determine the independent coordinates and velocities. The dependent velocities and accelerations in the Lagrange–D'Alembert principle are determined using the *velocity transformation,* as discussed in the remainder of this section (Shabana 2019).

Embedding Technique The technique of Lagrange multipliers, which is the basis of the augmented formulation, was originally introduced to handle systems subjected to non-holonomic constraints. However, this technique has been used in the MBS literature for both holonomic and non-holonomic systems. Using this approach, the dynamic equations of motion are formulated in terms of redundant coordinates, which are related by algebraic constraint functions. Consequently, constraint forces must appear in the equations of motion of the system.

Another approach that can be used to eliminate constraint forces and obtain a number of equations equal to the number of the system degrees of freedom is the *embedding technique.* In this alternate approach, algebraic constraint functions are used to write dependent variables in terms of independent variables. This leads to the definition of the *velocity transformation matrix*, which allows for systematically eliminating the constraint forces and obtaining a minimum number of equations of motion. The embedding technique has its roots in the *Lagrange–D'Alembert principle* $\left(\mathbf{M}\ddot{\mathbf{q}} - \mathbf{Q}_e - \mathbf{Q}_v + \mathbf{C}_{\mathbf{q}}^T\boldsymbol{\lambda}\right)^T \delta\mathbf{q} = 0$. This equation, as previously discussed in this chapter, is a statement of the virtual work principle $\delta W_i = \delta W_e + \delta W_c$, where $\delta W_i = \left(\mathbf{M}\ddot{\mathbf{q}} - \mathbf{Q}_v\right)^T \delta\mathbf{q}$ is the virtual work of the inertia forces of the system, $\delta W_e = \mathbf{Q}_e^T \delta\mathbf{q}$ is the virtual work of the applied forces, and $\delta W_c = -\left(\mathbf{C}_{\mathbf{q}}^T\boldsymbol{\lambda}\right)^T \delta\mathbf{q}$ is the virtual work of the constraint forces. As discussed before, when the dynamic equilibrium of the system is considered, one has $\delta W_c = -\left(\mathbf{C}_{\mathbf{q}}^T\boldsymbol{\lambda}\right)^T \delta\mathbf{q} = 0$, and the Lagrange–D'Alembert principle can be written as

$$\left(\mathbf{M}\ddot{\mathbf{q}} - \mathbf{Q}_e - \mathbf{Q}_v\right)^T \delta\mathbf{q} = 0 \tag{6.59}$$

For a virtual change in coordinates, the constraint functions lead to

$$\mathbf{C}_{\mathbf{q}}\delta\mathbf{q} = \mathbf{C}_{\mathbf{q}_i}\delta\mathbf{q}_i + \mathbf{C}_{\mathbf{q}_d}\delta\mathbf{q}_d = \mathbf{0} \tag{6.60}$$

This equation allows writing the virtual change in dependent coordinates in terms of the virtual change in independent coordinates as $\delta \mathbf{q}_d = -\left(\mathbf{C}_{\mathbf{q}_d}\right)^{-1} \mathbf{C}_{\mathbf{q}_i} \delta \mathbf{q}_i$, which can be used to write the virtual change in all coordinates in terms of the virtual change of independent coordinates as

$$\delta \mathbf{q} = \begin{bmatrix} \delta \mathbf{q}_i \\ \delta \mathbf{q}_d \end{bmatrix} = \begin{bmatrix} \mathbf{I} \\ -\left(\mathbf{C}_{\mathbf{q}_d}\right)^{-1} \mathbf{C}_{\mathbf{q}_i} \end{bmatrix} \delta \mathbf{q}_i = \mathbf{B}_{di} \delta \mathbf{q}_i \tag{6.61}$$

where $\mathbf{B}_{di}$ is the *velocity transformation matrix* defined as

$$\mathbf{B}_{di} = \begin{bmatrix} \mathbf{I} \\ -\left(\mathbf{C}_{\mathbf{q}_d}\right)^{-1} \mathbf{C}_{\mathbf{q}_i} \end{bmatrix} \tag{6.62}$$

One also has

$$\dot{\mathbf{q}} = \mathbf{B}_{di} \dot{\mathbf{q}}_i, \qquad \ddot{\mathbf{q}} = \mathbf{B}_{di} \ddot{\mathbf{q}}_i + \boldsymbol{\gamma}_i \tag{6.63}$$

where $\boldsymbol{\gamma}_i = \dot{\mathbf{B}}_{di} \dot{\mathbf{q}}_i$. Substituting Eqs. 61 and 63 into Eq. 59, one obtains $\left(\mathbf{M}\left(\mathbf{B}_{di}\ddot{\mathbf{q}} + \boldsymbol{\gamma}_i\right) - \mathbf{Q}_e - \mathbf{Q}_v\right)^T \mathbf{B}_{di} \delta \mathbf{q}_i = 0$. Because the elements of vector $\mathbf{q}_i$ are independent, their coefficients can be set equal to zero, leading to $\mathbf{B}_{di}^T \left(\mathbf{M}\left(\mathbf{B}_{di}\ddot{\mathbf{q}}_i + \boldsymbol{\gamma}_i\right) - \mathbf{Q}_e - \mathbf{Q}_v\right) = \mathbf{0}$, which upon rearranging the terms can be written as $\left(\mathbf{B}_{di}^T \mathbf{M} \mathbf{B}_{di}\right) \ddot{\mathbf{q}}_i = \mathbf{B}_{di}^T \left(\mathbf{Q}_e + \mathbf{Q}_v - \mathbf{M}\boldsymbol{\gamma}_i\right)$. This equation can be written as

$$\mathbf{M}_{ii} \ddot{\mathbf{q}}_i = \mathbf{B}_{di}^T \left(\mathbf{Q}_e + \mathbf{Q}_v - \mathbf{M}\boldsymbol{\gamma}_i\right) \tag{6.64}$$

where $\mathbf{M}_{ii} = \mathbf{B}_{di}^T \mathbf{M} \mathbf{B}_{di}$ is the generalized mass matrix associated with the independent coordinates. The number of equations in the preceding matrix equation is equal to the number of the system degrees of freedom n_d. Furthermore, no constraint forces appear in Eq. 64 because this equation is formulated in terms of the independent coordinates. In fact, one can show that $\mathbf{B}_{di}^T \mathbf{C}_{\mathbf{q}}^T \boldsymbol{\lambda} = \mathbf{0}$. This is clear from the definition of the velocity transformation matrix in Eq. 62, which leads to

$$\mathbf{C}_{\mathbf{q}} \mathbf{B}_{di} = \begin{bmatrix} \mathbf{C}_{\mathbf{q}_i} & \mathbf{C}_{\mathbf{q}_d} \end{bmatrix} \begin{bmatrix} \mathbf{I} \\ -\left(\mathbf{C}_{\mathbf{q}_d}\right)^{-1} \mathbf{C}_{\mathbf{q}_i} \end{bmatrix} = \mathbf{0} \tag{6.65}$$

That is, the product of the transpose of the velocity transformation matrix $\mathbf{B}_{di}$ and the constraint forces is always equal to zero.

Equation 64 is an alternate formulation that leads to a smaller number of equations as compared to the augmented formulation. However, Eq. 64 has a dense and highly nonlinear mass matrix $\mathbf{M}_{ii}$, and therefore, using sparse-matrix techniques with algorithms based on the embedding technique does not lead to significant computational savings. Nonetheless, when the independent coordinates $\mathbf{q}_i$ are selected to be joint variables for open-chain systems, one can write the system coordinates in terms of the independent coordinates using closed-form expressions. In this case of open-chain systems, using the Newton–Raphson iterative procedure at the position level to determine dependent coordinates can be avoided.

Example 6.6 The equations of motion of the wheel/rail system formulated in Example 6.5 using the augmented formulation can also be formulated using the embedding technique, which leads to a number of equations equal to the number of degrees of freedom of the system (Shabana et al. 2008). To this end, the constraints at the acceleration level, $\ddot{\mathbf{r}}_c^w = \mathbf{0}$, are used to write two dependent accelerations $\ddot{\mathbf{q}}_d$ in terms of the independent acceleration $\ddot{\mathbf{q}}_i$. In this wheel/rail example, the dependent accelerations $\ddot{\mathbf{q}}_d$ can be selected to be $\ddot{R}_x^w$ and $\ddot{R}_y^w$, while the independent acceleration $\ddot{\mathbf{q}}_i$ can be selected as $\ddot{\theta}^w$. Using the condition $\ddot{\mathbf{r}}_c^w = \mathbf{0}$ of Example 6.5, one can write

$$\ddot{\mathbf{q}}_d = \begin{bmatrix} \ddot{R}_x^w \\ \ddot{R}_y^w \end{bmatrix} = \begin{bmatrix} -r\cos\phi \\ r\sin\phi \end{bmatrix} \ddot{\theta}^w + \begin{bmatrix} r\dot{\theta}^w\dot{\phi}\sin\phi \\ r\dot{\theta}^w\dot{\phi}\cos\phi \end{bmatrix}$$

Using this equation, the vector of the system accelerations can be written in terms of independent acceleration in the form $\ddot{\mathbf{q}} = \mathbf{B}_{di}\,\ddot{\mathbf{q}}_i + \boldsymbol{\gamma}_i$, where in this case $\ddot{\mathbf{q}}_i$ reduces to the scalar $\ddot{\theta}^w$. One therefore has

$$\ddot{\mathbf{q}} = \begin{bmatrix} \ddot{\mathbf{q}}_d \\ \ddot{\mathbf{q}}_i \end{bmatrix} = \begin{bmatrix} \ddot{R}_x^w \\ \ddot{R}_y^w \\ \ddot{\theta}^w \end{bmatrix} = \begin{bmatrix} -r\cos\phi \\ r\sin\phi \\ 1 \end{bmatrix} \ddot{\theta}^w + \begin{bmatrix} r\dot{\theta}^w\dot{\phi}\sin\phi \\ r\dot{\theta}^w\dot{\phi}\cos\phi \\ 0 \end{bmatrix}$$

Using this equation, the velocity transformation matrix $\mathbf{B}_{di}$ and quadratic velocity vector $\boldsymbol{\gamma}_i$ are recognized as

$$\mathbf{B}_{di} = \begin{bmatrix} -r\cos\phi \\ r\sin\phi \\ 1 \end{bmatrix}, \qquad \boldsymbol{\gamma}_i = \begin{bmatrix} r\dot{\theta}^w\dot{\phi}\sin\phi \\ r\dot{\theta}^w\dot{\phi}\cos\phi \\ 0 \end{bmatrix}$$

The equations of motion of the system were obtained in Example 6.5 as $\mathbf{M}\ddot{\mathbf{q}} = \mathbf{Q}_e + \mathbf{Q}_c$, where the vectors and matrices in this equation were defined in Example 6.5. It is clear that

$$\mathbf{B}_{di}^T\mathbf{Q}_c = \begin{bmatrix} -r\cos\phi & r\sin\phi & 1 \end{bmatrix} \begin{bmatrix} N_x \\ N_y \\ N_x r\cos\phi - N_y r\sin\phi \end{bmatrix} = 0$$

Therefore, substituting $\ddot{\mathbf{q}} = \mathbf{B}_{di}\,\ddot{\mathbf{q}}_i + \boldsymbol{\gamma}_i$ into the equations of motion $\mathbf{M}\ddot{\mathbf{q}} = \mathbf{Q}_e + \mathbf{Q}_c$ and premultiplying by the transpose of the velocity transformation matrix $\mathbf{B}_{di}^T$, one obtains (Shabana et al. 2008)

$$\left(J^w + m^w r^2\right)\ddot{\theta}^w = M^w - r\left(F_x^w\cos\phi + F_y^w\sin\phi\right) - m^w g r\sin\phi$$

This scalar equation, which does not have any constraint forces, is the system equation of motion, which can be solved for the angular acceleration $\ddot{\theta}^w$. If no external forces and moments are applied to the wheel, and if the wheel is represented by a cylinder with a mass moment of inertia $J^w = m^w r^2/2$, the preceding equation, upon using

(Continued)

the kinematic relationship $r\dot{\theta}^w = (R-r)\dot{\phi}$ obtained in Example 6.5, reduces to $(3(R-r)/2)\ddot{\phi} + g\sin\phi = 0$.

The wheel/rail problem considered in this example can be used to demonstrate that the embedding technique has its roots in D'Alembert's principle. By treating inertia forces the same way as applied forces, one can eliminate constraint forces by equating the moments of these forces about the contact point. This leads to one scalar equation of motion because the system has one degree of freedom. This can be demonstrated using the free-body diagram shown in Figure 4. The moments of the applied force about contact point c is $M^w - r\left(F_x^w \cos\phi + F_y^w \sin\phi\right) - m^w gr\sin\phi$. The moment of the inertia forces about the same point is $J^w\ddot{\theta}^w - m^w\ddot{R}_x^w r\cos\phi + m^w\ddot{R}_y^w r\sin\phi$. By using the constraint equations at the acceleration level to write accelerations $\ddot{R}_x^w$ and $\ddot{R}_y^w$ in terms of independent acceleration $\ddot{\theta}^w$, it can be shown that the moment of the inertia forces about the contact point is $\left(J^w + m^w r^2\right)\ddot{\theta}^w$. Therefore, the same equation of motion obtained previously in this example using the embedding technique can be obtained by equating the moments of the inertia and applied forces. However, the embedding technique can be used as the basis for developing general MBS algorithms that are based on the connectivity conditions and do not require using free-body diagrams.

6.8 WHEEL/RAIL CONSTRAINT CONTACT FORCES

Wheel/rail contact formulations were discussed in Chapter 5. Two fundamentally different methods can be used to mathematically define wheel/rail interaction forces: the *constraint approach,* which does not allow for wheel/rail separation; and the *elastic approach,* which allows for wheel/rail separations. In the constraint approach, the wheel/rail normal contact force is determined as a reaction force. In the elastic approach, on the other hand, the wheel/rail contact force is defined using a compliant force model formulated using assumed stiffness and damping coefficients. The two approaches (constraint and elastic) considered in this book require the formulation and solution of a system of algebraic equations.

Wheel and Rail Contact Surfaces As discussed in the preceding chapters, formulating the three-dimensional wheel/rail contact problem requires the parameterization of the wheel and rail surfaces. As shown in Figure 5 and discussed in Chapter 5, the wheel and rail surface parameters can be written in a vector form as

$$\mathbf{s} = \begin{bmatrix} s_1^w & s_2^w & s_1^r & s_2^r \end{bmatrix}^T \tag{6.66}$$

where superscripts w and r refer, respectively, to the wheel and rail surfaces. In the preceding equation, s_1^w and s_2^w are, respectively, the wheel lateral and angular surface parameters; and s_1^r and s_2^r are, respectively, the rail longitudinal and lateral surface parameters, as shown in Figure 5. This surface-parameter description is used to capture the three-dimensional nature of the wheel/rail contact problem.

The locations of a potential contact point P^k, $k = w, r$, on the wheel and rail surfaces with respect to the wheel and rail coordinate systems can be written, respectively, as $\overline{\mathbf{u}}_P^k\left(s_1^k, s_2^k\right) = \left[x^k\left(s_1^k, s_2^k\right) \; y^k\left(s_1^k, s_2^k\right) \; z^k\left(s_1^k, s_2^k\right)\right]^T$, $k = w, r$, as discussed in Chapters 3–5. The global position of the potential contact point can be written as $\mathbf{r}_P^k = \mathbf{R}^k + \mathbf{A}^k\overline{\mathbf{u}}_P^k$, $k = w, r$.

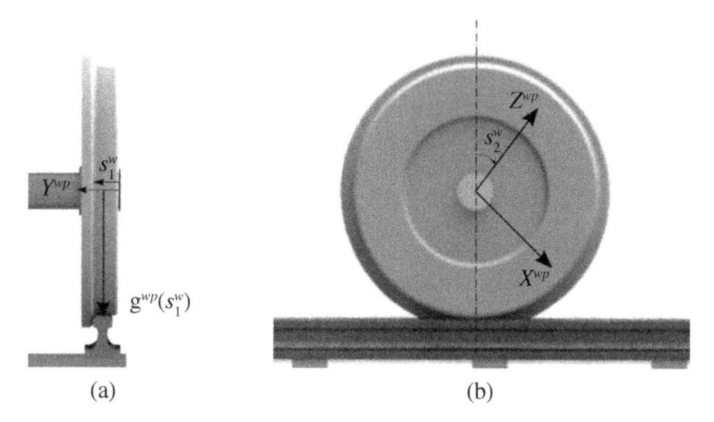

Figure 6.5 Wheel and rail surfaces.

The tangent and normal vectors to the wheel and rail surfaces can be defined in the body coordinate systems, respectively, as

$$\bar{\mathbf{t}}_1^k = \frac{\partial \bar{\mathbf{u}}_P^k}{\partial s_1^k}, \quad \bar{\mathbf{t}}_2^k = \frac{\partial \bar{\mathbf{u}}_P^k}{\partial s_2^k}, \quad \bar{\mathbf{n}}^k = \bar{\mathbf{t}}_1^k \times \bar{\mathbf{t}}_2^k, \quad k = w, r \tag{6.67}$$

These vectors can be defined in the global coordinate system as $\mathbf{t}_1^k = \mathbf{A}^k \bar{\mathbf{t}}_1^k$, $\mathbf{t}_2^k = \mathbf{A}^k \bar{\mathbf{t}}_2^k$ and $\mathbf{n}^k = \mathbf{A}^k \bar{\mathbf{n}}^k$, $k = w, r$.

Contact Constraint Approach In the contact constraint approach, as discussed in Chapter 5, wheel/rail separation is not allowed. In this case, the wheel is assumed to have five degrees of freedom with respect to the rail. The freedom of the wheel to move with respect to the rail along the normal to the surface at the contact point is eliminated. Because four geometric surface parameters are introduced as non-generalized coordinates that must be solved for in order to determine the location of the contact point, the elimination of one relative degree of freedom between the wheel and rail requires imposing the following five algebraic constraint equations for a *non-conformal contact j*

$$\mathbf{C}^j\left(\mathbf{q}^w, \mathbf{q}^r, \mathbf{s}^{wj}, \mathbf{s}^{rj}\right) = \begin{bmatrix} \left(\mathbf{r}_P^w - \mathbf{r}_P^r\right)^j \\ \left(\mathbf{t}_1^{w^T} \mathbf{n}^r\right)^j \\ \left(\mathbf{t}_2^{w^T} \mathbf{n}^r\right)^j \end{bmatrix} = \mathbf{0} \tag{6.68}$$

which can be written in an alternate and equivalent form, upon using the definition $\mathbf{r}_P^{wr} = \mathbf{r}_P^w - \mathbf{r}_P^r$, as

$$\mathbf{C}^j\left(\mathbf{q}^w, \mathbf{q}^r, \mathbf{s}^{wj}, \mathbf{s}^{rj}\right) = \begin{bmatrix} \left(\mathbf{t}_1^{r^T} \mathbf{r}_P^{wr}\right)^j \\ \left(\mathbf{t}_2^{r^T} \mathbf{r}_P^{wr}\right)^j \\ \left(\mathbf{n}^{r^T} \mathbf{r}_P^{wr}\right)^j \\ \left(\mathbf{t}_1^{w^T} \mathbf{n}^r\right)^j \\ \left(\mathbf{t}_2^{w} \cdot \mathbf{n}^r\right)^j \end{bmatrix} = \mathbf{0} \tag{6.69}$$

As discussed in Chapter 5, by using the contact constraint equations written in terms of the surface parameters $\mathbf{s} = \begin{bmatrix} s_1^w & s_2^w & s_1^r & s_2^r \end{bmatrix}^T$, the constraint functions that include the effect of all joints and specified motion trajectories as well as the contact constraints can be written in a vector form as $\mathbf{C}(\mathbf{q}, \mathbf{s}) = \mathbf{0}$, where in this case $\mathbf{s}$ contains the surface parameters associated with all the contacts in the system – that is, $\mathbf{s} = \begin{bmatrix} \mathbf{s}^{1^T} & \mathbf{s}^{2^T} & \cdots & \mathbf{s}^{n_{cc}^T} \end{bmatrix}^T$ – n_{cc} is the number of contacts, and $\mathbf{s}^j = \begin{bmatrix} s_1^{wj} & s_2^{wj} & s_1^{rj} & s_2^{rj} \end{bmatrix}^T$ is the vector of surface parameters associated with contact j. It follows that $\delta\mathbf{C} = \mathbf{C}_\mathbf{q}\delta\mathbf{q} + \mathbf{C}_\mathbf{s}\delta\mathbf{s} = \mathbf{0}$, where $\mathbf{C}_\mathbf{q}$ and $\mathbf{C}_\mathbf{s}$ are, respectively, the constraint Jacobian matrices associated with vectors $\mathbf{q}$ and $\mathbf{s}$. Therefore, one has $\boldsymbol{\lambda}^T(\mathbf{C}_\mathbf{q}\delta\mathbf{q} + \mathbf{C}_\mathbf{s}\delta\mathbf{s}) = 0$, where $\boldsymbol{\lambda}$ is the vector of Lagrange multipliers. Adding this equation to the equation defining the Lagrange–D'Alembert principle $\delta\mathbf{q}^T\left(\mathbf{M}\ddot{\mathbf{q}} - \mathbf{Q}_e - \mathbf{Q}_v\right) = 0$, previously obtained in this chapter, one has

$$\delta\mathbf{q}^T\left(\mathbf{M}\ddot{\mathbf{q}} + \mathbf{C}_\mathbf{q}^T\boldsymbol{\lambda} - \mathbf{Q}_e - \mathbf{Q}_v\right) + \delta\mathbf{s}^T\mathbf{C}_\mathbf{s}^T\boldsymbol{\lambda} = 0 \tag{6.70}$$

Using a procedure similar to the one discussed in Section 6.2 of this chapter, one can show that the preceding equation leads to

$$\mathbf{M}\ddot{\mathbf{q}} + \mathbf{C}_\mathbf{q}^T\boldsymbol{\lambda} = \mathbf{Q}_e + \mathbf{Q}_v, \qquad \mathbf{C}_\mathbf{s}^T\boldsymbol{\lambda} = \mathbf{0} \tag{6.71}$$

Differentiating the constraint equations $\mathbf{C}(\mathbf{q}, \mathbf{s}) = \mathbf{0}$ twice with respect to time, one obtains

$$\mathbf{C}_\mathbf{q}\ddot{\mathbf{q}} + \mathbf{C}_\mathbf{s}\ddot{\mathbf{s}} = \mathbf{Q}_d \tag{6.72}$$

where $\mathbf{Q}_d$ is the vector that absorbs terms that are not linear in accelerations. Combining the preceding two equations, one has the augmented form of the equations of motion in the case of the wheel/rail contact constraint approach as

$$\begin{bmatrix} \mathbf{M} & \mathbf{0} & \mathbf{C}_\mathbf{q}^T \\ \mathbf{0} & \mathbf{0} & \mathbf{C}_\mathbf{s}^T \\ \mathbf{C}_\mathbf{q} & \mathbf{C}_\mathbf{s} & \mathbf{0} \end{bmatrix} \begin{bmatrix} \ddot{\mathbf{q}} \\ \ddot{\mathbf{s}} \\ \boldsymbol{\lambda} \end{bmatrix} = \begin{bmatrix} \mathbf{Q}_e + \mathbf{Q}_v \\ \mathbf{0} \\ \mathbf{Q}_d \end{bmatrix} \tag{6.73}$$

There are important observations about the augmented form of the equations of motion of Eq. 73, which accounts for the wheel/rail contact constraint conditions. These observations are summarized as follows:

- Equation 73 can be solved for the second time derivatives of the generalized coordinates $\ddot{\mathbf{q}}$, the second time derivatives of the non-generalized coordinates $\ddot{\mathbf{s}}$, and the vector of Lagrange multipliers $\boldsymbol{\lambda}$. If the constraint functions are linearly independent and the mass matrix is nonsingular, one can show that the coefficient matrix in Eq. 73 has a full row rank and is nonsingular.
- It is clear from Eq. 73 that there are no inertia or applied forces associated with the vector of surface parameters $\mathbf{s}$. Therefore, the surface parameters are geometric parameters and can be considered *non-generalized coordinates*, as discussed in the literature (Shabana and Sany 2001).
- Equation 73 implies that the system differential equations of motion and the wheel/rail contact constraints must be solved simultaneously for the system generalized and non-generalized coordinates as well as the constraint forces. This is necessary to correctly account for couplings between the generalized and non-generalized coordinates.

In the augmented contact constraint formulation, the total vector of coordinates can be written as $\mathbf{p} = \begin{bmatrix}\mathbf{q}^T & \mathbf{s}^T\end{bmatrix}^T$.

- The system constraint Jacobian matrix can be written as $\mathbf{C_p} = \partial\mathbf{C}/\partial\mathbf{p} = \begin{bmatrix}\partial\mathbf{C}/\partial\mathbf{q} & \partial\mathbf{C}/\partial\mathbf{s}\end{bmatrix}$, which can be written as $\mathbf{C_p} = \begin{bmatrix}\mathbf{C_q} & \mathbf{C_s}\end{bmatrix}$, where $\mathbf{C_q} = \partial\mathbf{C}/\partial\mathbf{q}$ and $\mathbf{C_s} = \partial\mathbf{C}/\partial\mathbf{s}$. Using the system constraint Jacobian matrix $\mathbf{C_p}$ to identify the dependent and independent coordinates, the system degrees of freedom can include both generalized and non-generalized coordinates. In this case, the total vector of coordinates can be written as $\mathbf{p} = \begin{bmatrix}\mathbf{p}_i^T & \mathbf{p}_d^T\end{bmatrix}^T$, where $\mathbf{p}_i$ and $\mathbf{p}_d$ are, respectively, the vectors of independent and dependent coordinates. Both vectors can include generalized and non-generalized coordinates.
- Because four surface parameters are used for each contact, which is described using five algebraic equations, the constraint conditions for one contact lead to one independent reaction force. This force is the normal contact constraint force. Equation 73 shows that $\mathbf{C}_\mathbf{s}^T\boldsymbol{\lambda} = \mathbf{0}$ (Shabana and Sany 2001). In this equation, there are five Lagrange multipliers associated with five contact constraint equations. For a contact j, one has $\mathbf{C}_\mathbf{s}^{j^T}\boldsymbol{\lambda}^j = \mathbf{0}$, $j = 1, 2, \ldots, n_{cc}$. That is, each contact j introduces five independent contact constraints and five Lagrange multipliers. The equation $\mathbf{C}_\mathbf{s}^{j^T}\boldsymbol{\lambda}^j = \mathbf{0}$ for contact j represents a system of four scalar equations in five unknown Lagrange multipliers $\boldsymbol{\lambda}^j$. If the constraint equations are linearly independent, the rank of the 5×4 constraint Jacobian matrix $\mathbf{C}_\mathbf{s}^j$ associated with the surface parameters of contact j is 4, and therefore, the equation $\mathbf{C}_\mathbf{s}^{j^T}\boldsymbol{\lambda}^j = \mathbf{0}$ ensures that there is only one independent Lagrange multiplier. That's is, the five contact constraints, which eliminate the relative degree of freedom of the wheel to move with respect to the rail in the normal direction, can be used to determine only one independent constraint force. This constraint force is used in the augmented contact formulation to determine the wheel/rail normal contact force. This normal force is used with the wheel and rail geometry to determine the tangential creep forces, as discussed in Chapter 5. The creep forces acting on the wheel and rail can be entered into the equations of motion using vector $\mathbf{Q}_e$.

It is clear, therefore, that when the augmented contact constraint formulation is used, the normal contact force is obtained as a reaction force. Because this normal force is used to determine the tangential creep forces included in vector $\mathbf{Q}_e$ on the right-hand side of Eq. 73, the value of the Lagrange multipliers associated with the contact constraints from the previous time step can be stored and used to determine the tangential creep force. This approximation has proven to work well and allows avoiding the use of an iterative procedure to solve Eq. 73, which represents a large system of equations in the case of complex railroad vehicle systems.

Elimination of Surface Parameters Four of the contact constraint equations can be used to eliminate surface parameters (non-generalized coordinates) by writing them in terms of generalized coordinates. In this case, contact between the wheel and rail can be represented by one algebraic constraint equation imposed on the generalized coordinates. This constraint equation eliminates the relative motion between the wheel and rail in a direction normal to the contact surfaces at the contact point. For simplicity, the superscript j that refers to the contact number is dropped in the remainder of this section. To explain the procedure for eliminating the surface parameters, the constraint equations in Eq. 69 can be rearranged and written as

$$\mathbf{C}\left(\mathbf{q}^w, \mathbf{q}^r, \mathbf{s}^w, \mathbf{s}^r\right) = \begin{bmatrix} \mathbf{C}^d \\ C^n \end{bmatrix} = \begin{bmatrix} \mathbf{t}_1^{r^T} \mathbf{r}_P^{wr} \\ \mathbf{t}_2^{r^T} \mathbf{r}_P^{wr} \\ \mathbf{t}_1^{w^T} \mathbf{n}^r \\ \mathbf{t}_2^{w^T} \mathbf{n}^r \\ \mathbf{n}^{r^T} \mathbf{r}_P^{wr} \end{bmatrix} = \mathbf{0} \tag{6.74}$$

where

$$\left.\begin{aligned} \mathbf{C}^d\left(\mathbf{q}^w, \mathbf{q}^r, \mathbf{s}^w, \mathbf{s}^r\right) &= \left[\mathbf{t}_1^{r^T} \mathbf{r}_P^{wr} \quad \mathbf{t}_2^{r^T} \mathbf{r}_P^{wr} \quad \mathbf{t}_1^{w^T} \mathbf{n}^r \quad \mathbf{t}_2^{w^T} \mathbf{n}^r\right]^T = \mathbf{0}, \\ C^n\left(\mathbf{q}^w, \mathbf{q}^r, \mathbf{s}^w, \mathbf{s}^r\right) &= \mathbf{n}^{r^T} \mathbf{r}_P^{wr} = 0 \end{aligned}\right\} \tag{6.75}$$

Therefore, one has

$$\mathbf{C}_\mathbf{s}^d \delta\mathbf{s} = -\mathbf{C}_\mathbf{q}^d \delta\mathbf{q}, \quad \mathbf{C}_\mathbf{s}^d \dot{\mathbf{s}} = -\mathbf{C}_\mathbf{q}^d \dot{\mathbf{q}}, \quad \mathbf{C}_\mathbf{s}^d \ddot{\mathbf{s}} = -\mathbf{C}_\mathbf{q}^d \ddot{\mathbf{q}} + \overline{\boldsymbol{\gamma}}_C^d \tag{6.76}$$

in which $\mathbf{C}_\mathbf{s}^d = \partial \mathbf{C}^d / \partial \mathbf{s}$, $\mathbf{C}_\mathbf{q}^d = \partial \mathbf{C}^d / \partial \mathbf{q}$, and $\overline{\boldsymbol{\gamma}}_C^d = -\dot{\mathbf{C}}_\mathbf{s}^d \dot{\mathbf{s}} - \dot{\mathbf{C}}_\mathbf{q}^d \dot{\mathbf{q}}$. Assuming the case of nonconformal contact in which matrix $\mathbf{C}_\mathbf{s}^d$ is a nonsingular matrix, Eq. 76 leads to

$$\delta\mathbf{s} = -\left(\mathbf{C}_\mathbf{s}^{d^{-1}} \mathbf{C}_\mathbf{q}^d\right) \delta\mathbf{q}, \quad \dot{\mathbf{s}} = -\left(\mathbf{C}_\mathbf{s}^{d^{-1}} \mathbf{C}_\mathbf{q}^d\right) \dot{\mathbf{q}}, \quad \ddot{\mathbf{s}} = -\left(\mathbf{C}_\mathbf{s}^{d^{-1}} \mathbf{C}_\mathbf{q}^d\right) \ddot{\mathbf{q}} + \boldsymbol{\gamma}_C^d \tag{6.77}$$

where $\boldsymbol{\gamma}_C^d = \mathbf{C}_\mathbf{s}^{d^{-1}} \overline{\boldsymbol{\gamma}}_C^d$. Using Eq. 77, the second equation of Eq. 75 leads to

$$\left.\begin{aligned} \delta C^n &= C_\mathbf{s}^n \delta\mathbf{s} + C_\mathbf{q}^n \delta\mathbf{q} = \left(C_\mathbf{q}^n - C_\mathbf{s}^n \left(\mathbf{C}_\mathbf{s}^{d^{-1}} \mathbf{C}_\mathbf{q}^d\right)\right) \delta\mathbf{q} = 0, \\ \dot{C}_\mathbf{s}^n &= C_\mathbf{s}^n \dot{\mathbf{s}} + C_\mathbf{q}^n \dot{\mathbf{q}} = \left(C_\mathbf{q}^n - C_\mathbf{s}^n \left(\mathbf{C}_\mathbf{s}^{d^{-1}} \mathbf{C}_\mathbf{q}^d\right)\right) \dot{\mathbf{q}} = 0 \\ \ddot{C}_\mathbf{s}^n &= C_\mathbf{s}^n \ddot{\mathbf{s}} + C_\mathbf{q}^n \ddot{\mathbf{q}} + \left(\dot{C}_\mathbf{s}^n \dot{\mathbf{s}} + \dot{C}_\mathbf{q}^n \dot{\mathbf{q}} + C_\mathbf{s}^n \boldsymbol{\gamma}_C^d\right) = \left(C_\mathbf{q}^n - C_\mathbf{s}^n \left(\mathbf{C}_\mathbf{s}^{d^{-1}} \mathbf{C}_\mathbf{q}^d\right)\right) \ddot{\mathbf{q}} + \gamma_C^n = 0 \end{aligned}\right\} \tag{6.78}$$

where $\gamma_C^n = \left(\dot{C}_\mathbf{s}^n \dot{\mathbf{s}} + \dot{C}_\mathbf{q}^n \dot{\mathbf{q}} + C_\mathbf{s}^n \boldsymbol{\gamma}_C^d\right)$ is a vector that absorbs terms that are not linear in accelerations, $C_\mathbf{s}^n = \partial C^n / \partial \mathbf{s}$, and $C_\mathbf{q}^n = \partial C^n / \partial \mathbf{q}$. Using the preceding equation, the following algorithm can be used to eliminate surface parameters and represent the wheel/rail interaction using one constraint equation defined by C^n:

1. Given the configurations of the wheel and rail defined, respectively, by the generalized coordinates $\mathbf{q}^w$ and $\mathbf{q}^r$, the four constraint equations $\mathbf{C}^d$ of Eq. 75 can be solved iteratively using a Newton–Raphson algorithm to determine the four surface parameters $\mathbf{s}^w$ and $\mathbf{s}^r$. The wheel and rail coordinates $\mathbf{q}^w$ and $\mathbf{q}^r$ are assumed to be known from the numerical integration of the system equations of motion. In the Newton–Raphson iterations, the equation $(\partial \mathbf{C}^d / \partial \mathbf{s}) \Delta\mathbf{s} = -\mathbf{C}^d$ is solved for the Newton differences $\Delta\mathbf{s}$, which are used to update $\mathbf{s}$ until convergence is achieved. The result of this numerical procedure is equivalent to writing $\mathbf{s} = \mathbf{s}(\mathbf{q}^w, \mathbf{q}^r)$.
2. After determining the surface parameters, their derivatives can be determined using Eq. 77 in terms of the derivatives of the wheel and rail coordinates. Furthermore, because the surface parameters are numerically computed using wheel and rail coordinates, the Jacobian matrix of constraint C^n defined by differentiation with respect to the generalized coordinates can be assumed equal to $\left(C_\mathbf{q}^n - C_\mathbf{s}^n \left(\mathbf{C}_\mathbf{s}^{d^{-1}} \mathbf{C}_\mathbf{q}^d\right)\right)$. This is the Jacobian matrix that must be used in the position and velocity analysis steps.

3. At the acceleration analysis step, the Jacobian matrix of constraint C^n is again assumed to take the form $\left(C_{\mathbf{q}}^n - C_{\mathbf{s}}^n\left(\mathbf{C}_{\mathbf{s}}^{d^{-1}}\mathbf{C}_{\mathbf{q}}^d\right)\right)$, while the element in vector $\mathbf{Q}_d$ associated with constraint C^n is set equal to $-\gamma_C^n$.

The use of this algorithm can be analytically explained by writing the Lagrange–D'Alembert principle as (Shabana et al. 2008)

$$\delta\mathbf{q}^T\left(\mathbf{M}\ddot{\mathbf{q}} + \mathbf{C}_{\mathbf{q}}^{d^T}\boldsymbol{\lambda}^d + \mathbf{C}_{\mathbf{q}}^{n^T}\boldsymbol{\lambda}^n - \mathbf{Q}\right) + \delta\mathbf{s}^T\left(\mathbf{C}_{\mathbf{s}}^{d^T}\boldsymbol{\lambda}^d + \mathbf{C}_{\mathbf{s}}^{n^T}\boldsymbol{\lambda}^n\right) = 0 \tag{6.79}$$

where $\boldsymbol{\lambda}^d$ and $\boldsymbol{\lambda}^n$ are Lagrange multipliers associated with constraint equations $\mathbf{C}^d$ and $\mathbf{C}^n$, respectively. The first equation in Eq. 77 can be written as

$$\delta\mathbf{s} = \mathbf{B}^{dn}\delta\mathbf{q} \tag{6.80}$$

where $\mathbf{B}^{dn} = -\left(\mathbf{C}_{\mathbf{s}}^d\right)^{-1}\mathbf{C}_{\mathbf{q}}^d$. Therefore, Eq. 79 reduces to

$$\delta\mathbf{q}^T\left(\mathbf{M}\ddot{\mathbf{q}} + \left(\mathbf{C}_{\mathbf{q}}^n + \mathbf{C}_{\mathbf{s}}^n\mathbf{B}^{dn}\right)^T\boldsymbol{\lambda}^n - \mathbf{Q}\right) = 0 \tag{6.81}$$

The forces associated with contact constraints $\mathbf{C}^d$ are eliminated from this equation as a consequence of eliminating the surface parameters. This is the result of using the identity $\mathbf{C}_{\mathbf{q}}^{d^T}\boldsymbol{\lambda}^d + \mathbf{B}^{dn^T}\mathbf{C}_{\mathbf{s}}^{d^T}\boldsymbol{\lambda}^d = \mathbf{C}_{\mathbf{q}}^{d^T}\boldsymbol{\lambda}^d + \left(-\left(\mathbf{C}_{\mathbf{s}}^d\right)^{-1}\mathbf{C}_{\mathbf{q}}^d\right)^T\mathbf{C}_{\mathbf{s}}^{d^T}\boldsymbol{\lambda}^d = \mathbf{0}$. Also, as previously mentioned, the second and third equations in Eq. 77, $\dot{\mathbf{s}} = \mathbf{B}^{dn}\dot{\mathbf{q}}$ and $\ddot{\mathbf{s}} = \mathbf{B}^{dn}\ddot{\mathbf{q}} + \boldsymbol{\gamma}_C^d$, respectively, can be used to eliminate $\dot{\mathbf{s}}$ and $\ddot{\mathbf{s}}$ from the time derivatives of the constraint equation. Therefore, by eliminating the non-generalized coordinates (surface parameters) and constraint equations $\mathbf{C}^d$, one can write the augmented form of the equations of motion as

$$\begin{bmatrix} \mathbf{M} & \overline{\mathbf{C}}_{\mathbf{q}}^T \\ \overline{\mathbf{C}}_{\mathbf{q}} & \mathbf{0} \end{bmatrix}\begin{bmatrix} \ddot{\mathbf{q}} \\ \boldsymbol{\lambda} \end{bmatrix} = \begin{bmatrix} \mathbf{Q}_e + \mathbf{Q}_v \\ \overline{\mathbf{Q}}_d \end{bmatrix} \tag{6.82}$$

where $\overline{\mathbf{C}}_{\mathbf{q}}$ is the Jacobian matrix of all constraint equations $\overline{\mathbf{C}}$ after eliminating $\mathbf{s}$ and $\mathbf{C}^d$, $\boldsymbol{\lambda}$ is the vector of Lagrange multipliers associated with constraint equations $\overline{\mathbf{C}}$, and $\overline{\mathbf{Q}}_d$ is the vector that results from the differentiation of $\overline{\mathbf{C}}$ twice with respect to time: that is, $\overline{\mathbf{C}}_{\mathbf{q}}\ddot{\mathbf{q}} = \overline{\mathbf{Q}}_d$. Constraint functions $\overline{\mathbf{C}}$, constraint Jacobian matrix $\overline{\mathbf{C}}_{\mathbf{q}}$, and vector $\overline{\mathbf{Q}}_d$ are implicit functions of the surface parameters, and therefore, the coupling between the generalized coordinates and the non-generalized surface parameters is not ignored by eliminating $\mathbf{s}$ and $\mathbf{C}^d$.

The computer implementation of the procedure discussed in this section for the elimination of the surface parameters has shown that no clear computational advantage results from eliminating surface parameters $\mathbf{s}$ and constraint functions $\mathbf{C}^d$. This is mainly due to the significant mathematical operations required for this elimination, as demonstrated by the analysis presented in this section. Therefore, when the constraint contact formulation is used, it is recommended to use Eq. 73, which is written in terms of both the generalized and non-generalized coordinates.

6.9 WHEEL/RAIL ELASTIC CONTACT FORCES

The analysis presented in the preceding section shows that when the constraint contact formulation is used, there is no wheel/rail separation, and the wheel is assumed to remain in contact with the rail regardless of the magnitude of applied forces. The constraint contact formulation can provide insight and be more efficient in many simulation scenarios, particularly in the case of smooth dynamics and when the wheel remains in contact with the rail. Nonetheless, such a constraint contact formulation can have serious limitations when examining some important scenarios, including derailments. For this reason, it is important to have an implementation of another approach that allows for wheel/rail separation, as discussed in Chapter 5. In this approach, referred to as the *elastic contact formulation*, wheel/rail interaction forces are described using a compliant force model.

In the elastic contact formulation, no kinematic constraints are imposed on the relative motion between the wheel and rail, and consequently, the unconstrained wheel is assumed to have six degrees of freedom with respect to the rail if no other motion constraints are introduced. In this formulation, the compliant force model is written as the sum of elastic and damping forces expressed in terms of assumed stiffness and damping coefficients, respectively. Whereas in the contact constraint formulation, the location of the contact point is determined by solving constraint equations at the position level, in the elastic contact formulation, several methods can be used to determine the location of the wheel/rail contact point. These methods include *discrete nodal search*, the use of *lookup tables*, and solving a set of *algebraic equations*. In this section, the method of solving algebraic equations to determine the location of the contact point online is used to formulate wheel/rail contact forces based on the elastic contact approach. The implementation of the other two methods, nodal search and lookup tables, is straightforward, and the same force model described in this section can be used once the location of the wheel/rail contact point is determined.

As discussed in Chapter 5, the algebraic equations used in the elastic force formulation to determine the contact point online are the same as four of the equations used in the constraint contact formulation. These equations are the four algebraic equations given in the first equation of Eq. 75, which are reproduced here for convenience (Escalona 2002; Pombo and Ambrosio 2003; Shabana et al. 2005)

$$\mathbf{g} = \begin{bmatrix} \mathbf{t}_1^{r^T}\mathbf{r}_P^{wr} & \mathbf{t}_2^{r^T}\mathbf{r}_P^{wr} & \mathbf{t}_1^{w^T}\mathbf{n}^r & \mathbf{t}_2^{w^T}\mathbf{n}^r \end{bmatrix}^T = \mathbf{0} \tag{6.83}$$

where all the vectors that appear in this equation are the same as previously defined. As discussed in Chapter 5, given the wheel and rail configurations defined, respectively, by vectors $\mathbf{q}^w$ and $\mathbf{q}^r$ at a given point in time, the preceding nonlinear algebraic equations can be solved for the surface parameters $\mathbf{s}$. In this case, $\mathbf{g} = \mathbf{g}(\mathbf{s}) = \mathbf{0}$ is a system of four algebraic equations that can be solved for the four unknown surface parameters. Using the Newton–Raphson method, the system of algebraic equations $(\partial\mathbf{g}/\partial\mathbf{s})\Delta\mathbf{s} = -\mathbf{g}(\mathbf{s})$ is solved iteratively for the Newton-differences $\Delta\mathbf{s}$, which are used to update the surface parameters until convergence is achieved. The converged solution defines the surface parameters that can be used to determine the locations and velocities of the contact points on the wheel and rail using the equations

$$\left.\begin{aligned} \mathbf{r}_P^k &= \mathbf{R}^k + \mathbf{A}^k\overline{\mathbf{u}}_P^k, \\ \mathbf{v}_P^k &= \dot{\mathbf{r}}_P^k = \dot{\mathbf{R}}^k + \boldsymbol{\omega}^k \times \mathbf{u}_P^k, \quad k = w, r \end{aligned}\right\} \tag{6.84}$$

Using the first equation in Eq. 84, the distance (penetration) between surfaces along the normal is evaluated using the equation

$$\delta = \mathbf{r}^{wr^T}\mathbf{n}^r \tag{6.85}$$

In the case of wheel/rail penetrations, normal contact forces can be calculated using the assumptions of non-conformal contact and Hertz contact theory. The following expression for wheel/rail normal contact force F_n has been used in the literature (Shabana et al. 2004)

$$F_n = F_{ns} + F_{nd} = -k_{ns}\delta^{1.5} - \left(c_{nd}\left|\delta\right|\right)\dot{\delta} \tag{6.86}$$

where δ measures the penetration, F_{ns} is the normal Hertzian (elastic) contact force, F_{nd} is the normal damping force, k_{ns} is the Hertzian stiffness coefficient that depends on the surface curvatures and properties of the materials of the two bodies in contact, and c_{nd} is a damping coefficient. The time rate of penetration $\dot{\delta}$ is evaluated using the velocity equation in Eq. 84 as $\dot{\delta} \approx \left(\dot{\mathbf{r}}_P^w - \dot{\mathbf{r}}_P^r\right)^T \mathbf{n}^r = \dot{\mathbf{r}}_P^{wr^T} \mathbf{n}^r$. The absolute value of penetration $|\delta|$ is included in the expression of the normal damping force to ensure zero normal contact force when $\delta = 0$. Discussion of defining the time rate of penetration, using the equation $\dot{\delta} \approx \left(\dot{\mathbf{r}}_P^w - \dot{\mathbf{r}}_P^r\right)^T \mathbf{n}^r = \dot{\mathbf{r}}_P^{wr^T} \mathbf{n}^r$, was provided in the preceding chapter.

After determining the normal contact force F_n using the elastic contact formulation, this normal force can be used with material properties and the dimensions of the contact area to determine the tangential creep forces. The tangential creep forces are entered into the equations of motion using the vector of generalized applied forces $\mathbf{Q}_e$. It is clear from these force definitions that normal contact force determined using the elastic approach is not a constraint force. Therefore, the algebraic equations of Eq. 83 should not be viewed as constraints imposed on the motion of the system. These algebraic equations are not enforced at the velocity and acceleration levels, and there are no constraint forces or Lagrange multipliers associated with them. Instead, a compliant force model that allows for wheel/rail separation is used.

6.10 OTHER FORCE ELEMENTS

In addition to wheel/rail contact forces, other forces can have significant effects on the dynamics and stability of railroad vehicle systems. For example, suspension forces influence the vehicle critical speed; beyond this critical speed, the vehicle becomes unstable. Most rail vehicles have two types of suspensions: the *primary suspension,* which connects the wheelsets to the bogie frame or equalizer bar, and the *secondary suspension* used to connect the car body to the bogie. These suspension elements can be represented mathematically using combinations of spring-damper-actuator and friction elements. Furthermore, bearings are used to mount wheelsets on bogie frames, and bushings are widely used in developing virtual railroad models. Figure 6 shows examples of the force elements used in railroad vehicles. This section discusses the formulations of several force elements used in the virtual prototyping of railroad vehicle systems.

Spring-Damper-Actuator Element This force element can be used to connect two arbitrary bodies in the vehicle system using a spring, damper, and/or actuator. The stiffness and damping coefficients and actuator force can be represented using linear or nonlinear functions defined analytically or numerically using spline functions. Figure 7 shows two bodies i and j connected by the spring-damper-actuator force element. The attachment point on body i is denoted as P^i, while the attachment point on body j is P^j. The stiffness

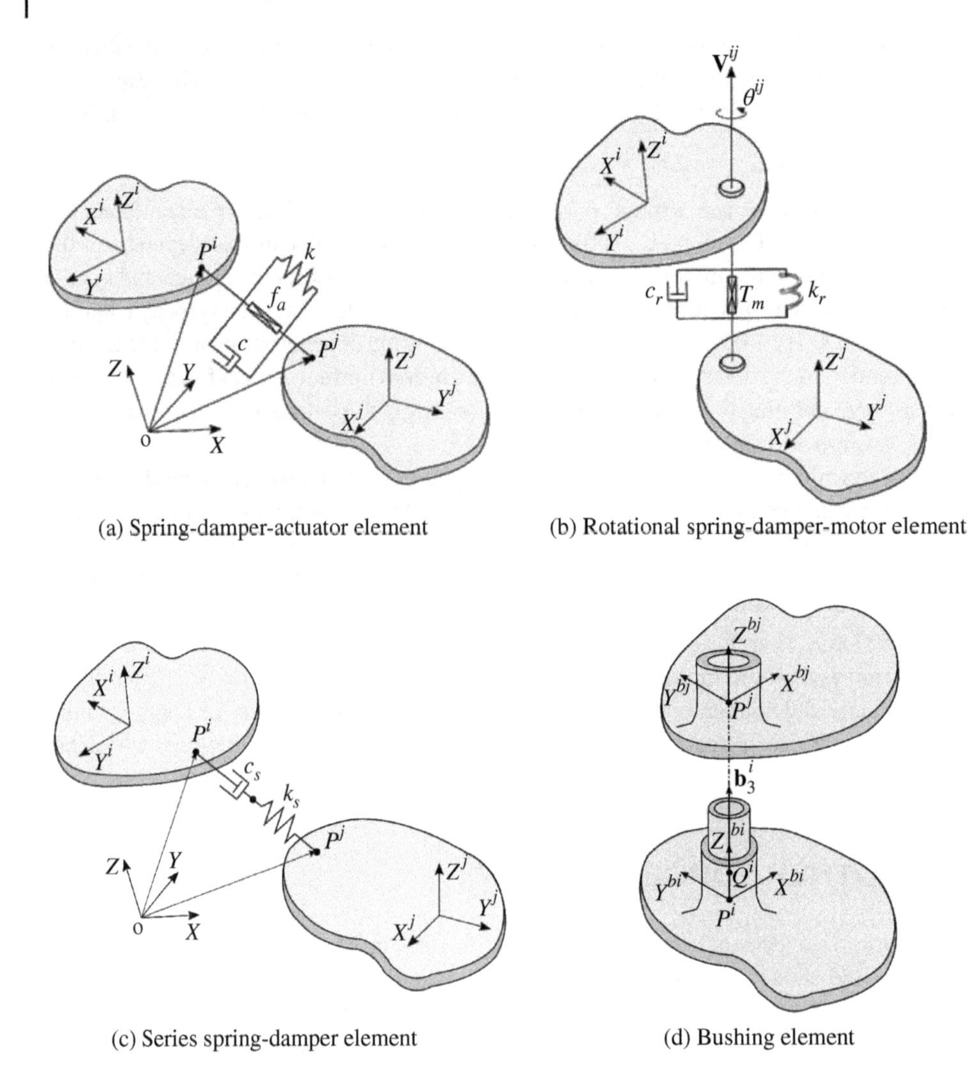

Figure 6.6 Force elements. (a) Spring-damper-actuator element; (b) rotational spring-damper-motor element; (c) series spring-damper element; (d) bushing element.

coefficient of the spring is assumed to be k, and its undeformed and current lengths are l_o and l, respectively. The damping coefficient is assumed to be c, and the actuator force is f_a. The force produced by this element can be written as

$$F_s = k\left(l - l_0\right) + c\dot{l} + f_a \tag{6.87}$$

This force can be used to define the force vector $\mathbf{F}_s^{ij} = F_s\hat{\mathbf{r}}_P^{ij}$, where $\hat{\mathbf{r}}_P^{ij} = \left(\mathbf{r}_P^i - \mathbf{r}_P^j\right) / \left|\mathbf{r}_P^i - \mathbf{r}_P^j\right|$ is a unit vector along the line connecting points P^i and P^j on bodies i and j, respectively.

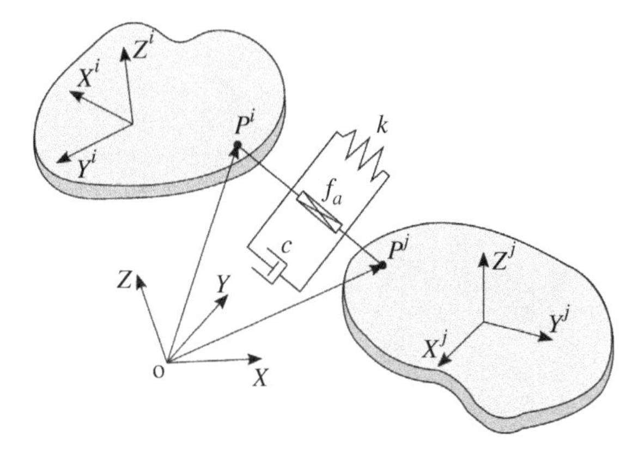

Figure 6.7 Spring-damper-actuator element.

Therefore, the virtual work of the spring-damper-actuator force element can be written as

$$\begin{aligned}\delta W_e^{ij} &= -\mathbf{F}_s^{ij^T}\delta\mathbf{r}_P^{ij} = -\left(F_s\hat{\mathbf{r}}_P^{ij}\right)^T\left(\delta\mathbf{r}_P^i - \delta\mathbf{r}_P^j\right)\\ &= -\left(F_s\hat{\mathbf{r}}_P^{ij}\right)^T\left(\mathbf{L}_P^i\delta\mathbf{q}^i - \mathbf{L}_P^j\delta\mathbf{q}^j\right)\end{aligned} \tag{6.88}$$

where $\mathbf{L}_P^k = \left[\mathbf{I} \quad -\mathbf{A}^k\tilde{\overline{\mathbf{u}}}_P^k\overline{\mathbf{G}}^k\right]$, $k = i, j$, and $\tilde{\overline{\mathbf{u}}}_P^k$ is the skew-symmetric matrix associated with vector $\overline{\mathbf{u}}_P^k$, which defines the position of point P^k, $k = i, j$, with respect to the coordinate system of body k. Using Eq. 88, the virtual work of the spring-damper-actuator force element can be written as $\delta W_e^{ij} = \mathbf{Q}_e^{i^T}\delta\mathbf{q}^i + \mathbf{Q}_e^{j^T}\delta\mathbf{q}^j$, where $\mathbf{Q}_e^i$ and $\mathbf{Q}_e^j$ are the generalized spring-damper-actuator forces associated, respectively, with the generalized coordinates of bodies i and j given by

$$\left.\begin{aligned}\mathbf{Q}_e^i &= -\mathbf{L}_P^{i^T}\left(F_s\hat{\mathbf{r}}_P^{ij}\right) = \begin{bmatrix}\left(\mathbf{Q}_e^i\right)_R\\ \left(\mathbf{Q}_e^i\right)_\theta\end{bmatrix} = -\begin{bmatrix}\mathbf{F}_s^{ij}\\ -\overline{\mathbf{G}}^{i^T}\tilde{\overline{\mathbf{u}}}_P^{i^T}\mathbf{A}^{i^T}\mathbf{F}_s^{ij}\end{bmatrix} = -\begin{bmatrix}\mathbf{F}_s^{ij}\\ \mathbf{G}^{i^T}\left(\mathbf{u}_P^i\times\mathbf{F}_s^{ij}\right)\end{bmatrix}\\ \mathbf{Q}_e^j &= \mathbf{L}_P^{j^T}\left(F_s\hat{\mathbf{r}}_P^{ij}\right) = \begin{bmatrix}\left(\mathbf{Q}_e^j\right)_R\\ \left(\mathbf{Q}_e^j\right)_\theta\end{bmatrix} = \begin{bmatrix}\mathbf{F}_s^{ij}\\ -\overline{\mathbf{G}}^{j^T}\tilde{\overline{\mathbf{u}}}_P^{j^T}\mathbf{A}^{j^T}\mathbf{F}_s^{ij}\end{bmatrix} = \begin{bmatrix}\mathbf{F}_s^{ij}\\ \mathbf{G}^{j^T}\left(\mathbf{u}_P^j\times\mathbf{F}_s^{ij}\right)\end{bmatrix}\end{aligned}\right\} \tag{6.89}$$

where $\mathbf{u}_P^k = \mathbf{A}^k\overline{\mathbf{u}}_P^k$, $k = i, j$, and $\mathbf{G}^k$ and $\overline{\mathbf{G}}^k$ are, respectively, the matrices that relate angular velocity vectors $\boldsymbol{\omega}^k$ and $\overline{\boldsymbol{\omega}}^k$ to the time derivatives of the orientation parameters, as previously discussed in this book. Since the spring-damper-actuator force formulation presented in this section is written in terms of the two matrices $\mathbf{G}^k$ and $\overline{\mathbf{G}}^k$, $k = i, j$, such a formulation can be used with any set of orientation parameters.

Rotational Spring-Damper-Motor Element The coefficients that enter into the definition of the rotational spring-damper-motor element can also be defined analytically or numerically as linear or nonlinear functions of the system coordinates and time. The torque

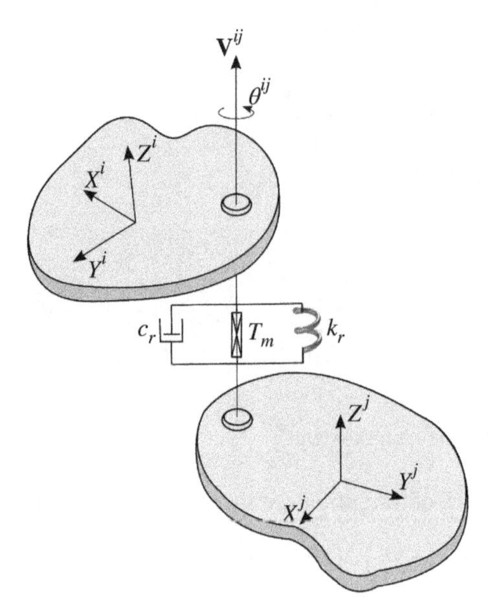

Figure 6.8 Rotational spring-damper-motor element.

produced by this force element when connecting two arbitrary bodies i and j in the system, as shown in Figure 8, is written in terms of the rotational spring stiffness coefficient k_r; current and initial relative angles between the two bodies θ^{ij} and θ_o^{ij}, respectively, about the element axis $\mathbf{v}^{ij}$; damping coefficient c_r; and motor torque T_m. The axis of rotation $\mathbf{v}^{ij}$ can be defined on body i as $\mathbf{v}^{ij} = \mathbf{A}^i\overline{\mathbf{v}}^{ij}$, where $\overline{\mathbf{v}}^{ij}$ is a constant vector defined in the coordinate system of body i. The torque produced by the rotational spring-damper-motor element about axis $\mathbf{v}^{ij}$ can be written as

$$T_s = k_r\left(\theta^{ij} - \theta_o^{ij}\right) + c_r\dot{\theta}^{ij} + T_m \tag{6.90}$$

The virtual work of this torque can be written as $\delta W_e^{ij} = -T_s\delta\theta^{ij}$, where $\delta\theta^{ij} = \mathbf{v}^{ij^T}\left(\delta\boldsymbol{\pi}^i - \delta\boldsymbol{\pi}^j\right)$, and $\delta\boldsymbol{\pi}^i$ and $\delta\boldsymbol{\pi}^j$ are virtual rotations about the axes of the Cartesian coordinate system, which can be written as $\delta\boldsymbol{\pi}^i = \boldsymbol{\omega}^i\delta t = \mathbf{G}^i\delta\boldsymbol{\theta}^i$ and $\delta\boldsymbol{\pi}^j = \boldsymbol{\omega}^j\delta t = \mathbf{G}^j\delta\boldsymbol{\theta}^j$. Therefore, the virtual work $\delta W_e^{ij} = -T_s\delta\theta^{ij}$ can be written as

$$\begin{aligned}\delta W_e^{ij} &= -T_s\delta\theta^{ij} = -T_s\mathbf{v}^{ij^T}\left(\delta\boldsymbol{\pi}^i - \delta\boldsymbol{\pi}^j\right)\\ &= -T_s\mathbf{v}^{ij^T}\left(\mathbf{G}^i\delta\boldsymbol{\theta}^i - \mathbf{G}^j\delta\boldsymbol{\theta}^j\right) = \mathbf{Q}_e^{i^T}\delta\mathbf{q}^i + \mathbf{Q}_e^{j^T}\delta\mathbf{q}^j\end{aligned} \tag{6.91}$$

where $\mathbf{Q}_e^i$ and $\mathbf{Q}_e^j$ are the generalized forces associated with the coordinates of bodies i and j, respectively, and defined as

$$\mathbf{Q}_e^i = \begin{bmatrix}\left(\mathbf{Q}_e^i\right)_R\\ \left(\mathbf{Q}_e^i\right)_\theta\end{bmatrix} = \begin{bmatrix}\mathbf{0}\\ -\mathbf{G}^{i^T}\left(T_s\mathbf{v}^{ij}\right)\end{bmatrix}, \quad \mathbf{Q}_e^j = \begin{bmatrix}\left(\mathbf{Q}_e^j\right)_R\\ \left(\mathbf{Q}_e^j\right)_\theta\end{bmatrix} = \begin{bmatrix}\mathbf{0}\\ \mathbf{G}^{j^T}\left(T_s\mathbf{v}^{ij}\right)\end{bmatrix} \tag{6.92}$$

It is clear from the definitions of the generalized forces in this equation that the rotational spring-damper-motor element does not produce generalized forces associated with the reference translations of the two bodies connected by this element.

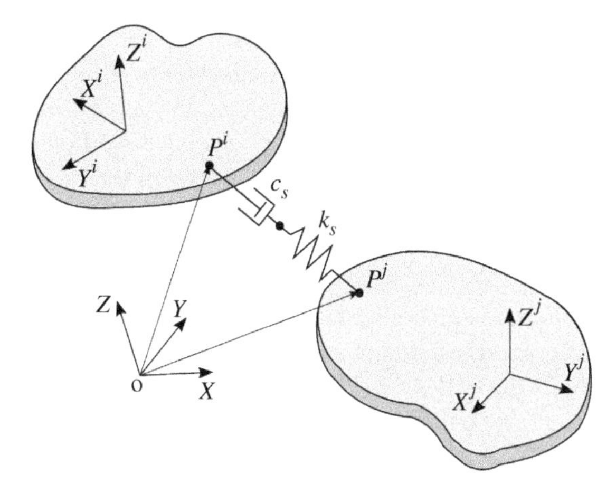

Figure 6.9 Series spring-damper element.

Series Spring-Damper Element The series spring-damper element, shown in Figure 9, consists of a spring and damper connected in series. As shown in Figure 9, the spring-damper can be used to connect two arbitrary bodies i and j. The attachment points on bodies i and j are assumed to be P^i and P^j, respectively. The spring stiffness and damping coefficients are assumed to be k_s and c_s, respectively. These coefficients can be defined analytically or numerically as linear or nonlinear functions of the coordinates. In the series spring-damper force element, widely used in railroad vehicle systems, the spring force is equal to the damping force: that is, $F_s = k_s d_s = c_s \dot{d}_d$, where d_s and d_d are, respectively, the spring deflection and relative displacement between the two ends of the damper, as shown in Figure 9. The total displacement of the element is the sum of these two displacements: that is, $d_t = d_s + d_d$. Differentiating this equation with respect to time, one obtains $\dot{d}_t = \dot{d}_s + \dot{d}_d$. In the special case where k_s is assumed to be constant, one has $\dot{F}_s = k_s \dot{d}_s$, and consequently, the equation $\dot{d}_t = \dot{d}_s + \dot{d}_d$ can be written as $\dot{d}_t = \dot{F}_s/k_s + F_s/c_s$, which leads to

$$\dot{F}_s + \left(\frac{k_s}{c_s}\right) F_s = k_s \dot{d}_t \tag{6.93}$$

The total element displacement d_t can be written in terms of the generalized coordinates of bodies i and j. This can be accomplished by defining the vector $\mathbf{r}_P^{ij} = \mathbf{r}_P^i - \mathbf{r}_P^j$, which can be used to define the relative velocity along the unit vector that connects the two points P^i and P^j. Equation 93 can be integrated numerically during dynamic simulations to define force F_s. Once this force is defined, a procedure similar to the one used for the translational spring-damper-actuator force element can be used to determine the generalized forces associated with the generalized coordinates of the two bodies connected by the series spring-damper element.

If k_s is not constant, one has $\dot{F}_s = k_s \dot{d}_s + \dot{k}_s d_s$, from which $\dot{d}_s = (\dot{F}/k_s) + (\dot{k}_s/k_s)\, d_s$. This equation can be written as $\dot{d}_s = (\dot{F}/k_s) + \left(\dot{k}_s/(k_s)^2\right) F_s$. In this case, $\dot{d}_t = \dot{d}_s + \dot{d}_d$ leads to the more general equation

$$\dot{F}_s + \left(\frac{\dot{k}_s c_s + (k_s)^2}{k_s c_s}\right) F_s = k_s \dot{d}_t \tag{6.94}$$

This equation reduces to Eq. 93 if k_s is constant.

Bushing Element The bushing element can be used to connect two arbitrary bodies i and j in the system, as shown in Figure 10. The attachment points on bodies i and j are assumed to be points P^i and P^j, respectively. An assumption is made that the relative translation and rotation between the two bodies at the bushing definition point are small. For the convenience of describing the relative rotations between the two bodies, two *bushing coordinate systems* on bodies i and j are introduced. These two coordinate systems can be selected such that their axes are assumed to be initially parallel. The bushing elements, often made of rubber materials and widely used in developing virtual prototyping models in a wide class of applications including railroad vehicle systems, produce restoring and damping forces in different translational and rotational directions. The bushing stiffness and damping coefficients can be described in the virtual models analytically or numerically using linear and/or nonlinear characteristics.

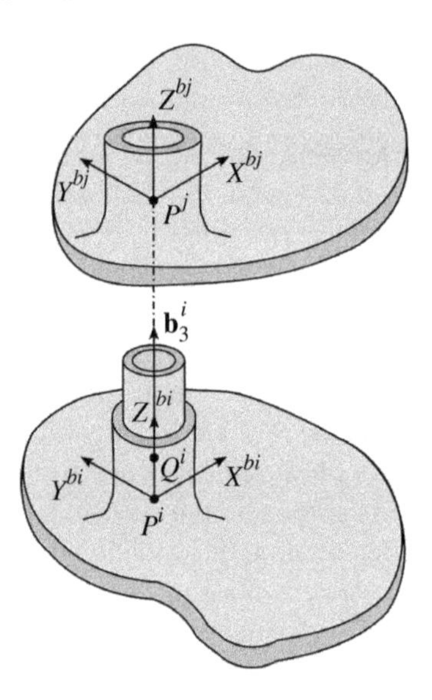

Figure 6.10 Bushing element.

To develop the mathematical model of the bushing element and associate the relative displacements and rotations between the two bodies connected by the bushing element with the bushing material properties, the *bushing coordinate systems* $X^{bi}Y^{bi}Z^{bi}$ and $X^{bj}Y^{bj}Z^{bj}$, which are assumed to be rigidly attached to bodies i and j connected by the bushing element, as shown in Figure 10, are used. Because the bushing coordinate system $X^{bi}Y^{bi}Z^{bi}$ is assumed to be attached to body i, this coordinate system can be defined using two points P^i and Q^i on body i. These two points can be used to define one of the axes of the bushing coordinate system in body i coordinate system $X^iY^iZ^i$ using the unit vector $\overline{\mathbf{b}}_3^i = \left(\overline{\mathbf{u}}_P^i - \overline{\mathbf{u}}_Q^i\right) / \left|\overline{\mathbf{u}}_P^i - \overline{\mathbf{u}}_Q^i\right|$, where $\overline{\mathbf{u}}_P^i$ and $\overline{\mathbf{u}}_Q^i$ are, respectively, the constant local position vectors of points P^i and Q^i. Using this axis, a simple procedure can be used to complete an orthogonal triad defined by the orthogonal matrix $\overline{\mathbf{A}}^{bi} = \left[\overline{\mathbf{b}}_1^i \;\; \overline{\mathbf{b}}_2^i \;\; \overline{\mathbf{b}}_3^i\right]$, where $\overline{\mathbf{b}}_1^i$ and $\overline{\mathbf{b}}_2^i$ are two unit vectors orthogonal to the unit vector $\overline{\mathbf{b}}_3^i$. The axes of the bushing coordinate system $X^{bi}Y^{bi}Z^{bi}$ can be defined in the global coordinate system using the transformation $\mathbf{A}^{bi} = \mathbf{A}^i\overline{\mathbf{A}}^{bi} = \left[\mathbf{b}_1^i \;\; \mathbf{b}_2^i \;\; \mathbf{b}_3^i\right]$, where $\mathbf{b}_k^i = \mathbf{A}^i\overline{\mathbf{b}}_k^i$, $k = 1, 2, 3$. The transformation matrix $\mathbf{A}^{bi}$

is used to define the relative displacements and rotations between the two bodies in the bushing coordinate system, thereby giving these displacements and rotations obvious meanings that can be associated with the bushing material properties, including the stiffness and damping in different directions.

To define, on body j, the bushing coordinate system $X^{bj}Y^{bj}Z^{bj}$, which has axes that are initially parallel to the axes of the bushing coordinate system $X^{bi}Y^{bi}Z^{bi}$ on body i, before the start of the simulation, matrix $\overline{\mathbf{A}}_o^{bj} = \mathbf{A}_o^{j^T}\mathbf{A}_o^{bi} = \begin{bmatrix}\overline{\mathbf{b}}_1^j & \overline{\mathbf{b}}_2^j & \overline{\mathbf{b}}_3^j\end{bmatrix}_o$ is defined, where subscript o refers to the initial configuration. Matrix $\overline{\mathbf{A}}_o^{bj}$ defines the axes of a coordinate system $X^{bj}Y^{bj}Z^{bj}$ rigidly attached to body j that coincide with the axes of the bushing coordinate system $X^{bi}Y^{bi}Z^{bi}$ rigidly attached to body i in the initial configuration. As body j moves with respect to body i, one can define the matrix $\mathbf{A}^{bj} = \mathbf{A}^j\overline{\mathbf{A}}_o^{bj}$ that defines the orientation of the coordinate system $X^{bj}Y^{bj}Z^{bj}$ in the global coordinate system.

The relative displacement and velocity between two points P^j and P^i can be defined, respectively, in the bushing coordinate system $X^{bi}Y^{bi}Z^{bi}$ as

$$\left.\begin{aligned}\mathbf{d}^{bji} &= \mathbf{A}^{bi^T}\mathbf{r}_P^{ji} = \mathbf{A}^{bi^T}\left(\mathbf{R}^j + \mathbf{A}^j\overline{\mathbf{u}}_P^j - \mathbf{R}^i - \mathbf{A}^i\overline{\mathbf{u}}_P^i\right)\\ \mathbf{v}^{bji} &= \mathbf{A}^{bi^T}\dot{\mathbf{r}}_P^{ji} = \mathbf{A}^{bi^T}\left(\dot{\mathbf{R}}^j + \boldsymbol{\omega}^j \times \mathbf{u}_P^j - \dot{\mathbf{R}}^i - \boldsymbol{\omega}^i \times \mathbf{u}_P^i\right)\end{aligned}\right\} \tag{6.95}$$

The relative rotations between the two bodies can be associated with the bushing axes using matrix $\mathbf{A}^{bji} = \mathbf{A}^{bi^T}\mathbf{A}^{bj}$. In the case of small rotations, matrix $\mathbf{A}^{bji}$ can be written in terms of small relative rotations β_1^{bji}, β_2^{bji}, and β_3^{bji}, which can be extracted from matrix $\mathbf{A}^{bji} = \mathbf{A}^{bi^T}\mathbf{A}^{bj}$, which can be computed in a straightforward manner if the coordinates of the two bodies are known from the numerical integration. Similarly, the components of the relative angular velocity vector can be defined in the bushing coordinate system as $\boldsymbol{\omega}^{bji} = \mathbf{A}^{bi^T}\left(\boldsymbol{\omega}^j - \boldsymbol{\omega}^i\right)$. The translational bushing stiffness and damping properties can, in general, be defined using the 3×3 stiffness and damping matrices $\mathbf{K}_t^b$ and $\mathbf{D}_t^b$. Similarly, the rotational bushing stiffness and damping coefficients can be defined using the 3×3 stiffness and damping matrices $\mathbf{K}_r^b$ and $\mathbf{D}_r^b$. The elements of the stiffness and damping matrices $\mathbf{K}_t^b$, $\mathbf{K}_r^b$, $\mathbf{D}_t^b$, and $\mathbf{D}_r^b$ can be determined using experimental measurements. Using these matrices, the bushing restoring and damping forces and moments can be defined in the bushing coordinate system $X^{bi}Y^{bi}Z^{bi}$, respectively, as

$$\overline{\mathbf{F}}_R^b = \mathbf{K}_t^b\mathbf{d}^{bji} + \mathbf{D}_t^b\mathbf{v}^{bji}, \qquad \overline{\mathbf{M}}_\alpha^b = \mathbf{K}_r^b\boldsymbol{\beta}^{bji} + \mathbf{D}_r^b\boldsymbol{\omega}^{bji} \tag{6.96}$$

where $\boldsymbol{\beta}^{bji} = \begin{bmatrix}\beta_1^{bji} & \beta_2^{bji} & \beta_3^{bji}\end{bmatrix}^T$ are the small relative rotations between the two bodies. The force and moment vectors of Eq. 96 can be defined, respectively, in the global coordinate system XYZ as $\mathbf{F}_R^b = \mathbf{A}^{bi}\overline{\mathbf{F}}_R^b$ and $\mathbf{M}_\alpha^b = \mathbf{A}^{bi}\overline{\mathbf{M}}_\alpha^b$. Using these force and moment vectors, the generalized bushing forces associated with the generalized coordinates of the two bodies can be written as

$$\left.\begin{aligned}\mathbf{Q}_R^i &= \mathbf{F}_R^b, &\quad \mathbf{Q}_\theta^i &= \mathbf{G}^{i^T}\left(\mathbf{M}_\alpha^b + \mathbf{u}_P^i \times \mathbf{F}_R^b\right)\\ \mathbf{Q}_R^j &= -\mathbf{F}_R^b, &\quad \mathbf{Q}_\theta^j &= -\mathbf{G}^{j^T}\left(\mathbf{M}_\alpha^b + \mathbf{u}_P^j \times \mathbf{F}_R^b\right)\end{aligned}\right\} \tag{6.97}$$

The generalized bushing forces defined in this equation can be added to the vector of the system generalized applied forces $\mathbf{Q}_e$.

6.11 TRAJECTORY COORDINATES

The absolute Cartesian coordinates used in this chapter to formulate the nonlinear dynamic equations of motion of a railroad vehicle system allow developing general computational MBS algorithms. This is particularly true when such algorithms are generalized for flexible body dynamics, an important generalization for strength and durability investigations that are routinely conducted for tank car strength and rail integrity. Other sets of coordinates, however, have been used in the literature to develop specialized railroad vehicle system software. While using specialized algorithms can make the implementation of flexible-body formulations more cumbersome, these specialized algorithms can be efficient in developing software for specific applications that require the use of fewer degrees of freedom for each body. Examples of these applications are in the important area of *longitudinal train dynamics* (LTD) in which the focus is on the longitudinal coupler and braking forces as well as energy consumption.

Trajectory Coordinates and Track Geometry This section presents the formulation of the equations of motion using *trajectory coordinates*, which are used to develop specialized railroad vehicle and LTD algorithms. As demonstrated in Chapter 3, trajectory and Cartesian coordinates are related by coordinate transformation, thereby allowing for systematically converting the equations of motion expressed in terms of Cartesian coordinates to equations expressed in terms of trajectory coordinates. As explained in Chapter 3 and shown in Figure 11, the general displacement of a rigid body i in the system can be described using the six trajectory coordinates

$$\mathbf{p}^i = \begin{bmatrix} s^i & y^{ir} & z^{ir} & \psi^{ir} & \phi^{ir} & \theta^{ir} \end{bmatrix}^T \tag{6.98}$$

where s^i is the arc length coordinate of the track space curve; y^{ir} and z^{ir} are, respectively, the lateral and vertical displacements of the origin O^i of the body coordinate system $X^iY^iZ^i$ relative to the trajectory body coordinate system $X^{ti}Y^{ti}Z^{ti}$ that follows the body, as shown in Figure 11; and ψ^{ir}, ϕ^{ir}, and θ^{ir} are, respectively, the *yaw*, *roll*, and *pitch* angles that define the orientation of the body coordinate system $X^iY^iZ^i$ with respect to the trajectory body coordinate system $X^{ti}Y^{ti}Z^{ti}$. As discussed in Chapter 3, if arc length parameter s^i is known, the location of the origin of the trajectory body coordinate system $X^{ti}Y^{ti}Z^{ti}$ can be defined and written as $\mathbf{R}^{ti} = \mathbf{R}^{ti}(s^i)$.

Euler Angles and Geometry As discussed in Chapters 3 and 4, another set of Euler angles, expressed as field variables in arc length parameter s^i, can be used to describe track geometry. Therefore, arc length parameter s^i is sufficient to define the orientation of the trajectory body coordinate system $X^{ti}Y^{ti}Z^{ti}$ in terms of three Euler angles $\psi^{ti}(s^i)$, $\theta^{ti}(s^i)$, and $\phi^{ti}(s^i)$ about the three axes Z^{ti}, $-Y^{ti}$, and $-X^{ti}$. These three Euler angles, used to define the orientation of the trajectory body coordinate system $X^{ti}Y^{ti}Z^{ti}$, are considered geometric field variables that only depend on the arc length parameter s^i and should be distinguished from the time-dependent yaw, roll, and pitch Euler angles $\psi^{ir}(t)$, $\phi^{ir}(t)$, and $\theta^{ir}(t)$ previously introduced to define the orientation of the body coordinate system with respect to its trajectory body coordinate system. As explained in Chapter 3, these geometry Euler angles can

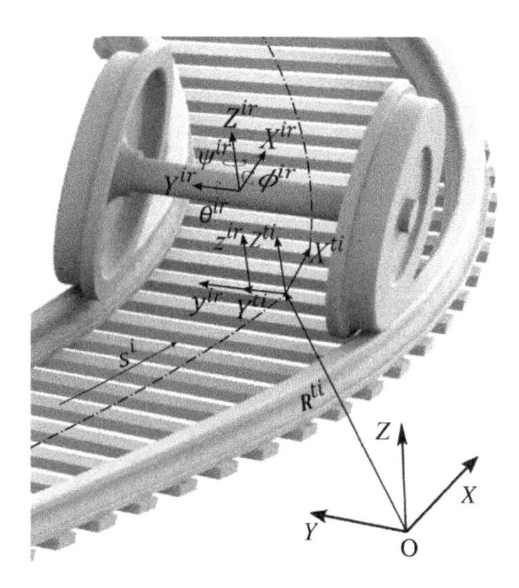

Figure 6.11 Trajectory coordinates.

be used to construct the following transformation matrix that defines the orientation of the trajectory body coordinate system $X^{ti}Y^{ti}Z^{ti}$:

$$\mathbf{A}^{ti} = \begin{bmatrix} \cos\psi^{ti}\cos\theta^{ti} & -\sin\psi^{ti}\cos\phi^{ti} + \cos\psi^{ti}\sin\theta^{ti}\sin\phi^{ti} & -\sin\psi^{ti}\sin\phi^{ti} - \cos\psi^{ti}\sin\theta^{ti}\cos\phi^{ti} \\ \sin\psi^{ti}\cos\theta^{ti} & \cos\psi^{ti}\cos\phi^{ti} + \sin\psi^{ti}\sin\theta^{ti}\sin\phi^{ti} & \cos\psi^{ti}\sin\phi^{ti} - \sin\psi^{ti}\sin\theta^{ti}\cos\phi^{ti} \\ \sin\theta^{ti} & -\cos\theta^{ti}\sin\phi^{ti} & \cos\theta^{ti}\cos\phi^{ti} \end{bmatrix} \tag{6.99}$$

If the geometry of the track space curve is given, the field-variable Euler angles $\psi^{ti} = \psi^{ti}(s^i)$, $\theta^{ti} = \theta^{ti}(s^i)$, and $\phi^{ti} = \phi^{ti}(s^i)$ can be determined for a given value of arc length parameter s^i. That is, vector $\boldsymbol{\theta}^{ti} = \begin{bmatrix}\psi^{ti}(s^i) & \theta^{ti}(s^i) & \phi^{ti}(s^i)\end{bmatrix}^T$ at a given s^i is assumed to be known from the specified track geometry. Determining $\boldsymbol{\theta}^{ti} = \begin{bmatrix}\psi^{ti}(s^i) & \theta^{ti}(s^i) & \phi^{ti}(s^i)\end{bmatrix}^T$ for a given value of s^i using the track data file produced by the track preprocessor was discussed in more detail in Chapter 4.

Position, Velocity, and Acceleration Vectors It was shown in Chapter 3 that, using the trajectory body coordinate system, the global position vector of the origin O^i of the body i coordinate system can be defined as

$$\mathbf{R}^i = \mathbf{R}^{ti}(s^i) + \mathbf{A}^{ti}(s^i)\bar{\mathbf{u}}^{ir} \tag{6.100}$$

where vector $\bar{\mathbf{u}}^{ir} = \begin{bmatrix}0 & y^{ir} & z^{ir}\end{bmatrix}^T$ is the position vector of reference point O^i with respect to the origin of the trajectory body coordinate system $X^{ti}Y^{ti}Z^{ti}$. As discussed in Chapter 3, because the location of point O^i in the longitudinal direction is assumed to be defined using arc length s^i, the first element of vector $\bar{\mathbf{u}}^{ir}$ is equal to zero. The absolute velocity and acceleration vectors of reference point O^i, respectively, can be obtained by differentiating Eq. 100 once and twice with respect to time. This leads to

$$\dot{\mathbf{R}}^i = \mathbf{L}^i\dot{\mathbf{p}}^i, \qquad \ddot{\mathbf{R}}^i = \mathbf{L}^i\ddot{\mathbf{p}}^i + \boldsymbol{\gamma}_R^i \tag{6.101}$$

where $\mathbf{L}^i = \left[\left(\left(\partial \mathbf{R}^{ti}/\partial s^i\right) + \left(\partial \mathbf{A}^{ti}/\partial s^i\right)\overline{\mathbf{u}}^{ir}\right) \quad \mathbf{a}_2^{ti} \quad \mathbf{a}_3^{ti} \quad \mathbf{0}\right]$ is a 3×6 matrix in which $\mathbf{a}_2^{ti}$ and $\mathbf{a}_3^{ti}$ are the second and third columns of the transformation matrix $\mathbf{A}^{ti}$ of Eq. 99, $\boldsymbol{\gamma}_R^i$ absorbs the terms that are not linear in generalized accelerations, and $\mathbf{0}$ is the 3×3 null matrix.

The absolute angular velocity vector of body i can be written, as explained in Chapter 3, as $\boldsymbol{\omega}^i = \boldsymbol{\omega}^{ti} + \boldsymbol{\omega}^{ir}$, which is the sum of the absolute angular velocity vector $\boldsymbol{\omega}^{ti} = \mathbf{G}^{ti}\dot{\boldsymbol{\theta}}^{ti}$ of the trajectory body coordinate system and vector $\boldsymbol{\omega}^{ir} = \mathbf{A}^{ti}\mathbf{G}^{ir}\dot{\boldsymbol{\theta}}^{ir}$, which defines the angular velocity of the body with respect to the trajectory body coordinate system. Therefore, the absolute angular velocity $\boldsymbol{\omega}^i$ of body i can be written in terms of the time derivatives of the trajectory coordinates as

$$\boldsymbol{\omega}^i = \mathbf{G}^{ti}\dot{\boldsymbol{\theta}}^{ti} + \mathbf{A}^{ti}\mathbf{G}^{ir}\dot{\boldsymbol{\theta}}^{ir} = \mathbf{H}^i\dot{\mathbf{p}}^i \tag{6.102}$$

where matrices $\mathbf{G}^{ti}$ and $\mathbf{G}^{ir}$ are, respectively, defined as

$$\mathbf{G}^{ti} = \begin{bmatrix} 0 & \sin\psi^{ti} & -\cos\psi^{ti}\cos\theta^{ti} \\ 0 & -\cos\psi^{ti} & -\sin\psi^{ti}\cos\theta^{ti} \\ 1 & 0 & -\sin\theta^{ti} \end{bmatrix}, \quad \mathbf{G}^{ir} = \begin{bmatrix} 0 & \cos\psi^{ir} & -\sin\psi^{ir}\cos\phi^{ir} \\ 0 & \sin\psi^{ir} & \cos\psi^{ir}\cos\phi^{ir} \\ 1 & 0 & \sin\phi^{ir} \end{bmatrix} \tag{6.103}$$

and $\mathbf{H}^i = \left[\left(\mathbf{G}^{ti}\left(\partial\boldsymbol{\theta}^{ti}/\partial s^i\right)\right) \quad \mathbf{0} \quad \mathbf{0} \quad \left(\mathbf{A}^{ti}\mathbf{G}^{ir}\right)\right]$ is a 3×6 matrix in which $\mathbf{0}$ is the three-dimensional zero vector. Differentiating Eq. 102 with respect to time, the absolute angular acceleration vector $\boldsymbol{\alpha}^i$ of body i can be expressed in terms of the second time derivatives of the trajectory coordinates as $\boldsymbol{\alpha}^i = \mathbf{H}^i\ddot{\mathbf{p}}^i + \boldsymbol{\gamma}_\alpha^i$, where $\boldsymbol{\gamma}_\alpha^i$ is a vector that absorbs terms that are not linear in accelerations (Sanborn et al. 2007; Sinokrot et al. 2008; Shabana et al. 2008). This equation that defines $\boldsymbol{\alpha}^i$ can also be used to write the absolute angular acceleration vector defined in the body coordinate system $X^iY^iZ^i$ as

$$\overline{\boldsymbol{\alpha}}^i = \mathbf{A}^{i^T}\boldsymbol{\alpha}^i = \overline{\mathbf{H}}^i\ddot{\mathbf{p}}^i + \overline{\boldsymbol{\gamma}}_\alpha^i \tag{6.104}$$

where $\overline{\mathbf{H}}^i = \mathbf{A}^{i^T}\mathbf{H}^i = \mathbf{A}^{ir^T}\mathbf{A}^{ti^T}\mathbf{H}^i$, $\overline{\boldsymbol{\gamma}}_\alpha^i = \mathbf{A}^{i^T}\boldsymbol{\gamma}_\alpha^i$, and $\mathbf{A}^{ir}$ is the matrix that defines the orientation of the body coordinate system $X^iY^iZ^i$ with respect to the trajectory body coordinate system $X^{ti}Y^{ti}Z^{ti}$. This matrix, as explained in Chapter 3, can be written in terms of the Euler angle generalized coordinates $\boldsymbol{\theta}^{ir} = \left[\psi^{ir} \quad \phi^{ir} \quad \theta^{ir}\right]^T$ as

$$\mathbf{A}^{ir} = \begin{bmatrix} \cos\psi^{ir}\cos\theta^{ir} - \sin\psi^{ir}\sin\phi^{ir}\sin\theta^{ir} & -\sin\psi^{ir}\cos\phi^{ir} & \cos\psi^{ir}\sin\theta^{ir} + \sin\psi^{ir}\sin\phi^{ir}\cos\theta^{ir} \\ \sin\psi^{ir}\cos\theta^{ir} + \cos\psi^{ir}\sin\phi^{ir}\sin\theta^{ir} & \cos\psi^{ir}\cos\phi^{ir} & \sin\psi^{ir}\sin\theta^{ir} - \cos\psi^{ir}\sin\phi^{ir}\cos\theta^{ir} \\ -\cos\phi^{ir}\sin\theta^{ir} & \sin\phi^{ir} & \cos\phi^{ir}\cos\theta^{ir} \end{bmatrix} \tag{6.105}$$

Acceleration Transformation Combining the second equation in Eqs. 101 and 104, one obtains (Sanborn et al. 2007; Sinokrot et al. 2008)

$$\begin{bmatrix} \ddot{\mathbf{R}}^i \\ \overline{\boldsymbol{\alpha}}^i \end{bmatrix} = \begin{bmatrix} \mathbf{L}^i \\ \overline{\mathbf{H}}^i \end{bmatrix}\ddot{\mathbf{p}}^i + \begin{bmatrix} \boldsymbol{\gamma}_R^i \\ \overline{\boldsymbol{\gamma}}_\alpha^i \end{bmatrix} \tag{6.106}$$

In this acceleration transformation, the absolute angular acceleration vector $\overline{\boldsymbol{\alpha}}^i$ is used instead of vector $\boldsymbol{\alpha}^i$ because the mass matrix associated with the absolute Cartesian accelerations $\begin{bmatrix} \ddot{\mathbf{R}}^{iT} & \overline{\boldsymbol{\alpha}}^{iT} \end{bmatrix}^T$ is constant in the Newton–Euler formulation of the equations of motion. Equation 106 can be used with the Newton–Euler equations to obtain the equations of motion in terms of trajectory coordinates without using any small-angle assumptions; therefore, these equations of motion include all the nonlinear effects that can be significant in some motion scenarios.

Trajectory-Coordinate Equations of Motion The form of the Newton–Euler equations given in Eq. 47 is used with the acceleration transformation of Eq. 106 to obtain the nonlinear equations of motion expressed in terms of trajectory coordinates. To this end, the absolute accelerations in Eq. 106, written in terms of trajectory accelerations, are substituted in Eq. 47, and Eq. 47 is premultiplied by the transpose of the coefficient matrix of $\ddot{\mathbf{p}}^i$ in Eq. 106. This leads to (Sanborn et al. 2007; Sinokrot et al. 2008; Shabana et al. 2008)

$$\mathbf{M}_{pp}^i \ddot{\mathbf{p}}^i = \mathbf{Q}_{pe}^i + \mathbf{Q}_{pv}^i \tag{6.107}$$

where

$$\left.\begin{aligned} \mathbf{M}_{pp}^i &= m^i \mathbf{L}^{i^T} \mathbf{L}^i + \overline{\mathbf{H}}^{i^T} \overline{\mathbf{I}}_{\theta\theta}^i \overline{\mathbf{H}}^i \\ \mathbf{Q}_{pe}^i &= \mathbf{L}^{i^T} \mathbf{F}_e^i + \overline{\mathbf{H}}^{i^T} \overline{\mathbf{M}}_e^i \\ \mathbf{Q}_{pv}^i &= -m^i \mathbf{L}^{i^T} \boldsymbol{\gamma}_R^i - \overline{\mathbf{H}}^{i^T} \left[\overline{\mathbf{I}}_{\theta\theta}^i \overline{\boldsymbol{\gamma}}_\alpha^i + \overline{\boldsymbol{\omega}}^i \times \left(\overline{\mathbf{I}}_{\theta\theta}^i \overline{\boldsymbol{\omega}}^i \right) \right] \end{aligned}\right\} \tag{6.108}$$

Mass matrix $\mathbf{M}_{pp}^i$ in this equation is not constant as it is the case when absolute Cartesian coordinates are used with the Newton–Euler equations of Eq. 47.

In the case of constrained motion, kinematic algebraic constraint functions can be formulated in terms of trajectory coordinates. In this case, the augmented form of the equations of motion can be written as (Shabana et al. 2008)

$$\begin{bmatrix} \mathbf{M}_{pp} & \mathbf{C}_{\mathbf{p}}^T \\ \mathbf{C}_{\mathbf{p}} & \mathbf{0} \end{bmatrix} \begin{bmatrix} \ddot{\mathbf{p}} \\ \boldsymbol{\lambda} \end{bmatrix} = \begin{bmatrix} \mathbf{Q}_{pe} + \mathbf{Q}_{pv} \\ \mathbf{Q}_d \end{bmatrix} \tag{6.109}$$

where $\mathbf{p} = \begin{bmatrix} \mathbf{p}^{1T} & \mathbf{p}^{2T} & \dots & \mathbf{p}^{n_b T} \end{bmatrix}^T$ is the vector of system trajectory coordinates, n_b is the total number of bodies in the system, $\mathbf{M}_{pp}$ is the system mass matrix associated with the trajectory coordinates, $\mathbf{Q}_{pe}$ is the vector of system applied forces associated with the trajectory coordinates, $\mathbf{Q}_{pv}$ is the vector of centrifugal and Coriolis forces, $\mathbf{C}_{\mathbf{p}}$ is the Jacobian matrix of the kinematic constraint equations, $\boldsymbol{\lambda}$ is the vector of Lagrange multipliers, and $\mathbf{Q}_d$ is the vector resulting from the differentiation of the system constraint equations twice with respect to time.

Trajectory Coordinate Constraints. Using trajectory coordinates allows developing simpler formulations for some types of motion constraints. For example, if it is desired to specify the forward velocity V of a vehicle along the track centerline, one can use the *constant forward velocity constraint*, which can be expressed in terms of trajectory coordinate s^i as (Shabana et al. 2008)

$$C = s^i - V\,t - s_0^i = 0 \tag{6.110}$$

where s_0^i is the initial arc length coordinate. The constraint condition of Eq. 110 ensures that the origin of the trajectory body coordinate system moves with a constant velocity $\dot{s}^i = V$

along the track centerline. The constraint of Eq. 110 is a *linear function* in s^i. This linearity cannot be achieved if absolute Cartesian coordinates are used.

Similar linear conditions can be developed when constraints are imposed on the yaw, roll, or pitch angles of the vehicle. For example, the constraint that specifies the yaw motion of a body with respect to the trajectory body coordinate system can be simply written in the case of the trajectory coordinates as

$$C = \psi^{ir} - \psi_0^{ir} = 0 \tag{6.111}$$

where ψ_0^{ir} is the yaw angle in the initial configuration. In general, one can specify any of the trajectory coordinates as a function of time using the constraint equation

$$C = p_k^i - f(t) = 0 \tag{6.112}$$

where p_k^i is the kth trajectory coordinate in vector $\mathbf{p}^i$ of Eq. 98, and $f(t)$ is a given function of time. Using the simple trajectory constraints of Eqs. 111 and 112 leads to constant elements in the constraint Jacobian matrix and allows for a straightforward elimination of dependent coordinates. Using trajectory coordinates to formulate other constraint equations is demonstrated by the following example.

Example 6.7 The wheelset shown in Figure 12 is denoted as body i and subjected to the following three kinematic constraint conditions: (i) the forward velocity of the wheelset along the track centerline is assumed to be g(t); (ii) the wheelset is assumed to rotate about its own axle with a constant angular velocity ω^i (pitch rotation); and (iii) the wheelset is assumed to have zero displacement along the Z^{ti} axis of the trajectory body coordinate system. The three constraint equations imposed on the motion of the wheelset can be written mathematically as

$$\mathbf{C}(\mathbf{p}, t) = \begin{bmatrix} C_1 \\ C_2 \\ C_3 \end{bmatrix} = \begin{bmatrix} s^i - g(t) \\ \theta^{ir} - \theta_0^{ir} - \omega^i t \\ z^{ir} - z_0^{ir} \end{bmatrix} = \mathbf{0}$$

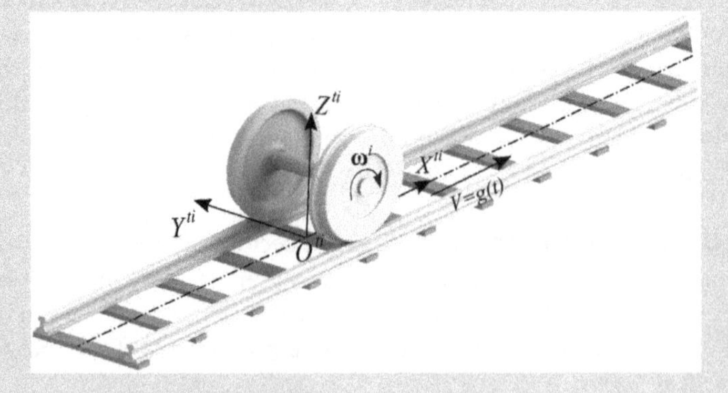

Figure 6.12 Trajectory coordinate constraints.

where θ_o^{ir} is the initial value of the pitch rotation and z_o^{ir} is the initial value of coordinate z^{ir}. Differentiating the constraint equations with respect to time, one obtains $\mathbf{C_p}\dot{\mathbf{p}} + \mathbf{C}_t = \mathbf{0}$, where the constraint Jacobian matrix and vector $\mathbf{C}_t$ are defined as

$$\mathbf{C_p} = \begin{bmatrix} 1 & 0 & 0 & 0 & 0 & 0 \\ 0 & 0 & 0 & 0 & 0 & 1 \\ 0 & 0 & 1 & 0 & 0 & 0 \end{bmatrix}, \qquad \mathbf{C}_t = \begin{bmatrix} -dg(t)/dt \\ -\omega^i \\ 0 \end{bmatrix}$$

The constraint equations at the acceleration level are defined as $\mathbf{C_p}\ddot{\mathbf{p}} = \mathbf{Q}_d$, where

$$\mathbf{Q}_d = \begin{bmatrix} d^2g(t)/dt^2 & 0 & 0 \end{bmatrix}^T$$

Using these kinematic relationships, the embedding technique can be used systematically to obtain three equations of motion for the wheelset. It is clear that the degrees of freedom of this system are y^{ir}, ψ^{ir}, and ϕ^{ir}: that is, the vector of independent coordinates is $\mathbf{p}_i = \begin{bmatrix} y^{ir} & \psi^{ir} & \phi^{ir} \end{bmatrix}^T$, and the vector of dependent coordinates is $\mathbf{p}_d = \begin{bmatrix} s^i & z^{ir} & \theta^{ir} \end{bmatrix}^T$. The constraint Jacobian matrix can be written in a partitioned form as $\mathbf{C_p} = \begin{bmatrix} \mathbf{C}_{\mathbf{p}_d} & \mathbf{C}_{\mathbf{p}_i} \end{bmatrix}$, where

$$\mathbf{C}_{\mathbf{p}_d} = \begin{bmatrix} 1 & 0 & 0 \\ 0 & 0 & 1 \\ 0 & 1 & 0 \end{bmatrix}, \qquad \mathbf{C}_{\mathbf{p}_i} = \begin{bmatrix} 0 & 0 & 0 \\ 0 & 0 & 0 \\ 0 & 0 & 0 \end{bmatrix}$$

It is clear from this partitioning that the system velocities can be written in terms of the independent velocities as

$$\dot{\mathbf{p}} = \begin{bmatrix} \dot{s}^i \\ \dot{y}^{ir} \\ \dot{z}^{ir} \\ \dot{\psi}^{ir} \\ \dot{\phi}^{ir} \\ \dot{\theta}^{ir} \end{bmatrix} = \begin{bmatrix} dg(t)/dt \\ \dot{y}^{ir} \\ 0 \\ \dot{\psi}^{ir} \\ \dot{\phi}^{ir} \\ \omega^i \end{bmatrix} = \begin{bmatrix} 0 & 0 & 0 \\ 1 & 0 & 0 \\ 0 & 0 & 0 \\ 0 & 1 & 0 \\ 0 & 0 & 1 \\ 0 & 0 & 0 \end{bmatrix} \begin{bmatrix} \dot{y}^{ir} \\ \dot{\psi}^{ir} \\ \dot{\phi}^{ir} \end{bmatrix} + \begin{bmatrix} dg(t)/dt \\ 0 \\ 0 \\ 0 \\ 0 \\ \omega^i \end{bmatrix}$$

Using this equation, the total vector of accelerations can be written in the form of the second equation in Eq. 63 as $\ddot{\mathbf{p}} = \mathbf{B}_{di}\ddot{\mathbf{p}}_i + \boldsymbol{\gamma}_i$, where

$$\mathbf{B}_{di} = \begin{bmatrix} 0 & 0 & 0 \\ 1 & 0 & 0 \\ 0 & 0 & 0 \\ 0 & 1 & 0 \\ 0 & 0 & 1 \\ 0 & 0 & 0 \end{bmatrix}, \qquad \boldsymbol{\gamma}^i = \begin{bmatrix} d^2g(t)/dt^2 \\ 0 \\ 0 \\ 0 \\ 0 \\ 0 \end{bmatrix}$$

Substituting $\ddot{\mathbf{p}} = \mathbf{B}_{di}\ddot{\mathbf{p}}_i + \boldsymbol{\gamma}_i$ in Eq. 107, given as $\mathbf{M}_{pp}^i\ddot{\mathbf{p}}^i = \mathbf{Q}_{pe}^i + \mathbf{Q}_{pv}^i$, and premultiplying by the transpose of the velocity transformation matrix $\mathbf{B}_{di}$, one obtains three equations of

(Continued)

motion from which the constraint forces are eliminated. These equations can be written in the form of Eq. 64 as

$$\mathbf{M}_{ii}\ddot{\mathbf{p}}_i = \mathbf{B}_{di}^T\left(\mathbf{Q}_{pe} + \mathbf{Q}_{pv} - \mathbf{M}_{pp}\boldsymbol{\gamma}_i\right)$$

where $\mathbf{M}_{ii} = \mathbf{B}_{di}^T\mathbf{M}_{pp}\mathbf{B}_{di}$.

6.12 LONGITUDINAL TRAIN DYNAMICS (LTD)

LTD algorithms are used in the analysis of railroad vehicle coupler and braking forces as well as energy consumption of long trains traveling for long distances. The train models can include 100 or 200 rail cars. For this reason, it is necessary to develop efficient algorithms in which fewer degrees of freedom are used for each rail car. In most LTD algorithms, a single-degree-of-freedom rail car is used in modeling long trains. This single degree of freedom is assumed to be the longitudinal coordinate s^i for a rail car i. This section discusses formulations that employ one or two degrees of freedom for the rail car.

Single-Degree-of-Freedom LTD Model When using a single degree of freedom for each rail car, it is assumed that the degree of freedom is the arc length coordinate s^i and all other coordinates are constrained. In this case, reference point O^i of the rail car does not deviate from the track centerline. Therefore, no other translational or rotational degrees of freedom are allowed except the motion in the longitudinal direction. Using these simplifying assumptions, one can write the equations of motion of each rail car i in the train in terms of the arc length coordinate s^i only. Since in this case $y^{ir} = y_o^{ir}$ and $z^{ir} = z_o^{ir}$, where subscript o refers to the initial configuration, the position of reference point O^i can be written in the track coordinate system, which is assumed to be fixed and to coincide with the global coordinate system, as $\mathbf{R}^i = \mathbf{R}^{ti}\left(s^i\right) + \mathbf{A}^{ti}\left(s^i\right)\overline{\mathbf{u}}_o^{ir}$, where $\overline{\mathbf{u}}_o^{ir} = \begin{bmatrix}0 & y_o^{ir} & z_o^{ir}\end{bmatrix}^T$ is a constant vector. In this case, the absolute velocity of the reference point can be written as

$$\dot{\mathbf{R}}^i = \dot{\mathbf{R}}^{ti}\left(s^i\right) + \dot{\mathbf{A}}^{ti}\left(s^i\right)\overline{\mathbf{u}}_o^{ir} = \mathbf{L}^i\dot{s}^i \tag{6.113}$$

where $\mathbf{L}^i = \left(\partial\mathbf{R}^{ti}/\partial s^i\right) + \left(\partial\mathbf{A}^{ti}/\partial s^i\right)\overline{\mathbf{u}}_o^{ir}$. Since all the rotations of the rail car with respect to the trajectory body coordinate system $X^{ti}Y^{ti}Z^{ti}$ are constrained – that is, ψ^{ir}, ϕ^{ir}, and θ^{ir} remain constant – the angular velocity of the rail car $\overline{\boldsymbol{\omega}}^i$ is equal to the angular velocity of its trajectory body coordinate system $\overline{\boldsymbol{\omega}}^{ti} = \overline{\mathbf{G}}^{ti}\dot{\boldsymbol{\theta}}^{ti}$. In this case, one has $\dot{\boldsymbol{\theta}}^{ti} = \left(\partial\boldsymbol{\theta}^{ti}/\partial s^i\right)\dot{s}^i$, and (Shabana et al. 2008)

$$\overline{\mathbf{G}}^{ti} = \begin{bmatrix} \sin\theta^{ti} & 0 & -1 \\ -\cos\theta^{ti}\sin\phi^{ti} & -\cos\phi^{ti} & 0 \\ \cos\theta^{ti}\cos\phi^{ti} & -\sin\phi^{ti} & 0 \end{bmatrix} \tag{6.114}$$

Therefore, the angular velocity of rail car i can be written as

$$\overline{\boldsymbol{\omega}}^i = \overline{\boldsymbol{\omega}}^{ti} = \overline{\mathbf{G}}^{ti}\dot{\boldsymbol{\theta}}^{ti} = \overline{\mathbf{G}}^{ti}\left(\frac{\partial\boldsymbol{\theta}^{ti}}{\partial s^i}\right)\dot{s}^i \tag{6.115}$$

By differentiating Eqs. 113 and 115 with respect to time, one obtains Eq. 106, which can be written in this case as

$$\begin{bmatrix} \ddot{\mathbf{R}}^i \\ \bar{\boldsymbol{\alpha}}^i \end{bmatrix} = \begin{bmatrix} \mathbf{L}^i \\ \overline{\mathbf{H}}^i \end{bmatrix} \ddot{s}^i + \begin{bmatrix} \bar{\boldsymbol{\gamma}}_R^i \\ \bar{\boldsymbol{\gamma}}_\alpha^i \end{bmatrix} \tag{6.116}$$

where

$$\left.\begin{aligned} &\overline{\mathbf{H}}^i = \overline{\mathbf{G}}^{ti}\left(\frac{\partial \boldsymbol{\theta}^{ti}}{\partial s^i}\right), \quad \bar{\boldsymbol{\gamma}}_R^i = \dot{\mathbf{L}}^i \dot{s}^i = \left(\frac{\partial^2 \mathbf{R}^{ti}}{\partial s^{i2}} + \left(\frac{\partial^2 \mathbf{A}^{ti}}{\partial s^{i2}}\right)\bar{\mathbf{u}}_o^{ir}\right)\left(\dot{s}^i\right)^2 \\ &\bar{\boldsymbol{\gamma}}_\alpha^i = \left(\left(\frac{\partial \overline{\mathbf{G}}^{ti}}{\partial s^i}\right)\left(\frac{\partial \boldsymbol{\theta}^{ti}}{\partial s^i}\right) + \overline{\mathbf{G}}^{ti}\left(\frac{\partial^2 \boldsymbol{\theta}^{ti}}{\partial s^{i2}}\right)\right)\left(\dot{s}^i\right)^2 \end{aligned}\right\} \tag{6.117}$$

Substituting Eq. 116 into the Newton–Euler equations given in Eq. 47, and premultiplying by the transpose of the coefficient vector of $\ddot{s}^i$ in Eq. 116, the equation of motion of the one-degree-of-freedom rail car can be written as (Sanborn et al. 2007; Sinokrot et al. 2008; Shabana et al. 2008)

$$M_{pp}^i \ddot{s}^i = Q_{pe}^i + Q_{pv}^i \tag{6.118}$$

where $M_{pp}^i = m^i \mathbf{L}^{i^T} \mathbf{L}^i + \overline{\mathbf{H}}^{i^T} \bar{\mathbf{I}}_{\theta\theta}^i \overline{\mathbf{H}}^i$, $Q_{pe}^i = \mathbf{L}^{i^T} \mathbf{F}_e^i + \overline{\mathbf{H}}^{i^T} \overline{\mathbf{M}}_e^i$, and $Q_{pv}^i = -m^i \mathbf{L}^{i^T} \bar{\boldsymbol{\gamma}}_R^i - \overline{\mathbf{H}}^{i^T}\left[\bar{\mathbf{I}}_{\theta\theta}^i \bar{\boldsymbol{\gamma}}_\alpha^i + \bar{\boldsymbol{\omega}}^i \times \left(\bar{\mathbf{I}}_{\theta\theta}^i \bar{\boldsymbol{\omega}}^i\right)\right]$.

Example 6.8 In the case of a rail car i negotiating a curve with zero grade and superelevation, one has $\theta^{ti} = \phi^{ti} = 0$, and the only nonzero track geometry angle is ψ^{ti}. In this case, one has $\mathbf{L}^i = \left(\partial \mathbf{R}^{ti}/\partial s^i\right) + \left(\partial \mathbf{A}^{ti}/\partial s^i\right)\bar{\mathbf{u}}_o^{ir} = \left(\partial \mathbf{R}^{ti}/\partial s^i\right) + \left(\partial \mathbf{A}^{ti}/\partial \psi^{ti}\right)\left(\partial \psi^{ti}/\partial s^i\right)\bar{\mathbf{u}}_o^{ir}$. Using Eq. 99 and the definition of $\bar{\mathbf{u}}_o^{ir}$ as $\bar{\mathbf{u}}_o^{ir} = \begin{bmatrix} 0 & y_o^{ir} & z_o^{ir} \end{bmatrix}^T$, one has $\left(\partial \mathbf{R}^{ti}/\partial s^i\right) = \begin{bmatrix} \cos\psi^{ti} & \sin\psi^{ti} & 0 \end{bmatrix}^T$ and $\left(\partial \mathbf{A}^{ti}/\partial \psi^{ti}\right)\bar{\mathbf{u}}_o^{ir} = -y_o^{ir}\begin{bmatrix} \cos\psi^{ti} & \sin\psi^{ti} & 0 \end{bmatrix}^T$. Therefore, $\mathbf{L}^i$ can be defined as (Shabana et al. 2008)

$$\mathbf{L}^i = \left(1 - y_o^{ir}\psi_s^{ti}\right)\begin{bmatrix} \cos\psi^{ti} \\ \sin\psi^{ti} \\ 0 \end{bmatrix}$$

where $\psi_s^{ti} = \left(\partial \psi^{ti}/\partial s^i\right)$. In this special case of the track geometry, one has

$$\overline{\mathbf{G}}^{ti} = \begin{bmatrix} 0 & 0 & -1 \\ 0 & -1 & 0 \\ 1 & 0 & 0 \end{bmatrix}$$

Therefore, the coefficient matrix $\overline{\mathbf{H}}^i = \overline{\mathbf{G}}^{ti}\left(\partial \boldsymbol{\theta}^{ti}/\partial s^i\right)$ in Eq. 116 can be written as

$$\overline{\mathbf{H}}^i = \overline{\mathbf{G}}^{ti}\left(\frac{\partial \boldsymbol{\theta}^{ti}}{\partial s^i}\right) = \begin{bmatrix} 0 \\ 0 \\ \psi_s^{ti} \end{bmatrix}$$

(Continued)

Since $\partial \overline{\mathbf{G}}^{ti}/\partial s^i = \mathbf{0}$, one can evaluate vectors $\overline{\boldsymbol{\gamma}}_R^i$ and $\overline{\boldsymbol{\gamma}}_\alpha^i$ of Eq. 117, respectively, as

$$\overline{\boldsymbol{\gamma}}_R^i = \left(\frac{\partial^2 \mathbf{R}^{ti}}{\partial s^{i2}} + \left(\frac{\partial^2 \mathbf{A}^{ti}}{\partial s^{i2}}\right)\overline{\mathbf{u}}_o^{ir}\right)(\dot{s}^i)^2$$
$$= -\left[\psi_s^{ti}\left(1 - y_o^{ir}\psi_s^{ti}\right)\begin{bmatrix}\sin\psi^{ti}\\ -\cos\psi^{ti}\\ 0\end{bmatrix} + y_o^{ir}\psi_s^{ti}\begin{bmatrix}\cos\psi^{ti}\\ \sin\psi^{ti}\\ 0\end{bmatrix}\right](\dot{s}^i)^2$$

and

$$\overline{\boldsymbol{\gamma}}_\alpha^i = \left(\left(\frac{\partial \overline{\mathbf{G}}^{ti}}{\partial s^i}\right)\left(\frac{\partial \boldsymbol{\theta}^{ti}}{\partial s^i}\right) + \overline{\mathbf{G}}^{ti}\left(\frac{\partial^2 \boldsymbol{\theta}^{ti}}{\partial s^{i2}}\right)\right)(\dot{s}^i)^2 = \begin{bmatrix}0\\ 0\\ \psi_{ss}^{ti}\end{bmatrix}(\dot{s}^i)^2$$

where $\psi_{ss}^{ti} = \partial^2\psi^{ti}/\partial s^{i2}$.

Using the matrices and vectors developed in this example, the mass matrix $M_{pp}^i = m^i\mathbf{L}^{i^T}\mathbf{L}^i + \overline{\mathbf{H}}^{i^T}\overline{\mathbf{I}}_{\theta\theta}^i\overline{\mathbf{H}}^i$ in Eq. 118 can be evaluated as (Shabana et al. 2008)

$$M_{pp}^i = m^i\mathbf{L}^{i^T}\mathbf{L}^i + \overline{\mathbf{H}}^{i^T}\overline{\mathbf{I}}_{\theta\theta}^i\overline{\mathbf{H}}^i$$
$$= m^i\left(1 - y_o^{ir}\psi_s^{ti}\right)^2 + i_{zz}\left(\psi_s^{ti}\right)^2$$

The scalars Q_{pe}^i, and Q_{pv}^i can be evaluated, respectively, as

$$Q_{pe}^i = \mathbf{L}^{i^T}\mathbf{F}_e^i + \overline{\mathbf{H}}^{i^T}\overline{\mathbf{M}}_e^i$$
$$= \left(1 - y_o^{ir}\psi_s^{ti}\right)\left(F_{ex}^i\cos\psi^{ti} + F_{ey}^i\sin\psi^{ti}\right) + \psi_s^{ti}\overline{M}_{ez}^i$$

and

$$Q_{pv}^i = -m^i\mathbf{L}^{i^T}\overline{\boldsymbol{\gamma}}_R^i - \overline{\mathbf{H}}^{i^T}\left[\overline{\mathbf{I}}_{\theta\theta}^i\overline{\boldsymbol{\gamma}}_\alpha^i + \overline{\boldsymbol{\omega}}^i \times \left(\overline{\mathbf{I}}_{\theta\theta}^i\overline{\boldsymbol{\omega}}^i\right)\right]$$
$$= \left(m^i y_o^{ir}\left(1 - y_o^{ir}\psi_s^{ti}\right) - i_{zz}\psi_s^{ti}\right)\psi_{ss}^{ti}(\dot{s}^i)^2$$

where in this special example of planar motion, the gyroscopic moment $\overline{\boldsymbol{\omega}}^i \times \left(\overline{\mathbf{I}}_{\theta\theta}^i\overline{\boldsymbol{\omega}}^i\right)$ is equal to zero, $\mathbf{F}_e^i = \left[F_{ex}^i \;\; F_{ey}^i \;\; F_{ez}^i\right]^T$ is the force vector defined in the global coordinate system XYZ, and $\overline{\mathbf{M}}_\alpha^i = \left[\overline{M}_{\alpha x}^i \;\; \overline{M}_{\alpha y}^i \;\; \overline{M}_{\alpha z}^i\right]^T$ is the moment vector defined in the rail car coordinate system $X^iY^iZ^i$. Substituting M_{pp}^i, Q_{pe}^i, and Q_{pv}^i into the equation of motion of Eq. 118, one obtains (Shabana et al. 2008)

$$\left(m^i\left(1 - y_o^{ir}\psi_s^{ti}\right)^2 + I_{zz}^i\left(\psi_s^{ti}\right)^2\right)\ddot{s}^i = \left(1 - y_o^{ir}\psi_s^{ti}\right)\left(F_{ex}^i\cos\psi^{ti} + F_{ey}^i\sin\psi^{ti}\right) + \psi_s^{ti}\overline{M}_{ez}^i$$
$$+ \left[m^i y_o^{ir}\left(1 - y_o^{ir}\psi_s^{ti}\right) - I_{zz}^i\psi_s^{ti}\right]\psi_{ss}^{ti}(\dot{s}^i)^2$$

If $y_o^{ir} = 0$, the preceding equation reduces to

$$\left[m^i + I_{zz}^i\left(\psi_s^{ti}\right)^2\right]\ddot{s}^i = F_{ex}^i\cos\psi^{ti} + F_{ey}^i\sin\psi^{ti} + \psi_s^{ti}\overline{M}_{ez}^i - I_{zz}^i\psi_s^{ti}\psi_{ss}^{ti}(\dot{s}^i)^2$$

Furthermore, if the track is assumed to be tangent, $\psi^{ti} = \psi_s^{ti} = \psi_{ss}^{ti} = 0$, and the preceding equation is further simplified to $m^i\ddot{s}^i = F_{ex}^i$.

Two-Degree-of-Freedom Rail Car Model The single-degree-of-freedom rail car model has been widely used in the development of LTD algorithms designed to efficiently solve problems of long freight trains that can have more than 100 rail cars. These problems focus on the longitudinal coupler and braking forces as well as energy consumptions. However, the simplified single-degree-of-freedom rail car model cannot be used to capture important phenomena such as wheelset hunting, which is the result of coupling between the lateral and yaw displacements of the wheelsets. Therefore, simplified two-degree-of-freedom models have been developed in the literature to study the hunting phenomenon as well as wheelset stability.

In the two-degree-of-freedom model discussed in this section, the lateral displacement y^{ir} and yaw angle ψ^{ir} are assumed to be the two independent coordinates of the wheelset. As in the case of the single-degree-of-freedom model, the two-degree-of-freedom model can be derived as a special case from the general trajectory coordinate formulation previously introduced in this chapter. This can be achieved by imposing proper kinematic constraint conditions on the other four trajectory coordinates s^i, z^{ir}, ϕ^{ir}, and θ^{ir}. In the general case of profiled wheels and rails, the displacement in the normal direction z^{ir} and roll angle ϕ^{ir} depend on the lateral displacement y^{ir} and yaw angle ψ^{ir}: that is, $z^{ir} = z^{ir}(y^{ir}, \psi^{ir})$ and $\phi^{ir} = \phi^{ir}(y^{ir}, \psi^{ir})$. If the forward velocity and pitch rotations of the wheelset are prescribed, one can write two constraints on s^i and θ^{ir}. These two constraint equations can be written, respectively, as $C_1 = s^i - s_o^i - Vt = 0$ and $C_2 = \theta^{ir} - \theta_o^{ir} - \omega_p t = 0$, where V and ω_p are, respectively, the specified forward velocity and pitch angular rotation of the wheelset, which are assumed in this section to be constant in order to develop a model that can be used in stability and hunting analysis at prescribed velocities. The other two constraints on z^{ir} and ϕ^{ir} can be written, respectively, as $C_3 = z^{ir} - f_1(y^{ir}, \psi^{ir}) = 0$ and $C_4 = \phi^{ir} - f_2(y^{ir}, \psi^{ir}) = 0$, where f_1 and f_2 are two functions that can be specified depending on the profile and track geometries. It is important to point out that in the case of general track geometry, these two functions also depend on the longitudinal arc length coordinate s^i. Other simplifying assumptions are also made in the literature. Since hunting stability is, for the most part, investigated in the case of tangent tracks, the two functions f_1 and f_2 are assumed in this section to have the forms $f_1 = f_1(y^{ir}, \psi^{ir})$ and $f_2 = f_2(y^{ir}, \psi^{ir})$, respectively: that is, these two functions are not dependent on other coordinates, including the arc length parameter s^i. The four constraint equations can then be written in vector form as

$$\mathbf{C}(\mathbf{p}, t) = \begin{bmatrix} C_1 \\ C_2 \\ C_3 \\ C_4 \end{bmatrix} = \begin{bmatrix} s^i - s_o^i - Vt \\ \theta^{ir} - \theta_o^{ir} - \omega_p t \\ z^{ir} - f_1(y^{ir}, \psi^{ir}) \\ \phi^{ir} - f_2(y^{ir}, \psi^{ir}) \end{bmatrix} = \mathbf{0} \tag{6.119}$$

The constraint equations at the velocity level can be written as

$$\dot{\mathbf{C}}(\mathbf{p}, t) = \mathbf{C}_{\mathbf{p}}\dot{\mathbf{p}} + \mathbf{C}_t = \begin{bmatrix} \dot{s}^i - V \\ \dot{\theta}^{ir} - \omega_p \\ \dot{z}^{ir} - (\partial f_1/\partial y^{ir})\,\dot{y}^{ir} - (\partial f_1/\partial \psi^{ir})\,\dot{\psi}^{ir} \\ \dot{\phi}^{ir} - (\partial f_2/\partial y^{ir})\,\dot{y}^{ir} - (\partial f_2/\partial \psi^{ir})\,\dot{\psi}^{ir} \end{bmatrix}$$

$$= \begin{bmatrix} 1 & 0 & 0 & 0 & 0 & 0 \\ 0 & 0 & 0 & 0 & 0 & 1 \\ 0 & -\left(\partial f_1/\partial y^{ir}\right) & 1 & -\left(\partial f_1/\partial \psi^{ir}\right) & 0 & 0 \\ 0 & -\left(\partial f_2/\partial y^{ir}\right) & 0 & -\left(\partial f_2/\partial \psi^{ir}\right) & 1 & 0 \end{bmatrix} \begin{bmatrix} \dot{s}^i \\ \dot{y}^{ir} \\ \dot{z}^{ir} \\ \dot{\psi}^{ir} \\ \dot{\phi}^{ir} \\ \dot{\theta}^{ir} \end{bmatrix} + \begin{bmatrix} -V \\ -\omega_p \\ 0 \\ 0 \end{bmatrix} = \mathbf{0} \tag{6.120}$$

In this equation, the 4×6 constraint Jacobian matrix $\mathbf{C_p}$ and the four-dimensional vector $\mathbf{C}_t$ can be recognized as

$$\mathbf{C_p} = \begin{bmatrix} 1 & 0 & 0 & 0 & 0 & 0 \\ 0 & 0 & 0 & 0 & 0 & 1 \\ 0 & -\left(\partial f_1/\partial y^{ir}\right) & 1 & -\left(\partial f_1/\partial \psi^{ir}\right) & 0 & 0 \\ 0 & -\left(\partial f_2/\partial y^{ir}\right) & 0 & -\left(\partial f_2/\partial \psi^{ir}\right) & 1 & 0 \end{bmatrix}, \quad \mathbf{C}_t = \begin{bmatrix} -V \\ -\omega_p \\ 0 \\ 0 \end{bmatrix} \tag{6.121}$$

The constraint equations at the acceleration level can be written as

$$\ddot{\mathbf{C}}(\mathbf{p}, t) = \mathbf{C_p}\ddot{\mathbf{p}} - \mathbf{Q}_d = \begin{bmatrix} \ddot{s}^i \\ \ddot{\theta}^{ir} \\ \ddot{z}^{ir} - (\partial f_1/\partial y^{ir})\ddot{y}^{ir} - (\partial f_1/\partial \psi^{ir})\ddot{\psi}^{ir} - D_{s1} \\ \ddot{\phi}^{ir} - \left(\partial f_2/\partial y^{ir}\right)\ddot{y}^{ir} - \left(\partial f_2/\partial \psi^{ir}\right)\ddot{\psi}^{ir} - D_{s_2} \end{bmatrix} \tag{6.122}$$

where $D_{sk} = (\partial^2 f_k/\partial y^{ir^2})(\dot{y}^{ir})^2 + (\partial^2 f_k/\partial \psi^{ir^2})(\dot{\psi}^{ir})^2 + 2(\partial^2 f_k/\partial y^{ir}\partial \psi^{ir})\dot{y}^{ir}\dot{\psi}^{ir}$, $k = 1, 2$, and vector $\mathbf{Q}_d$ is defined as

$$\mathbf{Q}_d = \begin{bmatrix} 0 & 0 & D_{s1} & D_{s2} \end{bmatrix}^T \tag{6.123}$$

The constraint Jacobian matrix $\mathbf{C_p}$ and vector $\mathbf{Q}_d$ can be used to obtain the augmented form of the equations of motion of this two-degree-of-freedom system. This augmented form, in the case of the trajectory coordinates, is defined by Eq. 109.

Alternatively, the embedding technique can be used to obtain two equations of motion that do not include any constraint forces. This can be achieved by writing dependent accelerations in terms of independent accelerations. It is clear from the definition of the constraint Jacobian matrix that the sub-Jacobians matrices associated with the dependent and independent coordinates $\mathbf{p}_d = \begin{bmatrix} s^i & z^{ir} & \phi^{ir} & \theta^{ir} \end{bmatrix}^T$ and $\mathbf{p}_i = \begin{bmatrix} y^{ir} & \psi^{ir} \end{bmatrix}^T$ can be written as

$$\mathbf{C}_{\mathbf{p}_d} = \begin{bmatrix} 1 & 0 & 0 & 0 \\ 0 & 0 & 0 & 1 \\ 0 & 1 & 0 & 0 \\ 0 & 0 & 1 & 0 \end{bmatrix}, \qquad \mathbf{C}_{\mathbf{p}_i} = \begin{bmatrix} 0 & 0 \\ 0 & 0 \\ -\left(\partial f_1/\partial y^{ir}\right) & -\left(\partial f_1/\partial \psi^{ir}\right) \\ -\left(\partial f_2/\partial y^{ir}\right) & -\left(\partial f_2/\partial \psi^{ir}\right) \end{bmatrix} \tag{6.124}$$

Therefore, the total vector of accelerations can be written in terms of the independent accelerations as

$$\ddot{\mathbf{p}} = \begin{bmatrix} \ddot{s}^i \\ \ddot{y}^{ir} \\ \ddot{z}^{ir} \\ \ddot{\psi}^{ir} \\ \ddot{\phi}^{ir} \\ \ddot{\theta}^{ir} \end{bmatrix} = \begin{bmatrix} 0 & 0 \\ 1 & 0 \\ \left(\partial f_1/\partial y^{ir}\right) & \left(\partial f_1/\partial \psi^{ir}\right) \\ 0 & 1 \\ \left(\partial f_2/\partial y^{ir}\right) & \left(\partial f_2/\partial \psi^{ir}\right) \\ 0 & 0 \end{bmatrix} \begin{bmatrix} \ddot{y}^{ir} \\ \ddot{\psi}^{ir} \end{bmatrix} + \begin{bmatrix} 0 \\ 0 \\ -D_{s_1} \\ 0 \\ -D_{s_2} \\ 0 \end{bmatrix} \tag{6.125}$$

This equation can be written as $\ddot{\mathbf{p}} = \mathbf{B}_{di}\ddot{\mathbf{p}}_i + \boldsymbol{\gamma}_i$, where velocity transformation matrix $\mathbf{B}_{di}$ and vector $\boldsymbol{\gamma}_i$ are given, respectively, by

$$\mathbf{B}_{di} = \begin{bmatrix} 0 & 0 \\ 1 & 0 \\ \left(\partial f_1/\partial y^{ir}\right) & \left(\partial f_1/\partial \psi^{ir}\right) \\ 0 & 1 \\ \left(\partial f_2/\partial y^{ir}\right) & \left(\partial f_2/\partial \psi^{ir}\right) \\ 0 & 0 \end{bmatrix} \tag{6.126}$$

and

$$\boldsymbol{\gamma}_i = \begin{bmatrix} 0 & 0 & -D_{s1} & 0 & -D_{s2} & 0 \end{bmatrix}^T \tag{6.127}$$

Using velocity transformation matrix $\mathbf{B}_{di}$ and vector $\boldsymbol{\gamma}_i$, a number of equations of motion equal to the number of the system degrees of freedom can be obtained and written, as previously explained, as

$$\mathbf{M}_{ii}\ddot{\mathbf{p}}_i = \mathbf{B}_{di}^T\left(\mathbf{Q}_{pe} + \mathbf{Q}_{pv} - \mathbf{M}_{pp}\boldsymbol{\gamma}_i\right) \tag{6.128}$$

where $\mathbf{M}_{ii} = \mathbf{B}_{di}^T\mathbf{M}_{pp}\mathbf{B}_{di}$, and the vector and matrices that appear in Eq. 128 are the same as those of Eq. 107.

6.13 HUNTING STABILITY

Hunting stability has been investigated in the literature using simplified two-degree-of-freedom models whose motion is governed by linear differential equations. These simplified models are obtained using the more general equations developed in the preceding section by making several assumptions. In hunting stability investigations, a tangent track is often used, and the assumption of small angles is made. By making these assumptions, one can show that the nonlinear equations of Eq. 128 reduce to a set of linear differential equations that can be used to explain important dynamic behaviors of the railroad vehicle system. Therefore, the equations used in this section can be considered a special form of the more general and nonlinear equations of Eq. 128.

Nonlinear Dynamic Equations Consider a wheelset i that consists of two wheels, as shown in Figure 13. In the case of a tangent track, the vector that defines the global position of origin O^i of the wheelset coordinate system $X^iY^iZ^i$ can be written as

$$\mathbf{R}^i = \begin{bmatrix} R_x^i & R_y^i & R_z^i \end{bmatrix}^T = \begin{bmatrix} s^i & y^{ir} & z^{ir} \end{bmatrix}^T \tag{6.129}$$

Because of this definition of vector $\mathbf{R}^i$, and because of the assumption of a tangent track used in the development presented in this section, it is expected that Eq. 107 will reduce to the Newton–Euler equations of Eq. 44. Using vector $\mathbf{R}^i$, the global position vector of a contact point on wheel $k = r, l$, where r and l refer, respectively, to the right and left wheel, can be written as $\mathbf{r}^{ik} = \mathbf{R}^i + \mathbf{A}^i\overline{\mathbf{u}}_P^{ki}$, where $\overline{\mathbf{u}}_P^{ki}$ is the position vector of contact point P with respect to wheel reference point; and $\mathbf{A}^i$ is the transformation matrix that defines the wheelset orientation in terms of the three Euler angles $\boldsymbol{\theta}^i = \begin{bmatrix} \psi^{ir} & \phi^{ir} & \theta^{ir} \end{bmatrix}^T$. Using the assumption of tangent track, matrix $\mathbf{L}^i$ of Eq. 101 reduces to $\mathbf{L}^i = \begin{bmatrix} \mathbf{I} & \mathbf{0} \end{bmatrix}$, matrix $\overline{\mathbf{H}}^i$ of Eq. 103 reduces to $\overline{\mathbf{H}}^i = \begin{bmatrix} \mathbf{0} & \overline{\mathbf{G}}^i \end{bmatrix}$, $\mathbf{A}^{ti} = \mathbf{I}$, $\mathbf{G}^{ti}$ is a constant matrix, $\boldsymbol{\omega}^{ti}$ is zero, and $\overline{\mathbf{G}}^i = \overline{\mathbf{G}}^{ir}$ that assumes the form of matrix $\overline{\mathbf{G}}^i$ defined in Eq. 21. Using these results, the vectors and matrices of Eq. 106 reduce to (Shabana et al. 2008)

$$\mathbf{B}^i = \begin{bmatrix} \mathbf{L}^i \\ \overline{\mathbf{H}}^i \end{bmatrix} = \begin{bmatrix} \mathbf{I} & \mathbf{0} \\ \mathbf{0} & \overline{\mathbf{G}}^i \end{bmatrix}, \qquad \boldsymbol{\gamma}^i = \begin{bmatrix} \boldsymbol{\gamma}_R^i \\ \overline{\boldsymbol{\gamma}}_\alpha^i \end{bmatrix} = \begin{bmatrix} \mathbf{0} \\ \dot{\overline{\mathbf{G}}}^i\dot{\boldsymbol{\theta}}^{ir} \end{bmatrix} \tag{6.130}$$

Therefore, one can write a matrix equation similar to Eq. 107 as

$$\mathbf{M}_{pp}^i\ddot{\mathbf{p}}^i = \mathbf{Q}_{pe}^i + \mathbf{Q}_{pv}^i \tag{6.131}$$

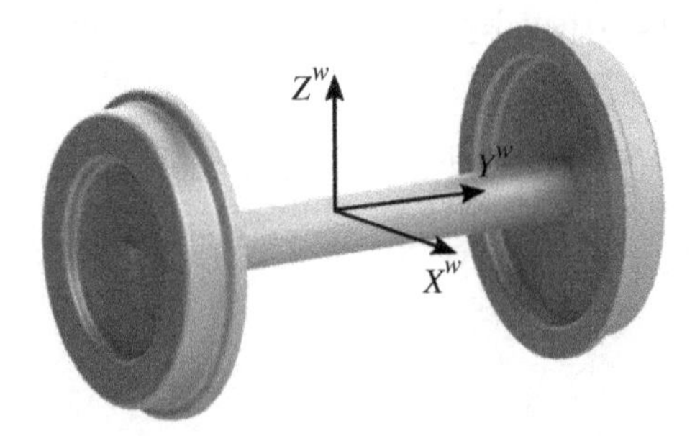

Figure 6.13 Wheelset model.

where

$$\left.\begin{aligned}
\mathbf{M}_{pp}^{i} &= m^{i}\mathbf{L}^{i^{T}}\mathbf{L}^{i} + \overline{\mathbf{H}}^{i^{T}}\overline{\mathbf{I}}_{\theta\theta}^{i}\overline{\mathbf{H}}^{i}\mathbf{M}_{p}^{w} = \begin{bmatrix} m^{i}\mathbf{I} & \mathbf{0} \\ \mathbf{0} & \overline{\mathbf{G}}^{i^{T}}\overline{\mathbf{I}}_{\theta\theta}^{i}\overline{\mathbf{G}}^{i} \end{bmatrix}, \\
\mathbf{Q}_{pe}^{i} &= \mathbf{L}^{i^{T}}\mathbf{F}_{e}^{i} + \overline{\mathbf{H}}^{i^{T}}\overline{\mathbf{M}}_{e}^{i} = \begin{bmatrix} \mathbf{F}_{e}^{i} \\ \overline{\mathbf{G}}^{i^{T}}\overline{\mathbf{M}}_{e}^{i} \end{bmatrix}, \\
\mathbf{Q}_{pv}^{i} &= -m^{i}\mathbf{L}^{i^{T}}\boldsymbol{\gamma}_{R}^{i} - \overline{\mathbf{H}}^{i^{T}}\left[\overline{\mathbf{I}}_{\theta\theta}^{i}\overline{\boldsymbol{\gamma}}_{\alpha}^{i} + \overline{\boldsymbol{\omega}}^{i} \times \left(\overline{\mathbf{I}}_{\theta\theta}^{i}\overline{\boldsymbol{\omega}}^{i}\right)\right] \\
&= \begin{bmatrix} \mathbf{0} \\ -\overline{\mathbf{G}}^{i^{T}}\left[\overline{\mathbf{I}}_{\theta\theta}^{i}\dot{\overline{\mathbf{G}}}^{i}\dot{\boldsymbol{\theta}}^{i} + \overline{\boldsymbol{\omega}}^{i} \times \left(\overline{\mathbf{I}}_{\theta\theta}^{i}\overline{\boldsymbol{\omega}}^{i}\right)\right] \end{bmatrix}
\end{aligned}\right\} \tag{6.132}$$

To obtain the two equations of motion that govern the hunting stability of the wheelset, Eqs. 126 and 127 are used and rewritten as

$$\mathbf{B}_{di} = \begin{bmatrix} (\mathbf{B}_{di})_{R} \\ (\mathbf{B}_{di})_{\theta} \end{bmatrix} = \begin{bmatrix} 0 & 0 \\ 1 & 0 \\ f_{1y} & f_{1\psi} \\ 0 & 1 \\ f_{2y} & f_{2\psi} \\ 0 & 0 \end{bmatrix}, \quad \boldsymbol{\gamma}_{i} = \begin{bmatrix} (\boldsymbol{\gamma}_{i})_{R} \\ (\boldsymbol{\gamma}_{i})_{\theta} \end{bmatrix} = \begin{bmatrix} 0 \\ 0 \\ -D_{s_1} \\ 0 \\ -D_{s_2} \\ 0 \end{bmatrix} \tag{6.133}$$

where the first derivatives $f_{1y} = \partial f_1/\partial y^{ir}, f_{2y} = \partial f_2/\partial \psi^{ir}, f_{1\psi} = \partial f_1/\partial y^{ir}$, and $f_{2\psi} = \partial f_2/\partial \psi^{ir}$; the second derivatives $f_{1yy} = \partial^2 f_1/\partial y^{ir^2}, f_{2yy} = \partial^2 f_2/\partial y^{ir^2}, f_{1\psi\psi} = \partial^2 f_1/\partial \psi^{ir^2}$, $f_{2\psi\psi} = \partial^2 f_2/\partial \psi^{ir^2}$, $D_{sk} = f_{kyy}(\dot{y}^{ir})^2 + f_{k\psi\psi}(\dot{\psi}^{ir})^2 + 2f_{ky\psi}\dot{y}^{ir}\dot{\psi}^{ir}, k = 1, 2$; and

$$\left.\begin{aligned}
(\mathbf{B}_{di})_{R} &= \begin{bmatrix} 0 & 0 \\ 1 & 0 \\ f_{1y} & f_{1\psi} \end{bmatrix}, \quad (\mathbf{B}_{di})_{\theta} = \begin{bmatrix} 0 & 1 \\ f_{2y} & f_{2\psi} \\ 0 & 0 \end{bmatrix}, \\
(\boldsymbol{\gamma}_{i})_{R} &= \begin{bmatrix} 0 \\ 0 \\ -D_{s1} \end{bmatrix}, \quad (\boldsymbol{\gamma}_{i})_{\theta} = \begin{bmatrix} 0 \\ -D_{s2} \\ 0 \end{bmatrix}
\end{aligned}\right\} \tag{6.134}$$

Using these definitions, the time derivatives of the Euler angles and the angular velocity vector can be written as

$$\begin{bmatrix} \dot{\psi}^{ir} \\ \dot{\phi}^{ir} \\ \dot{\theta}^{ir} \end{bmatrix} = \begin{bmatrix} \dot{\psi}^{ir} \\ \dot{y}^{ir}f_{2y} + \dot{\psi}^{ir}f_{2\psi} \\ \omega_{p} \end{bmatrix}, \quad \overline{\boldsymbol{\omega}}^{i} = \begin{bmatrix} -\dot{\psi}^{ir}\cos\phi^{ir}\sin\theta^{ir} + \left(\dot{y}^{ir}f_{2y} + \dot{\psi}^{ir}f_{2\psi}\right)\cos\theta^{ir} \\ \omega_{p} + \dot{\psi}^{ir}\sin\phi^{ir} \\ \dot{\psi}^{ir}\cos\phi^{ir}\cos\theta^{ir} + \left(\dot{y}^{ir}f_{2y} + \dot{\psi}^{ir}f_{2\psi}\right)\sin\theta^{ir} \end{bmatrix} \tag{6.135}$$

Assuming $i_{xx} = i_{zz}$ and the products of inertia of the wheelset are equal to zero, one has

$$\overline{\mathbf{G}}^i(\mathbf{B}_{di})_\theta = \begin{bmatrix} f_{2y}\cos\theta^{ir} & \alpha_1 \\ 0 & \sin\phi^{ir} \\ f_{2y}\sin\theta^{ir} & \alpha_2 \end{bmatrix}, \quad (\mathbf{B}_{di})_\theta^T \overline{\mathbf{G}}^{i^T} \overline{\mathbf{I}}_{\theta\theta}^i \overline{\mathbf{G}}^i (\mathbf{B}_{di})_\theta = \begin{bmatrix} i_{xx}^i (f_{2y})^2 & i_{xx}^i f_{2y} f_{2\psi} \\ i_{xx}^i f_{2y} f_{2\psi} & i_{\psi\psi}^i \end{bmatrix} \tag{6.136}$$

In this equation, $\alpha_1 = f_{2\psi}\cos\theta^{ir} - \cos\phi^{ir}\sin\theta^{ir}$, $\alpha_2 = f_{2\psi}\sin\theta^{ir} + \cos\phi^{ir}\cos\theta^{ir}$, and $i_{\psi\psi}^i = i_{yy}^i \sin^2\phi^{ir} + i_{xx}^i\left((f_{2y})^2 + \cos^2\phi^{ir}\right)$. Substituting Eq. 133 into Eq. 131, and premultiplying by the transpose of transformation matrix $\mathbf{B}_{di}$, one obtains two scalar differential equations of motion, which can be written in the form of Eq. 128 as

$$\mathbf{M}_{ii}\ddot{\mathbf{p}}_i = \mathbf{B}_{di}^T\left(\mathbf{Q}_{pe}^i + \mathbf{Q}_{pv}^i - \mathbf{M}_{pp}^i \boldsymbol{\gamma}_i\right) \tag{6.137}$$

where $\mathbf{p}_i = \begin{bmatrix} y^{ir} & \psi^{ir} \end{bmatrix}^T$ and

$$\left.\begin{aligned} \mathbf{M}_{ii} &= \mathbf{B}_{di}^T \mathbf{M}_{pp}^i \mathbf{B}_{di} = \begin{bmatrix} m^i\left(1 + (f_{1y})^2\right) + i_{xx}^i (f_{2y})^2 & f_{1y} f_{1\psi} + i_{xx}^i f_{2y} f_{2\psi} \\ f_{1y} f_{1\psi} + i_{xx}^i f_{2y} f_{2\psi} & (f_{1\psi})^2 + i_{\psi\psi}^i \end{bmatrix} \\ \mathbf{B}_{di}^T \mathbf{Q}_{pe}^i &= \begin{bmatrix} F_{ey}^i + F_{ez}^i f_{1y} + Q_{\theta 2}^i f_{2y} \\ F_{ez}^i f_{1\psi} + Q_{\theta 1}^i + Q_{\theta 2}^i f_{2\psi} \end{bmatrix}, \quad \mathbf{Q}_{pe}^i = \begin{bmatrix} F_{ex}^i & F_{ey}^i & F_{ez}^i & Q_{\theta 1}^i & Q_{\theta 2}^i & Q_{\theta 3}^i \end{bmatrix}^T \\ \mathbf{B}_{di}^T\left(\mathbf{Q}_{pv}^i - \mathbf{M}_{pp}^i \boldsymbol{\gamma}_i\right) &= -(\mathbf{B}_{di})_\theta^T \overline{\mathbf{G}}^{i^T}\left[\overline{\mathbf{I}}_{\theta\theta}^i \dot{\overline{\mathbf{G}}}^i \dot{\boldsymbol{\theta}}^i + \overline{\boldsymbol{\omega}}^i \times \left(\overline{\mathbf{I}}_{\theta\theta}^i \overline{\boldsymbol{\omega}}^i\right)\right] \\ &\quad + \begin{bmatrix} m^i f_{1y} D_{s_1} + i_{yy}^i f_{2y} D_{s_2} \\ m^i f_{1\psi} D_{s_1} + i_{yy}^i f_{2\psi} D_{s_2} \end{bmatrix} \end{aligned}\right\} \tag{6.138}$$

Equation 137, which is a nonlinear matrix equation, has two scalar equations that can be solved using direct numerical integration methods. Vector $\mathbf{Q}_{pe}^i$ in the preceding equation includes the wheel/rail normal and tangential contact forces, which can be nonlinear functions in creepages if nonlinear contact theory is used.

Creepage Definitions Vector $\mathbf{Q}_{pe}^i$ of Eq. 138 includes the effect of creep forces, which are expressed in terms of velocity creepages as explained in Chapter 5. For the simple wheelset model considered in this section, closed-form expressions can be obtained for creepages in terms of the prescribed velocities. Creepages at wheel/rail contact points can be conveniently defined by introducing an intermediate wheelset coordinate system that does not rotate with the wheelset about its Y^i axis. This coordinate system is defined by transformation matrix $\mathbf{A}^{ii}$, which is expressed in terms of two angles ψ^{ir} and ϕ^{ir} as

$$\mathbf{A}^{ii} = \begin{bmatrix} \cos\psi^{ir} & -\sin\psi^{ir}\cos\phi^{ir} & \sin\psi^{ir}\sin\phi^{ir} \\ \sin\psi^{ir} & \cos\psi^{ir}\cos\phi^{ir} & \cos\psi^{ir}\sin\phi^{ir} \\ 0 & \sin\phi^{ir} & \cos\phi^{ir} \end{bmatrix} \tag{6.139}$$

Using this transformation, the global position vector at contact point P^{ik} is given by $\mathbf{r}_P^{ik} = \mathbf{R}^i + \mathbf{A}^{ii}\overline{\mathbf{u}}_P^{iik}$, $k = 1, 2$, where $k = 1$ for the right contact, $k = 2$ for the left contact,

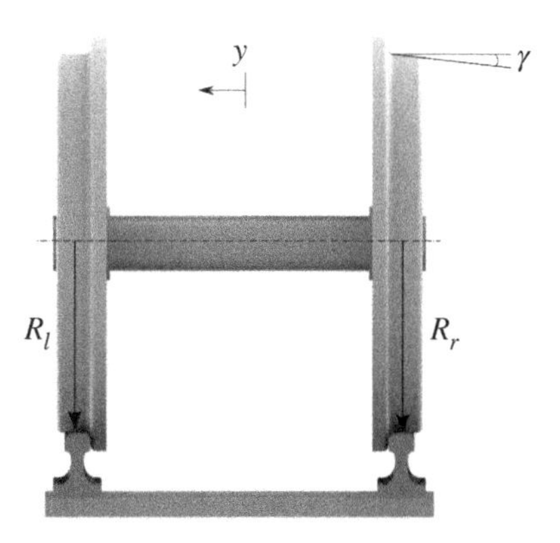

Figure 6.14 Right and left wheel rolling radius and conicity.

$\overline{\mathbf{u}}_P^{iik} = \begin{bmatrix} 0 & d^k & -r^k \end{bmatrix}^T$, d^k is the distance of the contact point from the origin of the wheelset coordinate system in the lateral direction, and r^k is the rolling radius at the contact point. For the right and left wheels, one has, respectively, $r^1 = R_r = r_0 - \gamma y^{ir}$ and $r^2 = R_l = r_0 + \gamma y^{ir}$, where r_o and γ are, respectively, the nominal rolling radius and the conicity, as shown in Figure 14. It follows that the absolute velocity of a contact point defined in the intermediate wheelset coordinate system can be written as $\mathbf{v}_P^{ik} = \mathbf{A}^{ii^T}\dot{\mathbf{R}}^i + \overline{\boldsymbol{\omega}}^{ii} \times \overline{\mathbf{u}}_P^{iik}$, where $\overline{\boldsymbol{\omega}}^{ii} = \overline{\mathbf{G}}^{ii}\dot{\boldsymbol{\theta}}^{ir}$, where $\overline{\mathbf{G}}^{ii}$ can be defined using $\overline{\mathbf{G}}^{i}$ by assuming $\theta^{ir} = 0$ since the intermediate wheelset coordinate system is not affected by the pitch rotation. This leads to

$$\overline{\mathbf{G}}^{ii} = \begin{bmatrix} 0 & 1 & 0 \\ \sin\phi^{ir} & 0 & 1 \\ \cos\phi^{ir} & 0 & 0 \end{bmatrix} \tag{6.140}$$

Consequently, one can write $\overline{\boldsymbol{\omega}}^{ii} = \begin{bmatrix} \dot{\phi}^{ir} & \dot{\theta}^{ir} + \dot{\psi}^{ir}\sin\phi^{ir} & \dot{\psi}^{ir}\cos\phi^{ir} \end{bmatrix}^T$. Using small-angle assumptions and neglecting higher-order terms, the angular velocity vector $\overline{\boldsymbol{\omega}}^{ii}$ reduces to $\overline{\boldsymbol{\omega}}^{ii} = \begin{bmatrix} \dot{\phi}^{ir} & \dot{\theta}^{ir} & \dot{\psi}^{ir} \end{bmatrix}^T$. Using this definition of the angular velocity and the small-angle assumption in transformation matrix $\mathbf{A}^{ii}$, one can show that velocity vector $\mathbf{v}_P^{ik} = \mathbf{A}^{ii^T}\dot{\mathbf{R}}^i + \overline{\boldsymbol{\omega}}^{ii} \times \overline{\mathbf{u}}_P^{iik}$ can be written as

$$\mathbf{v}_P^{ik} = \begin{bmatrix} v_{px}^{ik} \\ v_{py}^{ik} \\ v_{pz}^{ik} \end{bmatrix} = \begin{bmatrix} V \\ \dot{y}^{ir} - V\psi^{ir} \\ \dot{z}^{ir} \end{bmatrix} + \begin{bmatrix} -d^k\dot{\psi}^{ir} - r^k\dot{\theta}^{ir} \\ r^k\dot{\phi}^{ir} \\ d^k\dot{\phi}^{ir} \end{bmatrix} \tag{6.141}$$

As discussed in Chapter 5, longitudinal, lateral, and spin creepages are defined, respectively, as $\zeta_x^k = v_{px}^{ik}/V$, $\zeta_y^k = v_{px}^{ik}/V$, and $\zeta_s^k = \overline{\boldsymbol{\omega}}^{ii} \cdot \overline{\mathbf{n}}^{ik}/V$, where $\overline{\mathbf{n}}^{ik} = \begin{bmatrix} 0 & (-1)^{k+1}\gamma & 1 \end{bmatrix}^T / \sqrt{1 + (\gamma)^2}$, $k = 1, 2$, is a unit vector along the normal at the contact point. This unit normal vector is defined in the intermediate wheelset coordinate system. Therefore, by using Eq. 141, the following linear creepage expressions can be obtained:

$$\begin{bmatrix} \zeta_x^1 \\ \zeta_y^1 \\ \zeta_s^1 \end{bmatrix} = \begin{bmatrix} \left(\frac{d}{V}\right)\dot{\psi}^{ir} + \left(\frac{\gamma}{r_o}\right)y^{ir} \\ \frac{\dot{y}^{ir}}{V} + \frac{\left(r_o - \gamma y^{ir}\right)\dot{\phi}^{ir}}{V} - \psi^{ir} \\ \frac{\left(\dot{\psi}^{ir} + \gamma\,\dot{\theta}^{ir}\right)}{V} \end{bmatrix}, \quad \begin{bmatrix} \zeta_x^2 \\ \zeta_y^2 \\ \zeta_s^2 \end{bmatrix} = \begin{bmatrix} -\left(\frac{d}{V}\right)\dot{\psi}^{ir} - \left(\frac{\gamma}{r_o}\right)y^{ir} \\ \frac{\dot{y}^{ir}}{V} + \frac{\left(r_o + \gamma y^{ir}\right)\dot{\phi}^{ir}}{V} - \psi^{ir} \\ \frac{\left(\dot{\psi}^{ir} - \gamma\,\dot{\theta}^{ir}\right)}{V} \end{bmatrix} \tag{6.142}$$

In obtaining these linear creepage expressions, the higher-order term $y^{ir}\dot{\psi}^{ir}$ is neglected, and it is assumed that $d^1 = -d - y^{ir}$ and $d^2 = d - y^{ir}$, where d is half the gage. Furthermore, in the case of conical wheels, contact angle δ_c is the same as the conicity, and this leads to the definition of the normal vector $\overline{\mathbf{n}}^{ik}$.

Linear Creep Force Model As discussed in Chapter 5, in Kalker's linear creep theory, longitudinal, lateral, and spin creep forces can be written, respectively, in terms of creepages as $\overline{F}_{crx} = -c_{xx}\zeta_x$, $\overline{F}_{cry} = -c_{yy}\zeta_y - c_{ys}\zeta_s$, and $\overline{M}_{crs} = c_{sy}\zeta_y - c_{ss}\zeta_s$, where c_{xx}, c_{yy}, $c_{ys} = c_{sy}$, and c_{ss} are creep coefficients that are assumed to be functions of the material properties and dimensions of the contact area. These forces are determined in a contact frame, which can be different from the wheelset and global coordinate systems. If lateral displacement y^{ir} and yaw angle ψ^{ir} are assumed small, the creep coefficients can be assumed the same for the right and left contacts. In this case, one has only the generalized creep forces associated with coordinates y^{ir} and ψ^{ir}, defined, respectively, as $F_{ey}^i = \sum_{k=1}^2 \left(-c_{yy}\zeta_y^k - c_{ys}\zeta_s^k\right)$ and $Q_{\theta 1}^i = \sum_{k=1}^2 d(-1)^k c_{xx}\zeta_x^k + \left(c_{sy}\zeta_y^k - c_{ss}\zeta_s^k\right)$. The first term in generalized force $Q_{\theta 1}^i$ is a yaw moment attributed to longitudinal creep forces at the two contacts. Using Eq. 142, one can write the following expression for generalized creep forces:

$$\left.\begin{aligned} F_{ey}^i &= -\frac{2c_{yy}}{V}\left(1 + \frac{r_o\gamma}{d}\right)\dot{y}^{ir} + 2c_{11}\psi^{ir} - \frac{2c_{ys}}{V}\dot{\psi}^{ir} \\ Q_{\theta 1}^i &= \frac{2c_{ys}}{V}\left(1 + \frac{r_o\gamma}{d}\right)\dot{y}^{ir} - 2c_{ys}\psi^{ir} - \frac{2c_{ss}}{V}\dot{\psi}^{ir} - 2dc_{xx}\left(\frac{d}{V}\dot{\psi}^{ir} + \frac{\gamma}{r_o}y^{ir}\right) \end{aligned}\right\} \tag{6.143}$$

Linearized Dynamic Equations A linear set of equations that sheds light on wheelset dynamics and stability can be obtained from Eq. 137. Because yaw angle ψ^{ir} is small in most railroad vehicle applications, z^{ir} and ϕ^{ir} can be assumed to depend only on y^{ir}. Using this assumption, one has

$$\mathbf{B}_{di} = \begin{bmatrix} \left(\mathbf{B}_{di}\right)_R \\ \left(\mathbf{B}_{di}\right)_\theta \end{bmatrix} = \begin{bmatrix} 0 & 0 \\ 1 & 0 \\ f_{1y} & 0 \\ 0 & 1 \\ f_{2y} & 0 \\ 0 & 0 \end{bmatrix}, \quad \boldsymbol{\gamma}_i = \begin{bmatrix} \left(\boldsymbol{\gamma}_i\right)_R \\ \left(\boldsymbol{\gamma}_i\right)_\theta \end{bmatrix} = \begin{bmatrix} 0 \\ 0 \\ -f_{1yy}\left(\dot{y}^{ir}\right)^2 \\ 0 \\ -f_{2yy}\left(\dot{y}^{ir}\right)^2 \\ 0 \end{bmatrix} \tag{6.144}$$

and the matrices and vectors in Eq. 138 reduce to

$$\left.\begin{aligned}
\mathbf{M}_{ii} &= \mathbf{B}_{di}^{T}\mathbf{M}_{pp}^{i}\mathbf{B}_{di} = \begin{bmatrix} m^{i}\left(1+\left(f_{1y}\right)^{2}\right)+i_{xx}^{i}\left(f_{2y}\right)^{2} & 0 \\ 0 & i_{\psi\psi}^{i} \end{bmatrix} \\
\mathbf{B}_{di}^{T}\mathbf{Q}_{pe}^{i} &= \begin{bmatrix} F_{ey}^{i}+F_{ez}^{i}f_{1y}+Q_{\theta 2}^{i}f_{2y} \\ Q_{\theta 1}^{i} \end{bmatrix} \\
\mathbf{B}_{di}^{T}\left(\mathbf{Q}_{pv}^{i}-\mathbf{M}_{pp}^{i}\boldsymbol{\gamma}_{i}\right) &= -\left(\mathbf{B}_{di}\right)_{\theta}^{T}\overline{\mathbf{G}}^{iT}\left[\overline{\mathbf{I}}_{\theta\theta}^{i}\dot{\overline{\mathbf{G}}}^{i}\dot{\boldsymbol{\theta}}^{i}+\overline{\boldsymbol{\omega}}^{i}\times\left(\overline{\mathbf{I}}_{\theta\theta}^{i}\overline{\boldsymbol{\omega}}^{i}\right)\right] \\
&\quad + \begin{bmatrix} \left(m^{i}f_{1y}f_{1yy}+i_{yy}^{i}f_{2y}f_{2yy}\right)\left(\dot{y}^{ir}\right)^{2} \\ 0 \end{bmatrix}
\end{aligned}\right\} \tag{6.145}$$

Using the assumptions of small yaw and roll angles, a small change in the profile function, and small lateral displacements and velocities, and recalling that $i_{xx}^{i}=i_{zz}^{i}$ for a symmetric wheelset, one obtains the linearized equations of motion of the two-degree-of-freedom wheelset, which can be written as

$$\begin{bmatrix} m^{i} & 0 \\ 0 & i_{zz}^{i} \end{bmatrix}\begin{bmatrix} \ddot{y}^{ir} \\ \ddot{\psi}^{ir} \end{bmatrix} = \begin{bmatrix} F_{ey}^{i}+F_{ez}^{i}f_{1y}+Q_{\theta 2}^{i}f_{2y} \\ Q_{\theta 1}^{i}-i_{yy}^{i}\dot{\phi}^{ir}\dot{\theta}^{ir} \end{bmatrix} \tag{6.146}$$

If the assumption of pure rolling is made, the pitch angular velocity can be written as $\dot{\theta}^{ir}=V/r_{o}$, where V is the constant forward velocity of the wheelset and r_{o} is the nominal rolling radius. The applied forces and moments on the right-hand side of the preceding equations include the creep forces and moments. Force F_{ey}^{i} is the lateral creep force, and $Q_{\theta 1}^{i}$ is the spin creep moment. Substituting Eq. 143 into Eq. 146 and assuming zero applied forces except creep forces and moments and gravity forces, one obtains

$$\begin{bmatrix} m^{i} & 0 \\ 0 & i_{zz}^{i} \end{bmatrix}\begin{bmatrix} \ddot{y}^{ir} \\ \ddot{\psi}^{ir} \end{bmatrix}+\begin{bmatrix} d_{yy} & d_{y\psi} \\ d_{\psi y} & d_{\psi\psi} \end{bmatrix}\begin{bmatrix} \dot{y}^{ir} \\ \dot{\psi}^{ir} \end{bmatrix}+\begin{bmatrix} k_{yy} & k_{y\psi} \\ k_{\psi y} & k_{\psi\psi} \end{bmatrix}\begin{bmatrix} y^{ir} \\ \psi^{ir} \end{bmatrix}=\begin{bmatrix} 0 \\ 0 \end{bmatrix} \tag{6.147}$$

where

$$\left.\begin{aligned}
&d_{yy}=2c_{yy}\left(1+\left(r_{o}\gamma/d\right)\right)/V, \quad d_{y\psi}=2c_{ys}/V, \\
&d_{y\psi}=\left(-2c_{ys}\left(r_{o}\gamma/d\right)+\left(i_{yy}^{i}\gamma(V)^{2}\right)/r_{o}d\right)/V, \quad d_{\psi\psi}=2\left(c_{ss}+d^{2}c_{xx}\right)/V, \\
&k_{yy}=k_{w1}, \quad k_{y\psi}=-2c_{yy}, \quad k_{\psi y}=2dc_{xx}\gamma/r_{o}, \quad k_{\psi\psi}=2c_{ys}+k_{w2}
\end{aligned}\right\} \tag{6.148}$$

and k_{w1} and k_{w2} are stiffness coefficients resulting from accounting for gravity force. In the case of a suspended wheelset like the one shown in Figure 15, suspension spring and damper elements are used to connect the wheelset to the frame in the longitudinal and lateral directions. The preceding equation can be modified to account for the effect of suspension forces as

$$\begin{bmatrix} m^{i} & 0 \\ 0 & i_{zz}^{i} \end{bmatrix}\begin{bmatrix} \ddot{y}^{ir} \\ \ddot{\psi}^{ir} \end{bmatrix}+\begin{bmatrix} \overline{d}_{yy} & d_{y\psi} \\ d_{\psi y} & \overline{d}_{\psi\psi} \end{bmatrix}\begin{bmatrix} \dot{y}^{ir} \\ \dot{\psi}^{ir} \end{bmatrix}+\begin{bmatrix} \overline{k}_{yy} & k_{y\psi} \\ k_{\psi y} & \overline{k}_{\psi\psi} \end{bmatrix}\begin{bmatrix} y^{ir} \\ \psi^{ir} \end{bmatrix}=\begin{bmatrix} 0 \\ 0 \end{bmatrix} \tag{6.149}$$

where

$$\left.\begin{aligned}
&\overline{d}_{yy}=\left(2c_{yy}\left(1+\left(r_{o}\gamma/d\right)\right)+2d_{y}V\right)/V, \quad \overline{d}_{\psi\psi}=\left(2\left(c_{ss}+d^{2}c_{xx}\right)+2d_{x}(b)^{2}V\right)/V, \\
&\overline{k}_{yy}=k_{w1}+2k_{y}, \quad \overline{k}_{\psi\psi}=2c_{ys}+k_{w2}+2k_{x}b^{2}
\end{aligned}\right\} \tag{6.150}$$

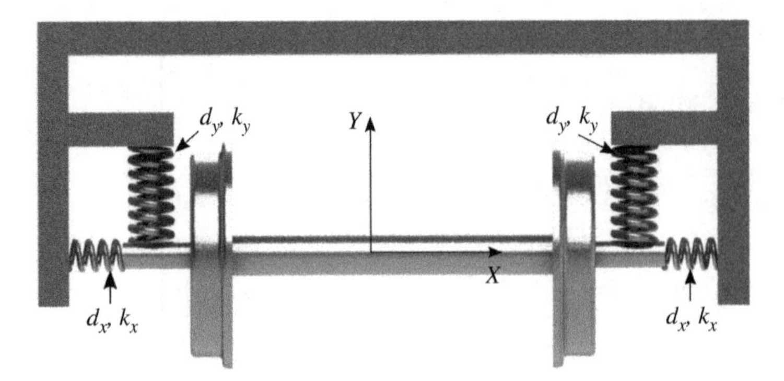

Figure 6.15 Suspended wheelset.

where d_x and d_y are suspension damping coefficients, and k_x and k_y are suspension stiffness coefficients, as shown in Figure 15. The effect of the suspension of the wheelset is examined in the remainder of this section.

Effect of the Primary Suspension Unsuspended wheelsets are unstable if the effects of spin creepage and gravity force are neglected. To investigate this instability, the primary suspension coefficients in Eq. 149 are assumed to be zero. For such models, the hunting frequency can be approximated using *Klingel's formula* (Klingel 1883), which is derived using kinematic equations without regard to the forces applied to the wheelset. This formula is given by $f = (V/2\pi)\sqrt{\gamma/(r_o d)}$. Figure 16 shows the results of an eigenvalue analysis based on the equations developed in this section, which account for the effect of creep forces. As shown in Figure 16, this comparative analysis leads to a solution that is in good agreement with the solution obtained using Klingel's formula.

While unsuspended wheelsets are unstable, suspended wheelsets have *critical speeds* that depend on suspension characteristics. For a given wheelset with specified suspension characteristics, there is a critical speed below which the wheelset is stable and above which

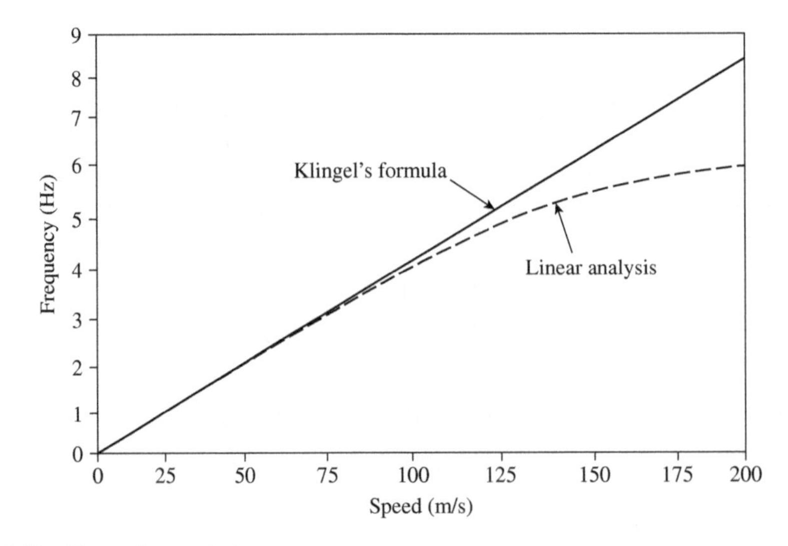

Figure 6.16 Eigenvalue analysis.

the wheelset is unstable. To examine the effect of the primary suspension on stability, the wheelset shown in Figure 15 is assumed to be connected to the frame with a primary suspension. This system can be unstable if the forward velocity V of the wheelset exceeds the critical speed. To investigate the effects of the primary suspension and conicity on the stability and critical speed of the wheelset, the data of Table 1 are used (Shabana et al. 2008). The root loci results of Figure 17 show that as the wheelset forward velocity V increases, the real part of the eigenvalues becomes positive, leading to unstable behavior. The effect of the wheel profile conicity γ on the critical speed is demonstrated by the results shown in Figure 18. In vibration theory, the ratio between the real part α and the imaginary part ω of

Table 6.1 Data for the suspended wheelset model.

Parameter description	Value
Wheelset mass m^i	1568 kg
Inertia moment i^i_{xx}	656 kg m^2
Inertia moment i^i_{yy}	168 kg m^2
Inertia moment i^i_{zz}	656 kg m^2
Longitudinal spring stiffness k_x	1.35×10^5 N/m
Lateral spring stiffness k_y	2.50×10^5 N/m
Longitudinal damping coefficient d_x	0 N/(m s)
Lateral damping coefficient d_y	0 N/(m s)
Distance between the longitudinal springs $2b$	1.8 m
Half gage distance d	1.435 m
Nominal rolling radius r_o	0.4566 m
Conicity γ	1/40

Source: Courtesy of Shabana, A.A., Zaazaa, K.E., and Sugiyama, H.

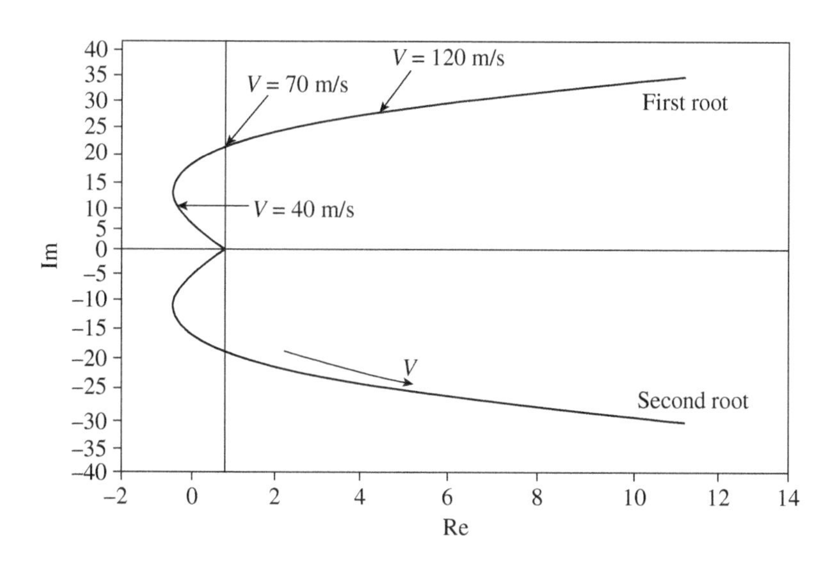

Figure 6.17 Root loci results.

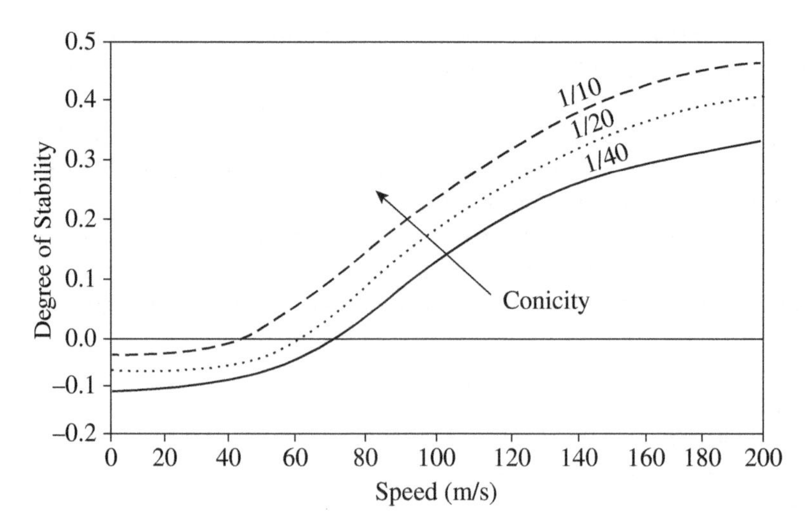

Figure 6.18 Conicity effect.

the eigenvalue defines the degree of stability. The critical speed is reached when the ratio α/ω is equal to zero, which is the case for a critically stable system (sustained oscillations). Because increasing the conicity causes the wheel/rail contact forces to add more energy to the system, such an increase can lead to unstable behavior. This instability can be clearly seen at low speeds by lowering the stiffness coefficients of the primary suspension.

6.14 MBS MODELING OF ELECTROMECHANICAL SYSTEMS

Electromechanical systems are integral parts of modern vehicle systems, including railroad systems. The equations that govern the electric circuits required to produce and control forces such as levitation forces can be systematically integrated with the constrained dynamic equations used in MBS railroad vehicle algorithms. In the preceding chapter, *maglev trains* were discussed, and an expression for the *magnetic levitation* force F_z was obtained as

$$F_z = F_z\left(d_z, i\right) = \frac{1}{P_p A_p}\left(\frac{n_{co}\, i\,(t)}{R_M}\right)^2 \tag{6.151}$$

where $d_z = d_z(t)$ is the distance between the pole faces; $A_p = l_p w_p$; l_p and w_p are, respectively, the length and width of the pole face; P_p is the permeability of the free space; n_{co} is the number of turns of the coil; i is the current; and $R_M = R_M(d_z)$ is the reluctance of the mutual flux. The force of Eq. 151 can be used to define magnetic levitation force vector $\mathbf{F}_{ml}$ in the global coordinate system. This force represents the magnetic force between the vehicle and the guide. Using this force vector, the generalized forces associated with vehicle v and guide g can be written, respectively, as

$$\mathbf{Q}_{ml}^{v} = \begin{bmatrix} \mathbf{F}_{ml} \\ \mathbf{G}^{v^T}\left(\mathbf{u}^{v} \times \mathbf{F}_{ml}\right) \end{bmatrix}, \qquad \mathbf{Q}_{ml}^{g} = -\begin{bmatrix} \mathbf{F}_{ml} \\ \mathbf{G}^{g^T}\left(\mathbf{u}^{g} \times \mathbf{F}_{ml}\right) \end{bmatrix} \tag{6.152}$$

where $\mathbf{u}^k$, $k = v, g$, is the position of the point of application of the force with respect to the origin of the respective body coordinate system. The formulations presented in this chapter also allow for using distributed magnetic levitation forces, which can be used to determine the generalized forces associated with the generalized coordinates of the vehicle and the guide. In the case of distributed forces, an integration over the area of application of the magnetic levitation force can be performed and used to determine the magnetic levitation generalized forces, which can be added to the equation of motion defined by Eq. 55. Current i and the distance between pole faces d_z depend on time, and therefore, they are, in general, functions of the system coordinates.

A more general approach that accounts for the time rate of change in the current combines electric circuit equations with vehicle dynamic equations of motion. Using this approach, power-supply voltage v_s can be written in terms of current i, circuit resistance R_c, and circuit inductance L_c as $v_s = R_c i(t) + d(L_c(t)i(t))/dt$, where, as defined in the preceding chapter, $L_c(t) = (n_{co})^2(R_L + R_M)/R_L R_M$, $R_L = 1/P_L$ is the reluctance of the leakage flux, and P_L is the permeance of the leakage flux. Power-supply voltage v_s can be written as

$$v_s = R_c i(t) + L_c(t)(di/dt) + i(t)\dot{L}_c(t) \tag{6.153}$$

Inductance L_c, which depends on air gap d_z, is defined as

$$L_c = L_c(d_z) = L_L + \frac{P_p(n_{co})^2 A_p}{2(d_z(t) + c_p)} \tag{6.154}$$

where P_p is the permeability of the free space, c_p is a constant defined in the preceding chapter, and $L_L = (n_{co})^2 P_L = (n_{co})^2/R_L$. Substituting Eq. 154 into Eq. 153 and rearranging the terms, one obtains

$$\left(L_L + \frac{P_p(n_{co})^2 A_p}{2(d_z(t) + c_p)}\right)\frac{di(t)}{dt} = \left(\frac{P_p(n_{co})^2 A_p \dot{d}_z(t)}{2(d_z(t) + c_p)^2} - R_c\right)i(t) + v_s \tag{6.155}$$

This is a first-order ordinary differential equation in current $i(t)$. It is a non-homogeneous equation because of the presence of power-supply voltage v_s on the right-hand side. Therefore, the value of the magnetic levitation force, which depends on air gap d_z, is determined by the solution of the preceding equation, which depends on the value of voltage v_s. As discussed in Chapter 5, stability issues must be taken into consideration in the design of maglev train systems. The magnetic levitation force decreases as air gap d_z increases, and this can lead to instability of the maglev suspension system. For this reason, a feedback control system is needed to achieve a stable electromagnetic suspension. The appropriate value of power-supply voltage v_s required to ensure stability can be determined using a feedback control system. To this end, the power-supply voltage is allowed to vary as a function of air gap d_z and its derivative $\dot{d}_z$ as $v_s(d_z, \dot{d}_z) = v_{so} + f(d_z, \dot{d}_z)$, where $f(d_z, \dot{d}_z)$ is a control function that can be evaluated if air gap d_z and its derivative $\dot{d}_z$ are measured, and v_{so} is the power voltage at the initial equilibrium configuration. Using the control function $f(d_z, \dot{d}_z)$, the value of the voltage can be adjusted to ensure that d_z remains constant and ensure stable behavior of the maglev suspension.

The electric current differential equation defined by Eq. 155 can be systematically integrated with the MBS constrained dynamic equations developed in this chapter.

To demonstrate the procedure used to achieve this integration, the case of a maglev train system consisting of n_s electromagnetic suspensions is considered. In this case, all the electric currents of all the suspension circuits can be written in vector form as $\mathbf{i} = \mathbf{i}(t) = \begin{bmatrix} i_1(t) & i_2(t) & \cdots & i_{n_s}(t) \end{bmatrix}^T$. Using Eq. 155 for each suspension circuit, one can write $d\mathbf{i}/dt = \mathbf{f}_{ms}(\mathbf{i}, \mathbf{q}, \dot{\mathbf{q}})$, where the function $\mathbf{f}_{ms}(\mathbf{i}, \mathbf{q}, \dot{\mathbf{q}})$ can be defined using Eq. 155, and $\mathbf{q}$ is the vector of the system generalized coordinates. The electric current differential equation $d\mathbf{i}/dt = \mathbf{f}_{ms}(\mathbf{i}, \mathbf{q}, \dot{\mathbf{q}})$ can be combined with the augmented form of the equations of motion defined by Eq. 55 to obtain the MBS equation, which can be used to model the maglev train system as

$$\begin{bmatrix} \mathbf{M} & \mathbf{C}_{\mathbf{q}}^T & \mathbf{0} \\ \mathbf{C}_{\mathbf{q}} & \mathbf{0} & \mathbf{0} \\ \mathbf{0} & \mathbf{0} & \mathbf{I} \end{bmatrix} \begin{bmatrix} \ddot{\mathbf{q}} \\ \boldsymbol{\lambda} \\ d\mathbf{i}/dt \end{bmatrix} = \begin{bmatrix} \mathbf{Q}(\mathbf{i}, \mathbf{q}, \dot{\mathbf{q}}) \\ \mathbf{Q}_d(\mathbf{q}, \dot{\mathbf{q}}) \\ \mathbf{f}_{ms}(\mathbf{i}, \mathbf{q}, \dot{\mathbf{q}}) \end{bmatrix} \tag{6.156}$$

where $\mathbf{I}$ is the $n_s \times n_s$ identity matrix, $\mathbf{Q} = \mathbf{Q}_e + \mathbf{Q}_v + \mathbf{Q}_{ml}$, and $\mathbf{Q}_{ml}$ is a vector that contains all the generalized magnetic levitation forces. The preceding equation can be solved for $\ddot{\mathbf{q}}$, $\boldsymbol{\lambda}$ and $d\mathbf{i}/dt$. Acceleration vector $\ddot{\mathbf{q}}$ and the time derivative of the electric current $d\mathbf{i}/dt$ can be integrated to determine generalized coordinates $\mathbf{q}$, generalized velocities $\dot{\mathbf{q}}$, and electric currents $\mathbf{i}$. It is important, however, to point out that, because of the lack of coupling in the coefficient matrix of the preceding equation between the derivatives of the electric currents and other unknown accelerations and Lagrange multipliers, it is more efficient to solve the following two equations separately instead of solving Eq. 156:

$$\begin{bmatrix} \mathbf{M} & \mathbf{C}_{\mathbf{q}}^T \\ \mathbf{C}_{\mathbf{q}} & \mathbf{0} \end{bmatrix} \begin{bmatrix} \ddot{\mathbf{q}} \\ \boldsymbol{\lambda} \end{bmatrix} = \begin{bmatrix} \mathbf{Q}(\mathbf{i}, \mathbf{q}, \dot{\mathbf{q}}) \\ \mathbf{Q}_d(\mathbf{q}, \dot{\mathbf{q}}) \end{bmatrix}, \qquad \frac{d\mathbf{i}}{dt} = \mathbf{f}_{ms}(\mathbf{i}, \mathbf{q}, \dot{\mathbf{q}}) \tag{6.157}$$

Clearly, the second equation in this equation does not require any matrix inversion or LU factorization, and therefore, there is no need to combine the two equations of Eq. 157 in the matrix equation of Eq. 156.

Chapter 7

PANTOGRAPH/CATENARY SYSTEMS

Pantograph/catenary systems are used to supply power for electrically operated trains. Most freight trains that travel for long distances are driven by diesel locomotives. However, such diesel-powered locomotives have a speed limit of approximately 238 km/h (148 mph); therefore, diesel-powered engines cannot be used for high-speed passenger trains that operate at speeds of 300 km/h or higher. Extensive experimentation is currently being conducted to significantly increase train speed to a level that minimizes travel time. Therefore, pantograph/catenary technology is necessary: it represents the only viable option, particularly as the demand for higher train speeds increases. Such technology is also more environmentally friendly as compared to reliance on fuel, due to its higher efficiency and lower emissions. Using pantograph/catenary systems is not limited to high-speed trains; such systems are also used for urban transportation systems such as trams, trolleys, and electrically powered buses that travel for short distances at lower speeds. Electrification systems are less costly, quieter, and more reliable and can produce the greater amount of power needed for higher speeds.

A *pantograph*, which is often mounted on the top of a train, tram, or bus, is used to collect power through contact with an overhead power line called a *catenary*. The use of overhead electric lines to provide power for transportation systems started in the late 1800s. Such technology was first proposed for urban transportation, including trams and trolleys, and later for buses; therefore, as previously mentioned, the use of this technology is not limited to pantograph/catenary systems for high-speed trains (Figure 1).

It is also important to point out that pantograph/catenary systems, although they are the most widely used method for providing power for high-speed trains, are not the only method for supplying electric power to rail transportation systems. A third rail, batteries, and ground-level power supplies are also used, particularly for urban rail transportation systems.

Electrically powered trains use both overhead AC and DC voltage. AC voltage is normally used with an overhead arrangement, while DC voltage is used with a third rail. The voltage used in some countries for electric train operations can reach 25 kV AC at 50–60 Hz. AC electric supplies are used with induction motors. However, because of the DC-motor advantage, an AC supply can be converted to DC voltage.

This chapter presents mathematical formulations that can be used for the virtual prototyping and design of pantograph/catenary systems. The focus of the chapter is on developing computational approaches that can be integrated with general multibody

Mathematical Foundation of Railroad Vehicle Systems: Geometry and Mechanics,
First Edition. Ahmed A. Shabana.

Figure 7.1 Pantograph/catenary system. Source: Vaccaro (2019).

system (MBS) algorithms for computer modeling and simulation of electrically operated railroad vehicle systems.

A pantograph can be modeled as a mechanical system that consists of interconnected rigid or flexible components whose motion equations can be developed using the MBS constrained dynamic formulations discussed in Chapter 6. Catenary wires, on the other hand, are flexible cables that experience large displacements as the result of pantograph/catenary contact and aerodynamic forces. These displacements can be described using the finite element (FE) *absolute nodal coordinate formulation* (ANCF), which allows for accurately capturing initial wire geometries. The structure of the pantograph/catenary system is first discussed, and nonlinear kinematic and dynamic governing equations are developed. To ensure the efficient and safe operation of pantograph/catenary systems, and to minimize loss of contact, which can lead to arcing, several important issues must be taken into consideration, including proper design of the system to maintain contact between the pantograph and catenary wire, the effect of aerodynamic forces, wear control and reduction, and temperature effects. Some of these topics are discussed in this chapter.

7.1 PANTOGRAPH/CATENARY DESIGN

As shown in Figure 1, a pantograph/catenary system consists of two main subsystems: the pantograph, which is mounted on the top of the vehicle and is used to collect power, and the electric cable, called a catenary, which transmits electric power through contact. A pantograph can be modeled as a mechanical system consisting of interconnected components, while the catenary, which consists of several cables, can be treated as a structural system that experiences large displacements as the result of contact with the pantograph and aerodynamic forces. The catenary and pantograph systems are described in this section in more detail, and some of the important issues related to their design are highlighted to provide a better understanding of the challenges encountered in the operation, design, performance evaluation, and computer simulations of these systems.

Pantograph Two types of pantograph systems are in common use: *single-arm* and *double-arm pantograph systems*. These two systems are shown in Figure 2. The single-arm pantograph system was invented in 1955 by Louis Faiveley. It is Z-shaped and sometimes referred to as a *half-pantograph*, and is more widely used because of its compactness, lighter weight, and response and reliability at higher speeds; in addition, it requires less power to control. Double-arm pantographs are also used because they are more fault-tolerant as compared to single-arm pantographs.

(a) (b)

Figure 7.2 Pantograph types. Sources: (a) Vaccaro (2019); (b) hpgruesen/Pixabay.

Figure 3 shows the main components of a single-arm pantograph system. As shown, a pantograph consists of interconnected components that include the thrust rod, lower arm, balance rod, upper arm, crossbar, plunger, and pan-head. The pantograph system is designed to maintain contact between the pan-head and the catenary contact wire. To maintain this contact and avoid separations – which can result in air gaps and arcing during the operation of a railroad vehicle – pneumatic actuators are used, to provide sufficient uplift forces. The uplift force should not be very high, to avoid component wear that can adversely affect the performance of the pantograph/catenary system. Springs are also used as passive controllers to improve the stability of the pantograph mechanism. Carbon strips are used on the pan-head to collect electric current from the catenary wire, which is transmitted to the pantograph system. The use of carbon material, as compared to copper, is preferred for the pan-head strip due to the undesirable higher friction force that results from copper/copper contact.

Pantograph Data In some computer simulation models developed for pantographs, the pantograph lower arm is connected to the car body with a revolute joint; the lower link is connected to the car body with a spherical joint and to the upper arm with a spherical joint; the lower arm is connected to the upper arm with a revolute joint; the plunger is connected to the upper arm with a revolute joint and the upper link with a spherical joint; and the

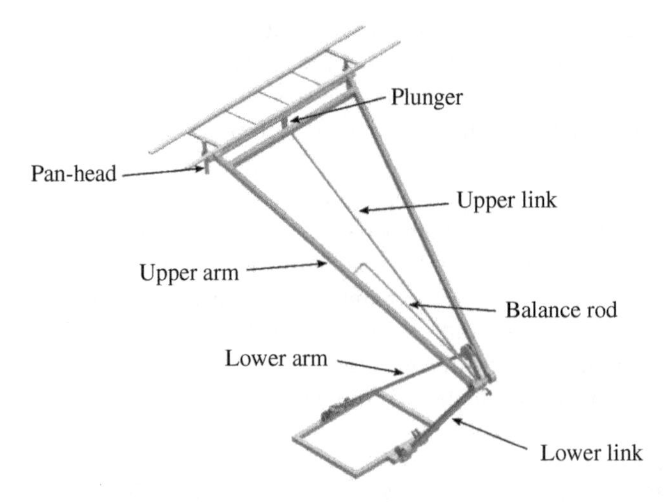

Figure 7.3 Single-arm pantograph.

lower arm is connected to the upper link with a spherical joint (Pappalardo et al. 2016; Daocharoenporn et al. 2019). In the model developed by Pappalardo et al. and Daocharoenporn et al., the lower arm mass is assumed to be 32.18 kg, and its mass moments of inertia are assumed to be 0.31, 10.43, and 10.65 kg·m^2. The upper arm mass is assumed to be 15.60 kg, and its mass moments of inertia are assumed to be 0.15, 7.76, and 7.86 kg·m^2. The lower link mass is assumed to be 3.10 kg, and its mass moments of inertia are assumed to be 0.05, 0.46, and 0.46 kg·m^2. The upper link mass is assumed to be 1.15 kg, and its mass moments of inertia are assumed to be 0.05, 0.48, and 0.48 kg·m^2. The plunger mass is assumed to be 1.51 kg, and its mass moments of inertia are assumed to be 0.07, 0.05, and 0.076 kg·m^2. The pan-head mass is assumed to be 9.50 kg, and its mass moments of inertia are assumed to be 1.59, 0.21, and 1.78 kg·m^2. When multiple pantographs are used to power high-speed trains, the minimum distance between two pantographs should not be less than 200 m. These data depend on the pantograph model used for specific train models and therefore should not be used with all simulation models. They are presented here for the purpose of providing examples of inertia data that can be used in computer simulation models.

Catenary Figures 4 and 5 show a catenary system used to provide electric power for the operation of railroad vehicles. Different names are used to refer to catenary systems, including *overhead line* and *overhead wire*. In addition to their use for electrically operated trains, catenary systems are used to supply the power required for the operation of urban vehicles such as trams, trolleys, and electric buses. In the case of intercity trains, catenary wires receive electricity from *feeder stations* located at different distances along the track. The feeder stations receive electricity from high-voltage electric grids. Electric current is collected by the train using the pantograph, which has a contact strip attached to the pan-head. The pan-head carbon strip maintains contact with the catenary by applying uplift force. The combined pantograph/catenary system is also referred to as a *current collection system*. Current collection systems allow electric current to flow through to the train and back

Figure 7.4 Catenary system. Source: Jackin/Adobe Stock.

to the feeder station through wheel/rail contact. This arrangement for the flow of current represents a closed electric circuit.

Catenary Structure The most common catenary system consists of *contact* and *messenger wires* connected by droppers that allow only tensile forces. The catenary system also includes supports, moving brackets, and steady arms (pull-off fittings). The contact wire provides electric power to the pantograph through contact with the pan-head, as previously mentioned. Because of the friction resulting from pantograph/catenary contact and the frequent trips normally made by trains, fatigue and wear of the catenary wires and pan-head of the pantograph are important design considerations. In addition to contact forces, the catenary can be subjected to significant cross-wind forces that result in severe oscillations. As a

Figure 7.5 Catenary system. Source: peuceta/Adobe Stock.

result of these forces, undesirable changes are possible in the geometry of the contact wire, which can negatively impact current-collection quality; these changes must be considered at the design stage.

To achieve high-quality, high-speed current collection, it is necessary to avoid significant deviations in the contact wire geometry, which can result from severe vibrations or large deformations. One of the methods used to avoid such changes in the contact wire geometry is to support the contact wire using a *messenger wire*, which prevents slanting and severe oscillations of the contact wire resulting from pantograph/catenary interaction forces and other forces such as aerodynamic forces. Nonetheless, both messenger and contact wires may experience nonlinear elastic deformations; therefore, it is necessary to examine their dynamic response using computer simulations. In the pantograph/catenary computer model developed by Seo et al. (2005), the mass density, cross-sectional area, and Young's modulus of the contact wire were assumed to be 3200 kg/m^3, 16.0 mm^2, and 9×10^9 N/m^2, respectively. Additional catenary data used in computer simulations were provided by Pappalardo et al. (2016) and Daocharoenporn et al. (2019).

The contact wire is connected to the messenger wire by vertical *droppers*, shown in Figure 5. The droppers provide additional stiffness that helps maintain system stability. The vertical droppers are designed to have high tensile stiffness but zero resistance to compressive forces. The steady horizontal arm is used to maintain the shape of the catenary such that pan-head wear is reduced. In developing computer models, the connection between the steady arm and the support can be modeled using a pin joint to allow for contact wire oscillations. As pointed out by Seo et al. (2005), the tension in the droppers between the contact wire and the messenger wire, span length of the support, gap, number of droppers, etc. have an effect on the vertical stiffness of catenary systems. In the computer simulations performed by Seo et al. (2005), the distance between two droppers is assumed to be 12.5 m. While the contact wire is modeled using ANCF finite elements, discrete springs and dampers are used to model the droppers. The stiffness and damping coefficients of the droppers are assumed to be 2.0×10^5 N/m and 100 N·s/m, respectively.

Catenary Geometry on Tangent Tracks Because the catenary and messenger wires are approximately straight lines between the supports, when a vehicle negotiates a tangent track, contact between the pantograph carbon strip used for current collection and the catenary contact wire can be localized; this results in a small contact area of high stress and nonuniform wear that can cause damage to the carbon strip. For this reason, the catenary is zigzagged over tangent track sections to achieve more uniform wear by allowing the contact wire to slide over the pantograph carbon strip. This zigzagging geometry is not required in curved track sections since as the vehicle negotiates a curve, the contact wire sweeps over the pantograph carbon strip. This issue is revisited when the pantograph/catenary wear problem is discussed in a later section of this chapter.

Catenary Pretension Pretension contributes to greater bending stiffness as a result of the coupling between axial and bending deformations. This additional geometric stiffening is necessary to maintain catenary stability and ensure stable contact with the pan-head. As the stiffness of the contact wire increases, the speed of the propagation of elastic waves resulting from pan-head/catenary interaction also increases to a level that is much higher than the vehicle operating speed. In addition to achieving stable pan-head/catenary contact, the speed of the propagation of the elastic waves is an important factor that must be considered. To avoid resonance, the frequency of the wire oscillations and the speed of the elastic waves must be much higher than the train operating speed, and this can be achieved using pretension applied to the catenary wires.

To make the pretension independent of weather conditions, including temperature variations, a constant pretension of 9000–20 000 N per wire is normally produced using balance weights or hydraulic tensioners. The weights are used to balance axial cable strains that cause pretension and increase the wire bending stiffness. When weights are used to produce wire pretension, the positions of the weights can change as the temperature changes. To prevent the weights from swaying, their motion is restricted by using sliding bars or tubes. Weight movements also impose a limit on the maximum length of the catenary wires. Due to temperature variations, the wires can expand or contract, and this deformation is proportional to the wire length and inversely proportional to the axial stiffness of the wire. For this reason, the maximum length of the wires between two pretension weight locations is limited to slightly less than 2 km. To prevent the catenary wires from sliding along the track as the result of the balance weights moving when the temperature varies, anchors fixed to the catenary poles or supports are placed close to the middle of the distance between the weight locations to constrain the motion of the catenary wire in the longitudinal direction of the track. Some other designs are also used that include stoppers to prevent the wires from sliding in the case of unexpected failure.

As previously mentioned, pretensioning the contact wire increases its stiffness and the speed of elastic wave propagation in the wire. The speed of elastic wave propagation depends not only on the contact wire stiffness but also on its inertia and material properties. It is recommended that the maximum train speed should not exceed 70% of the speed of elastic wave propagation in the contact wire. That is, for a train traveling at 300 km/h, the speed of wave propagation in the contact wire should be greater than 430 km/h. This is necessary to ensure safe operation of the train and avoid the loss of pantograph/catenary contact.

Voltage Specifications The voltage used for electrically operated trains depends on the country and region; different countries use different voltages for different rail transportation systems. Both AC and DC voltage are used. AC voltage is normally used with conductor motors. Although DC voltage systems are more commonly used with a third rail, due to the DC-motor advantage, an AC electric supply can be converted to a DC system. Some of the most commonly used voltages, shown in Table 1, were selected for international standardization (CENELEC 2007; IEC 2007). The voltage range allowed for standardized voltages depends on the number of trains collecting current and the distance of these trains from feed stations.

For example, the planned traction power supply system for a future California high-speed train will utilize a 25 kV and 60 Hz catenary and a negative (−25 kV) longitudinal feeder together with autotransformers placed approximately every 8 km (5 miles). The 2 × 25 kV – 60 Hz arrangement is used to reduce the number of feed stations (Hsiao 2010).

Table 7.1 Voltage standardization

Electrification system	Voltage				
	Min. non-permanent	Min. permanent	Nominal	Max. permanent	Max. non-permanent
600 V DC	400 V	400 V	600 V	720 V	800 V
750 V DC	500 V	500 V	750 V	900 V	1000 V
1500 V DC	1000 V	1000 V	1500 V	1800 V	1950 V
3 kV DC	2 kV	2 kV	3 kV	3.6 kV	3.9 kV
15 KV AC, 16.7 Hz	11 kV	12 kV	15 kV	17.25 kV	18 kV
25 KV AC, 50 Hz (EN 50163) and 60 Hz (IEC 60850)	17.5 kV	19 kV	25 kV	27.5 kV	29 kV

Source: CENELEC (2007); IEC (2007).

7.2 ANCF CATENARY KINEMATIC EQUATIONS

A large number of investigations have been devoted to studying catenary system dynamics and vibration. Some of these investigations are based on linear models, while others propose nonlinear models with varying degrees of complexity and assumptions. In some cases, the catenary is modeled using simple discrete spring-damper elements. Such simplified models do not account for the effect of distributed inertia and elasticity of catenary cables and do not accurately predict the cable stresses that are necessary for credible evaluation of cable integrity and durability. Such credible assessments can be made using continuum-based models that account for the distributed inertia and elasticity of catenary wires. Such continuum-based models allow for a more accurate evaluation of the speed of elastic wave propagation and the implementation of different material models; these models also account for geometric nonlinearities that result from large-amplitude oscillations.

Catenary design has a direct effect on the operation of electrically powered trains, as evidenced by the fact that pantograph/catenary contact is one of the main factors contributing to limiting train speeds. As a result of the forces acting on the catenary system, elastic waves are generated in the catenary cable. The train operating speed cannot exceed the speed of the propagation of catenary elastic waves due to safety considerations (Kumaniecka and Jacek 2008; Pappalardo et al. 2016). Therefore, it is recommended to use a continuum-based approach for the catenary model that can be systematically integrated with general MBS computational algorithms.

Catenary Models A catenary system consists of flexible wires that can be subjected to significant vibrations, including aerodynamics and contact forces. These vibrations can result in a loss of contact with the pan-head, leading to poor current collection that can adversely affect the performance of the pantograph/catenary system. As a result of catenary/pan-head contact, the contact wire can experience large nonlinear vibrations, which must be accounted for to develop realistic virtual prototyping models. Capturing the nonlinearity of catenary vibrations contributes to accurate prediction of the location of the contact point between the catenary wire and pan-head in computer simulations. In the literature, as previously mentioned, simplified models have been proposed; in some of these models, lumped parameters are used; other models use discrete springs that have no inertia. Other models include an analytical description of the catenary kinematics, including the Fourier sine expansion method. Some other approaches employ the finite element (FE) method using conventional beam elements that do not correctly capture large displacements as a result of the type of nodal coordinates used. In some studies of pantograph/catenary interaction, *co-simulation techniques* are used by using two different computer programs: one for modeling the pantograph and vehicle system, and an FE computer program for modeling the catenary system. When co-simulation techniques are used, communication between different software using different formulations and numerical procedures can be an issue.

In other investigations, continuum-based ANCF finite elements are used for modeling catenary flexibility (Seo et al. 2006; Lee and Park 2012). ANCF finite elements can be directly implemented in MBS computer programs, thereby eliminating the need to use co-simulations. The FE/ANCF approach is general and allows for developing a detailed catenary model that has a large number of degrees of freedom. This approach also allows for using nonlinear material models and conveniently describing catenary geometries. In this chapter, catenary equations of motion are developed using ANCF finite elements that are suited for nonlinear large displacement analysis. These elements lead to a constant mass matrix regardless of the magnitude of the catenary-wire displacements.

ANCF Finite Elements ANCF finite elements, introduced in Chapter 2 for the geometry description, also allow for developing accurate, general, flexible body models for predicting catenary vibrations and stresses. These continuum-based elements can capture both the axial and bending deformation of catenary cables, and also account for coupling between these two deformation modes. Two different three-dimensional ANCF finite elements are good candidates for developing catenary dynamic equations of motion: the *ANCF cable element,* which has a smaller number of degrees of freedom and does not capture the effect of shear deformation; and the fully parameterized *ANCF beam element* introduced in Chapter 2. The fully parameterized ANCF beam element captures the effect of shear deformation and the change in cross-section dimensions that results from cross-section stretch, axial, and bending displacements. That is, the effects of pretension and temperature on the change in cross-section dimensions can be examined using the ANCF fully parameterized beam element. Nonetheless, both the three-dimensional ANCF cable and fully parameterized beam elements capture the in-plane and out-of-plane bending deformations of the catenary. Furthermore, as a result of using ANCF position vector gradients as nodal coordinates, complex geometries can be modeled; these features allow for experimenting numerically with different catenary wire shapes. As previously mentioned in Chapter 2, the fully parameterized ANCF beam element was mainly used for the geometric description of the rail. In this chapter, ANCF finite elements are used to develop both kinematic and dynamic equations of motion of catenary wires.

ANCF Cable Element The three-dimensional ANCF cable element can be used to capture the extension and in- and out-of-plane bending deformations (Gerstmayr and Shabana 2006; Shabana 2018). However, this element does not allow for deformation of the cross-section since the element shape function matrix depends only one parameter. The ANCF cable element, shown in Figure 6, is assumed to have two nodal coordinates, and each node has six coordinates: three translation coordinates and three gradient coordinates that define the elements of the tangent vector at the node. Therefore, the vector of nodal coordinates at node k of an element j can be written as $\mathbf{e}^{jk} = \left[\mathbf{r}^{jk} \; \left(\mathbf{r}_x^{jk}\right)^T\right]^T, k = 1, 2,$ where $\mathbf{r}^{jk}$ is the global position vector of node k, $\mathbf{r}_x^{jk} = \partial\mathbf{r}^j/\partial x$ is the position-gradient vector evaluated at node k, and x is the element spatial coordinate in the longitudinal direction. Because the ANCF cable element has 6 coordinates at each node, the total number of element nodal coordinates is 12. Therefore, polynomials with 12 coefficients can be used as a starting point for developing the element displacement field. For the cable element, the following cubic polynomials can be used

$$\mathbf{r}^j = \mathbf{r}^j(x) = \begin{bmatrix} r_1^j \\ r_2^j \\ r_3^j \end{bmatrix} = \begin{bmatrix} a_0 + a_1 x + a_2(x)^2 + a_3(x)^3 \\ b_0 + b_1 x + b_2(x)^2 + b_3(x)^3 \\ c_0 + c_1 x + c_2(x)^2 + c_3(x)^3 \end{bmatrix} \tag{7.1}$$

where a_l, b_l, and c_l, $l = 0,1,2,3$, are the time-dependent polynomial coefficients that can be replaced by the element position and position-gradient nodal coordinates defined by vector $\mathbf{e}^j = \left[\left(\mathbf{e}^{j1}\right)^T \left(\mathbf{e}^{j2}\right)^T\right]^T = \left[\mathbf{r}^{j1^{\mathrm{T}}} \; \mathbf{r}_x^{j1^{\mathrm{T}}} \; \mathbf{r}^{j2^{\mathrm{T}}} \; \mathbf{r}_x^{j2^{\mathrm{T}}}\right]^{\mathrm{T}}$, which can also be written as $\mathbf{e}^j = \left[\mathbf{r}^{j^{\mathrm{T}}}(x = 0) \; \mathbf{r}_x^{j^{\mathrm{T}}}(x = 0) \; \mathbf{r}^{j^{\mathrm{T}}}(x = l) \; \mathbf{r}_x^{j^{\mathrm{T}}}(x = l)\right]^{\mathrm{T}}$. Using these definitions of the nodal coordinates, the cable element displacement field can be written as

$$\mathbf{r}^j = \mathbf{r}^j(x, t) = \mathbf{S}^j(x)\,\mathbf{e}^j(t) \tag{7.2}$$

In this equation, $\mathbf{S}^j(x)$ is the element *shape function matrix*, which can be written as

$$\mathbf{S}^j = \left[s_1\mathbf{I} \;\; s_2\mathbf{I} \;\; s_3\mathbf{I} \;\; s_4\mathbf{I}\right] \tag{7.3}$$

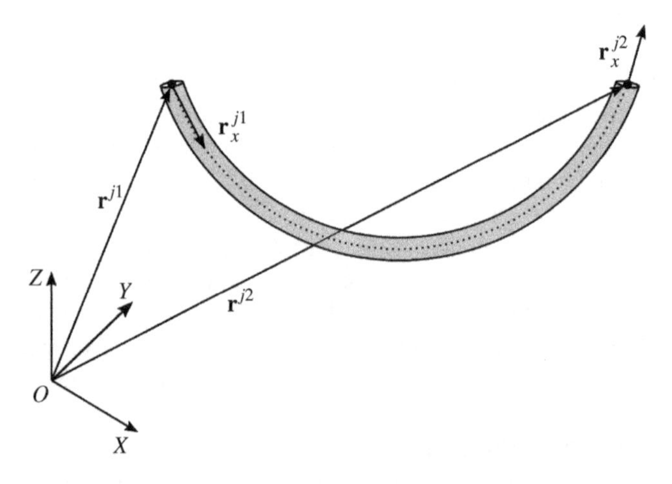

Figure 7.6 ANCF cable element.

where $\mathbf{I}$ is the 3×3 identity matrix,

$$\left.\begin{array}{ll} s_1 = 1 - 3\xi^2 + 2\xi^3, & s_2 = l\left(\xi - 2\xi^2 + \xi^3\right), \\ s_3 = 3\xi^2 - 2\xi^3, & s_4 = l\left(-\xi^2 + \xi^3\right) \end{array}\right\} \tag{7.4}$$

are the shape functions, $\xi = x/l$, and l is the length of the finite element.

The cable element described in this section is based on cubic interpolation; therefore, it is suited for modeling bending problems. Because it is a three-dimensional element, it captures both in- and out-of-plane bending deformations. The cubic interpolation allows for describing non-constant curvatures, a necessary feature for modeling bending problems. This description is also consistent with the basic bending-vibration equation of beams, which is a fourth-order partial differential equation. Using an interpolation lower than cubic imposes restrictions on the definition of the curvature within the element and can lead to undesirable discontinuity problems.

Furthermore, using position vector gradients as nodal coordinates allows for modeling initially curved cables and large amplitudes of vibration. The cable element displacement field, however, depends on only one spatial coordinate, the longitudinal coordinate x. Therefore, the deformation of the element cross-section cannot be accounted for when this element is used. In addition, the effect of shear deformation is not taken into consideration in the formulation of the displacement field of this element. Nonetheless, because this element has significantly fewer coordinates as compared to the fully parameterized ANCF beam element, it has proven to be very efficient for modeling catenary system vibrations and also does not suffer from the locking problems encountered when fully parameterized ANCF elements are used.

The ANCF displacement field can be used to describe arbitrarily large rigid-body displacements because it includes linear terms as part of the polynomial representation. Therefore, such an element displacement field leads to zero strain under arbitrary rigid-body displacements. Using this description eliminates the need to use an incremental solution procedure and allows for direct and efficient implementation in MBS computational algorithms. Furthermore, the ANCF displacement-field description allows for defining the absolute velocity and acceleration vectors as linear functions of the nodal velocities and accelerations. By differentiating the position equation $\mathbf{r}^j(x,t) = \mathbf{S}^j(x)\mathbf{e}^j(t)$ once and twice with respect to time, one obtains the absolute velocity and acceleration vectors, which can be written, respectively, as $\dot{\mathbf{r}}^j(x,t) = \mathbf{S}^j(x)\dot{\mathbf{e}}^j(t)$ and $\ddot{\mathbf{r}}^j(x,t) = \mathbf{S}^j(x)\ddot{\mathbf{e}}^j(t)$.

Fully Parameterized ANCF Beam Element The three-dimensional, two-node, fully parameterized ANCF beam element is shown in Figure 7. Each node has 12 coordinates that include the 3 components of the node global position vector and 9 components of the matrix of position vector gradients at the node. Therefore, the element has 24 degrees of freedom: 12 at each of the element nodes (Yakoub and Shabana 2001). This element was introduced in Chapter 2 and is used to define the rail geometry regardless of whether the rail is modeled as a flexible body. If x, y, and z are used as the element spatial coordinates, the geometry of the rail surface can be defined using the functional relationship $z = z(x,y)$. This relationship can be used for the geometry description, and it is not invoked in the analysis

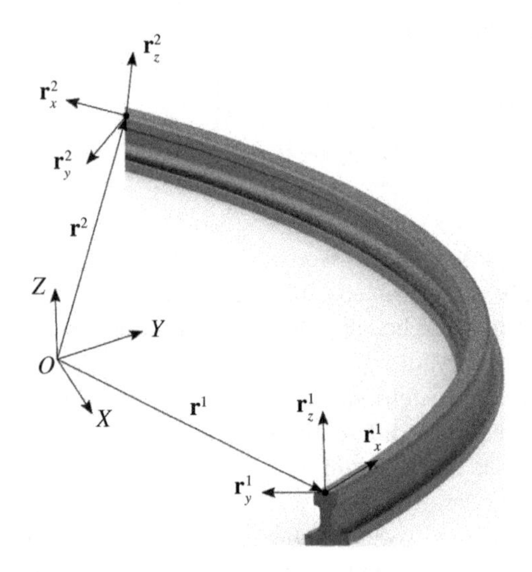

Figure 7.7 ANCF beam element.

of deformations. That is, during the deformation analysis step, the element spatial coordinates x, y, and z are assumed independent, while in the search for the contact point on the rail surface, for example, the functional relationship $z = z(x, y)$ is used. It is also important to mention that in a computer implementation, the solution algorithm can be designed to allow using different elements for the geometry and deformation analyses. For example, the ANCF fully parameterized element can be used to define the surface geometry, while the ANCF cable element can be used in the deformation analysis to achieve efficient computer simulations. In the case of wheel/rail contact, using the ANCF fully parameterized element is necessary to capture the surface geometry of the rail accurately. In the case of a pantograph/catenary system, the geometry is simpler, and using the ANCF fully parameterized element is not necessary unless the effects of deformation modes such as changing cross-section dimensions need to be evaluated.

Because the fully parameterized ANCF beam element has 24 nodal coordinates, the following cubic interpolation can be used for beam element j

$$\mathbf{r}^j = \begin{bmatrix} r_1^j \\ r_2^j \\ r_3^j \end{bmatrix} = \begin{bmatrix} a_0 + a_1x + a_2y + a_3z + a_4xy + a_5xz + a_6(x)^2 + a_7(y)^3 \\ b_0 + b_1x + b_2y + b_3z + b_4xy + b_5xz + b_6(x)^2 + b_7(y)^3 \\ c_0 + c_1x + c_2y + c_3z + c_4xy + c_5xz + c_6(x)^2 + c_7(y)^3 \end{bmatrix} \tag{7.5}$$

where a_l, b_l, and c_l, $l = 0, 1, \ldots, 7$, are the time-dependent polynomial coefficients that can be replaced by the element position and position-gradient nodal coordinates defined by vector

$$\mathbf{e}^j = \left[\left(\mathbf{e}^{j1}\right)^T \; \left(\mathbf{e}^{j2}\right)^T \right]^T = \left[\mathbf{r}^{j1^{\mathrm{T}}} \; \mathbf{r}_x^{j1^{\mathrm{T}}} \; \mathbf{r}_y^{j1^{\mathrm{T}}} \; \mathbf{r}_z^{j1^{\mathrm{T}}} \; \mathbf{r}^{j2^{\mathrm{T}}} \; \mathbf{r}_x^{j2^{\mathrm{T}}} \; \mathbf{r}_y^{j2^{\mathrm{T}}} \; \mathbf{r}_z^{j2^{\mathrm{T}}} \right]^{\mathrm{T}} \tag{7.6}$$

In this equation,

$$\mathbf{r}^{j1} = \mathbf{r}^{j}(0,0,0) = \begin{bmatrix} e_1 \\ e_2 \\ e_3 \end{bmatrix}, \quad \mathbf{r}_x^{j1} = \mathbf{r}_x^{j}(0,0,0) = \begin{bmatrix} e_4 \\ e_5 \\ e_6 \end{bmatrix},$$

$$\mathbf{r}_y^{j1} = \mathbf{r}_y^{j}(0,0,0) = \begin{bmatrix} e_7 \\ e_8 \\ e_9 \end{bmatrix}, \quad \mathbf{r}_z^{j1} = \mathbf{r}_z^{j}(0,0,0) = \begin{bmatrix} e_{10} \\ e_{11} \\ e_{12} \end{bmatrix} \tag{7.7}$$

are the coordinates of the first node of the element,

$$\mathbf{r}^{j2} = \mathbf{r}^{j}(l,0,0) = \begin{bmatrix} e_{13} \\ e_{14} \\ e_{15} \end{bmatrix}, \quad \mathbf{r}_x^{j2} = \mathbf{r}_x^{j}(l,0,0) = \begin{bmatrix} e_{16} \\ e_{17} \\ e_{18} \end{bmatrix},$$

$$\mathbf{r}_y^{j2} = \mathbf{r}_y^{j}(l,0,0) = \begin{bmatrix} e_{19} \\ e_{20} \\ e_{21} \end{bmatrix}, \quad \mathbf{r}_z^{j3} = \mathbf{r}_z^{j}(l,0,0) = \begin{bmatrix} e_{22} \\ e_{23} \\ e_{24} \end{bmatrix} \tag{7.8}$$

are the coordinates of the second node of the element, l is the element length, and $\mathbf{r}_\alpha^j = \partial \mathbf{r}^j/\partial\alpha, \alpha = x, y, z$. Using the conditions of Eqs. 7 and 8, the coefficients of the polynomials in Eq. 5 can be replaced by the vector of nodal coordinates of Eq. 6, which can be written using Eqs. 7 and 8 as $\mathbf{e}^j = \begin{bmatrix} e_1 & e_2 & \dots & e_{24} \end{bmatrix}^T$. Therefore, one can show that the displacement field of the fully parameterized ANCF beam element can be written as

$$\mathbf{r}^j = \mathbf{r}^j(\mathbf{x}, t) = \mathbf{S}^j(\mathbf{x})\,\mathbf{e}^j(t) \tag{7.9}$$

In this equation, $\mathbf{x} = \begin{bmatrix} x & y & z \end{bmatrix}^{\mathrm{T}}$ is the vector of spatial coordinates and $\mathbf{S}^j$ is the element shape function matrix, which can be written as

$$\mathbf{S}^j = \begin{bmatrix} s_1\mathbf{I} & s_2\mathbf{I} & s_3\mathbf{I} & s_4\mathbf{I} & s_5\mathbf{I} & s_6\mathbf{I} & s_7\mathbf{I} & s_8\mathbf{I} \end{bmatrix} \tag{7.10}$$

where the shape functions s_i, $i = 1, 2, \dots, 8$ are defined as (Yakoub and Shabana 2001)

$$\left.\begin{aligned} & s_1 = 1 - 3\xi^2 + 2\xi^3, \quad s_2 = l\left(\xi - 2\xi^2 + \xi^3\right), \quad s_3 = l(\eta - \xi\eta), \\ & s_4 = l(\varsigma - \xi\varsigma), \quad s_5 = 3\xi^2 - 2\xi^3, \quad s_6 = l\left(-\xi^2 + \xi^3\right), \\ & s_7 = l\xi\eta, \quad s_8 = l\xi\varsigma \end{aligned}\right\} \tag{7.11}$$

and $\xi = x/l$, $\eta = y/l$, and $\varsigma = z/l$.

Unlike the cable element, which has a displacement field written in terms of only one parameter x, the displacement field of the fully parameterized beam element is expressed in terms of three parameters x, y, and z. Therefore, the fully parameterized beam element is capable of capturing deformation modes that cannot be captured by the cable element. These deformations, as previously mentioned, include shear deformation and changing cross-section dimensions. This generality comes at the expense of higher computational cost as a result of the increase in the number of coordinates and the existence of high-frequency modes that couple different displacements. Therefore, the fully parameterized beam element is recommended for use in catenary problems in which the effects of these deformation modes are important and need to be investigated.

7.3 CATENARY INERTIA AND ELASTIC FORCES

The ANCF element displacement fields defined in the preceding section for the cable and fully parameterized elements can be used to develop the expressions of the element inertia and elastic forces. In the developments presented in this section, the catenary is assumed to be discretized using n_e finite elements. For both the gradient-deficient cable and fully parameterized beam elements, the displacement field for element j can be written as $\mathbf{r}^j = \mathbf{r}^j(\mathbf{x}, t) = \mathbf{S}^j(\mathbf{x})\mathbf{e}^j(t)$, which in the case of the cable element $\mathbf{x}$ reduces to a scalar – that is, $\mathbf{x} = x$ – and in the case of the fully parameterized beam element, $\mathbf{x}$ is a three-dimensional vector defined as $\mathbf{x} = \begin{bmatrix} x & y & z \end{bmatrix}^{\mathrm{T}}$. As a result of using the full parameterization for the beam element, a complete set of gradient vectors can be determined. This allows for using a more general approach for formulating the elastic forces at the expense of having high-frequency deformation modes and encountering locking problems in some applications.

Catenary Inertia Forces Using the displacement field $\mathbf{r}^j = \mathbf{r}^j(\mathbf{x}, t) = \mathbf{S}^j(\mathbf{x})\mathbf{e}^j(t)$, the absolute velocity and acceleration vectors of an arbitrary point on finite element j can be written, respectively, as $\mathbf{v}^j = \dot{\mathbf{r}}^j = \mathbf{S}^j\dot{\mathbf{e}}^j$ and $\mathbf{a}^j = \ddot{\mathbf{r}}^j = \mathbf{S}^j\ddot{\mathbf{e}}^j$, $j = 1, 2, \ldots, n_e$. The virtual displacement of the arbitrary point can be written as $\delta\mathbf{r}^j = \mathbf{S}^j\delta\mathbf{e}^j$. Therefore, the virtual work of the inertia forces of ANCF element j can then be defined as

$$\begin{aligned} \delta W_i^j &= \int_{V^j} \rho^j \mathbf{a}^{j^{\mathrm{T}}} \delta\mathbf{r}^j dV^j = \int_{V^j} \rho^j \left(\ddot{\mathbf{e}}^{j^{\mathrm{T}}} \mathbf{S}^{j^{\mathrm{T}}}\right) \left(\mathbf{S}^j \delta\mathbf{e}^j\right) dV^j \\ &= \left\{ \ddot{\mathbf{e}}^{j^{\mathrm{T}}} \int_{V^j} \rho^j \mathbf{S}^{j^{\mathrm{T}}} \mathbf{S}^j dV^j \right\} \delta\mathbf{e}^j, \quad j = 1, 2, \ldots, n_e \end{aligned} \tag{7.12}$$

where ρ^j and V^j are, respectively, the mass density and volume of the ANCF element. The preceding equation can be written as $\delta W_i^j = \left[\ddot{\mathbf{e}}^{j^{\mathrm{T}}} \mathbf{M}^j\right] \delta\mathbf{e}^j$, where $\mathbf{M}^j$ is the constant and symmetric mass matrix of ANCF element j defined as

$$\mathbf{M}^j = \int_{V^j} \rho^j \mathbf{S}^{j^{\mathrm{T}}} \mathbf{S}^j dV^j, \quad j = 1, 2, \ldots, n_e \tag{7.13}$$

Because the element mass matrix is constant, the quadratic velocity centrifugal and Coriolis inertia forces are equal to zero in this formulation. Given the dimensions and material properties of the catenary wires, the mass matrix in the preceding equation can be evaluated only once in advance of the dynamic simulation. In the case of the ANCF cable element, integration over the volume is not required, and the mass matrix can be evaluated by integrating over the element length. In this case, the cable-element mass matrix can be simply evaluated as $\mathbf{M}^j = \int_0^l \rho^j A^j \mathbf{S}^{j^{\mathrm{T}}} \mathbf{S}^j dx$, where A^j is the element cross-section area.

The virtual work of the inertia forces can then be written as $\delta W_i^j = \mathbf{Q}_i^{j^{\mathrm{T}}} \delta\mathbf{e}^j$, where $\mathbf{Q}_i^j$ is the vector of the generalized inertia forces associated with the element nodal coordinates. This vector can be written as

$$\mathbf{Q}_i^j = \mathbf{M}^j \ddot{\mathbf{e}}^j, \qquad j = 1, 2, \ldots, n_e \tag{7.14}$$

In obtaining this simple expression of the inertia forces, no assumptions are made except for the order of the interpolating polynomials. Therefore, this simple expression of the inertia forces can be used in the case of very large displacements and does not impose any restrictions on the amount of deformation or rotation within the element except for the restrictions imposed by the order of the interpolating polynomials. This ANCF feature is attributed to using position vector gradients as nodal coordinates. This choice of coordinates allows for describing arbitrary displacements, including general rigid-body displacements. The virtual work and vector of generalized inertia forces are used in the development of catenary equations of motion presented in this section.

Catenary Elastic Forces In the case of the fully parameterized ANCF beam element, one can evaluate all the position gradient vectors. In this case, a general continuum mechanics approach can be used in the formulation of the elastic forces. The general continuum mechanics approach for formulating stress forces is documented in standard texts on continuum mechanics and the theory of elasticity. Using this approach, the virtual work of the stresses of element j can be written in terms of the Green-Lagrange strain tensor $\boldsymbol{\varepsilon}^j$ and the second Piola-Kirchhoff stress tensor $\boldsymbol{\sigma}_{P2}^j$ as (Bonet and Wood 1997; Boresi and Chong 2000; Ogden 1984; Shabana 2018)

$$\delta W_s^j = -\int_{V^j} \boldsymbol{\sigma}_{P2}^j : \delta\boldsymbol{\varepsilon}^j dV^j, \qquad j = 1, 2, \ldots, n_e \tag{7.15}$$

The stress and strain tensors used in this equation are defined in the reference configuration. The virtual changes in the strains can be expressed in terms of the virtual changes of the position vector gradients as

$$\delta\boldsymbol{\varepsilon}^j = \frac{1}{2}\left[\left(\delta\mathbf{J}^{j^T}\right)\mathbf{J}^j + \mathbf{J}^{j^T}\left(\delta\mathbf{J}^j\right)\right], \qquad j = 1, 2, \ldots, n_e \tag{7.16}$$

where $\mathbf{J}^j = \left[\partial\mathbf{r}^j/\partial x \quad \partial\mathbf{r}^j/\partial y \quad \partial\mathbf{r}^j/\partial z\right]$ is the matrix of position vector gradients at an arbitrary point on element j. The second Piola-Kirchhoff stresses are related to the Green-Lagrange strains using the constitutive equations

$$\boldsymbol{\sigma}_{P2}^j = \mathbf{E}^j : \boldsymbol{\varepsilon}^j, \qquad j = 1, 2, \ldots, n_e \tag{7.17}$$

where $\mathbf{E}^j$ is the fourth-order tensor of elastic coefficients. Substituting the preceding two equations into Eq. 15, it can be shown that the virtual work of the stresses of ANCF element j can be written as

$$\begin{aligned}\delta W_s^j &= -\frac{1}{2}\int_{V^j}\left(\mathbf{E}^j : \boldsymbol{\varepsilon}^j\right) : \left[\left(\delta\mathbf{J}^{j^T}\right)\mathbf{J}^j + \mathbf{J}^{j^T}\left(\delta\mathbf{J}^j\right)\right] dV^j \\ &= -\mathbf{Q}_s^{j^T}\delta\mathbf{e}^j, \qquad j = 1, 2, \ldots, n_e\end{aligned} \tag{7.18}$$

where $\mathbf{Q}_s^j$ is the vector of the elastic forces associated with the nodal coordinates of ANCF element j.

The cable element is a gradient-deficient element, and therefore, a complete set of gradient vectors cannot be defined. For this reason, classical beam theory formulations can be used. In this case, the elastic forces of the cable element can be formulated using the virtual work or the strain energy. The virtual work of the elastic forces for the cable element can be written as

$$\delta W_s^j = \int_0^l E^j A^j \varepsilon_{11}^j \delta \varepsilon_{11}^j dx + \frac{1}{2}\int_0^l E^j I^j \kappa^j \delta \kappa^j dx, \qquad j = 1, 2, \ldots, n_e \tag{7.19}$$

where ε_{11}^j is the axial strain, E^j is the modulus of elasticity, A^j is the cross-section area, I^j is the second moment of area, and κ^j is the curvature. The elastic forces can also be evaluated using the strain energy of cable element j. The strain energy can be written as

$$U^j = \frac{1}{2}\int_0^l E^j A^j \left(\varepsilon_{11}^j\right)^2 dx + \frac{1}{2}\int_0^l E^j I^j \left(\kappa^j\right)^2 dx, \qquad j = 1, 2, \ldots, n_e \tag{7.20}$$

Using this expression of strain energy, the vector of generalized elastic forces associated with the element nodal coordinates can be defined as $\mathbf{Q}_s^j = -\left(\partial U^j / \partial \mathbf{e}^j\right)^T$. The axial strain ε_{11}^j can be defined using position-gradient vector $\mathbf{r}_x^j$ as $\varepsilon_{11}^j = (1/2)\left(\mathbf{r}_x^{jT}\mathbf{r}_x^j - 1\right)$. The curvature κ^j can be defined as the norm of the curvature vector obtained by differentiating unit tangent $\mathbf{r}_x^j / \left|\mathbf{r}_x^j\right|$ with respect to the arc length s, where $ds = \left|\mathbf{r}_x^j\right| dx$: that is, $\partial\left(\mathbf{r}_x^j / \left|\mathbf{r}_x^j\right|\right) / \partial s = \left[\partial\left(\mathbf{r}_x^j / \left|\mathbf{r}_x^j\right|\right) / \partial x\right] (\partial x / \partial s)$, which can be written as $\partial\left(\mathbf{r}_x^j / \left|\mathbf{r}_x^j\right|\right) / \partial s = \left[\partial\left(\mathbf{r}_x^j / \left|\mathbf{r}_x^j\right|\right) / \partial x\right]\left(1 / \left|\mathbf{r}_x^j\right|\right)$. Therefore, exact definitions of the axial strain and curvature can be used to develop the expression of the vector of the elastic forces of the ANCF cable element.

While the mass matrices of the ANCF cable and fully parameterized beam elements are constant and the generalized inertia forces are linear in the second derivatives of the nodal coordinates, the generalized elastic (stress) forces of both elements are highly nonlinear functions of the nodal coordinates. These elements, therefore, can capture the elastic nonlinearities due to large displacements of catenary wires. As a result of using the general continuum mechanics approach in the formulation of the elastic forces of the fully parameterized beam element, and as a result of using a complete set of gradient vectors that allow for the definition of all the components of the Green-Lagrange strain tensor, the resulting nonlinear elastic forces account for more deformation modes as compared to the gradient-deficient cable element. While these deformation modes can be important in many applications, they can also be a source of high frequencies that force the numerical integration routine to take smaller time steps, making computer simulations using fully parameterized beam-element models less efficient as compared to simulations based on cable-element models.

7.4 CATENARY EQUATIONS OF MOTION

In Chapter 6, the principle of virtual work in dynamics was derived using a system of particles. A continuum can be assumed to consist of an infinite number of particles, and therefore, the principle of virtual work is also applicable to flexible bodies. For example, in the case of n_p particles, the virtual work of the inertia forces can be written as $\delta W_i = \sum_{i=1}^{n_p} m^i \ddot{\mathbf{r}}^{iT} \delta \mathbf{r}^i$, where m^i is the particle mass and $\mathbf{r}^i$ is its global position vector. If a finite element j is assumed to consist of an infinite number of particles, the summation can be replaced by integration and m^i can be replaced by $\rho^j dV^j$, where dV^j is an infinitesimal volume. It follows that the virtual work of the inertia force of the finite element can be written as $\delta W_i = \int_{V^j} \rho^j \ddot{\mathbf{r}}^{j^T} \delta \mathbf{r}^j dV^j$, as previously used in this section.

It was shown in Chapter 6 that the principle of virtual work can be written as $\delta W_i = \delta W_e$, where δW_i is the virtual work of the system inertia forces and δW_e is the virtual work of the system applied forces. In the case of flexible bodies, the virtual work principle can account for the effect of internal elastic forces by including the virtual work of the system elastic forces δW_s. In the case of unconstrained motion, the equations of motion of the finite elements that form a catenary can be developed using the principle of virtual work defined as

$$\delta W_i = \delta W_s + \delta W_e \tag{7.21}$$

In the case of a flexible catenary, the virtual work δW_e of applied forces such as gravity, aerodynamics, and contact forces can be systematically developed. If, for example, an external force vector $\mathbf{F}^j$ acts at a point P defined by coordinates $\mathbf{x}_P$ on finite element j, the virtual work of this force vector can be written using the ANCF displacement field as

$$\delta W_e^j = \mathbf{F}^{j^T} \delta \mathbf{r}_P^j = \mathbf{F}^{j^T} \mathbf{S}^j \left(\mathbf{x}_P\right) \delta \mathbf{e}^j = \mathbf{Q}_e^{j^T} \delta \mathbf{e}^j \tag{7.22}$$

where $\mathbf{S}^j(\mathbf{x}_P)$ is a constant matrix that defines the element shape function matrix at point $\mathbf{x}_P$, and $\mathbf{Q}_e^j$ is the vector of generalized forces associated with the element nodal coordinates $\mathbf{e}^j$ as a result of the application of force vector $\mathbf{F}^j$. This vector of generalized forces is defined using the preceding equation as $\mathbf{Q}_e^j = \mathbf{S}^{j^T} \left(\mathbf{x}_P\right) \mathbf{F}^j$. The generalized forces associated with the element nodal coordinates can be systematically defined in cases of both concentrated and distributed external forces for both the fully parameterized and gradient-deficient finite elements discussed in this section. Therefore, one can write the virtual work of the applied forces acting on the catenary by summing up the virtual work of the forces of its finite elements as

$$\delta W_e = \sum_{j=1}^{n_e} \delta W_e^j = \sum_{j=1}^{n_e} \mathbf{Q}_e^{j^T} \delta \mathbf{e}^j \tag{7.23}$$

Using the expression for the virtual work of the inertia forces of the ANCF element obtained in the preceding section, the virtual work of the inertia forces of the catenary can be obtained by summing up the virtual work of the inertia forces of its ANCF elements as

$$\delta W_i = \sum_{j=1}^{n_e} \delta W_i^j = \sum_{j=1}^{n_e} \left(\mathbf{M}^j \ddot{\mathbf{e}}^j\right)^T \delta \mathbf{e}^j \tag{7.24}$$

Using the expression of the virtual work of the elastic forces of the finite element obtained in the preceding section, the virtual work of the elastic forces of the FE catenary model can be written as

$$\delta W_s = \sum_{j=1}^{n_e} \delta W_s^j = -\sum_{j=1}^{n_e} \mathbf{Q}_s^{j^T} \delta \mathbf{e}^j \tag{7.25}$$

Substituting the preceding three equations into the principle of virtual work of Eq. 21 yields

$$\sum_{j=1}^{n_e} \left(\mathbf{M}^j \ddot{\mathbf{e}}^j + \mathbf{Q}_s^j - \mathbf{Q}_e^j\right)^T \delta \mathbf{e}^j = 0 \tag{7.26}$$

This equation is equivalent to the following scalar equation, which is written in a more explicit form in terms of the element vectors and matrices:

$$\left\{\begin{bmatrix} \mathbf{M}^1 & \mathbf{0} & \dots & \mathbf{0} \\ \mathbf{0} & \mathbf{M}^2 & \dots & \mathbf{0} \\ \vdots & \vdots & \ddots & \mathbf{0} \\ \mathbf{0} & \mathbf{0} & \dots & \mathbf{M}^{n_e} \end{bmatrix}\begin{bmatrix} \ddot{\mathbf{e}}^1 \\ \ddot{\mathbf{e}}^2 \\ \vdots \\ \ddot{\mathbf{e}}^{n_e} \end{bmatrix} + \begin{bmatrix} \mathbf{Q}_s^1 \\ \mathbf{Q}_s^2 \\ \vdots \\ \mathbf{Q}_s^{n_e} \end{bmatrix} - \begin{bmatrix} \mathbf{Q}_e^1 \\ \mathbf{Q}_e^2 \\ \vdots \\ \mathbf{Q}_e^{n_e} \end{bmatrix}\right\}^T \begin{bmatrix} \delta\mathbf{e}^1 \\ \delta\mathbf{e}^2 \\ \vdots \\ \delta\mathbf{e}^{n_e} \end{bmatrix} = 0 \tag{7.27}$$

In the FE analysis, the element Boolean matrix $\mathbf{B}^j$ is introduced to define the element connectivity. This Boolean matrix is used to develop a standard procedure for the assembly of the elements by mapping the element nodal coordinates to catenary nodal coordinate vector $\mathbf{e}$. In this case, one has $\delta\mathbf{e}^j = \mathbf{B}^j\delta\mathbf{e}$ and $\ddot{\mathbf{e}}^j = \mathbf{B}^j\ddot{\mathbf{e}}$. Therefore, one can write Eq. 26 as $\left[\sum_{j=1}^{n_e}\left(\mathbf{M}^j\mathbf{B}^j\ddot{\mathbf{e}} + \mathbf{Q}_s^j - \mathbf{Q}_e^j\right)^T\mathbf{B}^j\right]\delta\mathbf{e} = 0$. In the case of unconstrained motion, the elements of vector $\mathbf{e}$ are independent, and therefore, their coefficients in this equation must be equal to zero. This leads to $\sum_{j=1}^{n_e}\left(\mathbf{B}^{j^T}\mathbf{M}^j\mathbf{B}^j\ddot{\mathbf{e}} + \mathbf{B}^{j^T}\mathbf{Q}_s^j - \mathbf{B}^{j^T}\mathbf{Q}_e^j\right) = \mathbf{0}$. Performing the summation in this equation, it can be shown that the catenary FE equations of motion can be written as

$$\mathbf{M}\ddot{\mathbf{e}} + \mathbf{Q}_s - \mathbf{Q}_e = \mathbf{0} \tag{7.28}$$

where $\mathbf{M} = \sum_{j=1}^{n_e}\mathbf{B}^{j^T}\mathbf{M}^j\mathbf{B}^j$ is the catenary symmetric mass matrix, $\mathbf{Q}_s = \sum_{j=1}^{n_e}\mathbf{B}^{j^T}\mathbf{Q}_s^j$ is the vector of catenary elastic forces, and $\mathbf{Q}_e = \sum_{j=1}^{n_e}\mathbf{B}^{j^T}\mathbf{Q}_e^j$ is the vector of the catenary applied forces. The mass matrix in the preceding equation remains constant and symmetric, and therefore, a Cholesky transformation can be used to obtain an identity generalized mass matrix, which has an optimum sparse matrix structure. Therefore, if an elastic contact formulation (ECF) is used to describe pantograph/catenary interaction, the solution for the catenary accelerations does not require an iterative LU factorization or finding the inverse of the mass matrix during the dynamic simulation. After solving for the ANCF Cholesky accelerations, the ANCF nodal accelerations of the catenary can be obtained using the Cholesky transformation (Shabana 2018).

7.5 PANTOGRAPH/CATENARY CONTACT FRAME

In the analysis presented in this chapter, it is assumed that catenary wires are modeled using ANCF finite elements, which allow for systematic definitions of the gradient vectors that enter into the formulation of pantograph/catenary contact forces. To develop a unified approach that can be used for both the gradient-deficient cable and fully parameterized beam elements, the contact conditions are formulated using a single position-gradient vector. Recall that the gradient-deficient ANCF cable element has only the longitudinal position-gradient vector $\mathbf{r}_x$, while the fully parameterized ANCF beam element has three position vector gradients $\mathbf{r}_x$, $\mathbf{r}_y$, and $\mathbf{r}_z$. Therefore, to develop a unified approach that allows for using either of the two ANCF elements, the contact equations are formulated using the longitudinal gradient vector $\mathbf{r}_x$, which is tangent to the element centerline. In the notations used in this section and the following sections, superscript c refers to the catenary; this is with the understanding that positions of arbitrary points on the ANCF finite elements are directly defined in the global coordinate system, and for such an ANCF catenary model, there is no need to introduce a local body coordinate system.

Pantograph/catenary forces can have both normal and tangential components that depend on the locations of the contact points between the pan-head and catenary. To properly define the components of these contact forces, a coordinate system $X^cY^cZ^c$, referred to as the *contact frame*, is introduced at the contact point, as shown in Figure 8. The orientation of this coordinate system in the global coordinate system is assumed to be defined by the orthogonal transformation matrix $\mathbf{A}^c = \begin{bmatrix}\mathbf{i}^c & \mathbf{j}^c & \mathbf{k}^c\end{bmatrix}$, where $\mathbf{i}^c, \mathbf{j}^c$, and $\mathbf{k}^c$ are orthogonal unit vectors along the axes of the coordinate system $X^cY^cZ^c$. The longitudinal axes of the contact frame $X^cY^cZ^c$can be defined by unit vector $\mathbf{i}^c$ along ANCF gradient vector $\mathbf{r}_x^c$ of the catenary wire, which comes into contact with the pan-head. The other two axes defined by unit vectors $\mathbf{j}^c$ and $\mathbf{k}^c$ can be determined as unit vectors perpendicular to ANCF gradient vector $\mathbf{r}_x^c$ at the contact point. These three orthogonal unit vectors can be defined as (Pappalardo et al. 2016)

$$\mathbf{i}^c = \frac{\mathbf{r}_x^c}{\|\mathbf{r}_x^c\|}, \quad \mathbf{j}^c = \frac{\mathbf{n}_1^c}{\|\mathbf{n}_1^c\|}, \quad \mathbf{k}^c = \frac{\mathbf{n}_2^c}{\|\mathbf{n}_2^c\|} \tag{7.29}$$

and vectors $\mathbf{r}_x^c$, $\mathbf{n}_1^c$, and $\mathbf{n}_2^c$ are

$$\mathbf{r}_x^c = \begin{bmatrix}(\mathbf{r}_x^c)_1 \\ (\mathbf{r}_x^c)_2 \\ (\mathbf{r}_x^c)_3\end{bmatrix}, \quad \mathbf{n}_1^c = \begin{bmatrix}-(\mathbf{r}_x^c)_1(\mathbf{r}_x^c)_2 \\ ((\mathbf{r}_x^c)_1)^2 + ((\mathbf{r}_x^c)_3)^2 \\ -(\mathbf{r}_x^c)_2(\mathbf{r}_x^c)_3\end{bmatrix}, \quad \mathbf{n}_2^c = \begin{bmatrix}-(\mathbf{r}_x^c)_3 \\ 0 \\ (\mathbf{r}_x^c)_1\end{bmatrix} \tag{7.30}$$

The norms of vectors $\mathbf{n}_1^c$ and $\mathbf{n}_2^c$ become zero if ANCF position gradient vector $\mathbf{r}_x^c$ is parallel to vector $\mathbf{j} = \begin{bmatrix}0 & 1 & 0\end{bmatrix}^T$. This leads to a singularity that can be avoided by using a different definition of the orthogonal vectors of Eq. 30. In this singular configuration, which is not encountered in the case of the catenary system, the following three orthogonal vectors can

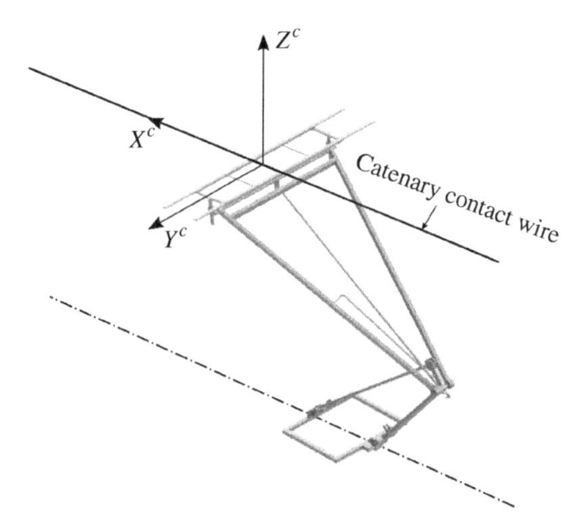

Figure 7.8 Pantograph/catenary contact.

be used instead of those defined by the preceding equation (Gere and Weaver 1965; Shabana 2019):

$$\mathbf{r}_x^c = \begin{bmatrix} 0 \\ (\mathbf{r}_x^c)_2 \\ 0 \end{bmatrix}, \quad \mathbf{n}_1^c = \begin{bmatrix} -(\mathbf{r}_x^c)_2 \\ 0 \\ 0 \end{bmatrix}, \quad \mathbf{n}_2^c = \begin{bmatrix} 0 \\ 0 \\ 1 \end{bmatrix} \tag{7.31}$$

It is clear from the preceding two equations that in both the non-singular and singular configurations, the pantograph/catenary contact frame $X^cY^cZ^c$ is defined using a single gradient vector, which is the longitudinal gradient vector $\mathbf{r}_x^c$ at the contact point. Therefore, such a definition of the contact frame can be used with both the ANCF cable element, which has only longitudinal gradient vector $\mathbf{r}_x$, and the fully parameterized ANCF beam element, which has three gradient vectors $\mathbf{r}_x$, $\mathbf{r}_y$, and $\mathbf{r}_z$.

As in the case of wheel/rail contact, two fundamentally different formulations can be used to predict pantograph/catenary contact forces: the *constraint contact formulation* (CCF) and the *elastic contact formulation* (ECF). In the constraint contact formulation, the pan-head is assumed to remain in contact with the catenary wire, and therefore, separations or penetrations are not allowed. In this case, the freedom of the pan-head to translate along the normal to the catenary wire is eliminated. The pan-head, however, can move relative to the catenary wire in the longitudinal and lateral directions. The normal contact force in the CCF approach is determined as a constraint (reaction) force. In the elastic contact formulation, on the other hand, pantograph/catenary separations and penetrations are allowed and no constraints are imposed on the motion of the pan-head with respect to the catenary wire. In this case, the pantograph/catenary normal contact force is described using a compliant force model with assumed stiffness and damping coefficients. The constraint and elastic contact formulations are described in the following two sections, respectively.

7.6 CONSTRAINT CONTACT FORMULATION (CCF)

The constraint contact formulation is recommended in simulation scenarios in which it is desirable to avoid the oscillations and high frequencies that result when the elastic contact formulation is used. Avoiding these oscillations and high frequencies allows for smoother solutions that can be used to shed light on dynamic behaviors and phenomena that can be difficult to observe or identify when high-frequency oscillations are superimposed on the solutions. For this reason, it is recommended to implement both constraint and elastic contact formulations in computer software used for nonlinear dynamic simulations of railroad vehicle systems.

Two CCF approaches can be used to formulate the constraint contact conditions between the pantograph head and catenary. In the first approach, the pan-head is not allowed to slide with respect to the catenary wire in the lateral direction, while in the second approach, lateral sliding is allowed. Both constraint formulations require using the concept of non-generalized coordinates, previously introduced in this book. As in the preceding section, it is assumed in the analysis presented in this section that the catenary wire is modeled using ANCF finite elements, while the pantograph is modeled as a system of interconnected bodies. The pan-head is assumed to be a rigid body whose configuration is defined using the Cartesian absolute coordinates $\mathbf{q}^p = \left[\mathbf{R}^{p^T} \quad \boldsymbol{\theta}^{p^T}\right]^T$ introduced in Chapter 3, where superscript p refers to the pan-head, $\mathbf{R}^p$ is the vector that defines the global position of the origin of the pan-head body coordinate system $X^pY^pZ^p$ with respect to the global coordinate system, and $\boldsymbol{\theta}^p$ is the set of rotational coordinates used to define the orientation of the pan-head body coordinate system. Using these coordinates, the global position of a point on the pan-head can be written as $\mathbf{r}^p = \mathbf{R}^p + \mathbf{A}^p\overline{\mathbf{u}}^p$, where $\overline{\mathbf{u}}^p$ is the vector that defines the position of the point with respect to the origin of the pan-head body coordinate system $X^pY^pZ^p$. The location of the origin of the coordinate system $X^pY^pZ^p$ is assumed to be the center of mass of the link to which the pan-head is attached. The global position vector of an arbitrary point on the catenary wire can simply be written as $\mathbf{r}^c = \mathbf{S}^c\mathbf{e}^c$, where superscript c refers to the catenary, and $\mathbf{S}^c$ and $\mathbf{e}^c$ are, respectively, the shape function matrix and vector of nodal coordinates of the ANCF element on which the contact point lies.

Longitudinal Relative Sliding If the effect of the relative lateral displacement between the pan-head and catenary wire is not considered, the constraint contact conditions can be represented by a *sliding joint* that describes the movement of the pan-head along the flexible catenary (Seo et al. 2006). The sliding joint kinematic conditions, which are used to model the relative motion between the rigid pan-head of the pantograph and the flexible catenary cable, can be solved to determine the location of the contact points on the pan-head and catenary wire. To determine the locations of the contact points, the arc length parameter s that defines the distance traveled by the pan-head with respect to the catenary wire is introduced. This arc length parameter, which is associated with the space curve that defines the centerline of the ANCF cable or fully parameterized beam-element catenary model, can be written in terms of the spatial longitudinal coordinate x of the finite element: that is, $s = s(x)$. Using this functional relationship, arc length parameter s is associated with the reference configuration, despite the fact that its value depends on the location of the point of contact between the pan-head and the catenary wire.

As a result of introducing arc length parameter s, three constraint equations are required to solve for s (one equation) and eliminate the two translational degrees of freedom of the pan-head with respect to the catenary wire in the lateral and normal directions (two equations). These constraint contact conditions can then be written as

$$\mathbf{C}^s = \mathbf{C}^s\left(\mathbf{q}^p, \mathbf{e}^c, s, t\right) = \mathbf{r}^p - \mathbf{r}^c = \mathbf{0} \tag{7.32}$$

where $\mathbf{r}^p = \mathbf{r}^p(\mathbf{q}^p, t)$ is the global position vector of contact point P on the pan-head, $\mathbf{r}^c = \mathbf{r}^c(\mathbf{e}^c, s, t)$ is the global position vector of contact point P on the catenary, $\mathbf{q}^p = \mathbf{q}^p(t)$ is the vector of generalized coordinates of the pantograph link to which the pan-head is attached, $\mathbf{e}^c = \mathbf{e}^c(t)$ is the vector of the nodal coordinates of the finite element on the catenary on which the contact point lies, and $s = s(t)$ is the non-generalized arc length coordinate used to determine the location of the contact point on the catenary cable.

The vectors in the three constraint conditions of Eq. 32 are defined in the global coordinate system. An equivalent formulation for the constraint conditions of Eq. 32 is to define the components of these vectors in the pantograph/catenary contact frame $X^cY^cZ^c$ introduced in the preceding section and whose axes are defined using the longitudinal ANCF position vector gradients $\mathbf{r}_x$, as explained in the preceding section. In this case, the preceding equation can be written in terms of vector components defined along the axes of the coordinate system $X^cY^cZ^c$ as (Pappalardo et al. 2016)

$$\mathbf{C}^s = \begin{bmatrix} \mathbf{i}^{c^T}(\mathbf{r}^p - \mathbf{r}^c) \\ \mathbf{j}^{c^T}(\mathbf{r}^p - \mathbf{r}^c) \\ \mathbf{k}^{c^T}(\mathbf{r}^P - \mathbf{r}^c) \end{bmatrix} = \mathbf{0} \tag{7.33}$$

where $\mathbf{i}^c = \mathbf{i}^c(\mathbf{e}^c, s, t)$, $\mathbf{j}^c = \mathbf{j}^c(\mathbf{e}^c, s, t)$, and $\mathbf{k}^c = \mathbf{k}^c(\mathbf{e}^c, s, t)$ are the three unit vectors along the axes of the coordinate system $X^cY^cZ^c$ defined on the catenary cable at the contact point, as previously explained. As explained in the preceding section, the definition of the pantograph/catenary contact frame is based on only one gradient vector instead of three gradient vectors. This allows for developing a general sliding joint formulation that can be used with both ANCF fully parameterized and gradient-deficient finite elements.

Longitudinal and Lateral Sliding The three constraint equations of Eq. 32 or, equivalently, Eq. 33 can be used to determine the arc length parameter s and eliminate two degrees of freedom of relative translation of the pan-head with respect to the catenary wire in the lateral and normal directions. For example, given the generalized coordinates of pan-head $\mathbf{q}^p$ and catenary nodal coordinates $\mathbf{e}^c$, the first equation $\mathbf{i}^{c^T}(\mathbf{r}^p - \mathbf{r}^c) = 0$ of Eq. 33 can be solved iteratively to determine arc length parameter s. This solution for s can be substituted into the last two equations of Eq. 33 to restrict pan-head motion with respect to the catenary wire in lateral and normal directions.

The same three equations of Eq. 32 or, alternatively, Eq. 33 can be used to develop a constraint contact formulation that allows for both longitudinal and lateral displacements of the pan-head with respect to the catenary wire. In this case, the lateral component of vector $\bar{\mathbf{u}}^p$ in the equation $\mathbf{r}^p = \mathbf{R}^p + \mathbf{A}^p\bar{\mathbf{u}}^p$ is allowed to vary and is treated as an unknown geometric parameter. Without any loss of generality, the coordinate system of the pan-head body can be selected such that local position vector $\bar{\mathbf{u}}^p$ can be written as $\bar{\mathbf{u}}^p = \begin{bmatrix} \bar{u}_1^p & y_l^p & \bar{u}_3^p \end{bmatrix}^T$, where $\bar{u}_1^p$ and $\bar{u}_3^p$ are assumed to be constants, while y_l^p is a parameter that can vary and can be determined by solving the constraint equations. In this case, two of the constraint equations can be used to solve for the parameters s and y_l^p, while the remaining constraint equation eliminates the freedom of the pan-head to translate with respect to the catenary wire along the normal direction. For example, given the generalized coordinates of pan-head $\mathbf{q}^p$ and catenary nodal coordinates $\mathbf{e}^c$, the first two equations $\mathbf{i}^{c^T}(\mathbf{r}^p - \mathbf{r}^c) = 0$ and $\mathbf{j}^{c^T}(\mathbf{r}^p - \mathbf{r}^c) = 0$ of Eq. 33 can be used to solve for s and y_l^p, while the third equation $\mathbf{k}^{c^T}(\mathbf{r}^p - \mathbf{r}^c) = 0$ is used to eliminate the freedom of the pan-head to move with respect to the catenary wire in the normal direction. By introducing the additional parameter y_l^p, lateral motion of the pan-head is allowed, and the constraints on the generalized coordinates are reduced to only one because two constraint equations are used to determine the unknown geometric parameters s and y_l^p.

Augmented Formulation and Embedding Technique The constraint contact conditions introduced in this section can be used with the augmented formulation or the embedding technique discussed in Chapter 6. In the augmented formulation, the constraint equations are combined with the differential equations of motion using the technique of

Lagrange multipliers. This approach allows for using non-generalized coordinates that represent geometric variables that have no associated inertia or applied forces. The normal pantograph/catenary contact forces in the augmented formulation can be evaluated using Lagrange multipliers as reaction forces.

In alternate augmented formulations, the geometric parameters can be systematically eliminated, leading to a number of constraint contact conditions equal to the number of degrees of freedom eliminated. In this second augmented formulation, the non-generalized geometric parameters do not appear in the equations of motion.

In the embedding technique, on the other hand, all the dependent generalized and non-generalized coordinates are eliminated, leading to a number of equations of motion equal to the number of system degrees of freedom. In general, a systematic procedure can be used to eliminate the geometric parameters to obtain a number of constraint equations equal to the number of degrees of freedom eliminated in the constraint contact formulation.

Elimination of Geometric Parameters To demonstrate the systematic procedure for eliminating geometric parameters, the case of the constraint formulation that only allows for relative sliding is considered (Pappalardo et al. 2016). A similar procedure can be used when relative lateral sliding between the pan-head and catenary contact wire is allowed. If lateral sliding is not allowed, one geometric parameter s is introduced, and the relative motion is governed by the constraints of Eq. 32 or Eq. 33. In this section, the procedure of eliminating arc length parameter s is demonstrated using Eq. 33. The first scalar equation in this equation, $\mathbf{i}^{c^T}(\mathbf{r}^c - \mathbf{r}^p)^T = 0$, is equivalent to the equation $C_e^s = (\mathbf{r}^c - \mathbf{r}^p)^T \mathbf{r}_x^c = 0$. The algebraic equation $C_e^s = (\mathbf{r}^c - \mathbf{r}^p)^T \mathbf{r}_x^c = 0$ can be used to eliminate arc length parameter s. Assuming that the generalized coordinates are known from the numerical integration of the equations of motion, the algebraic equation $C_e^s = (\mathbf{r}^c - \mathbf{r}^p)^T \mathbf{r}_x^c = 0$ can be considered a nonlinear equation in arc length parameter s. Using a Newton–Raphson algorithm, this equation can be used to determine s by iteratively solving the equation $\left(\partial C_e^s / \partial s\right) \Delta s = -C_e^s$, where Δs is the Newton difference, $\partial C_e^s / \partial s = (\partial \mathbf{r}^c / \partial s)^T \mathbf{r}_x^c + \mathbf{r}^{cp^T} \left(\partial \mathbf{r}_x^c / \partial s\right)$, and $\mathbf{r}^{cp} = \mathbf{r}^c - \mathbf{r}^p$. Having determined s, the location of the contact point on the catenary can be determined for a given position of the pan-head. Using this solution, one can write $\left(C_e^s\right)_{\mathbf{q}} \delta \mathbf{q} + \left(C_e^s\right)_s \delta s = 0$, where $\left(C_e^s\right)_{\mathbf{q}}$ is the row vector $\left(C_e^s\right)_{\mathbf{q}} = \partial C_e^s / \partial \mathbf{q} = \left[\partial C_e^s / \partial \mathbf{q}^p \quad \partial C_e^s / \partial \mathbf{e}^c\right]$, $\left(C_e^s\right)_s$ is the scalar $\left(C_e^s\right)_s = \partial C_e^s / \partial s$, and $\mathbf{q} = \left[\mathbf{q}^{p^T} \quad \mathbf{e}^{c^T}\right]^T$ is the vector of generalized coordinates of the pan-head body and ANCF catenary. Assuming that $\left(C_e^s\right)_s$ is different from zero, one has

$$
\left.
\begin{aligned}
\delta s &= -\left(\left(C_e^s\right)_{\mathbf{q}} / \left(C_e^s\right)_s\right) \delta \mathbf{q} \\
\dot{s} &= -\left(\left(C_e^s\right)_{\mathbf{q}} / \left(C_e^s\right)_s\right) \dot{\mathbf{q}} \\
\ddot{s} &= -\left(\left(C_e^s\right)_{\mathbf{q}} \ddot{\mathbf{q}} + \left(Q_e^s\right)_c\right) / \left(C_e^s\right)_s
\end{aligned}
\right\} \tag{7.34}
$$

where $\left(Q_e^s\right)_c$ is a vector that is quadratic in velocities, which arises from differentiating the constraint equation at the velocity level $\left(C_e^s\right)_{\mathbf{q}} \dot{\mathbf{q}} + \left(C_e^s\right)_s \dot{s} = 0$ with respect to time. This differentiation leads to $\left(C_e^s\right)_{\mathbf{q}} \ddot{\mathbf{q}} + \left(d\left(C_e^s\right)_{\mathbf{q}} / dt\right) \dot{\mathbf{q}} + \left(C_e^s\right)_s \ddot{s} + \left(d\left(C_e^s\right)_s / dt\right) \dot{s} = 0$. This equation defines $\left(Q_e^s\right)_c$ as $\left(Q_e^s\right)_c = \left(d\left(C_e^s\right)_{\mathbf{q}} / dt\right) \dot{\mathbf{q}} + \left(d\left(C_e^s\right)_s / dt\right) \dot{s}$. The value of s obtained using the

Newton–Raphson iterations and virtual change and the first and second time derivatives of s given by Eq. 34 can be used to eliminate the dependence of the second and third equations of Eq. 33 and their time derivatives on s and its time derivatives. These two equations can be written at the position level as

$$\mathbf{C}_m^s = \begin{bmatrix} \mathbf{j}^{c^T}\left(\mathbf{r}^p - \mathbf{r}^c\right) \\ \mathbf{k}^{c^T}\left(\mathbf{r}^P - \mathbf{r}^c\right) \end{bmatrix} = \mathbf{0} \tag{7.35}$$

The virtual change and the first and second time derivatives of these two equations can written, respectively, as

$$\left.\begin{aligned} \delta\mathbf{C}_m^s &= \left(\mathbf{C}_m^s\right)_{\mathbf{q}}\delta\mathbf{q} + \left(\mathbf{C}_m^s\right)_s\delta s = \mathbf{0} \\ \dot{\mathbf{C}}_m^s &= \left(\mathbf{C}_m^s\right)_{\mathbf{q}}\dot{\mathbf{q}} + \left(\mathbf{C}_m^s\right)_s\dot{s} = \mathbf{0} \\ \ddot{\mathbf{C}}_m^s &= \left(\mathbf{C}_m^s\right)_{\mathbf{q}}\ddot{\mathbf{q}} + \left(d\left(\mathbf{C}_m^s\right)_{\mathbf{q}}/dt\right)\dot{\mathbf{q}} + \left(\mathbf{C}_m^s\right)_s\ddot{s} + \left(d\left(\mathbf{C}_m^s\right)_s/dt\right)\dot{s} = \mathbf{0} \end{aligned}\right\} \tag{7.36}$$

where $\left(\mathbf{C}_m^s\right)_{\mathbf{q}} = \partial\mathbf{C}_m^s/\partial\mathbf{q}$ and $\left(\mathbf{C}_m^s\right)_s = \partial\mathbf{C}_m^s/\partial s$. Using the results of Eq. 34 with Eq. 36, the dependence of the constraint equations $\mathbf{C}_m^s$ and their derivatives on arc length parameter s can be eliminated, leading to

$$\left.\begin{aligned} \delta\mathbf{C}_m^s &= \left(\left(\mathbf{C}_m^s\right)_{\mathbf{q}} - \frac{1}{\left(C_e^s\right)_s}\left(\mathbf{C}_m^s\right)_s\left(C_e^s\right)_{\mathbf{q}}\right)\delta\mathbf{q} = \mathbf{0} \\ \dot{\mathbf{C}}_m^s &= \left(\left(\mathbf{C}_m^s\right)_{\mathbf{q}} - \frac{1}{\left(C_e^s\right)_s}\left(\mathbf{C}_m^s\right)_s\left(C_e^s\right)_{\mathbf{q}}\right)\dot{\mathbf{q}} = \mathbf{0} \\ \ddot{\mathbf{C}}_m^s &= \left(\left(\mathbf{C}_m^s\right)_{\mathbf{q}} - \frac{\left(\mathbf{C}_m^s\right)_s\left(C_e^s\right)_{\mathbf{q}}}{\left(C_e^s\right)_s}\right)\ddot{\mathbf{q}} - \left(\mathbf{Q}_m^s\right)_c = \mathbf{0} \end{aligned}\right\} \tag{7.37}$$

where $\left(\mathbf{Q}_m^s\right)_c$ is a vector that is quadratic in velocities and results from differentiating the constraint equations twice with respect to time. Because the vector of constraint functions $\mathbf{C}_m^s$ has two scalar functions, eliminating the arc length parameter leads to a number of constraint equations equal to the number the system degrees of freedom eliminated by the pantograph/catenary contact conditions. The details of the scalars, vectors, and matrices that appear in the equations used in this section are presented by Pappalardo et al. (2016). A similar procedure can also be used in the case of longitudinal and lateral sliding to eliminate the dependence on geometric parameters s and y_l^p and obtain one constraint equation expressed in terms of only the generalized coordinates, because one degree of freedom is eliminated in this case.

7.7 ELASTIC CONTACT FORMULATION (ECF)

The elastic contact formulation allows for pan-head/catenary penetrations and separations, and therefore, no degrees of freedom are eliminated when this formulation is used. Algebraic equations are used to solve for geometric parameters that define the location of the potential contact points on the pan-head and catenary wire. Distance equations are then used to check whether contacts occur between the pan-head and contact wire. Predicting possible separations between the pan-head and contact wire is important since such separations are the cause of arcing, which has an adverse effect on current-collection quality. This section discusses two cases in which lateral displacement of the pan-head relative to the contact wire is restricted or unrestricted, respectively. The equations that govern these two cases are similar to the equations used in the constraint contact formulation discussed in the preceding section.

Longitudinal Relative Sliding In this case, the pan-head is assumed to slide with respect to the catenary wire in the longitudinal direction, while the pan-head displacement relative to the catenary wire in the plane perpendicular to the longitudinal tangent is assumed to be restricted. The algebraic equations of Eq. 33 can be used as the basis for the elastic contact formulation discussed in this section. As in the CCF approach, arc length geometric parameter s is used to define the location of the contact point between the catenary and pan-head. The configuration of the pan-head is assumed to be known: that is, the vector of generalized coordinates $\mathbf{q}^p = \mathbf{q}^p(t)$ is assumed to be known from the numerical integration of the equations of motion of the system. Similarly, the vector of nodal coordinates $\mathbf{e}^c = \mathbf{e}^c(t)$ of the catenary is assumed to be known. Knowing these generalized coordinates, the first equation in Eq. 33, $\mathbf{i}^{c^T}(\mathbf{r}^p - \mathbf{r}^c) = 0$, can be used to determine the arc length parameter s that defines the location of the contact point. Recalling that $\mathbf{i}^c = \mathbf{r}_x^c / |\mathbf{r}_x^c|$, where $\mathbf{r}_x^c = \left[(\mathbf{r}_x^c)_1 \; (\mathbf{r}_x^c)_2 \; (\mathbf{r}_x^c)_3\right]^T$ is the longitudinal ANCF gradient vector tangent to the catenary wire at the potential contact point and $s = s(t)$, the algebraic equation $\mathbf{i}^{c^T}(\mathbf{r}^p - \mathbf{r}^c) = 0$ can be written in a more convenient form as (Kulkarni et al. 2017)

$$g(s) = \left(\mathbf{r}^c(s) - \mathbf{r}^p\right)^T \mathbf{r}_x^c(s) = 0 \tag{7.38}$$

Knowing the system generalized coordinates, this equation can be considered a nonlinear algebraic equation in arc length parameter s. Due to the nonlinearity of this equation, an iterative Newton–Raphson algorithm can be used to obtain the solution for arc length parameter s. In this Newton–Raphson iterative procedure, one must solve the equation $(\partial g/\partial s)\Delta s = -g$ for Newton-differenceΔs, where $g_s = \partial g/\partial s = (\partial \mathbf{r}^c/\partial s)^T \mathbf{r}_x^c + \mathbf{r}^{cp^T}(\partial \mathbf{r}_x^c/\partial s)$ and $\mathbf{r}^{cp} = \mathbf{r}^c - \mathbf{r}^p$. Assuming that $g_s \neq 0$, the solution of Eq. 38 defines arc length parameter s , which can be used to define the location of the potential contact point on the catenary wire $\mathbf{r}^c(s)$. Using the value of s, the last two equations in Eq. 33 can be used to define the following lateral and normal distances, respectively:

$$d_l = \mathbf{j}^{c^T}\left(\mathbf{r}^p - \mathbf{r}^c\right), \qquad d_n = \mathbf{k}^{c^T}\left(\mathbf{r}^P - \mathbf{r}^c\right) \tag{7.39}$$

These two distances along two orthogonal directions perpendicular to the longitudinal tangent can be used to check whether contact occurs between the catenary pan-head and contact wire. If d_1 and d_2 are smaller than certain tolerances, contact between the pan-head and catenary wire is assumed.

Longitudinal and Lateral Sliding If lateral motion of the pan-head with respect to the catenary wire is not restricted, the first two algebraic equations of Eq. 33 can be used to solve for geometric parameters s and y_l^p, where y_l^p defines the lateral relative displacement of the pan-head with respect to the catenary wire, as discussed in the preceding section. These two nonlinear algebraic equations can be written as

$$\left.\begin{aligned} g_1 = g_1\left(s, y_l^p\right) = \left(\mathbf{i}^c(s)\right)^T\left(\mathbf{r}^p\left(y_l^p\right) - \mathbf{r}^c(s)\right) = 0 \\ g_2 = g_2\left(s, y_l^p\right) = \left(\mathbf{j}^c(s)\right)^T\left(\mathbf{r}^p\left(y_l^p\right) - \mathbf{r}^c(s)\right) = 0 \end{aligned}\right\} \tag{7.40}$$

These two nonlinear algebraic equations can be solved iteratively using a Newton–Raphson algorithm to determine s and y_l^p. To this end, the following system of equations is constructed:

$$\begin{bmatrix} \partial g_1/\partial s & \partial g_1/\partial y_l^p \\ \partial g_2/\partial s & \partial g_2/\partial y_l^p \end{bmatrix} \begin{bmatrix} \Delta s \\ \Delta y_l^p \end{bmatrix} = -\begin{bmatrix} g_1 \\ g_2 \end{bmatrix} \tag{7.41}$$

The convergence of the Newton–Raphson iterations defines s and y_l^p, which can be substituted into the equation $\mathbf{k}^{c^T}\left(\mathbf{r}^P - \mathbf{r}^c\right)$ to calculate the distance

$$d_n = \left(\mathbf{k}^c\left(\mathbf{e}^c, s\right)\right)^T\left(\mathbf{r}^P\left(\mathbf{q}^P, y_l^p\right) - \mathbf{r}^c\left(\mathbf{e}^c, s\right)\right) \tag{7.42}$$

If d_n is smaller than a given tolerance, contact between the pan-head and catenary wire is assumed. If this condition is not satisfied, separation is assumed, and no interaction forces between the pan-head and catenary wire are applied.

Constraint and Elastic Formulations The two different cases of contact, without and with lateral sliding y_l^p of the pan-head relative to the contact wire, are considered in both the constraint and elastic contact formulations presented in this chapter. Capturing this lateral sliding is necessary for developing accurate pantograph/catenary wear models in many motion scenarios. Localized contact in a very small area leads to significant wear that can damage the carbon strip used for current collection. Such wear can be reduced if the contact wire slides laterally over the pan-head to avoid contact being confined to a small area on the pan-head. During curve negotiations, the contact wire sweeps over the pantograph carbon strip, so the contact point is not restricted to a localized region. In the case of a tangent track, the catenary is staggered to assume a zigzagging shape in order for the contact wire to slide laterally over the carbon strip and avoid having contact in a small region. Such lateral sliding contributes to reducing the wear of the carbon strip. Therefore, in investigations focused on wear, it is necessary to consider the effect of lateral sliding y_l^p to obtain realistic results that are consistent with actual catenary design.

Unlike the algebraic equations used in the constraint contact formulation described in the preceding section, the algebraic equations used in the elastic contact formulation do not eliminate degrees of freedom, and they do not represent motion constraints. The ECF algebraic equations are mainly used to determine the geometric parameters that define the locations of pan-head/catenary contact points. CCF algebraic equations eliminate degrees of freedom and must be satisfied at the position, velocity, and acceleration levels. When ECF algebraic equations are used, on the other hand, normal contact forces are defined using compliant force models with assumed stiffness and damping coefficients. In the case of separation, the compliant contact forces are assumed to be zero. When CCF algebraic equations are used, on the other hand, normal contact forces are determined as constraint (reaction) forces using the techniques described in Chapter 6. In this case, no assumption of stiffness and damping coefficients needs to be made. Therefore, while the two approaches employ similar algebraic equations, they are fundamentally different and require using different solution procedures, as is the case with wheel/rail contact formulations. The two contact formulations lead to a different number of degrees of freedom and a different number of system constraint forces.

7.8 PANTOGRAPH/CATENARY EQUATIONS AND MBS ALGORITHMS

The pantograph/catenary contact formulations presented in this chapter can be systematically integrated with general computational MBS algorithms that allow for developing complex railroad vehicle system models. These MBS algorithms also allow for modeling flexible bodies using the FE method and are based on fully nonlinear formulations of the system equations of motion, as discussed in Chapter 6. The development of such general computational algorithms for railroad vehicle systems requires introducing different sets of coordinates required for modeling components with different degrees of flexibility. Some components can be bulky and can be modeled as rigid bodies; other components are flexible but experience only small deformations that can lead to very high stresses, as is the case with stiff components; still other components, such as catenary wires, are very flexible and experience very large deformations. All these types of bodies, which have different degrees of flexibility, can exist in one railroad vehicle model. Different coordinates are used to describe the motion of rigid bodies, flexible bodies, and very flexible bodies.

Coordinate Types General MBS solution algorithms are designed to model systems that consist of interconnected bodies, which can be rigid or flexible. Flexible bodies that experience small deformations are often modeled in computational MBS algorithms using the *floating frame of reference* (FFR) formulation. The FFR formulation, which allows for coordinate reduction using component mode synthesis techniques, can be used to solve small-deformation problems efficiently. In the FFR formulation, a coordinate system is assigned to each flexible body. The configuration of the flexible body that experiences small deformations is defined by a set of reference coordinates that define the translation and orientation of the body coordinate system and a set of elastic coordinates that define the deformation of the body with respect to its coordinate system.

The vector of generalized coordinates used to describe the motion of rigid bodies and the reference motion of the body coordinate systems in the FFR formulation is denoted as $\mathbf{q}_r$. For both rigid bodies and the coordinate systems of the FFR bodies, the vector of generalized coordinates $\mathbf{q}_r$ includes Cartesian coordinates that define the global positions of the origins of the coordinate systems of the rigid and FFR bodies. The small elastic deformations of FFR bodies are described using the vector of elastic coordinates $\mathbf{q}_f$, which defines the deformations of FFR bodies with respect to their coordinate systems. The rigid-body and FFR formulation are suited for developing models of pantograph systems that consist of interconnected rigid and deformable bodies. Pantograph links can experience small deformations, and therefore, the FFR formulation is more suited for such systems. The small deformations of relatively stiff members can lead to very high stresses that can result in component failures.

In addition to the two sets of coordinates $\mathbf{q}_r$ and $\mathbf{q}_f$, other coordinate types need to be introduced to be able to develop realistic and more accurate virtual prototyping of railroad vehicle system models. As discussed in this chapter, the catenary system can be modeled using ANCF finite elements, which can be used to solve accurately nonlinear large displacement and large deformation problems. General computational MBS algorithms allow for introducing the ANCF coordinates $\mathbf{e}$ to develop catenary equations of motion without the need to use incremental-rotation or co-simulation techniques. Furthermore, in the augmented formulation of the equations of motion, the non-generalized geometric parameters $\mathbf{s}$, used to describe curve and surface geometries and define the locations of wheel/rail and pantograph/catenary contact points, enter into the formulation of the kinematic conditions if the constraint contact formulations are used to model pantograph/catenary and/or wheel/rail contacts.

MBS Formulation A general multibody system such as a railroad vehicle system may consist of rigid bodies, flexible bodies, and very flexible bodies subjected to kinematic motion constraints and external forces. The motion of the rigid bodies is governed by the Newton-Euler equations presented in Chapter 6. The flexible bodies are assumed to experience small deformations and can be efficiently modeled using the FFR formulation; while the very flexible bodies can be modeled using ANCF finite elements. Therefore, the vector of system generalized coordinates $\mathbf{p}$ used in MBS computational algorithms can be written as $\mathbf{p} = \begin{bmatrix} \mathbf{q}_r^T & \mathbf{q}_f^T & \mathbf{e}^T & \mathbf{s}^T \end{bmatrix}^T$, where $\mathbf{q}_r$ defines the reference coordinates of the rigid and FFR bodies, $\mathbf{q}_f$ defines the small deformations of the FFR bodies with respect to their body coordinate systems, $\mathbf{e}$ defines the vector of the system nodal coordinates of the ANCF finite elements used to describe the nonlinear dynamics and large displacements, and $\mathbf{s}$ is the vector of non-generalized coordinates or surface parameters used to describe the curve and surface geometries in the formulation of the wheel/rail and/or pantograph/catenary contact conditions. Using the augmented Lagrangian formulation, the system equations of motion can be written using the principle of virtual work in dynamics as (Shabana 2019)

$$\begin{bmatrix} \mathbf{M}_{rr} & \mathbf{M}_{rf} & \mathbf{0} & \mathbf{0} & \mathbf{C}_{\mathbf{q}_r}^T \\ \mathbf{M}_{fr} & \mathbf{M}_{ff} & \mathbf{0} & \mathbf{0} & \mathbf{C}_{\mathbf{q}_f}^T \\ \mathbf{0} & \mathbf{0} & \mathbf{M}_{ee} & \mathbf{0} & \mathbf{C}_{\mathbf{e}}^T \\ \mathbf{0} & \mathbf{0} & \mathbf{0} & \mathbf{0} & \mathbf{C}_{\mathbf{s}}^T \\ \mathbf{C}_{\mathbf{q}_r} & \mathbf{C}_{\mathbf{q}_f} & \mathbf{C}_{\mathbf{e}} & \mathbf{C}_{\mathbf{s}} & \mathbf{0} \end{bmatrix} \begin{bmatrix} \ddot{\mathbf{q}}_r \\ \ddot{\mathbf{q}}_f \\ \ddot{\mathbf{e}} \\ \ddot{\mathbf{s}} \\ \boldsymbol{\lambda} \end{bmatrix} = \begin{bmatrix} \mathbf{Q}_r \\ \mathbf{Q}_f \\ \mathbf{Q}_e \\ \mathbf{0} \\ \mathbf{Q}_c \end{bmatrix} \tag{7.43}$$

where subscripts r, f, e, and s refer, respectively, to reference, flexible, ANCF, and non-generalized coordinates; $\mathbf{M}_{rr}$ is the mass matrix associated with the reference motion of the rigid and FFR bodies; $\mathbf{M}_{ff}$ is the mass matrix associated with the FFR bodies that experience small deformations; $\mathbf{M}_{rf} = \mathbf{M}_{fr}^T$ is the mass matrix that defines the nonlinear dynamic inertia coupling between the reference motion and the small elastic deformations of the FFR bodies; $\mathbf{M}_{ee}$ is the mass matrix of the ANCF finite elements used to describe the large displacements of the flexible bodies; $\mathbf{C}_{q_r}$, $\mathbf{C}_{q_f}$, $\mathbf{C}_e$, and $\mathbf{C}_\mathbf{s}$ are the constraint Jacobian matrices associated with coordinates $\mathbf{q}_r$, $\mathbf{q}_f$, $\mathbf{e}$, and $\mathbf{s}$, respectively; λ is the vector of Lagrange multipliers, which can be used to define constraint forces; $\mathbf{Q}_r$, $\mathbf{Q}_f$, and $\mathbf{Q}_e$ are the generalized forces associated with the generalized coordinates $\mathbf{q}_r$, $\mathbf{q}_f$, and $\mathbf{e}$, respectively; and $\mathbf{Q}_c$ is a quadratic velocity vector, which results from differentiating the constraint equations twice with respect to time. This vector is defined by the constraint equation at the acceleration level $\ddot{\mathbf{C}} = \mathbf{C}_q\ddot{\mathbf{q}} + \mathbf{C}_s\ddot{\mathbf{s}} - \mathbf{Q}_c = \mathbf{0}$, where $\mathbf{C}$ is the vector of all constraint functions and $\mathbf{q} = \begin{bmatrix} \mathbf{q}_r^T & \mathbf{q}_f^T & \mathbf{e}^T \end{bmatrix}^T$. Because the ANCF mass matrix $\mathbf{M}_{ee}$ is constant, a Cholesky transformation can be used to obtain an identity inertia matrix associated with the ANCF Cholesky coordinates (Shabana 2018).

Because the FE catenary model can include a large number of nodal coordinates compared to the number of coordinates used to describe vehicle and pantograph dynamics, it is necessary to develop an efficient solution procedure for the equations used to form Eq. 43. It is not always recommended to combine all these equations in one matrix equation unless all the coordinates are kinematically coupled. When the elastic contact formulation is used to model pantograph/catenary contact, the catenary system is not kinematically coupled with the vehicle on which the pantograph is mounted. Therefore, the catenary system can be considered a structure subjected to pantograph contact, gravity, pretension, and cross-wind forces. In this case of an elastic contact formulation, $\mathbf{C}_\mathbf{e} = \mathbf{0}$ for the catenary, and the solution for the ANCF accelerations $\ddot{\mathbf{e}}$ can be obtained separately without the need to make this solution part of Eq. 43. Since the mass matrix $\mathbf{M}_{ee}$ is constant and its LU factors can be determined only once prior to the dynamic simulation, there is no need in this case to increase the size of the coefficient matrix in Eq. 43, which has to be solved at each time step. One can simply solve for the ANCF accelerations $\ddot{\mathbf{e}}$ using the equation $\ddot{\mathbf{e}} = \mathbf{M}_{ee}^{-1}\mathbf{Q}_e$. Therefore, once the forces $\mathbf{Q}_e$ are computed, the solution for the ANCF accelerations involves only vector–matrix multiplication, and the size of the coefficient matrix in Eq. 43 can be significantly reduced.

Pantograph/Catenary Contact Forces It is clear that the coefficient matrix in Eq. 43 is a sparse matrix. Therefore, Eq. 43 can be efficiently solved using sparse-matrix techniques to determine the second time derivatives of the coordinates $\mathbf{q}_r$, $\mathbf{q}_f$, $\mathbf{e}$, and $\mathbf{s}$ as well as the vector of Lagrange multipliers λ. The vector of Lagrange multipliers can be used to define the generalized constraint force as $\mathbf{F}_c = -\mathbf{C}_\mathbf{q}^T\boldsymbol{\lambda}$. When the pantograph/catenary constraint contact formulation is used, Lagrange multipliers can be used to determine the normal contact forces as constraint forces. Using these normal contact forces, the tangential forces, which can be defined using Coulomb's dry friction law or any other friction or damping model, can be defined and introduced to the dynamic formulation. To avoid using an iterative procedure to determine contact forces, which depend on Lagrange multipliers when the constraint contact formulation is used, Lagrange multipliers associated with these constraints can be stored from the previous time step and used to evaluate pantograph/catenary contact forces at the current time step. This approach has been found accurate since in most railroad vehicle system applications, a small time step is used in the numerical integration of the system equations of motion due to the high speed of rotation of the wheelsets and the high values of the contact forces. Furthermore, in the case of steady-state rolling and under normal operating conditions, significant changes in contact forces are not expected.

By using the values of Lagrange multipliers from the previous time step, the iterative solution of Eq. 43 can be avoided when the constraint contact formulation is used.

If the elastic contact formulation is used, on the other hand, pantograph/catenary contact forces can be determined using a compliant force model with assumed stiffness and damping coefficients. In this case, contact is not described using constraints, and consequently, no degrees of freedom are eliminated. When the elastic contact formulation is used to model pantograph/catenary interaction, the pan-head may lose contact with the catenary wire, and therefore, a unilateral force model must be used to allow for pantograph/catenary separation (Kulkarni et al. 2017). In this section, the development of this unilateral contact force model is demonstrated in the case of longitudinal sliding. A similar procedure can be used when both longitudinal and lateral sliding are considered.

In the case of longitudinal sliding only, the lateral and normal distances defined by Eq. 39, $d_l = \mathbf{j}^{c^T}(\mathbf{r}^p - \mathbf{r}^c)$ and $d_n = \mathbf{k}^{c^T}(\mathbf{r}^P - \mathbf{r}^c)$, can be used to check whether the pan-head and catenary contact wire come into contact. In these equations, $\mathbf{j}^c$ and $\mathbf{k}^c$ are the lateral and normal axes of the contact frame, and $\mathbf{r}^p$ and $\mathbf{r}^c$ are, respectively, the global position vectors of the potential contact points on the pan-head and catenary wire. If pan-head/catenary penetration occurs, one can define the compliant force vector $\mathbf{F}^{pc} = \begin{bmatrix} 0 & F_l^{pc} & F_n^{pc} \end{bmatrix}^T$, where $F_l^{pc} = k_l d_l + c_l \dot{d}_l$ and $F_n^{pc} = k_n d_n + c_n \dot{d}_n$ are, respectively, the lateral and normal components of the compliant force model, k_l and k_n are assumed stiffness coefficients, and c_l and c_n are assumed damping coefficients. The stiffness and damping coefficients used in this model can be constant, linear, or nonlinear functions of the pantograph/catenary coordinates. They can also be defined numerically using spline functions based on tabulated data obtained from force measurements. The virtual work of contact force $\mathbf{F}^{pc}$ can then be written as

$$\delta W^{pc} = -\left(F_l^{pc}(d_l, \dot{d}_l)\,\delta d_l + F_n^{pc}(d_n, \dot{d}_n)\,\delta d_n\right) \tag{7.44}$$

Using the distance equations $d_l = \mathbf{j}^{c^T}(\mathbf{r}^p - \mathbf{r}^c)$ and $d_n = \mathbf{k}^{c^T}(\mathbf{r}^P - \mathbf{r}^c)$, the virtual change in these distances can be written as $\delta d_k = ((\partial d_k/\partial \mathbf{q}^p)\delta \mathbf{q}^p + (\partial d_k/\partial \mathbf{e}^c)\delta \mathbf{e}^c)$, $k = l, n$. Therefore, the virtual work of the contact forces can be written as

$$\delta W^{pc} = -F_l^{pc}(d_l, \dot{d}_l)\left(\frac{\partial d_l}{\partial \mathbf{q}^p}\delta \mathbf{q}^p + \frac{\partial d_l}{\partial \mathbf{e}^c}\delta \mathbf{e}^c\right) - F_n^{pc}(d_n, \dot{d}_n)\left(\frac{\partial d_n}{\partial \mathbf{q}^p}\delta \mathbf{q}^p + \frac{\partial d_n}{\partial \mathbf{e}^c}\delta \mathbf{e}^c\right) \tag{7.45}$$

Using this equation, the generalized contact forces associated with the pan-head and catenary generalized coordinates can be written as

$$\left.\begin{aligned} \mathbf{Q}^p &= -\left(F_l^{pc}(d_l, \dot{d}_l)\left(\frac{\partial d_l}{\partial \mathbf{q}^p}\right) + F_n^{pc}(d_n, \dot{d}_n)\left(\frac{\partial d_n}{\partial \mathbf{q}^p}\right)\right), \\ \mathbf{Q}^c &= -\left(F_l^{pc}(d_l, \dot{d}_l)\left(\frac{\partial d_l}{\partial \mathbf{e}^c}\right) + F_n^{pc}(d_n, \dot{d}_n)\left(\frac{\partial d_n}{\partial \mathbf{e}^c}\right)\right) \end{aligned}\right\} \tag{7.46}$$

These unilateral generalized forces, which are assumed to be zero in the case of pantograph/catenary separation, can be introduced to the right-hand side of the equations of motion of Eq. 43 in generalized force vector $\mathbf{Q}_e$.

Once the normal force is defined using the constraint or elastic contact formulation, the tangential forces that oppose relative sliding can be introduced. In the case of dry friction, for example, tangential friction forces are a function of the resultant normal force F_{nr}^{pc} and the coefficient of friction μ. The resultant normal force can be defined in the case of longitudinal sliding as $F_{nr}^{pc} = \sqrt{\left(F_l^{pc}\right)^2 + \left(F_n^{pc}\right)^2}$. Using the Coulomb dry friction model, the tangential friction force can be defined as $\mathbf{F}_t^{pc} = -\mu F_{nr}^{pc}\mathbf{i}^c$ (Kulkarni et al. 2017). A similar procedure for developing the normal and tangential forces can be developed in the case of relative longitudinal and lateral sliding. In this case, the normal force has only one component since sliding is allowed in two different directions.

It is important to point out that pantograph/catenary contact forces are very small compared to wheel/rail contact forces. Therefore, pantograph/catenary contact force, which is necessary to avoid interruption of the electric power supply, normally does not have a significant effect on vehicle dynamics during steady-state motion. The expected static value of this force for high-speed trains ranges between 40 and 120 N (Hsiao 2010). Increasing the pantograph uplift force can contribute to a significant increase in the dynamic value of pantograph/catenary contact forces, as demonstrated in the literature (Pappalardo et al. 2016). Controlling the values of pantograph/catenary forces, however, is very important despite their small magnitudes to ensure the continuous, uninterrupted electric power supply required for train operation.

7.9 PANTOGRAPH/CATENARY CONTACT FORCE CONTROL

Providing the continuous supply of electric power necessary for the smooth operation of high-speed trains requires maintaining pantograph/catenary contact and ensuring system stability. This is a challenging problem due to the disturbances that can result from aerodynamics forces, rail car vibrations, and track irregularities. These disturbances can have an adverse effect on the current-collection system as the result of variation of the contact force between the pantograph strip and the catenary wire. For this reason, investigations have been conducted to study the effectiveness of control strategies that can be adopted to ensure the stability of pantograph/catenary contact and smooth train operation, particularly at high speeds. The design of effective control systems requires accurate modeling of train dynamics (Poetsch et al. 1997). The stability of the current-collection system depends on the dynamic behavior of the pantograph/catenary system and its response to disturbances. Pan-head/catenary contact is normally maintained using an uplift force exerted by actuators on the pantograph lower arm. The magnitude of the uplift force should not be very high to avoid high contact force between the pantograph strips and catenary wire that can lead to increased wear as a result of high friction forces. Low uplift forces, on the other hand, can lead to loss of contact, which increases the probability of electrical arcing. Arcing, which is the result of electric current flow in an air gap between the catenary contact wire and the pantograph contact strip, leads to electric sparks with intense light that can damage the contact wire and contact strip (Hsiao 2010). One solution to this problem is to use an *active control system* by placing an actuator between the pan-head and the plunger of the pantograph with the goal of reducing contact force variations (Pappalardo et al. 2016).

Pappalardo et al. (2016) presented a review of pantograph/catenary force control methods and proposed a contact force control strategy based on a continuum-based pantograph/catenary dynamic model that employs ANCF finite elements. This dynamic model, developed using the ANCF gradient-deficient cable elements, was used to test a method to control pantograph/catenary contact force variations. Because only one ANCF gradient vector is used in the formulation of pantograph/catenary contact conditions, as described in this chapter, the proposed control strategy can also be used with fully parameterized ANCF beam elements. A three-dimensional MBS model of a pantograph mounted on a train was developed using the nonlinear augmented MBS formulation defined by Eq. 43. Contact between the pantograph and catenary system was ensured using the constraint contact formulation, while an elastic contact formulation was used to predict wheel/rail contact forces. The standard deviation of the contact force was reduced without affecting its mean value. This was accomplished by using a control actuator placed between the pan-head and the plunger. Three types of control laws for the control action were proposed to improve contact quality in the transient and steady-state phases. The first control law is based on a feedback system, while the second and third control strategies are based on feedback plus feed-forward systems. To examine the effectiveness of the proposed control methods, Pappalardo et al. (2016) presented numerical results of train simulations with and without controllers. The results presented were used to evaluate the proposed new *dynamic damping* control method used with the continuum-based ANCF catenary model. Using this new formulation, three different control strategies were designed to control pantograph/catenary contact forces. These three strategies – a *derivative controller*, a *derivative/bang-bang controller*, and a *derivative/exponential controller* – were designed based on optimal control theory and the theory of virtual passive control. It was found that all three proposed controllers led to an improvement in the standard deviation of the contact force by more than 35% when compared to the uncontrolled system. The robustness of the control strategies was tested by using the controllers with a high uplift force. To address the practicality of the proposed control systems, their physical implementation using electro-rheological or magneto-rheological devices or general actuators was discussed.

7.10 AERODYNAMIC FORCES

Environmental conditions such as temperature and wind can have a significant impact on pantograph/catenary interaction forces. For example, high temperatures can alter the static equilibrium position and pretension in catenary wires, while cold weather conditions can result in the formation of ice on the wires, leading to undesirable deformations and poor contact quality. Aerodynamic forces, on the other hand, can result in severe vibrations of catenary cables as well as undesirable forces acting on the components of the pantograph that negatively influence its functional operation. Aerodynamic drag and lift forces can cause variations in contact forces, wear, and loss of contact. Wear can generate asymmetric drag and lift forces, leading to *galloping catenary motion* (Stickland Scanlon 2001; Stickland et al. 2003), while high cross-wind loads can also cause severe vehicle vibrations that directly influence pantograph/catenary interaction (Cheli et al. 2010). Aerodynamic drag and lift force components alter the uplift force exerted on the pantograph mechanism. For pantographs with multiple pan-heads, aerodynamic forces can cause an imbalance, resulting in faster wear of one of the collector strips (Carnevale et al. 2016; Pombo et al. 2009). Furthermore, boundary layer turbulence near the car body roof due to vortex shedding can excite the pantograph components, generate high frequencies, and increase the sparking level; this, in turn, can negatively impact the quality of current collection (Ikeda and Mitsumoji 2009). For these reasons, evaluating the effect of aerodynamic forces on pantograph/catenary dynamics is an important design consideration.

As pointed out by Kulkarni et al. (2017), in general, two main approaches can be used to evaluate the effect of the aerodynamic forces on pantograph/catenary systems. In the first approach, *computational fluid dynamics* (CFD) is used to calculate aerodynamic forces on pantograph system components and examine the contribution of these forces to the total uplift pantograph force (Carnevale et al. 2015). In the second approach, the drag and lift coefficients are obtained from experimental studies, and the aerodynamic forces are computed using the equation $\mathbf{F} = (0.5\,\rho C_a|\mathbf{v}^{wp}|^2)\mathbf{i}_a$, where $\mathbf{F}$ is the drag or lift force, ρ is the density of the fluid, C_a is the drag or lift pressure coefficient, $\mathbf{v}^{wp}$ is the relative velocity between the wind and pantograph component, and $\mathbf{i}_a$ is the unit vector in the direction of the drag or lift force (Pombo et al. 2009).

Aerodynamic Forces and ANCF Catenary Models Kulkarni et al. (2017) proposed an aerodynamic force formulation that can be used with the ANCF catenary models. It was shown that a simple aerodynamic force model could be integrated with the ANCF catenary and rigid body pantograph models to account for the cross-wind effect. The approximate drag and lift coefficients for pantograph components were obtained from previous investigations reported in the literature (Carnevale et al. 2016). To compute the aerodynamic forces on each body, a mesh data file, which defines the distribution of points at which aerodynamic forces are applied, is created for each rigid or flexible body in the model. Using this mesh data file, the effect of the moments of the aerodynamic forces can still be taken into consideration. The mesh data file contains 13 parameters for each data point selected: 3 Cartesian position coordinates that define the location of the point with respect to the body coordinate system, 9 direction cosines defining the orientation of a coordinate system for the data point with respect to the body coordinate system, and the area assigned to the data point. In the case of flexible catenaries, nodal locations are used to define the aerodynamic data points, and aerodynamic force is applied at the nodes of the flexible catenary wires. The aerodynamic drag and lift forces applied on a body i in the system can be defined as follows

$$\mathbf{F}_d^i = \left(\frac{1}{2}\rho\left(C_d A\right)\left|\mathbf{v}^{wi}\right|^2\right)\mathbf{i}_d, \quad \mathbf{F}_l^i = \left(\frac{1}{2}\rho\left(C_l A\right)\left|\mathbf{v}^{wi}\right|^2\right)\mathbf{i}_l \tag{7.47}$$

where $\mathbf{F}_d^i$ and $\mathbf{F}_l^i$ are the drag and lift forces, respectively; ρ is the fluid density; C_d and C_l are the drag and the lift pressure coefficients per unit area, respectively; A is the area on which the aerodynamic force is applied; $\mathbf{v}^{wi}$ is the vector of the relative velocity between the body and the wind; and $\mathbf{i}_d$ and $\mathbf{i}_l$ are unit vectors in the direction of the drag and the lift forces, respectively. The generalized forces associated with the generalized coordinates of body i as a result of aerodynamic drag force $\mathbf{F}_d^i$ and lift force $\mathbf{F}_l^i$ at point P can be defined, respectively, as $\mathbf{Q}_d^i = \sum_{i=1}^{n_p} \mathbf{J}_P^{iT}\mathbf{F}_{dP}^i$ and $\mathbf{Q}_l^i = \sum_{i=1}^{n_p} \mathbf{J}_P^{iT}\mathbf{F}_{lP}^i$, where n_p is the total number of points at which aerodynamic forces are applied and $\mathbf{J}_P^i$ is the Jacobian matrix of the position field defined by differentiation with respect to the vector of generalized coordinates of the body. In the case of a rigid body, one has $\mathbf{J}_P^i = \partial\mathbf{r}_P^i/\partial\mathbf{q}^i = \left[\mathbf{I} \;\; -\mathbf{A}_P^i\tilde{\overline{\mathbf{u}}}_P^i\overline{\mathbf{G}}^i\right]$; and in the case of an ANCF body $\mathbf{J}_P^i = \partial\mathbf{r}_P^i/\partial\mathbf{e}^i = \mathbf{S}_P^i$, where $\mathbf{r}_P^i$ is the absolute position vector of an arbitrary point P on the body, $\mathbf{q}^i$ is the vector of generalized coordinates of body i, $\mathbf{A}_P^i$ is the transformation matrix that defines the orientation of a point coordinate system on the body in the global coordinate system, $\tilde{\overline{\mathbf{u}}}_P^i$ is the skew-symmetric matrix associated with vector $\overline{\mathbf{u}}_P^i$ that defines the location of the point with respect to the coordinate system of the body on which the point is defined, and $\overline{\mathbf{G}}^i$ is the matrix that relates angular velocity vector $\overline{\boldsymbol{\omega}}^i$ defined in the body coordinate system to the time derivatives of the orientation coordinates $\dot{\boldsymbol{\theta}}^i$: that is, $\overline{\boldsymbol{\omega}}^i = \overline{\mathbf{G}}^i\dot{\boldsymbol{\theta}}^i$, and $\mathbf{S}_P^i$ is the ANCF shape function matrix defined at point P. By defining the aerodynamic forces at points on the bodies, the generalized aerodynamic forces associated with the generalized coordinates can be calculated, as previously discussed in this book. A more detailed discussion of the formulation of the aerodynamic forces of rigid and flexible bodies was presented by Kulkarni et al. (2017).

For design purposes, the National Electrical Safety Code (NESC) provides a formula for evaluating the wind velocity pressure p_v acting on an overhead line system. The NESC wind velocity pressure formula is defined as $p_v = c_{\rho a}c_{ve}c_g c_i c_f A_p (v^w)^2$ N/m^2, where $c_{\rho a} = 0.613$ is a velocity pressure numerical coefficient that accounts for the air mass density in the standard atmosphere, c_{ve} is the velocity pressure exposure coefficient, c_g is the gust response factor, c_i is the importance factor (assumed to be 1 in a pantograph/catenary system), c_f is the force coefficient shape factor, A_p is the projected wind area, and v^w is the basic wind speed. The catenary system should be designed such that contact wire displacement caused by cross-wind, with respect to the track centerline, should not exceed 0.4 m to avoid contact problems with the pantograph (Hsiao 2010).

7.11 PANTOGRAPH/CATENARY WEAR

The development of a general computational MBS algorithm for predicting pantograph/catenary wear allows incorporating the effects of vehicle vibration, wheel/rail contact forces, and track irregularities. Such an algorithm also allows for predicting the wear rate in the case of different motion scenarios that require using nonlinear models, including curve negotiations, accelerations, and braking (Daocharoenporn et al. 2019). In these motion scenarios, the effects of contact forces resulting from vehicle dynamics on the wear rates of the catenary wire can be significant. Severe environmental and operating conditions may cause arcing, high wear rates, or even failure of the pantograph/catenary system to provide the electric power necessary for train operation. Arcing, for example, may significantly increase the wear rate of the pantograph contact strip, leading to variations in contact forces and deterioration of the pantograph/catenary system performance.

Continuous localized contact between the pan-head and catenary wire can cause significant wear to the carbon strip inserted on the top of the pantograph and used for electric current collection. Between two overhead line supports, the contact wire segment is straight; therefore, during curve negotiations, the contact wire sweeps laterally over the entire carbon strip, resulting in uniform wear. When the vehicle negotiates a tangent track, it is also desirable to obtain such a pattern of contact in which the contact wire sweeps laterally over the surface of the carbon strip to achieve uniform wear instead of localized, more severe wear. Therefore, in the case of a tangent track, the contact catenary wire is slightly zigzagged around the centerline of the track to produce a lateral sweep that leads to uniform wear.

Several important factors can have a direct effect on pantograph/catenary wear, including the design of the contact wire, the pretension in the catenary cable, and the uplift force of the pantograph mechanism. For example, using a high uplift force can lead to a significant increase in the magnitude of the contact force, and this, in turn, can lead to an increase in the wear rate. Therefore, the wear rate can be reduced by controlling this uplift force of the pantograph mechanism. The wear rate can also be reduced by properly designing the pantograph/catenary system to better handle aerodynamic forces, staggering the contact wire, controlling the intensity of the collection current, using the proper materials for the contact wire and contact strip to reduce friction, and properly adjusting the pretension of the catenary cables (Bucca and Collina 2009; Bucca and Collina 2015; Daocharoenporn et al. 2019). A wear model that accounts for electric and mechanical effects such as the electric arcing effect, the *Joule effect* of electrical current, and the effect of contact friction forces can be developed (Bucca and Collina 2015). This section presents a brief summary of the investigation conducted by Daocharoenporn et al. (2019).

Brief Literature Review Studies of pantograph/catenary systems have shown that *electromechanical phenomena* have a significant effect on contact between the pan-head and catenary wire. He et al. (1998) used several contact strips with different material properties to estimate the friction coefficients. It was found that the value of the friction coefficients is highest in the case of unlubricated copper-to-copper contact. The optimal value of the friction coefficient was found to be in the range of 0.2–0.24 in the case of carbon-to-copper contact. He et al. (1998) also concluded that plastic deformation of soft asperities is the main factor that causes wear of the contact catenary wires in the case of metal-to-metal sliding. Kubo and Kato (1998, 1999), who investigated the effect of electrical discharge or arcing on the wear rates of copper-impregnated contact strips, concluded that the wear rate is proportional to the electrical discharge energy. One of the main conclusions drawn from the study performed by Kubo and Kato (1998, 1999) is the adverse effect of heat energy resulting from arcing. This generated heat energy is the cause of several temperature-related effects such as oxidation of carbon, separation of impregnated copper particles, melting, and evaporation. Temperatures during pantograph/catenary system operations can reach a very high level (200–550 °C) at which using grease becomes less effective. The arc discharge process was further investigated by Chen et al. (2013), who concluded that arcing has a very adverse effect on current-collection equipment. It was also found that the pan-head wear rate is approximately proportional to the discharge energy on a logarithmic scale. A similar conclusion was reported in the investigation performed by Kubota et al. (2013) for copper alloy–impregnated, carbon fiber–reinforced carbon composite collector strips. Ding et al. (2011) used a pure carbon contact strip and a copper contact wire in their investigation. They concluded that the friction coefficient is significantly higher if the current does not flow in the contact wire, and the friction coefficient decreases as the electrical current and normal force increase. Ding et al. (2011) found that the effect of sliding speed on the friction coefficient is insignificant, justifying the use of the dry friction tangential force model, which does not depend on the magnitude of the relative velocity. As pointed out by Daocharoenporn et al. (2019), a literature review revealed that a large number of investigations and research efforts have been devoted to understanding the wear mechanism of railroad vehicle current-collection systems. Some of these investigations were devoted to optimizing current-collection system design, which is influenced by several interdependent parameters.

Wear Models The wear formulation developed by Bucca and Collina (2015) was adopted by Daocharoenporn et al. (2019) to study the wear characteristics of a copper contact wire in a curved-track motion scenario. Daocharoenporn et al. (2019) integrated this wear model in a MBS computational algorithm that allowed for developing a detailed vehicle model based on the approaches previously discussed in this book. In the MBS railroad vehicle model developed by Daocharoenporn et al. (2019), the catenary wires were modeled using ANCF finite elements, and the pantograph system was modeled as a mechanism consisting of interconnected links. The model also included a detailed vehicle model with two bogies; each bogie had two wheelsets. A three-dimensional wheel/rail contact formulation was employed in the detailed MBS model used to study wear.

As previously mentioned, the tribological conditions generated during pantograph/catenary contact are the result of electrical and mechanical effects. It has been shown in the literature that the wear mechanism in a pantograph/catenary system is mainly governed by the following three factors: a *mechanical factor* resulting from sliding friction, a *Joule effect* resulting from the flow of high-intensity current in catenary wires, and *electrical arcs* resulting from pan-head/catenary separations. Because these factors do not contribute to wear independently, the development of a robust, credible wear model that can be used in computer simulations may require experimental testing. In such experimental testing, the relationship between the effects of these wear factors as well as their separate contributions can be evaluated by varying, for example, the materials of the pan-head and contact wire, the magnitude of the uplift force, the relative sliding velocity, and the electric current intensity (Bucca and Collina 2015). The *normal wear rate* N_{wr}, defined as the volume worn out (measured in mm^3) per unit distance (measured in km) was computed in the simulation scenarios performed by Daocharoenporn et al. (2019). In these simulation scenarios, a pure copper contact wire and a Kasperowski type of contact strip were used. The Kasperowski contact strip has the carbon material part encased in a copper casing from three sides. The normal wear rate N_{wr} of the copper contact line is given by (Bucca and Collina 2015)

$$N_{wr} = k_1 \left(\frac{1}{2} \left(1 + \frac{I_c}{I_0} \right) \right)^{-\alpha} \left(\frac{F_m}{F_0} \right)^{\beta} \frac{F_m}{H} + k_2 \frac{R_c \left(F_m \right) \cdot I_c^2}{H \cdot V} (1 - u) + k_3 u \frac{V_a \cdot I_c}{V \cdot H_m \cdot \rho} \quad (7.48)$$

The parameters in this equation are defined in Table 2, which also shows the numerical values used by Daocharoenporn et al. (2019). Using this equation, the effect of individual parameters on the normal wear rate N_{wr} can be evaluated separately by varying these parameters and keeping other parameters constant. Such a parametric study allows for a better understanding of the wear mechanism. The electrical contribution to the wear for a given current value is proportional to electrical contact resistance R_c, which depends on mean contact force F_m. Bucca et al. (2011) gave an explicit expression for this resistance based on extensive experimentation. This expression is defined as $R_c = R_c \left(F_m \right) = 0.013 + 0.09e^{-(F_m - 14)/11}$.

By examining Eq. 48, it can be shown that the contact wire normal wear rate N_{wr} decreases as electric current increases at high values of the mean contact force. An opposite trend

Table 7.2 Catenary wear parameters.

Symbol	Description	Value
F_m	Mean value of contact force F_n^{pc} (N)	—
k_1	Weight of the mechanical contribution to the N_{wr} value	22.4
k_2	Weight of the electrical contribution to the N_{wr} value	10.3
k_3	Weight of the contribution due to electrical arcs to the N_{wr} value	0.4
α	Coefficient of the dependence of the mechanical contribution on the electrical current	4.5
β	Coefficient of nonlinear dependence of the mechanical contribution on the mean value of the contact force	1.8
I_0	Reference value of electrical current (A)	500
I_c	Nominal electrical current during tests (A)	—
F_0	Reference value of contact force (N)	90
H	Hardness of material (N/mm^2)	700
R_c	Electrical contact resistance between strip and wire (Ω)	—
V	Sliding speed during tests (m/s)	—
u	Decimal fraction value of the percentage of contact loss	—
V_a	Electrical arc voltage (V)	50
H_m	Latent heat of fusion for copper (kJ/kg)	205
ρ	Density of copper (kg/m^3)	8940

Source: Courtesy of Daocharoenporn, S., Mongkolwongrojn, M., Kulkarni, S., and Shabana, A.A.

Table 7.3 Pantograph/catenary interaction criteria.

Pantograph parameter	Acceptable values
Mean contact force	$F_m = 0.00097(V)^2 + 70\text{N}$
Standard deviation	$\sigma_{max} < 0.3F_m$
Maximum contact force	$F_{max} < 350\text{N}$
Maximum catenary wire uplift at steady arm	$d_{up} \leq 120\,\text{mm}$
Maximum pantograph vertical amplitude	$\triangle_z \leq 80\,\text{mm}$
Percentage of real arcing	$NQ \leq 0.2\%$

occurs if the value of the mean contact force is below 90 N. For electric current values above 300 A, wear rates are lower as the contact force increases; this was attributed to the mechanism of *current lubrication* (Bucca and Collina 2009). On the other hand, for lower values of electric current, the wear rate increases as the mean contact force increases. This type of wear is characterized by an abrasive wear mechanism and dry friction because the temperatures in the contact region are not high enough to melt the metals in contact. The contribution of arcs is proportional to the power generated by the arcs, which depends on the electric voltage, the speed of the train, and the gap between the pan-head and contact wire. These findings demonstrate the interdependency of the electrical and mechanical contributions to the complex wear mechanism of the catenary contact wire (Daocharoenporn et al. 2019). The EN50367 and EN50317 standards provide criteria for electric current-collection systems of railroad vehicles. The limits shown in Table 3 were used by Daocharoenporn et al. to validate the results obtained using their computer model.

Appendix

CONTACT EQUATIONS AND ELLIPTICAL INTEGRALS

Chapter 5 discussed Hertz contact theory, which is widely used in the analysis of wheel/rail contact. For a given normal load and material properties, this theory can be used to determine the dimensions of the contact ellipse as a function of the wheel and rail surface geometries. Hertz theory, which assumes that the dimensions of the contact area are small in comparison with the dimensions of the two solids in contact, is based on the elastic half-space assumptions used to determine stress distribution in solids (Boussinesq 1885; Cerruti 1882; Love 1944). The development of Hertz theory requires the evaluation of the elliptical integrals whose values depend on the dimensions of the contact ellipse. Section A.1 presents the derivation of some of the equations required for the development of Hertz theory, and Section A.2 is devoted to elliptical integrals. The equations presented in this appendix are a summary of the equations presented in the two appendices of a previous book (Shabana et al. 2008).

A.1 DERIVATION OF THE CONTACT EQUATIONS

In the elastic half-space, shown in Figure A.1, point P is assumed to lie inside the contact area A_c resulting from load applications. By choosing the proper coordinate system and without any loss of generality, the coordinates of point P can be assumed to be $[x_P\, y_P\, 0]^T$. If Q is another arbitrary point inside the solid with coordinates $\left[x\ y\ z\right]^T$, as shown in Figure A.1, distance l_{PQ} between points P and Q can be written as

$$l_{PQ} = \sqrt{\left(x_P - x\right)^2 + \left(y_P - y\right)^2 + z^2} \tag{A.1}$$

The distributions of normal pressure p and tangential tractions τ_x and τ_y in contact area A_c are assumed to be functions of x_P and y_P. Using these assumptions, the following potential functions can be defined

$$\left.\begin{aligned}\overline{\Phi}_x &= \int_{A_c} \tau_x\left(x_P, y_P\right) h_P dA_c\\ \overline{\Phi}_y &= \int_{A_c} \tau_y\left(x_P, y_P\right) h_P dA_c\\ \overline{\Phi}_p &= \int_{A_c} p\left(x_P, y_P\right) h_P dA_c\end{aligned}\right\} \tag{A.2}$$

Mathematical Foundation of Railroad Vehicle Systems: Geometry and Mechanics,
First Edition. Ahmed A. Shabana.

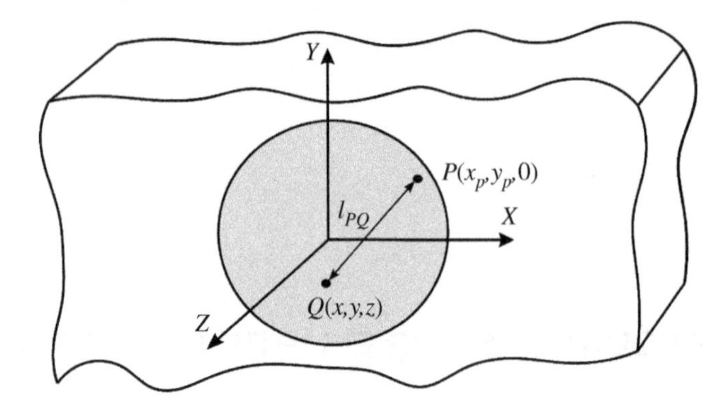

Figure A.1 Elastic half-space.

where $h_P = z\ln(l_{PQ}+z) - l_{PQ}$ and $dA_c = dx_P dy_P$. It follows that

$$\left.\begin{aligned}
\Phi_x &= \frac{\partial \overline{\Phi}_x}{\partial z} = \int_{A_c} \tau_x\left(x_P, y_P\right) \ln\left(l_{PQ}+z\right) dA_c \\
\Phi_y &= \frac{\partial \overline{\Phi}_y}{\partial z} = \int_{A_c} \tau_y\left(x_P, y_P\right) \ln\left(l_{PQ}+z\right) dA_c \\
\Phi_p &= \frac{\partial \overline{\Phi}_p}{\partial z} = \int_{A_c} p\left(x_P, y_P\right) \ln\left(l_{PQ}+z\right) dA_c
\end{aligned}\right\} \tag{A.3}$$

Using the definitions of the preceding equations, it can be shown that the elastic displacement of an arbitrary point Q can be written as (Love 1944)

$$\left.\begin{aligned}
u_x &= \frac{1}{4\pi G}\left[2\frac{\partial \Phi_x}{\partial z} - \frac{\partial \Phi_z}{\partial x} + 2\nu\frac{\partial \overline{\Phi}_s}{\partial x} - z\frac{\partial \Phi_s}{\partial x}\right] \\
u_y &= \frac{1}{4\pi G}\left[2\frac{\partial \Phi_y}{\partial z} - \frac{\partial \Phi_z}{\partial y} + 2\nu\frac{\partial \overline{\Phi}_s}{\partial y} - z\frac{\partial \Phi_s}{\partial y}\right] \\
u_z &= \frac{1}{4\pi G}\left[2\frac{\partial \Phi_z}{\partial z} + (1-2\nu)\,\Phi_s - z\frac{\partial \Phi_s}{\partial z}\right]
\end{aligned}\right\} \tag{A.4}$$

where G and ν are, respectively, the modulus of rigidity and Poisson ratio; $\overline{\Phi}_s = \left(\partial \overline{\Phi}_x/\partial x\right) + \left(\partial \overline{\Phi}_y/\partial y\right) + \left(\partial \overline{\Phi}_p/\partial z\right)$; and $\Phi_s = (\partial \Phi_x/\partial x) + (\partial \Phi_y/\partial y) + (\partial \Phi_p/\partial z)$ is the *Boussinesq function.* If the effect of friction is neglected and the pressure $p(x_P, y_P)$ is assumed to be normal to the contact surface, one has $\overline{\Phi}_x = 0,\ \ \Phi_x = 0,\ \ \overline{\Phi}_y = 0$ and $\Phi_y = 0$. Using these assumptions in Eqs. 2–4, one has

$$\left.\begin{aligned}
\overline{\Phi}_s &= \Phi_p = \textstyle\int_{A_c} p\left(x_P, y_P\right) \ln\left(l_{PQ}+z\right) dA_c \\
\Phi_s &= \textstyle\int_{A_c} p\left(x_P, y_P\right) \left(\frac{1}{l_{PQ}}\right) dA_c
\end{aligned}\right\} \tag{A.5}$$

and

$$\left.\begin{aligned} u_x &= -\frac{1}{4\pi G}\left[(1-2\nu)\frac{\partial\overline{\Phi}_s}{\partial x} + z\frac{\partial\Phi_s}{\partial x}\right] \\ u_y &= -\frac{1}{4\pi G}\left[(1-2\nu)\frac{\partial\overline{\Phi}_s}{\partial y} + z\frac{\partial\Phi_s}{\partial y}\right] \\ u_z &= \frac{1}{4\pi G}\left[2(1-\nu)\Phi_s - z\frac{\partial\Phi_s}{\partial z}\right] \end{aligned}\right\} \tag{A.6}$$

For an arbitrary point on the surface that has a z-coordinate equal to zero – that is, $z=0$ – the preceding equation leads to

$$\left.\begin{aligned} u_x\big|_{z=0} &= -\frac{1-2\nu}{4\pi G}\left(\frac{\partial\overline{\Phi}_s}{\partial x}\right)\Bigg|_{z=0} \\ u_y\big|_{z=0} &= -\frac{1-2\nu}{4\pi G}\left(\frac{\partial\overline{\Phi}_s}{\partial y}\right)\Bigg|_{z=0} \\ u_z\big|_{z=0} &= \frac{1-\nu}{2\pi G}\left(\frac{\partial\overline{\Phi}_s}{\partial z}\right)\Bigg|_{z=0} \end{aligned}\right\} \tag{A.7}$$

In the case of an elliptical contact area with semi axes that have dimensions a and b, the applied normal pressure can be written as

$$p(x,y) = p_0\left[1-\left(\frac{x}{a}\right)^2+\left(\frac{y}{b}\right)^2\right]^n \tag{A.8}$$

where n is a given number and the area of contact can be defined by the ellipse equation $(x/a)^2+(y/b)^2=1$. Using potential theory, the Boussinesq function (Eq. A.5) for a general point in the solid is given by

$$\Phi_s(x,y,z) = \frac{\Gamma(n+1)\Gamma(1/2)}{\Gamma(n+3/2)}p_0 ab\int_{\gamma_1}^{\infty}\frac{\left(1-\dfrac{(x)^2}{(a)^2+w}-\dfrac{(y)^2}{(b)^2+w}-\dfrac{(z)^2}{w}\right)^{\left(n+\frac{1}{2}\right)}}{\sqrt{\left((a)^2+w\right)\left((b)^2+w\right)w}}\,dw \tag{A.9}$$

In this equation, Γ is the gamma function and γ_1 is the positive root of the equation $((x)^2/((a)^2+\gamma))+((y)^2/((b)^2+\gamma))+(z)^2/\gamma=1$.

In Hertz theory, $n=1/2$. In this special case, Φ_s of Eq. 9 reduces to

$$\Phi_s(x,y,z) = \frac{1}{2}\pi p_0 ab\int_{\gamma_1}^{\infty}\frac{\left(1-\dfrac{(x)^2}{(a)^2+w}-\dfrac{(y)^2}{(b)^2+w}-\dfrac{(z)^2}{w}\right)}{\sqrt{\left((a)^2+w\right)\left((b)^2+w\right)w}}\,dw \tag{A.10}$$

For an arbitrary point on the surface, $z = 0$; and the preceding equation reduces to

$$\Phi_s(x,y,z) = \frac{1}{2}\pi p_0 ab \int_{\gamma_1}^{\infty} \frac{\left(1 - \frac{(x)^2}{(a)^2 + w} - \frac{(y)^2}{(b)^2 + w}\right)}{\sqrt{\left((a)^2 + w\right)\left((b)^2 + w\right)w}} dw \tag{A.11}$$

Using this equation and Eq. 7, the z-displacement can be written as

$$u_z = \frac{1-(v)^2}{\pi E}\left(L_e - M_e(x)^2 - N_e(y)^2\right) \tag{A.12}$$

where E is Young's modulus, and

$$\left.\begin{aligned}
M_e &= \frac{\pi p_0 ab}{2}\int_0^{\infty}\frac{dw}{\sqrt{\left((a)^2+w\right)^3\left((b)^2+w\right)w}} = \frac{\pi p_0 b}{(e)^2(a)^2}\left[K_e - E_e\right]\\
N_e &= \frac{\pi p_0 ab}{2}\int_0^{\infty}\frac{dw}{\sqrt{\left((a)^2+w\right)\left((b)^2+w\right)^3 w}} = \frac{\pi p_0 b}{(e)^2(a)^2}\left[\left(\frac{a}{b}\right)^2 E_e - K_e\right]\\
L_e &= \frac{\pi p_0 ab}{2}\int_0^{\infty}\frac{dw}{\sqrt{\left((a)^2+w\right)\left((b)^2+w\right)w}} = \pi p_0 b K_e
\end{aligned}\right\} \tag{A.13}$$

where E_e and K_e are complete elliptical integrals of argument $e = \sqrt{1-(a/b)^2}$ for $b > a$, as will be explained in the following section. Because the pressure is assumed to have an elliptical distribution (Eq. 8), the normal force F_{ns} can be defined as $F_{ns} = 2p_0\pi ab/3$, as discussed in Chapter 5.

A.2 ELLIPTICAL INTEGRALS

In Chapter 5, complete elliptical integrals were used in the development of the Hertz theory of contact. These integrals are defined as

$$\left.\begin{aligned}
B_e &= \int_0^{\pi/2}\frac{\cos^2 w}{\sqrt{1-(e)^2\sin^2 w}}dw = \int_0^{\pi/2}\frac{\cos^2 w}{\sqrt{\cos^2 w-(g)^2\sin^2 w}}dw,\\
C_e &= \int_0^{\pi/2}\sin^2 w\cos^2 w\left(1-(e)^2\sin^2 w\right)^{-\frac{3}{2}}dw,\\
D_e &= \int_0^{\pi/2}\frac{\sin^2 w}{\sqrt{1-(e)^2\sin^2 w}}dw,\quad E_e = \int_0^{\pi/2}\sqrt{1-(e)^2\sin^2 w}\,dw,\\
K_e &= \int_0^{\pi/2}\frac{1}{\sqrt{1-(e)^2\sin^2 w}}dw
\end{aligned}\right\} \tag{A.14}$$

Table A.1 Elliptical integrals.

$g = b/a$	B_e	C_e	D_e	E_e	K_e	$(e)^2$
0	1.0	$-2+\ln(4/g)$	$-1+\ln(4/g)$	1.0	$+\ln(4/g)$	1.00
0.1	0.9889	1.7351	2.7067	1.0160	3.6956	0.99
0.2	0.9686	1.1239	2.0475	1.0505	3.0161	0.96
0.3	0.9451	0.8107	1.6827	1.0965	2.6278	0.91
0.4	0.9205	0.6171	1.4388	1.1507	2.3593	0.84
0.5	0.8959	0.4863	1.2606	1.2111	2.1565	0.75
0.6	0.8719	0.3929	1.1234	1.2763	1.9953	0.64
0.7	0.8488	0.3235	1.0138	1.3456	1.8626	0.51
0.8	0.8267	0.27060	0.9241	1.4181	1.7508	0.36
0.9	0.8055	0.22925	0.8491	1.4933	1.6545	0.19
1.0	$\pi/4 = 0.7864$	$\pi/16 = 0.19635$	$\pi/4 = 0.7864$	$\pi/2 = 1.5708$	$\pi/2 = 1.5708$	0.00

Source: Courtesy of Kalker, J.J.

where $e = \sqrt{1-(a/b)^2}$, $b > a$ and $g = a/b$. The elliptical integrals presented in the preceding equations are not totally independent because they are related by the equations

$$\left.\begin{aligned} &K_e = 2D_e - (e)^2C_e, \quad E_e = \left(2-(e)^2\right)D_e - (e)^2C_e, \quad B_e = D_e - (e)^2C_e, \\ &D_e = \left(K_e - C_e\right)/(e)^2, \quad B_e = K_e - D_e, \quad C_e = \left(D_e - B_e\right)/(e)^2 \end{aligned}\right\} \tag{A.15}$$

The values of the elliptical integral as a function of the ratio of the contact ellipse semi axes $g = a/b$ are shown in Table A.1.

Bibliography

1. ACC. (2017). Competitive switching: making the switch to a more competitive, healthy freight rail system. Freight Rail Reform. https://www.freightrailreform.com/making-the-switch-to-a-more-competitive-healthy-freight-rail-system.
2. Andersson, C. and Abrahamsson, T. (2002). Simulation of interaction between a train in general motion and a track. *Vehicle System Dynamics* 38: 433–455.
3. APTA Press Task Force. (2007). Standard for wheel flange angle for passenger equipment. The American Public Transportation Association, Report # APTA PR-M-S-015-06.
4. Arnold, M. (1996). Numerical problems in the dynamical simulation of wheel-rail systems. *Zeitschrift Fur Angewandte Mathematik Und Mechanik* 76: 151–154.
5. Atkinson, K.E. (1978). *An Introduction to Numerical Analysis*. Wiley.
6. Bailey, J.R. and Wormley, D.N. (1992). A comparison of analytical and experimental performance data for a two-axle freight car. *ASME Journal of Dynamic Systems, Measurement, and Control* 114: 141–147.
7. Baumgarte, J. (1972). Stabilization of constraints and integrals of motion. *Computer Methods in Applied Mechanics and Engineering* 1: 1–16.
8. Bavinck, H. and Dieterman, H.H. (1996). Wave propagation in a finite cascade of masses and springs representing a train. *Vehicle System Dynamics* 26: 45–60.
9. Berg, M. (1998). A non-linear rubber spring model for rail vehicle dynamics analysis. *Vehicle System Dynamics* 30: 197–212.
10. Berghuvud, A. (2002). Freight car curving performance in braked conditions. *Proceedings of the Institution of Mechanical Engineers, Part F: Journal of Rail and Rapid Transit* 216: 23–29.
11. Berzeri, M., Sany, J.R., and Shabana, A.A. (2000). Curved track modeling using the absolute nodal coordinate formulation. Technical report MBS00-4-UIC. Department of Mechanical Engineering, University of Illinois at Chicago.
12. Bing, A.J., Shaun, R.B., and Henderson, B. (1996). Design data on suspension systems of selected rail passenger cars. FRA report DOT/FRA/ORD-96-01.
13. Blader, F. and Klauser, P. (1989). User's manual for NUCARS, version 1.0. Technical report R-734. Association of American Railroads.
14. Blader, F.B. (1989). A review of literature and methodologies in the study of derailments caused by excessive forces at the wheel/rail interface. AAR technical report R-717. AAR Technical Center, Chicago.

Mathematical Foundation of Railroad Vehicle Systems: Geometry and Mechanics,
First Edition. Ahmed A. Shabana.

15. Bocciolone, M., Caprioli, A., Cigada, A., and Collina, A. (2007). A measurement system for quick rail inspection and effective track maintenance strategy. *Mechanical Systems and Signal Processing* 21: 1242–1254.

16. Bocciolone, M., Resta, F., Rocchi, D., and Collina, A. (2006a). Pantograph aerodynamic effects on the pantograph–catenary interaction. *Vehicle System Dynamics* 44 (1): 560–570.

17. Bocciolone, M., Caprioli, A., Cigada, A., and Collina, A. (2007). A measurement system for quick rail inspection and effective track maintenance strategy. *Mechanical Systems and Signal Processing* 21: 1242–1254.

18. Bocciolone, M., Cheli, F., Corradi, R., and Collina, A. (2008a). Crosswind action on rail vehicles: Wind tunnel experimental analyses. *Journal of Wind Engineering and Industrial Aerodynamics* 96: 584–610.

19. Bocciolone, M., Resta, F., Rocchi, D., Tosi, A., and Collina, A. (2006b). Pantograph aerodynamic effects on the pantograph–catenary interaction. *Vehicle System Dynamics* 44 (1): 560–570.

20. Bocciolone, M., Cheli, F., Corradi, R., Muggiasca, S., and Tomasini, G. (2008b). Cross-wind action on rail vehicles: wind tunnel experimental analyses. *Journal of Wind Engineering and Industrial Aerodynamics* 96 (5): 584–610.

21. Bonet, J. and Wood, R.D. (1997). *Nonlinear Continuum Mechanics for Finite Element Analysis.* Cambridge University Press.

22. Boocock, D. (1969). The steady-state motion of railway vehicles on curved track. *Journal of Mechanical Engineering Science* 2: 556–566.

23. Boresi, A.P. and Chong, K.P. (2000). *Elasticity in Engineering Mechanics*, 2e. Wiley.

24. Bosso, N., Gugliotta, A., and Zampieri, N. (2015). Strategies to simulate wheel-rail adhesion in degraded conditions using a roller-rig. *Vehicle System Dynamic* 53 (5): 619–634.

25. Bosso, N. and Zampieri, N. (2013). Real-time implementation of a traction control algorithm on a scaled roller rig. *Vehicle System Dynamics* 51 (4): 517–541.

26. Boussinesq, J. (1885). *Application des Potentials 'al'étude de l'équilibre et du mouvement des solides élastiques.* Paris: Gauthier-Villars.

27. Bracciali, A., Cascini, G., and Ciuffi, R. (1998). Time domain model of the vertical dynamics of a railway track up to 5 kHz. *Vehicle System Dynamics* 30: 1–15.

28. British Railway Board. (1996). Wheelset tread standards & gauging. MT/288 report, issue 2, revision A.

29. Bruni, S., Facchinetti, A., Kolbe, M., and Massat, J-P., (2011). Hardware-in-the-loop testing of pantograph for homologation. *Proceedings 9th World Congress on Railway Research, 22–26 May 2011.* Lille, France, PB003802.

30. Bruni, S., Ambrosio, J., Carnicero, A., Cho, Y.H., Finner, L., Ikeda, M., and Zhang, W. (2015). The results of the pantograph–catenary interaction benchmark. *Vehicle System Dynamics* 53 (3): 412–435.

31. Bucca, G. and Collina, A. (2009). A procedure for the wear prediction of collector strip and contact wire in pantograph–catenary system. *Wear* 266 (1): 46–59.

32. Bucca, G. and Collina, A. (2015). Electromechanical interaction between carbon-based pantograph strip and copper contact wire: a heuristic wear model. *Tribology International* 92: 47–56.

33. Bucca, G., Collina, A., Manigrasso, R., Mapelli, F., and Tarsitano, D. (2011). Analysis of electrical interferences related to the current collection quality in pantograph–catenary interaction. *Proceedings of the Institution of Mechanical Engineers, Part F: Journal of Rail and Rapid Transit* 225 (5): 483–500.
34. Business, Wire (2017). Axalta launches new high performance, protective industrial rail car coatings. https://www.businesswire.com/news/home/20171018005982/en/Axalta-Launches-New-High-Performance-Protective-Industrial.
35. Carnevale, M., Facchinetti, A., Maggiori, L., and Rocchi, D. (2016). Computational fluid dynamics as a means of assessing the influence of aerodynamic forces on the mean contact force acting on a pantograph. *Proceedings of the Institution of Mechanical Engineers, Part F: Journal of Rail and Rapid Transit* 230 (7): 1698–1713.
36. Carter, F.W. (1926). On the action of a locomotive driving wheel. *Proceedings of the Royal Society of London Series A* 112: 151–157.
37. Cerruti, V. (1882). Roma, Acc. Lincei, Mem. Fis. Mat.
38. Cheli, F., Ripamonti, F., Rocchi, D., and Tomasini, G. (2010). Aerodynamic behaviour investigation of the new EMUV250 train to cross wind. *Journal of Wind Engineering and Industrial Aerodynamics* 98 (4): 189–201.
39. Chen, G.X., Yang, H.J., Zhang, W.H., Wang, X., Zhang, S.D., and Zhou, Z.R. (2013). Experimental study on arc ablation occurring in a contact strip rubbing against a contact wire with electrical current. *Tribology International* 61: 88–94.
40. Cole, C. and Sun, Y.Q. (2006). Simulated comparisons of wagon coupler systems in heavy haul trains. *Proceedings of the Institution of Mechanical Engineers, Part F: Journal of Rail and Rapid Transit* 220: 247–256.
41. Cole, C., Spiryagin, M., and Sun, Y.Q. (2013). Assessing wagon stability in complex train systems. *International Journal of Rail Transportation* 1 (4): 193–217.
42. Cole, C., Spiryagin, M., Wu, Q., and Sun, Y.Q. (2017). Modelling, simulation and applications of longitudinal train dynamics. *Vehicle System Dynamics* 55 (10): 1498–1571.
43. Collina, A. and Bruni, S. (2002). Numerical simulation of pantograph-overhead equipment interaction. *Vehicle System Dynamics* 38 (4): 261–291.
44. Cooperrider, N.K. (1991). Ride quality assessment for a 6-axle locomotive and a heavy truck. ASME Paper No. RTD-Vol.4: 153-160.
45. Cooperrider, N.K., Law, E.H., Hull, R., Kadala, P.S., and Tuten, P.S. (1975). Analytical and experimental determination of nonlinear wheel/rail geometric constraints. Report FRA-OR&D 76-244 (PB-252 290).
46. Cooperrider, N.K., Hedrick, J.K., Law, E.H., and Malstrom, C.W. (1976). The application of quasi-linearization techniques to the prediction of nonlinear railway vehicle response. In: *Proceedings of the IUTAM Symposium*, 314–325.
47. Coulomb, C.A. (1785). Theorie des Machines Simples. In: *Memoire de Mathematique et de Physique de l'Academie Royale*, 161–342.
48. Cummings, S. (2018). A new wheel profile for North American freight railroads: AAR-2A. Heavy Haul Seminar, May 2–3. http://www.wheel-rail-seminars.com/archives/2018/hh-papers/presentations/HH03.pdf.
49. Daocharoenporn, S., Mongkolwongrojn, M., Kulkarni, S., and Shabana, A.A. (2019). Prediction of the pantograph/catenary wear using nonlinear multibody system dynamic algorithms. *ASME Journal of Tribology*, accepted for publication.

50. Datoussaid, S., Verlinden, O., Wenderloot, L., and Conti, C. (1998). Computer-aided analysis of urban railway vehicles. *Vehicle System Dynamics* 30: 213–227.

51. De Pater, A.D. (1974). The propagation of waves in a semi-infinite chain of material points and springs representing a long train. *Vehicle System Dynamics* 3: 123–140.

52. De Pater, A.D. (1988). The geometrical contact between track and wheel-set. *Vehicle System Dynamics* 17: 127–140.

53. Diana, G., Fossati, F., and Resta, F. (1998). High speed railway: collecting pantographs active control and overhead lines diagnostic solutions. *Vehicle System Dynamics* 30: 69–84.

54. Dietz, S., Netter, H., and Sachau, D. (1988). Fatigue life prediction of a railway bogie under dynamic loads through simulation. *Vehicle System Dynamics* 29: 385–402.

55. Ding, T., Chen, G.X., Wang, X., Zhu, M.H., Zhang, W.H., and Zhou, W.X. (2011). Friction and wear behavior of pure carbon strip sliding against copper contact wire under AC passage at high speeds. *Tribology International* 44 (4): 437–444.

56. Diomin, Y.V. (1994). Stabilization of high-speed railway vehicles. *Vehicle System Dynamics* 23: 107–114.

57. Do Carmo, M.P. (1976). *Differential Geometry of Curves and Surfaces*. Prentice Hall.

58. Duff, I.S., Erisman, A.M., and Reid, J.K. (1986). *Direct Methods for Sparse Matrices*. Oxford: Clarendon Press.

59. Duffek, W. (1982). Contact geometry in wheel rail vehicles. In: *Proceedings of Contact Mechanics and Wear of Rail/Wheel Systems*, 161–181. Vancouver: University of WaterLoo.

60. Dukkipati, R.V. (1998). Dynamics of wheelset on roller rig. *Vehicle System Dynamics* 30 (6): 409–430.

61. Dukkipati, R.V. (2000). *Vehicle Dynamics*. New Delhi: CRC Press.

62. Dukkipati, R.V. and Amyot, J.R. (1988). *Computer-Aided Simulation in Railway Dynamics*. New York: Mercel-Dekker.

63. Dukkipati, R.V. and Dong, R. (1999). Impact loads due to wheel flats and shells. *Vehicle System Dynamics* 31: 1–22.

64. Dukkipati, R.V., Swamy, S.N., and Osman, M.O.M. (1991). Independently rotating wheel systems for railway vehicle – a survey of the state of the art. *The Archives of Transport* III: 297–330.

65. Durali, M. and Shadmehri, B. (2003). Nonlinear analysis of train derailment in severe braking. *ASME Journal of Dynamic Systems, Measurement, and Control* 125: 48–53.

66. Eich-Soellner, E. and Fhuhrer, C. (1998). *Numerical Methods in Multibody Dynamics*. Teubner, B. G. GmbH.

67. Elkins, J.A. (1992). Prediction of wheel rail interaction – the state-of-the-art. *Vehicle System Dynamics* 20: 1–27.

68. Elkins, J.A. and Carter, A. (1993). Testing and analysis techniques for safety assessment of rail vehicles: the state-of-the-art. *Vehicle System Dynamics* 22: 185–208.

69. Elkins, J.A. and Eikhoff, B.M. (1982). Advances in nonlinear wheel/rail force prediction methods and their validation. *ASME Journal of Dynamic Systems, Measurement, and Control* 104: 133–142.

70. Elkins, J.A. and Gostling, R.J. (1977). A general quasi-static curving theory for railway vehicles. In: *Proceedings of 2nd IUTAM Symposium, Vienna*, 388–406.

71. Elkins, J. and Wu, H. (2000). New criteria for flange climb derailment. IEEE/ASME Joint Rail Conference, Newark, New Jersey, April 4–6.
72. CENELEC. (2007). Railway applications. Supply voltages of traction systems. EN 50163.
73. Endlicher, K.O. and Lugner, P. (1990). Computer simulation of the dynamical curving behavior of a railway bogie. *Vehicle System Dynamics* 19: 71–95.
74. Escolona, J.L. (2002). Elastic Contact Personal Communications.
75. Fisette, P. and Samin, C. (1991). Lateral dynamics of a light railway vehicle with independent wheels. In: *Proceedings of the 12th IAVSD Symposium on the Dynamics of Vehicles on Road and Tracks, Lyon*, 157–171.
76. Fries, R.H., Cooperrider, N.K., and Law, E.H. (1981). Experimental investigation of freight car lateral dynamics. *ASME Journal of Dynamic Systems, Measurement, and Control* 103: 201–210.
77. Garcia de Jalon, J. and Bayo, E. (1994). *Kinematic and Dynamic Simulation of Multibody Systems: The Real-Time Challenge*. Springer-Verlag.
78. Garg, V.K. and Dukkipati, R.V. (1988). *Dynamics of Railway Vehicle Systems*. New York: Academic Press.
79. Gere, J.M. and Weaver, W. (1965). *Analysis of Framed Structures*. New York: Van Nostrand.
80. Gerstmayr, J. and Shabana, A.A. (2006). Analysis of thin beams and cables using the absolute nodal co-ordinate formulation. *Nonlinear Dynamics* 45 (1–2): 109–130.
81. Gilchrist, A.O. (1998). The long road to solution of the railway hunting and curving problems. *Proceedings of the Institution of Mechanical Engineers, Part F: Journal of Rail and Rapid Transit* 212: 219–226.
82. Gilchrist, A.O. and Brickle, B.V. (1976). A re-examination of the proneness to derailment of a railway wheel-set. *Journal of Mechanical Engineering Science* 18: 131–141.
83. Gioboata, D., Constantin, G., Abalaru, A., Stanciu, D., Logofatu, C., and Ghionea, I. (2017). Method and system for measurementof railway wheels rolling surface in re-shaping process. *Proceedings in Manufacturing Systems* 12: 113–118.
84. Gladwell, G.M.L. (1980). *Contact Problem in the Classical Theory of Elasticity*. Alphen aan den Rijn: Sijthogg and Noordhof.
85. Goetz, A. (1970). *Introduction to Differential Geometry*. Addison Wesley.
86. Goldsmith, W. (1960). *Impact – The Theory and Physical Behaviour of Colliding Solids*. London, UK: Edward Arnold LTD.
87. Grassie, S.L. (1993). Dynamic modeling of the track. In: *Rail Quality and Maintenance for Modern Railway Operation* (eds. J.J. Kalker, D. Cannon and O. Orringer). Dordrecht: Kluwer.
88. Greenwood, D.T. (1988). *Principle of Dynamics*, 2e. Prentice Hall.
89. Guida, D. and Pappalardo, C.M. (2014). Development of a closed-chain multibody model of a high-speed railway pantograph for hybrid motion/force control of the pantograph/catenary interaction. *International Journal of Mechanical Engineering and Industrial Design* 3 (5): 45–85.
90. Hamid, A., Rasmussen, K., Baluja, M., and Yang, T-L. (1983). Analytical description of track geometry variations. FRA report DOT/FRA/ORD-83/03.1.

91. Handoko, Y.A. (2006). Investigation of the dynamics of railway bogies subjected to traction/braking torque. Ph.D. dissertation. Central Queensland University, Australia.
92. Handoko, Y. and Dhanasekar, M. (2006). An inertial reference frame method for the simulation of the effect of longitudinal force to the dynamics of railway wheelsets. *Nonlinear Dynamics* 45: 399–425.
93. Handoko, Y., Xia, F., and Dhanasekar, M. (2004). Effect of asymmetric brake shoe force application on wagon curving performance. *Vehicle System Dynamics Supplement* 41: 113–122.
94. Haque, I., Law, E.H., and Cooperrider, N.K. (1979). User's manual for program for calculation of Kalker's linear creep coefficients. FRA report FRA/ORD-78/71.
95. He, D.H., Manory, R.R., and Grady, N. (1998). Wear of railway contact wires against current collector materials. *Wear* 215 (1–2): 146–155.
96. Hedrick, J.K. and Arslan, A.V. (1979). Nonlinear analysis of rail vehicle forced lateral response and stability. *ASME Journal of Dynamic Systems, Measurement, and Control* 101: 230–237.
97. Heller, R. and Cooperrider, N.K. (1977). Users' manual for asymmetric wheel/rail contact characterization program. FRA report FRA/ORD-78/05.
98. Hertz, H. (1882). Über die berührung fester elastische Körper und über die Harte. Verhandlungen des Vereins zur Beförderung des Gewerbefleisses, Leipzig.
99. Hesser, K. (1998). Progress in railway mechanical engineering: passenger and commuter rail vehicles and components. ASME International Mechanical Engineering Congress & Exposition (Rail Transportation), Anaheim, California.
100. Hsiao, M. (2010). OCS requirements. TM 3.2.1, prepared by PB for the California High Speed Rail Authority.
101. Huan, R.H., Pan, G.F., and Zhu, W.Q. (2012). Dynamics of pantograph–catenary system considering local singularities of contact wire with critical wavelengths. In: *Proceedings of the 1st International Workshop on High-Speed and Intercity Railways*, 319–333. Berlin, Heidelberg: Springer.
102. Huang, C., Zeng, J., Luo, G., and Shi, H. (2018). Numerical and experimental studies on the car body flexible vibration reduction due to the effect of car body-mounted equipment. *Proceedings of the Institution of Mechanical Engineers, Part F: Journal of Rail and Rapid Transit* 232: 103–120.
103. Huston, R.L. (1990). *Multibody Dynamics*. Butterworth-Heinemann.
104. Huston, R.L. (1993). Flexibility effects on multibody systems. NATO-Advanced Study Institute on Computer Aided Analysis of Rigid and Flexible Mechanical Systems, Troia, Portugal.
105. IEC. (2007). Railway applications – Supply voltages of traction systems. IEC 60850, 3e.
106. Ikeda, M. and Mitsumoji, T. (2009). Numerical estimation of aerodynamic interference between panhead and articulated frame. *Quarterly Report of RTRI* 50 (4): 227–232.
107. Ikeda, M., Mitsumoji, T., Sueki, T. and Takaishi, T. (2012). Aerodynamic noise reduction of a pantograph by shape-smoothing of panhead and its support and by the surface covering with porous material. In: *Noise and Vibration Mitigation for Rail Transportation Systems: Notes on Numerical Fluid Mechanics and Multidisciplinary Design 118* (eds. T. Maeda, P.-E. Gautier, C. Hanson, B. Hemsworth, J. Nelson, B. Schulte-Werning, D. Thompson and P. Vos), 419–426. Tokyo: Springer.

108. Institut Mines-Telecom. (2019). A mini revolution in railway catenaries. https://blogrecherche.wp.imt.fr/en/2019/10/03/a-minor-revolution-in-railway-catenary.

109. Iwnicki, S. (1999). *The Manchester Benchmarks for Rail Vehicle Simulation, Supplement to Vehicle System Dynamics*. Taylor & Francis.

110. Iwnicki, S.D. (1998). Manchester benchmarks for rail vehicle simulation. *Vehicle System Dynamics* 30 (3-4): 295–313.

111. Iwnicki, S.D. (Ed.), (1999). *The Manchester Benchmarks for Rail Vehicle Simulation. Swets & Zeitlinger*. Lisse, the Netherlands.

112. Iwnicki, S.D. and Wickens, A.H. (1998). Validation of a MATLAB railway vehicle simulation using a scale roller rig. *Vehicle System Dynamics* 30: 257–270.

113. Jacobson, B. and Kalker, J.J. (2001). *Rolling Contact Phenomena*. Springer.

114. Jahnke, F.E. (1943). *Tables of Functions*. Dover Publications.

115. Jaschinski, A. (1990). On the application of similarity laws to a scaled railway bogie model. Doctoral thesis. Delft University of Technology.

116. Jeambey, J. (1998). Improving high speed stability of freight cars with hydraulic dampers. ASME International Mechanical Engineering Congress & Exposition (Rail Transportation), Anaheim, California.

117. Johnson, K.L. (1958a). The effect of spin upon the rolling motion of an elastic sphere upon a plane. *ASME Journal of Applied Mechanics* 25: 332–338.

118. Johnson, K.L. (1958b). The effect of a tangential contact force upon the rolling motion of an elastic sphere on a plane. *ASME Journal of Applied Mechanics* 25: 339–346.

119. Johnson, K.L. (1985). *Contact Mechanics*. Cambridge University Press.

120. Jung, S.P., Kim, Y.G., Paik, J.S., and Park, T.W. (2012). Estimation of dynamic contact force between a pantograph and catenary using the finite element method. *ASME Journal of Computational and Nonlinear Dynamics* 7 https://doi.org/10.1115/1.4006733.

121. Kalker, J.J. (1967). On the rolling contact of two elastic bodies in the presence of dry friction. PhD thesis. Delft University, Netherlands.

122. Kalker, J.J. (1973). Simplified theory of rolling contact. *Delft Progress Report* 1: 1–10.

123. Kalker, J.J. (1979). Survey of wheel-rail rolling contact theory. *Vehicle System Dynamics* 8 (4): 317–358.

124. Kalker, J.J. (1980). Review of wheel-rail rolling contact theories. In: *The General Problem of Rolling Contact* (eds. L. Browne and N.T. Tsai). The 1989 ASME Winter Annual Meeting, Chicago, IL, AMD-Vol.40, 77–92.

125. Kalker, J.J. (1982). A fast algorithm for the simplified theory of rolling contact. *Vehicle System Dynamics* 11: 1–13.

126. Kalker, J.J. (1986). Wheel-rail wear calculations with the program contact. In: *Contact Mechanics and Wear of Rail/Wheel Systems II: Proceedings of the International Symposium Held at the University of Rhode Island, Kingston, RI, July 8–11, 1986* (eds. G.M.L. Gladwell, H. Ghonem and J. Kalousek), 3–26. Waterloo, ON: University of Waterloo Press.

127. Kalker, J.J. (1990). *Three-Dimensional Elastic Bodies in Rolling Contact*. Dordrecht: Kluwer.

128. Kalker, J.J. (1991). Wheel-rail rolling contact theory. *Wear* 144: 243–261.

129. Kalker, J.J. (1994). Consideration of rail corrugation. *Vehicle System Dynamics* 23: 3–28.

130. Kalker, J.J. (1996). Discretely supported rails subjected to transient loads. *Vehicle System Dynamics* 25: 71–88.

131. Karnopp, D. (2004). *Vehicle Stability*. Marcel Dekker.

132. Kassa, E., Andersson, C., and Nielsen, J. (2006). Simulation of dynamic interaction between train and railway turnout. *Vehicle System Dynamics* 44 (3): 247–258.

133. Kerr, A.D. (2000). On the determination of the rail support modulus *k*. *International Journal of Solids and Structures* 37: 4335–4351.

134. Kerr, A.D. and El-Sibaie, M.A. (1987). On the new equations for the lateral dynamics of rail-tie structure. *ASME Journal of Dynamic Systems, Measurement, and Control* 107: 180–185.

135. Khulief, Y.A. and Shabana, A. (1986). Dynamics of multi-body systems with variable kinematic structure, ASME Paper No. 85-DET-83. *ASME Journal of Mechanisms, Transmissions, and Automation in Design* 108 (2): 167–175.

136. Khulief, Y.A. and Shabana, A.A. (1987). A continuous force model for the impact analysis of flexible multi-body systems. *Mechanism and Machine Theory* 22 (3): 213–224.

137. Kik, W. (1992). Comparison of the behavior of different wheelset-track models. *Vehicle System Dynamics* 20: 325–339.

138. Kik, W. and Piotrowski, J. (1996). A fast approximation method to calculate normal load at contact between wheel and rail and creep forces during rolling. 2nd Mini Conference on Contact Mechanics and Wear of Rail/Wheel System.

139. Klapas, D., Benson, F.A., Hackam, R., and Evison, P.R. (1988). Wear in simulated railway overhead current collection systems. *Wear* 126 (2): 167–190.

140. Klingel, W. (1883). Über den Lauf von Eisenbahnwagen auf Gerarder Bahn. *Organ für die Fortschritte des Eisenbahnwesens* 20: 113–123, Table XXI.

141. Knothe, K. and Bohm, F. (1999). History of stability of railway and road vehicle. *Vehicle System Dynamics* 31: 283–323.

142. Knothe, K.L. and Grassie, S.L. (1993). Modeling of railway track and vehicle/track interaction at high frequencies. *Vehicle System Dynamics* 22: 209–262.

143. Knothe, K. and Stichel, S. (1994). Direct covariance analysis for the calculation of creepages and creep-forces for various bogie on straight track with random irregularities. *Vehicle System Dynamics* 23: 237–251.

144. Knudsen, C., Feldberg, R., and Jaschinski, A. (1991). Non-linear dynamic phenomenon in the behavior of a railway wheelset model. *Nonlinear Dynamics* 2: 389–404.

145. Kono, H., Suda, Y., Yamaguchi, M., Yamashita, H., Yanobu, Y., and Tsuda, K. (2005). Dynamic analysis of the vehicle running on turnout at high speed considering longitudinal variation of rail profiles. IDETC/CIE 2005, ASME 2005 International Design Engineering Technical Conference & Computers and Information in Engineering Conference, September 24–28, Long Beach, California.

146. Kreyszig, E. (1991). *Differential Geometry*. Dover Publications.

147. Kubo, S. and Kato, K. (1998). Effect of arc discharge on wear rate of Cu-impregnated carbon strip in unlubricated sliding against Cu trolley under electric current. *Wear* 216 (2): 172–178.

148. Kubo, S. and Kato, K. (1999). Effect of arc discharge on the wear rate and wear mode transition of a copper-impregnated metallized carbon contact strip sliding against a copper disk. *Tribology International* 32 (7): 367–378.

149. Kubota, Y., Nagasaka, S., Miyauchi, T., Yamashita, C., and Kakishima, H. (2013). Sliding wear behavior of copper alloy impregnated C/C composites under an electrical current. *Wear* 302 (1–2): 1492–1498.

150. Kulkarni, S., Pappalardo, C.M., and Shabana, A.A. (2017). Pantograph/catenary contact formulations. *ASME Journal of Vibration and Acoustics* 139: 011010-1–011010-12.

151. Kumaniecka, A. and Jacek, S. (2008). Dynamics of the catenary modelled by a periodical structure. *Journal of Theoretical and Applied Mechanics* 46 (4): 869–878.

152. Kumaniecka, A. and Snamina, J. (2008). Dynamics of the catenary modelled by a periodical structure. *Journal of Theoretical and Applied Mechanics* 46: 869–878.

153. Kwak, B.M. (1991). Complementarity problem formulation of three-dimensional friction contact. *ASME Journal of Applied Mechanics* 58: 134–140.

154. Law, E.H. and Cooperrider, N.K. (1974). A survey of railway vehicle dynamics research. *ASME Journal of Dynamic Systems, Measurement, and Control* 96 (2): 132–146.

155. Lee, J.H. and Park, T.W. (2012). Development of a three-dimensional catenary model using the cable elements based on absolute nodal coordinate formulation. *Journal of Mechanical Science and Technology* 26 (12): 3933–3941.

156. Lee, T.W. and Wang, A.C. (1983). On the dynamics of intermittent motion mechanisms, part I: dynamic model and response. *ASME Journal of Mechanisms, Transmissions, and Automation in Design* 105: 534–540.

157. Lee, H.W., Kim, K.C., and Lee, J. (2006). Review of maglev train technology. *IEEE Transactions on Magnetics* 42: 1917–1925.

158. Liao, W.H. and Wang, D.H. (2003). Semiactive vibration control of train suspension systems via magnetorheological dampers. *Journal of Intelligent Material Systems and Structures* 14 (161): 161–172.

159. Lieh, J. and Haque, I. (1991). A study of the parametrically excited behavior of passenger and freight railway vehicles using linear models. *ASME Journal of Dynamic Systems, Measurement, and Control* 113: 336–338.

160. Ling, H. and Shabana, A.A. (2020). Euler angles and numerical representation of the railroad track geometry. *Acta Mechanica* (in press).

161. Litvin, F.L. (1994). *Gear Geometry and Applied Theory*. Englewood Cliffs, NJ: Prentice Hall.

162. Liu, Z. (2015). Numerical study on multi-pantograph railway operation at high speed. Licentiate thesis. KTH Royal Institute of Technology, Stockholm, Sweden. https://www.diva-portal.org/smash/get/diva2:856402/FULLTEXT01.pdf.

163. Lodec Jinshu. (2016). Aluminium innovation in high-speed trains. http://lodecjinshu.com/en/innovation-aluminium-trains.

164. Love, A.E.H. (1944). *A Treatise on the Mathematical Theory of Elasticity*, 4e. Dover.

165. Luo, Y., Yin, H., and Hua, C. (1996). The dynamic response of railway ballast to the action of trains moving at different speeds. *Proceedings of the Institution of Mechanical Engineers, Part F: Journal of Rail and Rapid Transit* 210: 95–101.

166. Mace, S., Pena, R., Wilson, N., and DiBrito, D. (1996). Effects of wheel-rail contact geometry on wheelset steering forces. *Wear* 191: 204–209.

167. Malvezzi, M., Presciani, P., Allotta, B., and Toni, P. (2003). Probabilistic analysis of braking performance in railways. *Proceedings of the Institution of Mechanical Engineers, Part F: Journal of Rail and Rapid Transit* 217: 149–165.

168. Manory, R. and Sinkis, H. (2000). A sliding wear tester for overhead wires and current collectors in light rail systems. *Wear* 239 (1): 10–20.

169. Marquis, B. and Grief, R. (2011). Application of Nadal limit in the prediction of wheel climb derailment. Paper JRC2011-56064. ASME/ASCE/IEEE Joint Rail Conference, Pueblo, Colorado, March 16–18.

170. Massat, J.P., Laine, J.P., and Bobillot, A. (2006). Pantograph–catenary dynamics simulation. *Vehicle System Dynamics* 44 (1): 551–559.

171. Matsudaira, T. (1960). Paper awarded prize in the competition sponsored by Office of Research and Experiment (ORE) of the International Union of Railways (UIC), Utrecht. ORE-Report RP2/SVA-C9 UIC.

172. Matsumoto, A., Sato, A., Ohno, Y., Suda, H., Nishimura, Y., Tanimoto, R., and Oka, M. (1999). Compatibility of curving performance and hunting stability of railway bogie. *Vehicle System Dynamics Supplement* 33: 740–748.

173. Mayville, R., Rancatone, R., and Tegeler, L. (1998). Investigation and simulation of lateral bucling in trains. ASME International Mechanical Engineering Congress & Exposition (Rail Transportation), Anaheim, California.

174. McClanachan, M., Handoko, Y., Dhanasekar, M., Skerman, D., and Davey, J. (2004). Modeling freight wagon dynamics. *Vehicle System Dynamics Supplement* 41: 438–447.

175. McGonigal, R.S. (2006). Grades and curves. Trains. http://trn.trains.com/railroads/abcs-of-railroading/2006/05/grades-and-curves.

176. McPhee, J.J. and Anderson, R.J. (1996). A model reduction procedure for the dynamic analysis of rail vehicles subjected to linear creep forces. *Vehicle System Dynamics* 25: 349–367.

177. Meijaard J.P. (1991). Dynamics of mechanical systems; algorithms for a numerical investigation of the behaviour of non-linear discrete models. PhD thesis. Delft University of Technology.

178. Meijaard, J.P. and De Pater, A.D. (1989). Railway vehicle systems dynamics and chaotic vibrations. *International Journal of Non-Linear Mechanics* 24 (1): 1–17.

179. Melzer, F. (1994). Symbolic computations in flexible multibody systems. In: *Proceedings of NATO-Advanced Study Institute on Computer Aided Analysis of Rigid and Flexible Mechanical Systems*, vol. II, 365–381. Troia, Portugal.

180. Meng, Q., Heineman, J., and Shabana, A.A. (2005). A longitudinal force model for multibody railroad vehicle system applications. DETC2005/MSNDC-84058). ASME International Design Engineering Technical Conferences and Computer and Information in Engineering Conference, Long Beach, California, September 24–28.

181. Miyamoto, T., Suda, Y., and Ishida, H. (2006). Dynamics simulation of railway vehicles on vibrated track upon seismicity. Third Asian Conference on Multibody Dynamics, Tokyo, Japan, August 1–4.

182. Murase, S., Tanaka, A., Takahashi, Y., and Sekine, Y. (1988). Rail vehicle dynamics of pitch caused by phase difference inputs. In: *Proceedings of the ASME Rail Transportation Winter Conference, Chicago, Illinois*, 45–51.

183. Nadal, M.J. (1908). Locomotives a vapeur. In: *Collection Encyclopédie Scientifique*, vol. 186. Paris: Bibliothèque de Mécanique Appliquée et Génie.

184. Nagase, K., Wakabayashi, Y., and Sakahara, H. (2002). A study of the phenomenon of wheel climb derailment: results of basic experiment using model bogies. *Proceedings of the Institution of Mechanical Engineers, Part F: Journal of Rail and Rapid Transit* 216: 237–247.

185. Nakanishi, T., Yin, X.G., and Shabana, A.A. (1996). Dynamics of multibody tracked vehicles using experimentally identified modal parameters. *ASME Journal of Dynamic Systems, Measurement, and Control* 118: 449–507.

186. Netter, H., Schupp, G., Rulka, W., and Schroeder, K. (1998). New aspects of contact modeling and validation within multibody system simulation of railway vehicles. *Vehicle System Dynamics Supplement* 28: 246–269.

187. Newland, D.E. and Cassidy, R.J. (1975). Suspension and structure: some fundamental design considerations for railway vehicles. *Railway Engineering Journal* 4 (2): 4–26.

188. Nielsen, J.C.O. (1994). Dynamic interaction between wheel and track – a parametric search towards an optimal design of rail structures. *Vehicle System Dynamics* 23: 115–132.

189. Nikravesh, P.E. (1988). *Computer-Aided Analysis of Mechanical Systems*. New Jersey: Prentice Hall.

190. Nilsson, C.M., Jones, C.J.C., Thompson, D.J., and Ryue, J. (2009). A waveguide finite element and boundary element approach to calculating the sound radiated by railway and tram rails. *Journal of Sound and Vibration* 321: 813–836.

191. O'Connor, D.N., Eppinger, S.D., Seering, W.P., and Wormley, D.N. (1997). Active control of a high-speed pantograph. *ASME Journal of Dynamic Systems, Measurement, and Control* 119: 1–4.

192. O'Donnell, W. (1993). Experimental and numerical investigation of the pitch and bounce response of a railroad vehicle. M.S. thesis. Department of Mechanical Engineering, University of Illinois at Chicago.

193. Ogden, R.W. (1984). *Non-linear Elastic Deformations*. Dover Publications.

194. Oura, Y., Yoshifumi, M., and Hiroki, N. (2008). Railway electric power feeding systems. *Railway Technology Today* 16 (3): 48–58.

195. Ozaki, T. and Shabana, A.A. (2003a). Treatment of constraints in complex multibody systems, part I: methods of constrained dynamics. *International Journal for Multiscale Computational Engineering* 1 (2&3): 235–252.

196. Ozaki, T. and Shabana, A.A. (2003b, 2003). Treatment of constraints in complex multibody systems, part II: application to tracked vehicles. *International Journal for Multiscale Computational Engineering* 1 (2&3): 253–276.

197. Pappalardo, C.M., Patel, M.D., Tinsley, B., and Shabana, A.A. (2016). Contact force control in multibody pantograph/catenary systems. *Proceedings of the Institution of Mechanical Engineers, Part K: Journal of Multi-body Dynamics* 230 (4): 307–328.

198. Park, T.J., Han, C.S., and Jang, J.H. (2003). Dynamic sensitivity analysis for the pantograph of a high-speed rail vehicle. *Journal of Sound and Vibration* 266 (2): 235–260.

199. Pascal, J.P. (1993). About multi-Hertzian-contact hypothesis and equivalent conicity in the case of S1002 and UIC60 analytical wheel/rail profiles. *Vehicle System Dynamics* 22: 57–78.

200. Pascal, J.P. and Sauvage, G. (1993). The available methods to calculate the wheel/rail forces in non Hertizian contact patches and rail damaging. *Vehicle System Dynamics* 22: 263–275.

201. Pereira, M. and Ambrosio, J. (eds.) (1994). *Computer-Aided Analysis of Rigid and Flexible Mechanical Systems*. Dordrecht: Kluwer.

202. Pfeiffer, F. and Glocker, C. (1996). *Multibody Dynamics with Unilateral Contacts*. New York: Wiley.

203. Poetsch, G., Evans, J., Meisinger, R., Kortum, W., Baldauf, W., Veitl, A., and Wallaschek, J. (1997). Pantograph/catenary dynamics and control. *Vehicle System Dynamics* 28: 159–195.

204. Polach, O. (1999). A fast wheel-rail forces calculation computer code. *Vehicle System Dynamics Supplement* 33: 728–739.

205. Polach, O. (2001). Influence of locomotive tractive effort on the forces between wheel and rail. *Vehicle System Dynamics Supplement* 35: 7–22.

206. Polach, O. (2005). Creep forces in simulation of traction vehicles running on adhesion limit. *Wear* 258: 992–1000.

207. Pombo, J. and Ambrosio, J. (2003). A wheel/rail contact model for rail guided vehicles dynamics. ECCOMAS Thematic Conference on Advances in Computational Multibody Dynamics, Lisboa, Portugal, July 1–4.

208. Pombo, J. and Ambrósio, J. (2012). Influence of pantograph suspension characteristics on the contact quality with the catenary for high speed trains. *Computers and Structures* 110 (111): 32–42.

209. Pombo, J., Ambrósio, J., Pereira, M., Rauter, F., Collina, A., and Facchinetti, A. (2009). Influence of the aerodynamic forces on the pantograph–catenary system for high-speed trains. *Vehicle System Dynamics* 47 (11): 1327–1347.

210. Pop, K. and Schiehlen, W. (1993). *Fahrzeugdynamik*. Stuttgart: Teubner.

211. Press, W.H., Teukolsky, S.A., Vetterling, W.T., and Flannery, B.P. (1992). *Numerical Recipes in Fortran*, 2e. Cambridge University Press.

212. Profilidis, V.A. (2000). *Railway Engineering*. Cambridge University Press.

213. Rabinowicz, E. (1995). *Friction and Wear of Material*. Wiley.

214. Railsystem. (2015). High-speed rail. www.railsystem.net/high-speed-rail.

215. Rathod, C. and Shabana, A.A. (2006a). Geometry and differentiability requirements in multibody railroad vehicle dynamic formulations. *Nonlinear Dynamics* 47 (1): 249–261.

216. Rathod, C. and Shabana, A.A. (2006b). Rail geometry and Euler angles. *ASME Journal of Computational and Nonlinear Dynamics* 1 (3): 264–268.

217. Raymond, G.P. and Cai, Z. (2005). Dynamic track support loading from heavier and faster train sets. *Transportation Research Record* 1381: 53–59.

218. Riches, E. (1988). Will maglev lift off? *IEE Review* 34: 427–430.

219. RifTek Sensors & Instruments. (2008). Laser wheel profilometer, user's manual, IKP-5, IKP-5R series. www.riftek.com.

220. Ripke, B. and Knote, K. (1995). Simulation of high frequency wagon-track interaction. *Vehicle System Dynamics Supplement* 24: 72–85.

221. Roberson, R.E. and Schwertassek, R. (1988). *Dynamics of Multibody Systems.* Springer-Verlag.

222. Rosenberg, R.M. (1977). *Analytical Dynamics of Discrete Systems.* Plenum Press.

223. Samin, J.C. and Neuve, L. (1984). A multibody approach for dynamic investigation of rolling systems. *Ingenieur Archiv* 54: 1–15.

224. Sanborn, G., Heineman, J., and Shabana, A.A. (2007). A low computational cost non-linear formulation for multibody railroad vehicle systems. Paper DETC2007-34522. ASME Design Engineering Technical Conferences, Las Vegas, Nevada, September 4–7.

225. Schmid, R., Endlicher, K.O., and Lugner, P. (1994). Computer simulation of the dynamical behavior of a railway bogie passing a switch. *Vehicle System Dynamics* 23: 481–499.

226. Schupp, G. (1996). Different contact models for wheel-rail systems: a comparison. Internal report IB 515-96-22. Institute for Robotics and System Dynamics, DLR-Oberpfaffenhofen, Germany.

227. Schupp, G. (2003). Simulation of railway vehicles: necessities and application. *Mechanics Based Design of Structures and Machines* 31: 297–314.

228. Schupp, G., Weidemann, C., and Mauer, L. (2004). Modelling the contact between wheel and rail within multibody system simulation. *Vehicle System Dynamics* 41 (5): 349–364.

229. Schwab, A.L. (2002). Dynamics of flexible multibody systems. PhD thesis. Delft University of Technology.

230. Schwab, A.L. and Meijaard, J.P. (2002). Two special finite elements for modelling rolling contact in a multibody environment. In: *Proceedings of the First Asian Conference on Multibody Dynamics*, ACMD'02 (31 July to 2 August 2002), Iwaki, Fukushima, Japan (eds. N. Shimizu et al.), 386–391. The Japan Society of Mechanical Engineering.

231. Schwab, A.L. and Meijaard, J.P. (2003). Dynamics of flexible multibody systems with non-holonomic constraints: a finite element approach. *Multibody System Dynamics* 10: 107–123.

232. Schwartz, B. (1988). An analytical study of the bounce motion of a freight car model in response to profile irregularities. ASME Rail Transportation Conference, Chicago.

233. Schwarz, B. (1991). Wheel climb and rollover potential due to excess elevation and curvature, ASME Paper No. RTD. 4: 83–91.

234. Schwertassek, R. and Wallrap, O. (1999). *Dynamik flexibler Mehrkorper Systeme.* Germany: Vieweg.

235. Seering, W., Armbruster, K., Vesely, C., and Wormley, D. (1991). Experimental and analytical study of pantograph dynamics. *ASME Journal of Dynamic Systems, Measurement, and Control* 113: 591–605.

236. Seo, J.H., Sugiyama, H., and Shabana, A.A. (2005). Three dimensional large deformation analysis of the multibody pantograph/catenary systems. *Nonlinear Dynamics* 42: 199–215.

237. Seo, J.H., Kim, S., Jung, I., Park, T., Mok, J., Kim, Y., and Chai, J. (2006). Dynamic analysis of a pantograph–catenary system using absolute nodal coordinates. *Vehicle System Dynamics* 44 (8): 615–630.

238. Shabana, A.A. (1986). Dynamics of inertia-variant flexible systems using experimentally identified parameters. *ASME Journal of Mechanisms, Transmissions, and Automation in Design* 108: 358–366.

239. Shabana, A.A. (1996a). *Theory of Vibration: An Introduction*, 2e. Springer Verlag.

240. Shabana, A.A. (1996b). An absolute nodal coordinate formulation for the large rotation and deformation analysis of flexible bodies. Technical report MBS96-1-UIC. Department of Mechanical Engineering, University of Illinois at Chicago.

241. Shabana, A.A. (1997). Flexible multibody dynamics: review of past and recent development. *Multibody System Dynamics* 1: 189–222.

242. Shabana, A.A. (2005). *Dynamics of Multibody Systems*, 3e. Cambridge, UK: Cambridge University Press.

243. Shabana, A.A. (2010). *Computational Dynamics*, 3e. New York: Wiley.

244. Shabana, A.A. (2018). *Computational Continuum Mechanics*, 3e. Chichester, UK: Wiley.

245. Shabana, A.A. (2019). *Dynamics of Multibody Systems*, 5e. Cambridge, UK: Cambridge University Press.

246. Shabana, A.A. and Ling, H. (2019). Noncommutativity of finite rotations and definitions of curvature and torsion. *ASME Journal of Computational and Nonlinear Dynamics* 14: 091005-1–091005-10.

247. Shabana, A.A. and Rathod, C. (2007). Geometric coupling in the wheel/rail contact formulations: a comparative study. *Proceedings of the Institution of Mechanical Engineers, Part K: Journal of Multi-body Dynamics* 221: 147–160.

248. Shabana, A.A. and Sany, J.R. (2001). An augmented formulation for mechanical systems with non-generalized coordinates: application to rigid body contact problems. *Nonlinear Dynamics* 24: 183–204.

249. Shabana, A.A., Hussien, H.A., and Escalona, J.L. (1998). Application of the absolute nodal coordinate formulation to large rotation and large deformation problems. *ASME Journal of Mechanical Design* 120: 188–195.

250. Shabana, A.A., Berzeri, M., and Sany, J.R. (2001). Numerical procedure for the simulation of wheel/rail contact dynamics. *ASME Journal of Dynamic Systems, Measurement, and Control* 123 (2): 168–178.

251. Shabana, A.A., Zaazaa, K.E., Escalona, L.J., and Sany, J.R. (2004). Development of elastic force model for wheel/rail contact problems. *Journal of Sound and Vibration* 269: 295–325.

252. Shabana, A.A., Tobaa, M., Sugiyama, H., and Zaazaa, K.E. (2005). On the computer formulations of the wheel/rail contact. *Nonlinear Dynamics* 40: 169–193.

253. Shabana, A.A., Tobaa, M., Marquis, B., and El-Sibaie, M. (2006). Effect of the linearization of the kinematic equations in railroad vehicle system dynamics. *ASME Journal of Computational and Nonlinear Dynamics* 1: 25–34.

254. Shabana, A.A., Zaazaa, K.E., and Sugiyama, H. (2008). *Railroad Vehicle Dynamics: A Computational Approach*. Boca Raton, FL: CRC Press, Taylor & Francis Group.

255. Shabana, A.A., Hamper, M.B., and O'Shea, J.J. (2013). Rolling condition and gyroscopic moments in curve negotiations. *ASME Journal of Computational and Nonlinear Dynamics* 8: 0111015-1–011015-10.

256. Shampine, L. and Gordon, M. (1975). *Computer Solution of ODE: The Initial Value Problem*. San Francisco, CA: Freeman.

257. Sharma, V., Sneed, W., and Punwani, S. (1984). Freight equipment environmental sampling test-description and results. ASME Rail Transportation Spring Conference, Chicago.

258. Shen, G. and Goodall, R. (1997). Active yaw relaxation for improved bogie performance. *Vehicle System Dynamics* 28: 273–289.

259. Shen, G. and Pratt, I. (2001). The development of railway dynamics modelling and simulation package to cater for current industrial trends. *Proceedings of the Institution of Mechanical Engineers, Part F: Journal of Rail and Rapid Transit* 215: 167–178.

260. Shen, Z.Y., Hedrick, J.K., and Elkins, J.A. (1983). A comparison of alternative creep-force models for rail vehicle dynamic analysis. *Vehicle System Dynamics* 12: 79–87.

261. Shen, G., Ayasse, J.B., Chollet, H., and Pratt, I. (2003). A unique design method for wheel profiles by considering the contact angle function. *Proceedings of the Institution of Mechanical Engineers, Part F: Journal of Rail and Rapid Transit* 217, Part F: 25–30.

262. Shi, H. and Wu, P. (2016a). A nonlinear rubber spring model containing fractional derivatives for use in railroad vehicle dynamic analysis. *Proceedings of the Institution of Mechanical Engineers, Part F: Journal of Rail and Rapid Transit* 230: 1745–1759.

263. Shi, H. and Wu, P. (2016b). Flexible vibration analysis for car body of high-speed EMU. *Journal of Mechanical Science and Technology* 30: 55–66.

264. Shi, J., Wei, Q.C., and Zhao, Y. (2007). Analysis of dynamic response of the high-speed EMS maglev vehicle/guideway coupling system with random irregularity. *Vehicle System Dynamics* 45 (12): 1077–1095.

265. Shi, H., Wang, J., Wu, P., Song, C., and Teng, W. (2018). Field measurements of the evolution of wheel wear and vehicle dynamics for high-speed trains. *Vehicle System Dynamics* 56: 1187–1206.

266. Shikin, E.V. and Plis, A.I. (1995). *Handbook on Splines for User*. Boca Raton, FL: CRC Press.

267. Shust, W.C., Elkins, J.A., Kalay, J.A., and El-Sibaie, M. (1997). Wheel climb derailment tests using AAR's track loading vehicle. Report R-910. Association of American Railroads.

268. Simeon, B., Fuhrer, C., and Rentrop, P. (1991). Differential algebraic equations in vehicle system dynamics. *Surveys on Mathematics for Industry* 1: 1–37.

269. Sinha, P.K. (1987). *Electromagnetic Suspension: Dynamics & Control*. London: P. Peregrinus.

270. Sinokrot, T., Nakhaeinejad, M., and Shabana, A.A. (2008). A velocity transformation method for the nonlinear dynamic simulation of railroad vehicle systems. *Nonlinear Dynamics* 51: 289–307.

271. Smith, C.A.M. and Bowler, E.H. (1981). A curved track simulator. XV Pan American Railway Congress, Mexico.

272. Smith, R.. (2019). A railroad switch on the Schynige Platte-Bahn rack railway in Switzerland. https://www.reddit.com/r/InfrastructurePorn/comments/8xbvq3/a_railroad_switch_on_the_schynige_plattebahn_rack.

273. Spiryagin, M., Wu, Q., and Cole, C. (2017). International benchmarking of longitudinal train dynamics simulators: Benchmarking questions. *Vehicle System Dynamics* 55 (4): 450–463.

274. Stichel, S. (1999). On freight wagon dynamics and track deterioration. *Proceedings of the Institution of Mechanical Engineers, Part F: Journal of Rail and Rapid Transit* 213: 243–254.

275. Stickland, M.T. and Scanlon, T.J. (2001). An investigation into the aerodynamic characteristics of catenary contact wires in a cross-wind. *Proceedings of the Institution of Mechanical Engineers, Part F: Journal of Rail and Rapid Transit* 215 (4): 311–318.

276. Stickland, M.T., Scanlon, T.J., Craighead, I.A., and Fernandez, J. (2003). An investigation into the mechanical damping characteristics of catenary contact wires and their effect on aerodynamic galloping instability. *Proceedings of the Institution of Mechanical Engineers, Part F: Journal of Rail and Rapid Transit* 217 (2): 63–71.

277. Suda, Y. and Anderson, R.J. (1994). High speed stability and curving performance of longitudinal unsymmetric trucks with semi-active control. *Vehicle System Dynamics* 23: 29–52.

278. Sugiyama, H. and Shabana, A.A. (2006). Trajectory coordinate constraints in multibody railroad vehicle systems. Third Asian Conference on Multibody Dynamics, Tokyo, Japan, August 1–4.

279. Sugiyama, H., Escalona, J.L., and Shabana, A.A. (2003). Formulation of three-dimensional joint constraints using the absolute nodal coordinates. *Journal of Nonlinear Dynamics* 31: 167–195.

280. Sun, Y.Q. and Dhanasekar, M. (2001). A dynamic model for the vertical interaction of the rail track and wagon system. *International Journal of Solids and Structures* 39: 1337–1359.

281. Swanson, J. (2019). Railroad workers need to fight for workers control over trains, safety. *The Militant* 83 (25) https://themilitant.com/2019/07/06/railroad-workers-needto-fight-for-workers-control-over-trains-safety.

282. Sweet, L.M. and Garrison-Phelan, P. (1980). Identification of wheel/rail creep coefficients from steady state and dynamic wheelset experiments. In: *The General Problem of Rolling Contact* (eds. L. Browne and N.T. Tsai), The ASME 1980 Winter Annual Meeting, Chicago, IL, AMD-Vol. 40, 77–92.

283. Sweet, L.M. and Karmel, A. (1981). Evaluation of time-duration dependent wheel load criteria for wheel climb derailment. *ASME Journal of Dynamic Systems, Measurement, and Control* 103: 219–227.

284. Szolc, T. (1998). Simulation of bending-torsional-lateral vibrations of the railway wheelset-track system in the medium frequency range. *Vehicle System Dynamics* 30: 473–508.

285. Thompson, D. (2009). *Railway noise and vibration: Mechanisms, modelling and means of control*. Elsevier. Oxford, UK, 2009.

286. Thompson, D.J. and Jones, C.J.C. (2000). A review of the modelling of wheel/rail noise generation. *Journal of Sound and Vibration* 231: 519–536.

287. Tse, Y. and Martin, G. (1975). Flexible body railroad freight car, vol. I. Technical report R-199. Association of American Railroads.

288. Tur, M., García, E., Baeza, L., and Fuenmayor, F.J. (2014). A 3D absolute nodal coordinate finite element model to compute the initial configuration of a railway catenary. *Engineering Structures* 71: 234–243.

289. Tuten, J.M., Law, E.H., and Coperrider, N.K. (1979). Lateral stability of freight cars with axles having different wheel profiles and asymmetric loading. *ASME Journal of Engineering for Industry* 101: 1–16.

290. US Department of Transportation, Federal Railroad Administration, Office of Safety. (2004). Track safety standards, part 213, subpart G. Distributed by the Railway Educational Bureau.

291. US Department of Transportation, Federal Railroad Administration, Office of Safety. (2005). Track safety standards, part 213, subpart A to F. Distributed by the Railway Educational Bureau.

292. Vaccaro, A. (2019). Take the E-train? MBTA mulling electric locomotives. *Boston Globe*. https://www.bostonglobe.com/metro/2019/03/21/take-train-mbta-mullingelectric-locomotives/kWekh87ZPI7IKKr3EjmIUJ/story.html.

293. Valtorta, D., Zaazaa, K.E., Shabana, A.A., and Sany, J.R. (2001). A study of the lateral stability of railroad vehicles using a nonlinear constrained multibody formulation. ASME International Mechanical Engineering Congress and Exposition, New York.

294. Vermeulen, P.J. and Johnson, K.L. (1964). Contact of nonspherical bodies transmitting tangential forces. *ASME Journal of Applied Mechanics* 31: 338–340.

295. Wang, W., Suda, Y., Komine, H., Sato, Y., Nakai, T., and Shimokawai, Y. (2006). Transition curve negotiating simulation of railway vehicle for air suspension control to prevent wheel load reduction. Third Asian Conference on Multibody Dynamics, Tokyo, Japan, August 1–4.

296. Warren, P., Samavedam, G., Tsai, T., and Gomes, J. (1998). Passenger vehicle dynamic performance – tests and analysis correlation. ASME International Mechanical Engineering Congress & Exposition (Rail Transportation), Anaheim, California.

297. Wehage, R.A. (1980). Generalized coordinate partitioning in dynamic analysis of mechanical systems. PhD thesis. University of Iowa, Iowa City.

298. Weidemann, C., Mauer, L., and Arnold, M. (2006). Improving the calculation speed for time domain integration of complex railway vehicles. Third Asian Conference on Multibody Dynamics, Tokyo, Japan, August 1–4.

299. Weinstock, H. (1984). Wheel climb derailment criteria for evaluation of rail vehicle safety. Paper 84-WA/RT-1. ASME Winter Annual Meeting.

300. White, R.C., Limbert, J.K., Hedrick, K., and Cooperrider, N.K. (1978). Guidewaysuspension tradeoffs in rail vehicle systems. Report DOT-OS-50107. US Department of Transportation, Washington, DC.

301. Wickens, A.H. (1965). The dynamic stability of railway vehicle wheelsets and bogies having profiled wheels. *International Journal of Solids and Structures* 1: 319–341.

302. Wickens, A.H. (1986). Non-linear dynamics of railway vehicles. *Vehicle System Dynamics* 15: 289–301.

303. Wickens, A.H. (1988). Stability optimization of multi-axle railway vehicles possessing perfect steering. *ASME Journal of Dynamic Systems, Measurement, and Control* 110: 1–7.

304. Wickens, A.H. (1991). Steering and stability of bogie: vehicle dynamics and suspension design. *Proceedings of the Institution of Mechanical Engineers, Part F: Journal of Rail and Rapid Transit* 205: 109–122.

305. Wickens, A.H. (1996). Static and dynamic instabilities of bogie railway vehicles with linkage steered wheelsets. *Vehicle System Dynamics* 26: 1–16.

306. Wickens, A. (1998). Key note address. *Vehicle System Dynamics* 30 (3–4): 193–195.

307. Wickens, A.H. (2003). *Fundamental of Rail Vehicle Dynamics*. Swets & Zeitlinger.

308. Wilson, N., Shu, X., and Kramp, K. (2004). Effect of independently rolling wheels on flange climb derailment. ASME International Mechanical Engineering Congress.

309. Wilson, N., Wu, H., Tournay, H., and Urban, C. (2009). Effects of wheel/rail contact patterns and vehicle parameters on lateral stability. *Vehicle System Dynamics* 48 (S1): 487–503.

310. Wilson, N.G., Fries, R., Witte, M., Haigermoser, A., Wrang, M., Evans, J., and Orlova, A. (2011). Assessment of safety against derailment using simulations and vehicle acceptance tests: A worldwide comparison of the state of the art assessment methods. Proceedings 22nd IAVSD Symposium, 14–19 August 2011, Manchester, UK, CD-ROM.

311. Wu, T.X. and Brennan, M.J. (1998). Basic analytical study of pantograph–catenary system dynamics. *Vehicle System Dynamics* 30: 443–456.

312. Wu, T.X. and Thompson, D.J. (2002). A hybrid model for the noise generation due to railway wheel flats. *Journal of Sound and Vibration* 251: 115–139.

313. Wu, H. and Elkins, J. (1999). Investigation of wheel flange climb derailment criteria. Report R-931. Association of American Railroads.

314. Wu, W.X., Brickle, B.V., Smith, J.H., and Luo, R.K. (1998). An investigation into stick-slip vibrations on vehicle/track systems. *Vehicle System Dynamics* 30: 229–236.

315. Wu, Q., Cole, C., Luo, S., and Spiryagin, M. (2014). A review of dynamics modelling of friction draft gear. *Vehicle System Dynamics* 52: 733–758.

316. Wu, X., Cai, W., Chi, M., Lai, W., Shi, H., and Zhu, M. (2015). Investigation of the effects of sleeper-passing impacts on the high-speed train. *Vehicle System Dynamics* 53: 1902–1917.

317. Wu, Q., Spiryagin, M., and Cole, C. (2016). Longitudinal train dynamics: an overview. *Vehicle System Dynamics* 54: 1688–1714.

318. Yabuno, H., Okamoto, T., and Aoshima, N. (2001). Stabilization control for the hunting motion. *Vehicle System Dynamics Supplement* 35: 41–55.

319. Yakoub, R.Y. and Shabana, A.A. (2001). Three dimensional absolute nodal coordinate formulation for beam elements: implementation and applications. *ASME Journal of Mechanical Design* 123 (4): 614–621.

320. Yang, G. (1994). Aspects in modeling a railway vehicle on an arbitrary track, ASME Rail Transportation, RTD. 8: 31–36.

321. Yang, Y.B. and Yau, J.D. (2011). An iterative interacting method for dynamic analysis of the maglev train guideway/foundation soil system. *Engineering Structures* 33 (3): 1013–1024.

322. Yau, J.D. (2009). Response of a maglev vehicle moving on a series of guideways with differential settlement. *Journal of Sound and Vibration* 324 (3–5): 816–831.

323. Yau, J.D. (2010). Aerodynamic vibrations of a maglev vehicle running on flexible guideways under oncoming wind actions. *Journal of Sound and Vibration* 329 (10): 1743–1759.

324. Yokoyama, N. (2009). Research and development toward wear reduction of current collecting system. *JR East Technical Review* 13: 50–54.

325. Zaazaa, K.E. (2003) Elastic force model for wheel/rail contact in multibody railroad vehicle systems. PhD thesis. Department of Mechanical Engineering, University of Illinois at Chicago.

326. Zboiński, K. (1998). Dynamical investigation of railway vehicles on a curved track. *European Journal of Mechanics - A/Solids* 17 (6): 1001–1020.

327. Zboinski, K. (1999). The importance of imaginary forces and kinematic-type nonlinearities for the description of railway vehicle dynamics. *Proceedings of the Institution of Mechanical Engineers, Part F: Journal of Rail and Rapid Transit* 213: 199–210.

328. Zboinski, K. and Dusza, M. (2006). Development of the method and analysis for non-linear lateral stability of railway vehicles in a curved track. *Vehicle System Dynamics* 44 (sup 1): 147–157.

329. Zhang, Y., El-Sibaie, M., and Lee, S. (2004). FRA track quality indices and distribution characteristics. AREMA Annual Conference, Nashville.

330. Zhai, W. (2019). *Vehicle–track coupled dynamics: Theory and application*. Springer, Singapore.

331. Zhai, W.M., Cai, C.B., and Guo, S.Z. (1996). Coupling model of vertical and lateral vehicle/track interactions. *Vehicle System Dynamics* 26 (1): 61–79.

332. Zheng, X.J., Wu, J.J., and Zhou, Y.H. (2000). Numerical analyses on dynamic control of five-degree-of-freedom maglev vehicle moving on flexible guideways. *Journal of Sound and Vibration* 235 (1): 43–61.

333. Zhou, L. and Shen, Z.Y. (2013). Dynamic analysis of a high-speed train operating on a curved track with failed fasteners. *Journal of Zhejiang University Science A* 14 (6): 447–458.

334. Zhou, D., Tian, H.Q., Zhang, J., and Yang, M.Z. (2014). Pressure transients induced by a high speed train passing through a station. *Journal of Wind Engineering and Industrial Aerodynamics* 135: 1–9.

335. Zobory, I. (1997). Prediction of wheel/rail profile wear. *Vehicle System Dynamics* 28: 221–259.

Index

Mathematical Foundation of Railroad Vehicle Systems: Geometry and Mechanics,
First Edition. Ahmed A. Shabana.

h

i

j

q

r

S